Sustainable Community Health

"If we want to achieve the goal of sustainable community health, it is imperative to address the broad and multi-level context in which these communities are situated. Understanding and integrating the context (or many contexts) of a community is a challenging task, and this timely volume tackles this issue head-on. Drawing on many disciplines and perspectives, this text both integrates strong theoretical perspectives and relies on evidence and pragmatic implementational realities. In doing so, it provides a thoughtful and practical guide for how to build and sustain communities that are active agents in their own health and well-being. I recommend this to anyone seeking to better understand and/or implement community health programs."

—Joshua M. Smyth, *Distinguished Professor of Biobehavioral Health and Medicine, Pennsylvania State University, USA*

"This comprehensive yet highly readable text integrates health and behavioral theory with practical strategies, resulting in a roadmap for improving the health of communities in ways that prioritize sustainability. The approaches detailed by Mpofu and colleagues are grounded in principles of social justice, and intended to redress health inequities that have persisted for centuries. This volume is essential reading for anyone working to promote community health in an increasingly complicated and interconnected world."

—Elizabeth R. Bertone-Johnson, *Professor of Epidemiology and Chair of Health Promotion and Policy, University of Massachusetts, Amherst, USA*

"This volume, *Sustainable Community Health: Systems and Practices in Diverse Settings*, offers the most comprehensive review to date of a wide range of programs aiming to understand and support community actors in promoting health and reducing disparities worldwide. This volume offers a breadth and depth of evidence and practice based experience in four areas: sustainable health foundations, policies and practices, indicators and outcomes, and sustainable health in populations with vulnerability. This trans-disciplinary work addresses a wide range of subject matter, from substance use safety or prevention, to infectious disease control, to quality inter-personal relationships, and many other issues from around the world. In the age of COVID-19, this work provides a valuable contribution to the science of community health promotion and how it can contribute to long-term recovery and sustainability in the wake of the pandemic."

—W. Dounglas Evans, *Professor of Prevention and Community Health and Global Health and Director of the Public Communication and Marketing Program, George Washington University, USA*

"This edited book on *Sustainable Community Health* is extremely comprehensive, in-depth, and research-based, with an international, global approach. This book is interactive with thoughtful questions and discussions that make the chapters engaging and reader-friendly. This is a key resource book for not only researchers but community providers, and policy makers. It'll be your 'one-stop-shop,' 'go-to' sourcebook for understanding, implementing, and strengthening community health centers in culturally appropriate and meaningful ways. This book is so timely and up-to-date, including critical discussions on COVID-19, which has challenged community health systems around the world."

—Susan Chuang, *Professor of Family Relations and Nutrition, University of Guelph, Ontario, Canada*

Elias Mpofu
Editor

Sustainable Community Health

Systems and Practices in Diverse Settings

palgrave
macmillan

Editor
Elias Mpofu
University of North Texas
Denton, TX, USA

University of Sydney
Sydney, NSW, Australia

University of Johannesburg
Johannesburg, South Africa

ISBN 978-3-030-59686-6 ISBN 978-3-030-59687-3 (eBook)
https://doi.org/10.1007/978-3-030-59687-3

Cover illustration: agefotostock / Alamy Stock Photo

This Palgrave Macmillan imprint is published by the registered company Springer Nature Switzerland AG.
The registered company address is: Gewerbestrasse 11, 6330 Cham, Switzerland

I dedicate this book volume to my father, Mr. Machina Denhere, for his lifelong commitment to facilitating community partnerships for the betterment of humanity. My father spoke five languages. I learned a lot from working with him on community projects since I was a toddler, and grew to appreciate his work even more over the years, which matured me in many ways. I took a big leaf from him on developing sustainable community relationships based on respect for diversity and unlocking of community assets for the wellbeing of member partners.

Foreword: A Timely and Compelling Volume

This volume on *Sustainable Community Health: Systems and Practices in Diverse Settings* is both timely and compelling. As this volume goes to press in the midst of a global pandemic, it is impossible to ignore the importance of effective healthcare systems. While the pandemic is horrific in its effects on people and economies, this crisis also creates space for new thinking on how to deliver health care going forward. The scope of the pandemic also illuminates vividly how far most nations, and particularly the United States, are from having an effective system of health care. When the history of this pandemic is written, effective healthcare systems will be one of the major components of an effective response, one with less morbidity and mortality. All nations, as well as smaller governmental units, should be seeking improved systems. I believe that this volume presents an especially compelling approach for delivering effective health care.

A pandemic also makes clear the importance of global models and cost-effective approaches for health care. Many in the United States boast of the quality of our health care, yet for too many, it is unaffordable. Further, the high cost of health care is no guarantee of quality. The trend in the United States toward increasing ownership of healthcare practices by private equity has raised serious concerns about the sustainability of quality care if profit is the dominant motive. Already too many in the

United States cannot afford the health care delivered in most care settings. What good is care if the population cannot access it?

For the majority world, the health care provided in much of the minority world is beyond reach financially. In a pandemic, where the health of one is the health of all, unaffordable health care is useless, however outstanding scientifically and technically. Thinking globally, we must have systems that will enable all people to stay healthy. The community-based approach to health care described in this volume is affordable, and therefore sustainable.

This volume also makes clear that some community values can lead people to rejecting health care, even when it is scientifically based and high quality. In the United States, an example, are those who distrust vaccines, choosing not to vaccinate their children, thus leaving the entire population vulnerable to infectious diseases. Religious beliefs as well as other values can impede the uptake of health care. When the community is engaged with care, there is an opportunity to surface fears and beliefs, and potentially address them. We have seen many such examples globally in the treatment of Ebola and other viral diseases. It is unquestionably challenging to work with a resistant community to embrace effective treatments, especially when their values or fears lead them to reject such care. This volume discusses the approaches needed to partner with communities to provide sensitive and effective care that is embraced by community members.

Finally, this volume embraces the importance of people taking responsibility for their own wellness, or self-efficacy for health. The importance and power of individual investment in being healthy are well established. However, most of us need some support to achieve our personal health goals. Effectiveness is much more likely when supported by the community, and the results can be much more powerful. The health of the public requires all to participate, as we have seen with the current pandemic. This volume provides many useful guidelines for achieving healthy individuals within healthy communities.

This book presents a remarkable approach to sustainable healthcare systems that assumes that, "communities achieve environmental, economic, and social health sustainability under their own impetus and in partnership with local, state, and federal agencies" (preface). This approach considers communities as health systems, with their wellbeing influencing the health of individuals within the community.

The structure of this volume lends itself to the goal of explicating the evidence for the approach. The four sections take us through the foundations of health sustainability, essential policies, and practices that provide structure for health sustainability, metrics for quality care improvement for community health sustainability, and a concluding section on special populations requiring attention to achieve community health sustainability. This volume is intended to educate and to serve as a handbook on how to achieve community health sustainability. The learning goals are articulated clearly with novel tools including some available online, an increasingly essential asset for learning across the world. The implementation goals are aided with lots of discussion boxes with examples in every chapter so that readers can apply their learning. Questions are also offered in every chapter, and each has an extensive reference list for further exploration of information. This thorough approach increases the accessibility of the material, and the ultimate success of the learning.

The authors of chapters in this volume represent five continents on the global, about fifteen disciplines across the entire spectrum relevant to healthcare systems, with records of accomplishment in community health practices that lead to wellness rather than simply the reduction in disease or illness. The rich and diverse backgrounds of the authors bring remarkable strength to this volume, grounding the material in global experience.

This voluminous book will serve all who wish to move toward sustainable health care, an ambitious goal needed in these fraught times. I am grateful to the Editor, Elias Mpofu, together with the many chapter authors for so ably preparing this volume so that communities everywhere globally can develop effective healthcare systems.

Ann Arbor, MI, USA
St. Joseph, MI, USA

Anne C. Petersen

Preface and Overview of the Book

This book advances the emerging sustainable health science by framing community health as an agentic practice in which communities are primary actors and benefactors of their own health. Little usable information is currently available to educators, policymakers, development agencies, and social service providers on concepts and approaches to sustainable community health. We are in a pandemics age, and this book is timely in proposing innovative approaches to sustainable community health systems resilient to the vagaries of opportunistic environmental, economic, and social injustices in a globalized world. This book volume provides usable information for designing and implementing sustainable community health for addressing health disparities, health equity, and social justice in diverse communities and populations.

Approach Applying a trans-disciplinary approach, this book examines the interdependence between environmental, economic, and social justice pillars of community health. The 42 contributing authors have outstanding expertise in the science of community health. Most have established records of accomplishments in community health practices aimed at wellness rather than merely the amelioration of symptoms of disease or illness. They come from diverse health backgrounds, including applied gerontology, behavior analysis, community medicine, epidemiology, environmental science, internal medicine, kinesiology, health services administration, medical sociology, mental health, nutrition and diet, political science, and rehabilitation and health services.

Timeliness This trans-disciplinary book is timely in view of the importance of sustainable community health systems to the wellbeing of populations. Sustainable community health systems are accessible, inclusive, cost-effective, efficient, and responsive to the health needs of local populations. There is presently a pressing need for sustainable health systems for protecting present health rights, while ensuring the wellbeing of future generations. Moreover, health education, research, and practice well into the twenty-first century will necessarily center on sustainable community health systems for implementation, addressing the health rights of individuals and communities. This book volume provides a much-needed grounding in sustainable community wellness policies and practices that would explain present community health while also ensuring wellness in the long-term.

Unique Contributions This book is seminal in providing a resource on sustainable community health concepts, procedures, and practices for addressing health disparities, inequity, and social justice for wellbeing by partner communities. It assumes the perspective that sustainable community health is about investment into present health as well as that of future generations. Moreover, this book volume assumes the premise that communities achieve environmental, economic, and social health sustainability under their own impetus and in partnership with local, state, and federal agencies for health for all. In this partnership view of health, communities are themselves health systems and their wellbeing qualities affect the health of individuals and the collective alike. This book chapters provide practical ways to enhance the sustainable health of communities adopting strategies for investment in health futures, addressing known as well as emergent population health issues with proactive health policy formulation, implementation, and evaluation. Each of this book chapters discusses the cultural, professional, and legal practice influences on health systems sustainability, in the context of interdisciplinary practices. All the chapters identify research and practice issues for which the evidence would advance the specific sustainable health science approaches.

Organization

This book comprises four sections and 18 chapters, addressing sustainable health foundations, policies and practices, indicators and outcomes, and health in populations with vulnerability. The chapters in Part I, "Foundations of Sustainable Community Health," provides a firm grounding in the social, economic, and environmental pillars of sustainable health systems, beginning with the theoretical and methodological trends in sustainable community health (Chap. 1), the significance of health disparities to social sustainability of community health systems (Chap. 2), the importance of environmental sustainability to livable communities (Chap. 3), and the economics of sustainable community health (Chap. 4). The chapters from this section collectively speak to the importance of a balanced approach to harnessing the social, economic, and environmental pillars of sustainable health systems, engaging communities in health decision making and actions. On completing the readings from this section, you will be well acquainted with the three pillars of sustainable community health, how they intermix in health for all practices.

Part II, "Policies and Practices in Sustainable Community Health," reflects the fact that sustainable community health is an inter-disciplinary science that brings together community wellbeing enablers that span disparate knowledge domains and competencies. These include nutritional practices to grow healthy communities (Chap. 4), community substance use safety (Chap. 6), community mental health wellbeing (Chap. 7), and community epidemiologic approaches (Chap. 8). The chapters from this section consider evidence-based example programs that utilized community assets and resourcing approaches for sustainable health for health promotion. After completing the chapters from this section, you will have heightened knowledge of community wellness approaches rooted in participatory, preventive approaches grounded in community wellness policies and practices. Part III, "Indicators and Outcomes of Sustainable Community Health," canvases a broad range of established and emerging technologies tracking the health of community populations for wellness interventions. The lead chapter addresses community-oriented quality care improvement, considering value-based care aimed at reducing costs and improving the quality of care, and non-value-based approaches aimed to identify and address the nonmedical determinants of health (Chap. 9). This is followed by the chapters on community health informatics for screening for medical and nonmedical needs (Chap. 10), telehealth services utilization in low resource settings, and taking into account digital divide constraints (Chap. 11), and metrices for communicable and noncommunicable diseases (Chap. 12). Overall, the chapters from this section on indicators and outcomes of sustainability propose digital age applications foregrounding social, economic, and environmental justice considerations. On completing the chapter readings, you will have comprehensive knowledge on current and emerging technologies for measuring progress sustainable community health for the design and implementation of futuristic health systems.

Part IV, "Sustainable Community Health in Populations," examines trends in the evidence on sustainable community health in diverse populations and settings with aging, disability, lifestyle, cultural heritage, and geo-social strata inter-sectionalities. The individual chapters in this section examine long-term wellbeing approaches with older adults (Chap. 13), people with intellectual and developmental disabilities (Chap. 14), neurodivergent people (Chap. 15), people with obesity and metabolic conditions (Chap. 16), and indigenous community people (Chap. 17). The chapters consider lived and prospective health in diverse populations, applying wellness approaches for inclusive community health systems. Upon completion of the readings from this section, you will be able to consider alternative evidence-based approaches to participatory community health and wellbeing by populations with vulnerability, which would work in the long-term.

Part V, "Epilogue," is a single chapter section on the futures of sustainable community health (Chap. 18). The chapter reflects on the key themes addressed in this book volume, premised on health-in-all and for all approaches to sustainability community health. The concluding chapter of this book volume identifies environmental sustainability to provide for economic and social sustainability of community health systems aimed to address health disparities, equity, and social justice. On reading this chapter, you will result with a personal theory on what you perceive to be the futures of sustainable community health, based on the evidence presented in this book volume and your own personal persuasions.

Target Audience

Senior undergraduate and graduate students of the public or population health sciences, including community psychology, counseling psychology, public health, medical anthropology, social policy, rehabilitation counseling, and related disciplines, intend this book for use. Health policy and service providers in the private and public sectors and international aid agencies will find this book an invaluable resource for their health promotion and development work in the international community.

Pedagogy

All chapters in this book volume follow a similar presentation format, beginning with a brief overview and statement of five to six learning objectives to anticipate the contents of the chapter. The opening chapter sets the stage for the rest of this book volume, presenting a conceptual model on the pillars of sustainable community health and for addressing health disparities, equity, and social justice. It also defines the key concepts on sustainable community health systems that are used across the chapters of this book volume. The principal 16 chapters examine the relevant historical issues, theories, and empirical data, emphasizing critical analysis and application of knowledge for designing the pillars of sustainable community health systems. The concluding chapter invites the reader to put everything together, reflecting on the key lessons from this book volume.

Critical Learning Resources and Activities

The book chapters include critical learning activity assignments for readers including discussion boxes, research boxes, self-check questions, case illustrations and studies, field-based experiential learning exercises, internet resources, and key terms and concepts.

Discussion Boxes Discussion boxes describe a current hot issue, dilemma, or controversy pertinent to a chapter theme or concept. They encourage you to demonstrate an understanding of the diversity of views and potential solutions in addressing community health equity, disparities, and social justice issues.

Research Boxes Research boxes describe illustrative research on pertinent community health equity, disparities, and social justice concepts or practices, thereby highlighting the evidence base for the practices. Research box tasks prompt you to examine documented scientific research evidence for specific community health practices, and the implications of the evidence for the development of programs to address community health equity, disparities, and social justice.

Case Illustrations and Studies The case illustrations and studies guide you to reflect on the translation of health equity, disparities, and social justice concepts with specific population and in specific community contents. They profile unique expression or elaboration of community health equity, disparities, and social justice practices by identified programs or consumers of the programs.

Self-check and Discussion Questions The self-check questions assist your self-monitoring of understanding of the chapter key concepts of the chapter and their application to sustainable health policies and practices. The study questions are aligned to the learning objectives from the beginning of the each chapter.

Field-Based Experiential Exercises These are for inviting you to participate in community health actions for wellbeing. The reflective learning to result would enable you to engage in community health practices in support of the quality of health they aspire.

Exploring Internet Resources This book chapters list online resources on the specific topic for further learning about related research, practice, and policy issues. The list of online resources is selected for ease of navigation and supporting the critical learning outcomes from accessing the resources.

Key Terms and Concepts The key terms and concepts used in the text are listed in the glossary, referencing the pages where they are defined and used in the text.

Denton, TX, USA
May 2020

Elias Mpofu

Acknowledgments

I would like to acknowledge several people for their assistance in the preparation of this book volume: Chelsea Anderson, Maidei Machina, Qiwei Li, and Tinashe Dune. Chelsea Anderson and Maidei Machina proof-edited the entire pre-publication volume, which tremendously assisted in the timely completion and submission for publication. Their thorough proof editing was a huge help to the contributing authors and to me as editor of this book volume. Qiwei Li assisted with logistical support including the manuscript file documentation and formatting of the pre-publication copy. Qiwei Li also helped with the design and construction of several graphics included in this volume. Tinashe Dune assisted with the copy editing of some of this book chapters, creating the filing platform and correspondence with some of the contributing authors assisting their successful pre-publication records.

I want to thank all the 42 authors who contributed to this book volume for their instrumental knowledge of the science and practice of community health systems. Their expertise is self-evidence across all contributions. I also want to thank Madison Alums, Senior Editor at Palgrave/Macmillan, and others associated with Springer International for assisting this book progression from its contract, to manuscript preparation, and successful publication.

Contents

Notes on Contributors

Theresa Abah is an assistant professor at the University of Sacramento, CA, and with extensive healthcare management and geriatrics expertise. She has a strong record of accomplishment of work experience in health promotion and education, vaccine security, disease surveillance, and health systems strengthening. Her research is centered on health policy development for health systems, strengthening, and applying interdisciplinary approaches.

Martha I. Arrieta is Associate Professor of Medicine and Associate Director/Director of Research of the Center for Healthy Communities at the University of South Alabama College of Medicine. Arrieta's community-engaged scholarship and epidemiology and has focused on health equity issues in immigrant populations, disaster preparation and response, diabetes, HIV/AIDS, and several other areas. She has championed the research apprenticeship as a model to achieve health equity.

Ganesh Baniya is a doctoral candidate at the University of North Texas. His primary research interests include the impact of adverse childhood experiences on health screening behaviors, applying trauma-informed approaches in primary care, and strengthening organizational structures to implement trauma-informed care. He is also interested in the application of geographic information systems on health, the role of location in the distribution of disease, chronic diseases among seniors, osteoporosis health among seniors, minority health, and interventions that benefit elderly people to maintain health.

Abraham David Benavides is an associate professor in the Department of Public Administration at the University of North Texas. His research interests include local government, human resources, cultural competency, ethics and leadership, immigration, wellness programs, and livable communities for the aging population. His articles have appeared in the *Journal of State and Local Government Review, the Journal of Public Affairs Education, International Journal of Public Administration, Public Administration Quarterly, the Journal of Public Management and Social Policy*, and others.

Charles P. Bernacchio is Professor of Counselor Education at the University of Southern Maine. He successfully developed a nationally accredited Clinical Rehabilitation Counseling graduate program. He is directing his third national Long-Term Rehabilitation Counseling Training program that pilots graduate vocational rehabilitation (VR) counseling online for VR staff in Maine, New Hampshire, and Vermont. Bernacchio is also piloting the use of tele-rehabilitation technology to deliver vocational rehabilitation services in rural regions.

M. Harvey Brenner is Associate Professor of Global Health at the University of North Texas Health Sciences Center. He has been a leading researcher in the use and development of the econometric method as applied to health at the community, national, and international levels. His studies, including for the Joint Economic Committee of the U.S. Congress, have demonstrated the short- and long-term impact of economic change on mortality—especially economic development, recessions, pandemic, and unemployment.

Clare Brock is Assistant Professor of Political Science at Texas Woman's University, where she teaches classes in food politics and public policy. Clare Brock's research is grounded in the areas of public policy, lobbying and interest group behavior, and food politics. She also researches on the construction and framing of agricultural policy and on the intersection of race, ethnicity, and food politics.

Kendall R. Brune is a senior manager and healthcare executive at the Meharry Medical College, TN, and with over 30 years' experience in healthcare. He is a licensed administrator whose professional life has centered on developing, building, and operating post-acute care facilities for minority communities. He also has extensive experience in working with tribal nations in the United States in the area of healthcare services.

Li-Wu Chen is a professor in the Department of Health Services Research and Administration at the College of Public Health, the University of Nebraska Medical Center (UNMC), NE, USA. He is also the Director of the UNMC Center for Health Policy Analysis and Rural Health Research and was the Deputy Director of the RUPRI Center for Rural Health Policy Analysis (funded by the Federal Office of Rural Health Policy) from 2004 to 2010. Chen's health services research focus spans health policy analysis, rural health research, health economics, and public health services. Chen has engaged many local health departments in identifying, designing, implementing, disseminating, and translating research on critical issues related to public health systems.

Andrew M. Colombo-Dougovito is Assistant Professor of Sport Pedagogy and Motor Behavior and Certified Adapted Physical Educator (CAPE) in the Department of Kinesiology, Health Promotion, and Recreation within the College of Education at the University of North Texas (UNT). He is the Director of the Physical Activity and Motor Skills (UNT-PAMS) program and a Faculty Liaison to the Kristin Farmer Autism Center. His research expertise focuses on the lifespan physical activity behaviors and motor skill development patterns of autistic individuals, potential socio-environmental facilitators and barriers to physical activity, the overarching benefits of physical activity in relation to quality of life and health outcomes in autistic individuals.

Veronica Cortez is a master's student in rehabilitation counseling at the University of North Texas. She is also a teaching assistant in the Department of Rehabilitation and Health Services. Cortez's research interests include addiction, collegiate substance use, substance use among women, and sustainable treatment models.

Carrie E. Crook is a rising third-year student in the joint MD-MPH program at Tulane University. She attended the University of Pennsylvania for her undergraduate studies with a major in Health and Societies. Her research interests center around disparities in maternal–infant mortality.

Errol D. Crook is the Abraham A. Mitchell Professor and Chair of the Department of Internal Medicine and Director of the Center for Healthy Communities at the University of South Alabama College of Medicine. Crook's research has focused on health disparities in kidney and cardiovascular disease with an emphasis on hypertension, diabetes, and the social determinants of health.

Kathleen Davis is Assistant Professor of Nutrition and Food Sciences at Texas Woman's University, where she teaches classes in nutrition and sustainability of food systems. Kathleen Davis's research focus is pediatric nutrition, including drivers and prevention of overweight and obesity, as well as how developmental disability impacts nutrition in children.

Suzanna Rocco Dillon is Associate Professor of Adapted Physical Activity and a Certified Adapted Physical Educator (CAPE) in the School of Health Promotion and Kinesiology at Texas Woman's University at Denton (TWU). Previously, she served as President and Legislative Chair for the National Consortium for Physical Education for Individuals with Disabilities (NCPEID). Dillon is a faculty scholar in the Sherrill Teaching and Research Lab and a member of the Pioneers in Teaching and Learning Academy at TWU. Her research interests center on inclusive physical education for children and youth with autism spectrum disorder (ASD) as well as the use of evidence-based practices for individuals with ASD within inclusive and segregated physical activity settings that lead to improved outcomes for individuals on the autism spectrum.

Kirsteen Edereka-Great is a health services research doctoral student at the University of North Texas and a teaching fellow/health services research assistant at the Department of Rehabilitation & Health Services. Her academic research explores health disparities, health (in)equity, social (in)justice, mental health- substance use disorder, health literacy and health communication, chronic disease management and comorbidities, disabilities, behavioral health, and aging-related-issues.

Crystal Fernandez is a doctoral student at the University of North Texas. She has worked with individuals with disabilities and their families for the past ten years across a variety of settings, including public schools, post-secondary education, and clinical settings. Crystal's research interests include assessment and design of instructional programs for children with disabilities, practitioner training, and parent advocacy.

Jeewani Anupama Ginige is Senior Lecturer in Health Informatics in the School of Computer, Data and Mathematical Sciences (SoCDMS) at Western Sydney University, Australia. Her research expertise includes using suitable technologies to reach communities in rural and remote settings, and developing IT systems for clinical trial management.

Roma Stovall Hanks is Professor and Chair, Department of Sociology, Anthropology, and Social Work at the University of South Alabama (USA) and Director, USA Center for Generational Studies and USA Programs in Gerontology. She is Director of Community Engagement of the Center for Healthy Communities. Hanks' research interests focus on social gerontology and evaluation of community-engaged activities focusing on community empowerment and health disparities. Her work has included evaluation of community health advocate programs, life history analysis, social networks, and social support to prevent and manage illness.

Heath Harllee is a health services research doctoral student at the University of North Texas. Formerly a service member with the United States Army, his interests are on improving the aging services as well as that of the family members of older adults.

Elizabeth Houck is a doctoral student at the University of North Texas. She has worked for ten years in clinical practice with children and adolescents with autism spectrum disorder (ASD) and intellectual and developmental disabilities (IDD). A significant portion of Elizabeth's clinical work has focused on developing the repertoires needed for people with IDD and ASD to safely and successfully access community healthcare. Elizabeth's research interests include the assessment and treatment of challenging behavior for people with intellectual disabilities.

Stanley Ingman is Emeritus Professor of Applied Gerontology at the University of North Texas. He holds degrees in Botany from Miami University, Rural Sociology from Ohio State University, and Medical Sociology from University of Pittsburgh. As Director of the Center for Public Service at the University of North Texas, he created the healthy neighborhood program in 1995 and the Sustainable Communities Review in 1997. Ingman is the President of Future Without Poverty, Inc., an NGO dedicated to reducing poverty using sustainable practices and employing the four E's of sustainability–improving enterprises, environmental preservation, sustainable oriented education, and finally, empowering citizens to bring about reform.

Chisom Nmesoma Iwundu is Assistant Professor of Public Health in the Department of Rehabilitation and Health Services, at the University of North Texas, Denton, and formerly a postdoctoral fellow in a partnership program supported by the University of Houston and MD Anderson Cancer Center. Her research is focused on understanding the social determinants of health that account for disparities in health and health risk behaviors among marginalized groups, such as homeless adults.

Laura M. Keyes an AICP certified planner, holds a position of lecturer and undergraduate program coordinator for the Nonprofit Leadership Studies and Urban Policy and Planning degrees at the University of North Texas. Her research interests include aging policy, age-friendly community design, community development, and nonprofit management. Keyes provides consulting services to communities on creating age-friendly places. She recently published her research specific to aging policy and several other scholarly journal articles and textbook chapters on public administration.

Rebekah Knight is a certified professional gerontologist recently retired from the University of North Texas. She has 30 years in the healthcare administration field that includes hospital management, Long-Term care administration, and owned her own rehabilitation facility. Her areas of research interest include commodification of aging, healthcare administration policy and practice, wellness education for the older population, health department research and program structure, and federal regulatory policy.

Qiwei Li is a post-doctoral fellow at the John Hopkins University, MD, USA. He received his doctorate degree of Health Services Research from the University of North Texas with a concentration on rehabilitation science and gerontology. His research interests are in the areas of applying the International Classification of Functioning, Health, and Disability (ICF) in successful aging among older adults with disabilities and chronic conditions.

Lu Liang is an assistant professor in the Department of Geography and the Environment at the University of North Texas. She received her doctoral degree in ecosystem science from the University of California, Berkeley. She has authored or co-authored 48 scientific papers and book chapters, including publication in high impact journals such as *The Lancet, PLOS Neglected Tropical Diseases, Environment International,* and *Remote Sensing of Environment.* She is a member of two *Lancet* health committees, and her work has been cited 3883 times.

April Linden is a doctoral student at the University of North Texas. She has worked with young children diagnosed with autism spectrum disorder for the past ten years and has trained staff and parents on behavior analytic strategies and tactics to improve lives for children and their families. April's primary research interests are in staff training, approach-based intervention, and cultural diversity.

Maidei Machina is a senior occupational therapist with the Geriatric Medicine Department at Westmead Hospital, Sydney, Australia. She has extensive clinical experience providing occupational therapy services to clients in acute, sub-acute, and community settings. In 2020, she led the RACE & RACE-Extend Telehealth Initiative that integrated the use of telecommunications technology with community-dwelling geriatric clients in the Western Sydney Local Health District. Maidei has several research interests that include falls prevention and management, the use of telehealth as a service delivery model for outpatient occupational therapy services, and investigating the value and application of interdisciplinary approaches in sustainable aging.

Bradley McDaniels is assistant professor in the Department of Rehabilitation and Health Services at the University of North Texas. He completed a Post-Doctoral Research Fellowship at Virginia Commonwealth University in the Department of Physical Medicine and Rehabilitation in the School of Medicine. McDaniels' research focus includes psychosocial adaptation, quality of life, employment, and the role of meaning and purpose for individuals with Parkinson's disease.

Ami Moore is an associate professor in the Department of Rehabilitation and Health Services at the University of North Texas. She is the public health undergraduate coordinator and the Director of the Center for Psychosocial Health Research. Moore's areas of expertise include health disparities, social demography, HIV/AIDS prevention, and public health education and health promotion.

Elias Mpofu is Professor of Health Services Research at the University of North Texas, Honorary Professor of Health Sciences at the University of Sydney, Australia, and distinguished visiting professor at the University of Johannesburg, South Africa. Elias Mpofu is editor of *Community-Oriented Health Services: Practices across Disciplines* (Springer) and several other community health-related titles. Contributing to these works are over 100 authors internationally renowned in community action for health and wellbeing. In the past 20 years, Mpofu has proposed and tested community-centric approaches to health and wellness in diverse settings and populations. His ongoing work in this area is focused on how community identities influence people's decisions for wellness.

Seth Oppong is an associate professor at the Department of Psychology, at the University of Botswana. Prior to, he was a Carnegie Scholar at the University of Ghana, Legon and an international visiting research scholar to North Carolina State University, Raleigh, NC, USA. Oppong's research interests revolve around how to increase the relevance of psychological science and psychotechnology in non-Western societies, underpinned by his focus on indigenization, history-philosophy-and-metatheory of psychology, and workplace indigenous health.

Valerie Pacino is a PhD student in the Department of Health Services Research and Administration in the College of Public Health at the University of Nebraska Medical Center, NE, USA. Her research interests include the social determinants of mental health, the lifelong consequences of adverse childhood experiences, and the ways in which structural violence perpetuates inequity and makes us all less healthy.

David Palm is an associate professor in the Department of Health Services Research and Administration in the College of Public Health at the University of Nebraska Medical Center, NE, USA. Earlier, he was the Director of the Office of Community and Rural Health in the Nebraska Department of Health and Human Services. While in this position, he worked with small rural hospitals on quality improvement activities. His research interests include public health services and systems, care coordination between public health and primary care, the quality of the community health needs assessments and implementation plans of the nonprofit hospitals, and the allocation of community benefit expenditures by nonprofit hospitals.

Gayle Prybutok is an assistant professor in Health Services Administration in the Department of Rehabilitation and Health Services, College of Health and Public Service, at the University of North Texas. She served as the Coordinator for the Health Services Administration Master's Degree. She has had an extensive career in healthcare administration and has authored over 20 journal articles. She is the Editor of the *International Journal Health Care Quality Assurance* and serves on the Editorial Board of the *International Journal of Electronic Healthcare.*

Sonia Redwine is the Director of the Collegiate Recovery Program at the University of North Texas. She also serves as a public health adjunct instructor for the University of North Texas. She holds a BSc degree in Health from Texas A&M University and a MSc degree in Public Health from the University of Texas Health Science Center. Her professional experience includes improving health and wellbeing at individual and community levels in community-based and higher education settings, with an emphasis on behavioral health.

Danielle Resiak is a rehabilitation counsellor working in a health education officer role at the Uniting Medically Supervised Injecting Centre in Sydney, Australia. Danielle is a fellow of the Higher Education Teaching Academy, coordinator, lecturer, and tutor at the University of Sydney

Faculty of Medicine and Health, and workshop facilitator at La Trobe University, School of Psychology & Public Health. Danielle is an advocate for vulnerable and marginalized members of community whose research addresses harm minimization community-based services for people who inject drugs.

Solymar Rivera-Torres is a second-year PhD student in the Department of Rehabilitation and Health Services, the University of North Texas. Solymar's research interests encompass the areas of leisure activity participation among independent living facilities, older adults, and opioid use disorder among Native American Indians.

Diana Kuo Stojda completed her doctoral training in epidemiology from the University of Louisville. Stojda has extensive experience working in clinical healthcare settings as well as public health research. She is interested in health disparities and health outcomes in marginalized populations as well as environmental health, perinatal and pediatric health, and chronic disease.

Justin R. Watts is Assistant Professor of Rehabilitation and Health Services at the University of North Texas. Watts' research is focused on child maltreatment and its connection to issues in social, psychological and emotional development as well as chronic disease and disability in adulthood. Watts also conducts research related to co-occurring trauma and substance use disorders and the impact of these disorders on functioning and other mental health conditions across the lifespan.

Josephine F. Wilson is Professor of Population and Public Health Sciences in Substance Abuse Resources and Disability Issues Program, Boonshoft School of Medicine at Wright State University. She is the author of dozens of publications in the areas of substance abuse, addiction, eating disorders, disability, and child abuse and neglect, as well as two textbooks on behavioral neuroscience. Wilson's research focus is related to substance abuse in diverse populations including populations with disabilities, leading a team of IT experts that has developed innovative web-based products to utilize online behavioral assessment instruments.

List of Figures

List of Tables

Part I

Foundations of Sustainable Community Health

1

Concepts and Models in Sustainable Community Health

Elias Mpofu

Introduction

Health is a cornerstone of sustainable human development. Communities are health systems and the foundation for the wellbeing of societies across the globe. Health systems refer to "all the activities whose primary purpose is to promote, restore, and maintain health" (World Health Organization [WHO], 2020, p. 5). A health system is more than a health care system by its "focus on the ultimate outcome of interest — that is, the population's health and each individual's health — and not only on the formal system of care designed primarily to treat illness" (Fineberg, 2012, p. 1020). The perspective of this book volume is that communities are diverse in their health needs, solutions, and potentialities, — meaning

E. Mpofu (✉)
University of North Texas, Denton, TX, USA

University of Sydney, Sydney, NSW, Australia

University of Johannesburg, Johannesburg, South Africa
e-mail: Elias.Mpofu@unt.edu

E. Mpofu (ed.), *Sustainable Community Health*,
https://doi.org/10.1007/978-3-030-59687-3_1

they are in essence health systems. As health systems, the activities in which communities engage can influence the health of the community members. Communities are also open health learning systems in that they enact and change their health-related practices under their own impetus and/or from partnerships with external agencies (e.g., researchers, policy makers, other communities). Therefore, the activities that communities enact or adopt as health systems have consequences for the sustainable health and wellbeing of both individuals and the collective community.

Learning Objectives

By the end of the chapter, the reader should be able to:

1. Define sustainable community health and its pillars.
2. Differentiate between sustainable health, sustainable community health, and sustainable community health systems.
3. Discuss the relative social, economic, and environmental sustainability determinants to designing and implementing community health systems.
4. Consider approaches to sustainable community health that are responsive to health disparities, equity, and social justice.
5. Propose ways to optimize the social, economic, and environmental pillars for resilient community health.

Sustainable health is premised on long-term wellbeing of populations, advantaged by wellness and health-for-all policies and practices. While sustainable health care has its roots in providing high-quality medical services with financial and environmental cost effectiveness ensuring affordability and performance (Schroeder, Thompson, Frith, & Pencheon, 2012), sustainable community health goes beyond the disease management approach to encompass wellness actions for, and by community members, in so far as all human activities have health consequences.

The diversity in communities, their health needs, and assets suggests that sustainable community health is best understood as a product of a health system. Moreover, the concept of sustainable community health systems captures the plurality of communities as health systems. Sustainable community health systems comprise all the elements for enhancing population and individual health, including family, cultural, inter-agency and inter-sectoral relationships, natural environmental safety, financing, and technologies for health. They are "designed for people to attain their highest possible health and to provide timely care for the specific needs of the members and with affordability, fairness and equity for members and providers" (Fineberg, 2012).

Sustainability and health are intertwined in that "…health is automatically improved by commitment to sustainability" while "…sustainability provides a framework within which health gains and reductions in health disparities are possible and greatly facilitated" (Guidotti, 2018, p. 357). Indeed, in the United Nations Development Programme (UNDP, 2014), sustainable community health is the third of 17 *Sustainable Development Goals* (SDGs) to "…ensure healthy lives and promote well-being for all at all ages"; and this goal permeates all the other SGDs. The SGD specifies 13 priority targets to ensure health for all, including investing in reproductive and child health, control of communicable and noncommunicable diseases, addiction disorders, environmental health, and achieving universal health care. In addition, national governments are committed to tobacco control, vaccines and medicines, health financing and workforce development, and global health risk preparedness. Implementation of sustainable community health approaches would meet the SDG health goals and empower communities and subpopulations of communities themselves to become active agents for their wellness and health for all.

Health systems around the globe face significant challenges in their sustainability secondary to a wide range of factors including an aging global population, virus pandemics, ever-increasing costs from managing chronic diseases in the face of budget cuts, the need to acquire new generation health technologies, persistent scarcities in the health workforce, and environment degradation (WHO, 2013a, 2013b, 2016, 2020). Moreover, a 20-year gap in life expectancy at birth persists between the most developed countries compared to the lower placed developing countries, indicating the significant health care burdens faced by developing countries, which they would be unable to overcome without investing in sustainable health systems (WHO, 2013a, 2013b). With investment in sustainable health systems, provision of universal health coverage, reduction in neonatal and maternal mortality, and control of infectious diseases, the life expectancy gap between the most developed and least developed countries could be eliminated by the year 2035 (Jamison et al., 2013). Increases in healthy life expectancies are premised on economic development (Chang, Robinson, Hammitt, & Resch, 2017) and implementation of progressive economic regimens in the developing regions of the world. This would assist in eradicating extreme poverty and promoting wellbeing (Summers, 2015), which would narrow the life expectancy gap between developed and developing countries. Implementation of community health systems adopting a health-for-all approach would make for sustainability, with huge cost savings from communities invested in their own wellness.

All countries share a view of health as a universal right, which is central to people's life situations, social goals, and community membership. Not only is good health an indisputable human right, but it is also an important contributor to economic growth by way of economic productivity and environmental protection and support. The flow-on effects of good population health include a reduction in expenditure on treating illness, and overall community social cohesion as community members enjoy good health at all ages. A country's progress toward sustainable development closely relates to national health indicators (Borgonovi, Adinolfi, Palumbo, & Piscopo, 2018), which at the local level provide the data on community health and wellbeing (Cummins, Mpofu, & Machina, 2015). Community-level data and the strategies for their use for wellness promotion would make for sustainable health for all. A bottom-up approach in which a country's sustainable development progress reflected the diversity in community wellbeing would enable the timely and targeted resourcing of communities for sustainable health (Bago d'Uva, Van Doorslaer, Lindeboom, & O'Donnell, 2008).

Traditional health systems prioritizing hospitalization are unsustainable in managing the high prevalence and mortality from avoidable health conditions, the bulk of which are generated at the community and health care systems levels (Leff & Montalto, 2004; Verjan, Augusto, Xie, & Buthion, 2013; WHO, 2007). Not surprisingly, health systems around the globe constructed on managing disease in individuals lack in sustainability qualities and are characterized by the fragility and unpredictability of their ability to deliver health in the future (Mpofu, 2015). Patient care strategies focused on enabling adaption in medical care to align with the patient's social circumstances have greater impact when implemented with community-level strategies that focus on improving health and well-being at the local community level through multi-sectoral partnerships (Gottlieb, Fichtenberg, Alderwick, & Adler, 2019). In addition, "many improvement initiatives fail to sustain to a point where their full benefits can be realized" (Lennox, Maher, & Reed, 2018, p. 1), and in part from a lack of in-built sustainability. The concept of participation for sustainable community health extends to communities as active health agents rather than as passive consumers of health services.

Defining Sustainable Community Health

Sustainable health systems is an emerging science, and presently with evolving definitions for its referents (Lennox, Maher, & Reed, 2018). Presently, "there is no consensus with regard to either the definition of the term or the factors that characterize a sustainable healthcare system" (Fischer, 2015, p. 294) and "not everyone means the same thing when they speak about sustainable health care" (Muzyka et al., 2012, p. 1). There is an emerging consensus on sustainable health systems to the rest of the three pillars: social, economic, and environmental. These pillars define people's social health needs (social pillar), economic capabilities (economic pillar), and environmental affordances (ecological pillar) (Alliance for Natural Health, 2010; Jameton & McGuire, 2002). The three pillars of sustainable health systems are social oriented (patient care and satisfaction, health employee job satisfaction, work wellness, etc.), economics oriented (reduced medical bills, health care pollution, rural area, resource conservation, etc.) and environmental oriented (waste management, handling of chemical substances, recycling, green technologies, etc.) (Marimuthu & Paulose, 2016). Implementation of the three pillars of sustainable health involves collaborative partnerships between consumers and providers in resource sharing such as: finances and environmental energy use, the utilization of enabling technologies for addressing community health (Coiera & Hovenga, 2007; Faezipour & Ferreira, 2013), and employing evidence-based health decision making and prevention-oriented approaches (Popa & Ştefan, 2019; Stefan, Popa, & Dobrin, 2016).

Ideally, these three pillars work in tandem for the common goal of achieving sustainable health in populations, resulting with environmental and human systems that protect and promote the health of the present global population as well as the health and wellbeing of generations in the future (Faezipour & Ferreira, 2013; Mintzberg, 2015; Schroeder, 2013; Tsasis, Agrawal, & Guriel, 2019). A focus on only one dimension of health system sustainability would be less optimal (Borgonovi, Adinolfi, Palumbo, & Piscopo, 2018). Thus, in viewing health through a sustainable development lens, one must consider the interrelated issues of economy, environment, health, wellbeing, and social justice (Borgonovi et al., 2018; Mohrman, Shani, & McCracken, 2012), with the goal to "meet the needs of the present without compromising the ability to meet future needs" (Roberts & World Health Organization, 1998, p. 5).

The perspective of this book assumes that the social, economic, and environmental pillars of health systems work together in complex and often less understood ways to influence community health disparities, and to promote health equity and social justice (see Fig. 1.1 for a conceptual model). The conceptual model presented in Fig. 1.1 is premised on

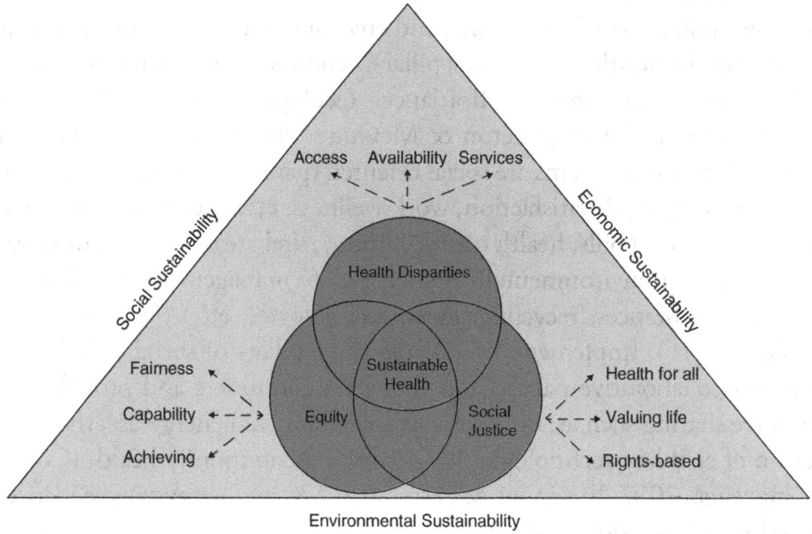

Fig. 1.1 Conceptual model on sustainable community health systems

the fact that community health systems differ on a spectrum in their presence and impact on population health regarding (1) disparities in availability, accessibility, and type of services; (2) equity in fairness, capabilities, and achievement; and (3) social justice in regard to promoting health for all, and valuing life and health as human right. Explaining the conceptual model from the outside in, and at the risk of oversimplification, the economic pillar would influence health disparities relatively more, while the social and environmental pillars would influence health equity and social justice more.

In real terms, the social, economic, and environmental pillars of community health are interactive in their influence of community health disparities, equity, and social justice. For instance, while economic sustainability would influence community health disparities in terms of access, availability, and services, economic resources would also enable community partnerships for health (social pillar) and protections of the natural environment (environmental pillar). Similarly, the social sustainability would influence health social justice (inclusive of health for all, and rights-based health) and equity, valuing the present health assets of the community, and treating with fairness their present health capabilities and aspirations. Social sustainability for health social justice also rests on environmental justice (as in sustainable natural environment practices by the community). Environmental sustainability would ensure economic sustainability through the use of renewable energy systems and minimize environmental degradation known to increase health system costs for the community. Moreover, environmental sustainability would also promote health equity and social justice through the community's investment in proactive wellness and health for all for mutually best human-to-nature interactions.

Health disparities from limited accessibility and availability of health resources, as well as from the types of services provided, are a major hurdle to sustainable community health (Braveman, 2006; Woolf & Braveman, 2011). For instance, there is a socioeconomic gradient effect on community health in that rural and low socioeconomic communities experience health disparities from poorer access to health care services (Woolf & Braveman, 2011; see also Chap. 2, this volume) and lack of availability of health-sustaining nutrition than comparatively advantaged communities (see also Chap. 5 this volume). Moreover, socioeconomically deprived communities are health systems with a poorer quality of service profile, creating avoidable health service disparities for the community members compared to relatively advantaged communities (McGrail & Humphreys, 2015; see also Chap. 17, this volume). The lived physical environment of a community has a huge impact on the health and wellbeing of the members, and those communities with exposure to hazardous pollutants are at elevated risk for chronic illnesses (Denny & Davidson, 2012; Marrone, 2007; Mitchell & Popham, 2008; Shostak, 2013; see also Chap. 3, this volume).

When communities are denied essential health services, they are denied health social justice, often by systematic powers they perceive to lack control over (Meara & Greenberg, 2015). These health social injustices manifest differently among and within community segments across the socioeconomic and lived environmental gradient so that disadvantaged communities experience denial of health rights and services that provide for health for all. Communities experiencing health inequities are denied fair access to achieve optimal health within their capabilities (Marmot & Commission on Social Determinants of Health, 2007; Sen, 2002; Solomon & Orridge, 2014). Fairness is when a health system is "without discrimination or disparities to all individuals and families, regardless of age, group identity, or place, and also to partner health professionals, institutions, and businesses supporting and delivering care" (Fineberg, 2012, p. 1020). From denials, communities experience a lack of health equity from being unfairly exposed to avoidable unhealthy living, disvaluing of their present as well as prospective health achievements and capabilities. Similarly, community health equity is possible by addressing the health disparities of access, availability, and services from a rights-based approach (see also Marmot, 2017; Peter, 2001).

Undeniably, there are socioeconomic gradients within and between communities, which explains both their achieved and aspired health (Marmot, 2017). Some communities (or their population segments) would come off as more health and wellbeing capable than others. For instance, depending on the health capabilities and resourcing of each community (social, economic, and environmental), a community as health system would be relatively more successful in employing its resources and practices for reducing health disparities, enhancing health equity, and promoting health social justice. In balance, the social, economic, and environmental pillars of health make for sustainable community health through their synergetic rather than linear influences on health disparities, social justice, and equity (Adam & de Savigny, 2012). Moreover, to be sustainable, health systems must be adaptable in how they intermingle their social, economic, and environmental resources to optimize population health since lived health is dynamic and evolving from responding to changing health needs and demographics; changing policy and practice environment; scientific discoveries; and emerging technologies. The major challenge for sustainable community health practices is to describe, explain, design, implement, monitor, and evaluate how the three pillars of health systems create a wide variety of lived health disparities, equity and social justice outcomes for the community members.

Social Sustainability

Health and wellbeing are both states of being and social outcomes. Social sustainability is about the ways and means by which relationship factors influence health outcomes (McKenzie, 2004). For that reason, the social sustainability of community health is improved by closer working relationships and alliances with public health agencies, other agencies in non-health sectors (such as transportation and housing support), and local community agencies to reduce health disparities, increase health equity, improve care coordination, and engage community members in their own health care. Social sustainability of community health is also premised on acceptability and co-ownership of the health system by the constituent communities, including individuals, families, public health service providers, and community welfare organizations.

As noted previously, communities are themselves health systems and their health capabilities affect the individuals who are member dwellers. Healthy communities have populations that participate in the design and implementation of their own health promotion, maintenance, or sustenance through value co-creation with their health care systems (Beirão,

Patrício, & Fisk, 2017; Russo, Moretta Tartaglione, & Cavacece, 2019; see also Chap. 9, this volume). Community health value co-creation occurs when community members and providers share resources to address health need gaps they perceive to result in sustainable health gains (Beirão et al., 2017; Frow, McColl-Kennedy, & Payne, 2016). Interventions aimed to improve people's attention to their health, in addition to providing greater availability of resources that deliver health protection, encourage people to participate in health management activities. Subsequently, people develop essential habits of caring about their health over the long term, and health promotion aimed at vulnerable populations makes for social sustainability of community health programs.

Greater Availability of Resources to Deliver Health Protection Social health sustainability is likely to be achieved with universal health coverage and long-term strategic perspectives and innovativeness in health systems quality management and institutional accountability (Fischer, 2015; Saviano, Bassano, & Calabrese, 2010). Universal health coverage typically includes access to basic health care services, mostly funded by general tax revenues or some type of reissuance scheme ensuring equitable access to health services regardless of socioeconomic status. Universal health coverage schemes are in recognition of the fact that health is a basic human right and everyone should have equal opportunity to achieve their full health potential. In many jurisdictions around the globe, huge inequities prevail in access to quality health care services or the resources to sustain optimal health. This is particularly the case in low-resource settings where social injustices derive health inequities (Ruger, 2004). Built-in long-term strategic perspective and innovativeness of community health systems is premised on efficiency and effectiveness to meet the health needs of the local community (Saviano et al., 2010) based on life expectancy at birth and fertility rate (Momete, 2016). Greater availability of resources to deliver community health protections also depends on a reliable supply of health care providers, access to medical care, and lower out-of-pocket expenses for community members in their diversity.

People with Essential Habits of Caring About Their Health over the Long Term The empowerment of communities to self-manage their health is requisite to health system sustainability. Sustainable community health is associated with cost reduction practices of improving community health literacy, improving the ability of community members to communicate their health needs, and increasing community members' engagement in prevention activities contributing to their health system (Russo, Moretta Tartaglione, & Cavacece, 2019; WHO, 2020; see also Chap. 6, this volume).

As an example community health empowerment approach, primary health care services for reducing community health disparities and increasing health equity and social justice seek to focus on disease prevention and health promotion, improving nutrition, enabling access to clean water and sanitation, and encouraging people to participate in health management as a collective responsibility (Fischer, 2015; Russo et al., 2019; see also Chaps. 3 and 7, this volume). If people do not perceive a link between the purposes of a health system and how it sustains their wellbeing, they will have a lower acceptance of the system and be less persuaded to follow its health guidelines (Mohrman, Shani, & McCracken, 2012; see also Chap. 17, this volume). The use of community navigators or community health advocates is another proven way to engage communities in their own health care (Borgonovi et al., 2018; Momete, 2016). Many community health partnerships fail because of the cultural divide and the tension that exists between their designated medical health support agencies and nonmedical community-based organizations (see also Chap. 9, this volume). For example, there may be differences in language or approaches between a medical health agency and a nonmedical community-based organization (individual treatment vs. whole populations) to supporting low-income populations (Alderwick & Gottlieb, 2019; Mills, 2014). Family buy-in with regard to treatment and care options is a major factor in community-centric health care systems (Boutin-Foster & Charlson, 2001). Often, this is potentiated when the families partner with a provider who communicates with them, listening to any suggestions or concerns that might have regarding treatment care (Gluyas, 2015) (see Discussion Box 1.1).

Discussion Box 1.1: Importance of Family Co-value Creation in Health Care

Consider the case of Mr. Scaramanga, a 62-year-old under outpatient care for hypertension and osteoarthritis. He was retired from work and lived with his wife. Mr. Scaramanga was fully independent and capable of driving and taking care of his personal, domestic, and community activities of daily living. He had three adult children who lived nearby and were supportive. Mr. Scaramanga received an elective total knee replacement, and one day post-surgery, his right leg had become pale and with the pulse difficult to palpate. Subsequent investigation revealed that Mr. Scaramanga had an ischemic limb with a thrombus in the popliteal artery for which he received an urgent right popliteal artery angioplasty and stenting. He also received right lower limb fasciotomy for a compartment syndrome complication and anticoagulation for new-onset right-sided facial droop, hemiparesis, and aphasia that showed two days after from deep vein thrombus. A home-based treatment care team looked after Mr. Scaramanga post-discharge to implement his rehabilitation plan for ongoing right upper and lower limb weakness, dysphagia, mild receptive and moderate expressive aphasia, and higher-level cognitive deficits.

During subsequent consultation, Mr. Scaramanga's wife and daughter expressed significant anger at the orthopedic team, reporting that they believed a prolonged tourniquet had caused Mr. Scaramanga's complications during his initial operation. They wanted Mr. Scaramanga to be admitted for in-patient rehabilitation as soon as possible, perceiving the home-based care team to be providing suboptimal care about which they wanted to make a formal complaint. However, Mr. Scaramanga did not speak for himself, likely due to his aphasia and presenting cognitive deficits. The family objected to not having met with the primary team looking after Mr. Scaramanga, and complained that no one had been able to talk to them after any of the surgeries or engage them as partners in his rehabilitation plan.

What Do You Think?

1. How would you apply value co-creation in treatment care to resolve the impasse between the Scaramanga family and the care providers?
2. What would be the potentialities and limits of value co-creation as a sustainable community health care practice?

When people perceive to achieve health and wellbeing from their health system, they are likely to engage in value co-creation with the health care system, practicing health behaviors and preserving the assets for the sustainable health of future generations (Borgonovi et al., 2018;

Garcés, Ródenas, & Sanjosé, 2003; Keat, Whiteley, & Abercrombie, 1994). However, while improving people's health awareness is a best sustainability practice, engaging communities in routinely attending to their wellness is even more important to health system efficiency and cost-effectiveness. Excessive out-of-pocket expense payments create socioeconomic vulnerability, making it more difficult for people to participate in their health self-management (Callander, Corscadden, & Levesque, 2017; Himmelstein, Thorne, Warren, & Woolhandler, 2009; Saksena, Xu, Elovainio, & Perrot, 2010). The constraints to people engaging in self-management of their own health behavior would be particularly prevalent in societies without universal health care, and foster a (mis) perception by the general population that high-cost private health provides higher quality of care than the public health system (Angeli, Ishwardat, Jaiswal, & Capaldo, 2018; Borgonovi et al., 2018).

Health Promotion Aimed at Vulnerable Populations Community-level interventions for addressing health disparities, health inequity, and social injustice affecting vulnerable populations and communities would make for sustainable health as compared to individual-level interventions (United Nations Development Programme [UNDP], 2014; see Chap. 2, this volume). Kindig and Stoddart (2003) defined population health as "the health outcomes of a group of individuals, including the distribution of such outcomes within the group" (p. 3). This definition goes beyond distribution of health outcomes of geographic regions (e.g., nations, states, and communities), to encompass minority statuses such as by socioeconomic status, sex and gender, neurodiversity, and aging. Historically marginalized populations with disempowerment from income, education, employment, social support, and culture disparities also have inordinate exposure to adverse physical environments (air and water pollution and a lack of sanitation), increasing their health vulnerabilities as well as marginal resources to manage their health (see also Chap. 3, this volume). They experience *health inequities* or the unfair and avoidable differences in health status (WHO, 2008), harming their health abilities and achievements over the long term.

Poverty is a major barrier that keeps people from self-managing their health, and the consequent ill health they experience exacerbates their financial situation making it less likely they would access health care services they require (Wagstaff, 2002). Yet, few countries have reliable universal health care systems and the world's private sector health market aims to provide for economically well-off consumers, ignoring a vast majority of the less well-off, estimated at over four billion people (Prahalad, 2009). Community health partnerships would fail with a cultural divide and tension between primary health care providers and community members (Alderwick & Gottlieb, 2019; Cummins et al., 2015).

Community health epidemiological approaches aimed to address the prevalence of long-term diseases (Bigdeli, Shroff, Godin, & Ghaffar, 2018; see also Chap. 8, this volume) would proactively address the needs of populations with vulnerability, applying community-centric concepts (Garcés, Ródenas, & Sanjosé, 2003; Zaidi & Morgan, 2017; see also Chaps. 14 and 15, this volume). Community action coalitions would address community health disparities, health equity, and social justice in sustainable ways. Sustainable community health programs lead to health equity and equality or the opportunity for all to live to their full health potential.

Despite diversity in approaches, sustainable community health systems are defined by the following qualities: (1) coordinated care across the continuum of care (e.g., primary care, hospital, and the community), (2) a broader scope in services for the health of populations, (3) use of community epidemiological approaches to best target health support services to population segments, (4) use of quality of care and health informatics data for effective referral system and data sharing between health care providers and non-health care organizations, (5) use of hospitalization at home, and (6) strong multisector partnerships between the public, not-for-profit, and for-profit private organizations to address community medical and nonmedical patient needs. Sustainable community health is enhanced with use of appropriate measures to track the health of populations influenced by the interplay between their social, economic, and environmental conditions for the design of evidence-based or best-practice interventions (Glasgow, Goldstein, Ockene, & Pronk, 2004).

Economic Sustainability

Economic sustainability is critical to a health system, and this has become quite apparent with the repeat economic crises over the past decade which have put a strain on public budgets, while health care expenditures continue to increase exponentially (Karanikolos et al., 2013; Liaropoulos & Goranitis, 2015; Pencheon, 2015). Changing population demographics investments in technology and infrastructure, medical products, and wages are associated with exponential growing public health expenditures (Garcés, Ródenas, & Sanjosé, 2003; Popescu, Militaru, Cristescu, Vasilescu, & Maer Matei, 2018). However, simply cutting expenses without improving the efficiency and effectiveness of public health systems and their responsiveness to current projected future needs would only make for inferior health services (see also Chap. 4, this volume). The economic sustainability of community health follows from realistic shared expenditures and the use of prospective control systems to prevent wastage. Moreover, investment in health technologies and prevention of misappropriations of funds enhances the economic sustainability of health systems.

Share Expenditures and Control Systems Economic sustainability of community health systems is ensured with a high share of health expenditures in the gross domestic product (GDP) on health, a commensurate high share of employment in the health sector, longer working life or later age retirement (Popescu et al., 2018), and reducing reduced risks to sustainability. The economic sustainability of community health is also associated with broader use of alternative care setting arrangements such as hospitalization-at-home practices and use of state-of-the-art health technologies.

Health Expenditures In general, consistent, higher tax dollars on health would provide for sustainable community health. However, the relationship is far from perfect (Conference Board of Canada [CBC], 2012; Organization for Economic Cooperation and Development [OECD],

2013; Jaswal, 2013; National Health Expenditure Accounts [NHEA], 2019). For instance, while Canada spends 12% of gross domestic product annually on health care, the majority of Canadians have less-than-optimal health outcomes (Jaswal, 2013). Similarly, the annual health care spending of the United States of $3.5 trillion has not resulted in significant changes in population health (NHEA, 2019). The United States compares poorly with other developed countries in rankings on infant mortality, premature mortality, and life expectancy (OECD, 2013; Ridic, Gleason, & Ridic, 2012). By contrast, Italy, Japan, Switzerland, and the United Kingdom have significantly improved population health outcomes (CBC, 2012). Japan has the lowest health care spending per capita and one of the highest life expectancies (CBC, 2012; Kontis, Bennett, Mathers, Li, Foreman, & Ezzati, 2017). This goes to say that gross health expenditure alone, while important to population health, is not singularly the solution to sustainable health. As previously noted, health systems operate dynamically, with input from their economic, social, and environmental systems.

The United States has the highest-paid physicians and high costs of medications among the OECD countries (OECD, 2013; Mossialos, Wenzl, Osborn, & Sarnak, 2016), inflating the country's health care expenditure costs (Fineberg, 2012). This inflation in personal health costs affects vulnerable populations such as those living in rural areas, older adults, people with disabilities, women, and racial/ethnic minorities (Hobijn & Lagakos, 2005; see also Chap. 2, this volume).

In the United States, new innovative reimbursement and health care delivery models have provided the foundation for innovation at the state and community level (see Discussion Box 1.2). These models make it easier for providers and community partners to link clinical and community approaches to health because they are based on the concept of value (Ryan, Riley, Abrams, & Nocon, 2015; see also Chap. 9, this volume).

Discussion Box 1.2: Impact of Locational and Service Area Differential Reimbursement on Sustainability of Hospitals in the United States

The Medicare reimbursement rates for hospitals across America are based on location and population of the surrounding area of the hospital. Therefore, hospitals in higher populated areas receive a larger percentage of reimbursement than hospitals in lower populated areas. The thought behind these statistical calculations is that costs are higher in urban areas, so health care providers and health care entities need to make more profit. However, in many rural areas, low population densities make it very difficult to maintain a sufficient volume of patients to make this model economically sustainable. In addition, many health care and community-based organizations in rural areas also have a difficult time recruiting and retaining health professionals and other staff.

As a result, small community hospitals across the United States are closing their doors due to lack of reimbursement to be sustainable, leaving regions of the country without hospital care.

What Do You Think?

1. What are possible solutions to the financial insolvency of rural hospitals in the United States?
2. What partnerships would be needed to address the financial insolvency of a hospital system of which you are aware?

Value is based on improvements in individual and population health outcomes, as well as the cost of delivering those outcomes. Value-based payment systems reimburse hospitals and providers based on patient health outcomes and shift the incentive from quantity of care to quality of care.

Control-Based Approaches In-built efficiencies for economic sustainability of health systems are associated with the adoption of control-based approaches to performance management that generate the evidence from implementation to feedback into the systems for effectiveness and efficiency services, cutting down on costs wastage (Saviano, Bassano, Piciocchi, Di Nauta, & Lettieri, 2018) and extensive use of hospitalization at home practices (Verjan, Augusto, Xie, & Buthion, 2013).

Implementation of effective control-based approaches in managing health systems maximizes use of available resources at the organizational level for economic sustainability (Porter, 2010), increasing value of the health care service to both consumers and providers (Adinolfi, Starace, & Palumbo, 2016; Harris, Green, & Elshaug, 2017). With these measures in place, patients, families, employers, and government have lower health expenditures, which would make more resources available for long-term investments in preparedness for new diseases or pandemics, cutting-edge health care research addressing the needs of vulnerable segments of the society, and adoption of new health care technologies (Fineberg, 2012; Palozzi, Brunelli, & Falivena, 2018). Moreover, investments in community health emergency preparedness, containment through rapid testing capabilities, mitigation to contain community spread and resolution, or deceleration tracking of successful interventions are core to sustainable health systems.

Misappropriations of health budgets can occur by omission or commission. For instance, misappropriation by omission could include failure to provide resources to address information gaps between physicians and patients, and between suppliers (medical devices and drug producers) and buyers (public health care organizations and agencies) (Bae, Masud, Kaium, & Kim, 2018), or not providing for affordable health despite adequate financial resources. Misappropriation by commission would involve unfairly advantaging special interests by deliberately sustaining unwanted complexity of the health system to make it difficult to analyze information for identifying avoidable financial leaks in the health system, shifting responsibilities and roles to cream the system, and implementing weak risk prevention and control systems along the supply chain (Kickbusch, Allen, & Franz, 2016). Either way, misappropriations compromises the sustainability of health systems by degrading institutional accountability to all the stakeholders and the community (Palumbo & Manna, 2018; Previtali & Cerchiello, 2018).

Moreover, while health inequity is a global phenomenon, resource scarcity, neglect, and misappropriation of resources significantly amplify its social injustices. This in part is because misappropriation of health funding compromises quality of patient care (Muldoon et al., 2011; Pinzón-Flórez et al., 2015; Savedoff & Hussmann, 2006) and loses the confidence of providers and the workforce, stifling care innovations for inclusive health care systems. Moreover, damage to the economic pillar has long-term pervasive effects on the quality of health systems institutions and the populations they serve. Regrettably, ordinary community members with high exposure to high health inequity may not be aware of the depth and extent of deprivation they are exposed to, which would result in avoidable morbidity and mortality.

Hospitalization-at-Home Practices Hospitalization-at-home practices aim to provide care in the patient's home setting. Such practices include outpatient services and/or the provision of remote clinical care in the home using telemedicine to provide timely ongoing care (Leff & Montalto, 2004; Voudris & Silver, 2018; see also Chap. 11, this volume). Hospitalization-at-home is cost-effective for treating chronic illnesses, obviating the need for re-hospitalization, and results in family carers acquiring dependable knowledge for addressing the health needs of the patient in a naturalistic setting (Verjan et al., 2013). It also reduces the risk of acquiring health care-associated infections secondary to in-patient hospitalization, which may be costly to treat (Glance, Stone, Mukamel, & Dick, 2011; Graves, 2004). Use of hospitalization-at-home also reduces costs for medical care facilities allowing them to focus on critical care cases requiring intensive medical care for which a home care setting would not be equipped to handle. With the rapid evolution of the Internet of Medical Things (IoMT) with device interconnectivity, hospitalization-at-home will increasingly be a preferred in-community sustainable care arrangement. However, presently, device interconnectivity is very proprietary, with little interoperability, limiting the rapid implementation of hospitalization-at-home requiring device interoperability. Nonetheless, as the IoMT becomes more widely adopted, homes of the future would have in-built hospitalization-at-home suites allowing for some critical health care needs to be met which are presently provided with hospitalization, such as pandemic community spread mitigations.

Technology Adoption Health technologies are rapidly available as tools by which communities can improve on their health outcomes, often at a lower cost to the citizens with wider adoption, distribution, and appropriate use. Use of innovative health technology holds high promise for the sustainability of health systems through savings from efficiency and effectiveness (Iandolo, Vito, Fulco, & Loia, 2018; Paris, Slawomirski, Colbert, Delaunay, & Oderkirk, 2017; see Chaps. 10 and 11, this volume). Increasingly, smart technologies are used to generate and manage health records for individuals within communities (Lo Presti, Testa, Marino, & Singer, 2019; see also Chap. 10, this volume). For example, Apple is developing fully medicalized iPhone sensors capable of measuring blood pressure, body fat, and heart function in real time, in addition to advising users of their immediate health care needs. These devices are intended for users to view, manage, and share medical records with health providers, increasing patient control over their own health data. Alphabet, the parent company to Google, is testing the utility of intelligent application to predict risk for patient mortality, in addition to another technology for processing dummified patients' records for prospecting for emerging diseases with a view to preventive illness as well as initiate care service planning. Sage Bionetworks is experimenting with an iPhone application to predict the risk for Parkinson's disease from measuring tremors while performing mundane tasks and long before other symptoms appear. CityHealth is mining health data from low-income communities for its use to predict and provide health care services to residents on Medicaid. Health insurance industry consortia such as Amazon/Berkshire Hathaway and JP Morgan Chase Health Care Company are developing smart technologies to sell drugs online. Moreover, a slew of so-called digiceuticals (Downes, Horigan, & Teixeira, 2019; Khirasaria, Singh, & Batta, 2020) (in contract to pharmaceuticals) smart technologies use sensors to provide real-time guidance to people, managing chronic illness symptoms by recognizing triggers to significant events requiring preventive actions or medical assistance.

The fact is that "digiceuticals" data can be compromised and pose an unfathomable risk to communities and individuals, for which risk they would not have consented (Cushing, 2013; Leetaru, 2016). In addition, the development and deployment of these digital health technologies are on the presumption that individuals can self-manage their health data or that the tech and digiceuticals organizations know what is suitable for the persons and communities whose data they hold. The fact is that the average person or community member is limited in capabilities to self-manage sensitive health data (Adjerid, Peer, & Acquisti, 2017), which could raise costs of health services long term from unintended neglect by would-be consumers. Health insurance organizations would not provide reimbursement for the use of health technologies with no regulatory approval, which would be an unwanted cost on individuals and communities. Thus, while innovative health technology adoption and implementation may be cost-effective along the value chain to health professionals, hospital managers, consumers, and the health care organization, they are not without disadvantages. Improving access to care and the quality of services using health technology comes with careful discretion by health system managers to avoid economic risk from acquiring technologies that do not do what they claim to do and are not value for money (Palozzi, Brunelli, & Falivena, 2018). Moreover, digiceuticals organizations would not invest in technologies needed by minority communities if they do not perceive the prospects to generate significant profits; while health systems organizations contend with potential conflicts of interest in the procurement and cost of proprietary health technologies they are involved in developing.

Environmental Sustainability

The health of nature is critically important to achieving sustainable community health due to the interdependency between nature and the health of humans (Tsasis, Agrawal, & Guriel, 2019). Degradations of the natural environment, mostly man-made, contribute substantially to the environmental unsustainability of health systems, with adverse effects of the health of populations (WHO, 2013a, 2013b) (see also Discussion Box 1.3). For instance, air pollution associates with escalating respiratory and cardiovascular conditions in populations, while climate change contributes to more frequent catastrophic flooding, insect-borne diseases, droughts, and wildfires (Selhub & Logan, 2012; WHO, 2013a, 2013b). Pollutants and contaminates exacerbate the prevalence of communicable and noncommunicable diseases (see Chap. 12, this volume). These natural environmental degradations compromise the economic sustainability of community health by increasing the burden of health care (Jameton & Pierce, 2001). In fact, health care organizations have an unenviable reputation for being harmful to their natural environment from generating "all existing classes of waste" (Carnero, 2015, p. 8270). Past and ongoing affronts to the natural environment are associated with increased health care, mostly from failures by societies to protect the natural environment ultimately damage the sustainability of their health systems, which would harm population health, increasing both the demand and cost of health care (Jameton & Pierce, 2001).

Moreover, proactively addressing matters concerning the biophysical and chemical environment affecting the health of populations such as energy use efficiencies (Pencheon, 2015) and the institutionalization of environmental protection concerns (Fischer, 2015) makes for sustainable community health. The ways societies create livable communities, attentive to environmentally friendly procurement supply chains, travel and transportation, waste management, the built environment, and partnerships for environmental protections are critical to sustainable health systems (Pencheon, 2015; Stuart & Adams, 2007; see also Chap. 3, this volume).

Discussion Box 1.3: Infectious Agents from Human–Animal Interactions

The way humans relate to animals is associated with a long history of catastrophic pandemics that have killed millions of people (Torrey &Yolken, 2005), notably the bubonic plague of 1347–1351 from rat infestation which converged into three pandemics through the early twentieth century (Dean, Kraurer, & Schmid, 2019). Recently the COVID-19 pandemic, which may have resulted from transmission of infectious agents among animals piled in cages on top of each other at a wet food market in China, is transmitting from human to human at a phenomenal rate. This is not to discount the fact that since the domestication of animals about 10,000 years ago, human–animal interactions have had both positive health effects on humans in a myriad of ways, including physical, mental, and nutritional health. The approximately 4500 mammal species each host many of one million known types of bacteria and viruses. Almost all domesticated animals are linked to an infectious agent or derivation from related family of viruses: whooping cough related to bacterium from pigs, measles from related viruses in cows, tuberculosis from related mycobacteria in goats, measles from related viruses in cows, typhoid fever from bacterium in chickens. Ducks carry the influenza virus that killed 20 million people in 1918–1920. The list is long. Human–animal interactions are increasingly close and personal, risking infectious agent transmissions, which then become human-to-human. Herd immunity or the population threshold for exposed individuals resulting in a decline in risk for new infections is still an inexact science (John & Samuel, 2000), and there are presently very few successful vaccines against new human immune-deficiency viruses related to infectious diseases. By the year 2060, about two-thirds of the world's population will be living in urban areas. The changing human habitats toward urbanization and the use of rapid mass transit systems interlinking global communities create opportunities for international corporations in combating pandemics. However, it also risks the quick transmission of viruses for which the global community may be ill-prepared to contain and mitigate without the risk of catastrophic failures in health systems, economies, and human–nature ecosystems.

What Do You Think?

1. Suggest ways by which human–animal interactions could minimize the risks for infectious agent transmissions in a globalized world economy?
2. How might the natural environmental pillars of sustainable health be leveraged for a safer world from communicable diseases?
3. In what ways would reducing risk of communicable diseases also mitigate the risk for noncommunicable diseases? Discuss with reference to given examples.

Negotiating the Social, Economic, and Environmental Sustenance Pillars

Sustainable health systems have to contend with complexities from the interactions between the social, economic, and environmental pillars (Armitage et al., 2009; Marcus, Kurucz, & Colbert, 2010). For their long-term function, community health systems need to show evidence of providing consistent, quality services to the community within the prevailing financial and environmental health conditions. They must demonstrate both efficiency and cost-effectiveness in how they deliver services to the community (economic pillar; see also Chap. 4, this volume). They also must be accessible to the community across segments of social (dis) advantage and with a focus on health promotion through value co-creation with the community (social pillar) for health self-management (Russo et al., 2019; see Chaps. 2 and 9, this volume). At the same time, they must provide for health services with natural environment protection, minimizing waste, which would risk population health (i.e., environmental pillar).

Health systems' political environment influences the permutations that occur in their implementation (Borgonovi et al., 2018; Muntaner et al., 2011). With health as a human right orientation, state and federal governments would provide universal care as a health social justice practice than without a human rights perspective. In general, health as a human right political environment would come with contingent financial resources for universal coverage. What is more, the overall costs of health systems with universal coverage do not rise because of the increased use of services (Lu & Hsiao, 2003). The ways to which human populations perceive to have guaranteed health protections might make for lower health system costs from people accessing health promotion services, thus actively engaging in their health management. With universal health coverage, there also would be a healthy relationship with the natural environment from engaging in less environmental pillage for short-term economic gains to afford health care. This effect may occur from people's understanding of the health benefits of the natural environment and engaging in pro-natural environment actions translating into health care cost saving in the long-term from the lower prevalence of health conditions resulting from environmental degradation (Selhub & Logan, 2013).

While state and federal governments around the globe would endorse health as a human rights, the actions that they take in translating that conviction into sustainable communities vary enormously. Some developed countries such as the United States, with enormous wealth and health expenditures, do not have universal health coverage, and have very steep health care costs on citizens, resulting in marginal population health status. Other countries like Japan, Canada, Germany, Australasia, and the Nordic countries with universal health care have admirable population health overall (OECD, 2013). The historical-political environment in which health systems function determine the commitment to health social justice, financing, and natural environment protections (Di Nardo, Saulle, & La Torre, 2010; Ridic, Gleason, & Ridic, 2012).

While there is consensus about the importance of the three pillars to sustainable health, it remains an open question as to their older or relative importance in influencing sustainable health systems. One view proposes that the economic pillar would exert more influence on sustainable health systems and that underfunded health systems are inherently unsustainable (Fineberg, 2012; Previtali & Cerchiello, 2018). Admittedly, the socioeconomy is fundamental to health and wellbeing (Link & Phelan, 2005). Nevertheless, as previously noted, the association between health expenditure and health outcomes is at best modest. Another view, vested in social sustainability, proposes participatory health by communities as in value co-creation in the health system and access or availability to all critically important. An emerging view is that attending to the health of the natural environment would make for population health from the health promotion benefits of mindful living with nature, and the reduction in health systems costs that follow. Each of these views has merit and undeniable assets for sustainable community health.

Summary and Conclusion

Sustainable community health rests on three interlinked pillars: social, economic, and environmental health and the enabling tools for them. While good health is a universal human pursuit, community health disparities persist in the absence of policies and practices for sustainable health. Programs and initiatives aimed to address the nexus between health disparities, health equity, and health social justice latently promote sustainable community health for all. Sustainable community health depends on many factors, including economic vitality, education, the environment, and community safety for all. Utilizing sustainable health approaches would not merely treat illnesses, but also enable collaborations to address nonmedical needs with health implications, embracing a model of health and wellness. For instance, sustainable health approaches target social needs, including housing, transportation, and food insecurity, as well as interpersonal violence and toxic stress, without medicalizing nonmedical issues.

Each of the social, economic, and environmental pillars of sustainable community health influences all human interactions for health in complex ways. The increasing demand for high-quality health care for all with efficiency to optimize on financial resources while protecting the environment calls for adoption of sustainable health policies and practices. The sustainable health of populations is improved not just by the medical care they receive, but by the social and physical environment in which people live, grow, and work as well as by the economic viability of their community health systems. Ultimately, human decisions influence the relative contribution of each of the health sustainability pillars toward population's health outcomes in their diversity. The decisional choices may be unbalanced and influenced by local contexts, while still with similar underpinnings in social, economic, and environment. This chapter assumes the perspective that the natural environment is fundamental to the health and wellbeing of the planet earth and all its inhabitants. Health systems that protect the natural environment would make for economic and social sustenance.

Self-Check Questions

1. Define sustainable community health and its pillars.
2. Differentiate between sustainable health, community health, and sustainable community health systems.
3. What is the relative advantage of social, economic, and environmental sustainability determinants to designing and implementing community health systems?
4. What approaches to sustainable community health would be responsive to health disparities, equity, and social justice.
5. How would you optimize social, economic, and environmental pillars for resilient community health?
6. Evaluate the role of the natural environment to sustainable community health.

Discussion Questions

1. How may sustainable community health vary between developed and developing nations? Explain as differences.
2. Of the pillars that define sustainable community health, which would you prioritize in your specific setting?
3. Propose and justify a pillar for sustainable health systems different from or related to those considered in this chapter. Justify your nomination.

Field-Based Experiential Exercises

1. Prepare a sustainable community health checklist to use on a community you are familiar with. Report on your preliminary findings, suggesting the strengths and potential of the health system.
2. Identify a nonmedical agency with work that has health and wellbeing implications. Interview a leader at the agency regarding health and wellbeing outcomes from their work. Share your summary with them.

Online Resources

1. Strategic steps to Sustainability in health care https://www.healthcarefacilitiestoday.com/posts/Strategic-steps-to-sustainability-in-healthcare%2D%2D13629
2. Connecting sustainability to health care mission https://www.healthcaredesignmagazine.com/architecture/connecting-sustainability-healthcare-mission/
3. Environmental sustainability in hospitals: The value of efficiency http://www.hpoe.org/Reports-HPOE/ashe-sustainability-report-FINAL.pdf
4. Planning for sustainability https://www.ruralhealthinfo.org/toolkits/rural-toolkit/5/sustainability-planning

References

Adam, T., & de Savigny, D. (2012). Systems thinking for strengthening health systems in LMICs: Need for a paradigm shift. *Health Policy and Planning, 27*(suppl_4), iv1–iv3.

Adinolfi, P., Starace, F., & Palumbo, R. (2016). Health outcomes and patient empowerment: The case of health budgets in Italy. *Journal of Health Management, 18*(1), 117–133.

Adjerid, L., Peer, E., & Acquisti, A. (2017). Beyond the privacy paradox: Objective versus relative risk in privacy decision making. *Management and Information Systems Quarterly*. Retrieved on January 26 from https://papers.ssrn.com/sol3/papers.cfm?abstract_id=2765097

Alderwick, H., & Gottlieb, L. M. (2019). Meanings and misunderstandings: A social determinants of health lexicon for health care systems. *The Milbank Quarterly, 97*(2).

Alliance for Natural Health (ANH). (2010). *Sustainable healthcare—Working towards the paradigm shift*. White Paper. Dorking, UK: Alliance for Natural Health.

Angeli, F., Ishwardat, S. T., Jaiswal, A. K., & Capaldo, A. (2018). Socio-cultural sustainability of private healthcare providers in an Indian slum setting: A bottom-of-the-pyramid perspective. *Sustainability, 10*(12), 4702.

Armitage, D. R., Plummer, R., Berkes, F., Arthur, R. I., Charles, A. T., Davidson-Hunt, I. J., … McConney, P. (2009). Adaptive co-management for social–ecological complexity. *Frontiers in Ecology and the Environment, 7*(2), 95–102.

Bae, S. M., Masud, M., Kaium, A., & Kim, J. D. (2018). A cross-country investigation of corporate governance and corporate sustainability disclosure: A signaling theory perspective. *Sustainability, 10*(8), 2611.

Bago d'Uva, T., Van Doorslaer, E., Lindeboom, M., & O'Donnell, O. (2008). Does reporting heterogeneity bias the measurement of health disparities? *Health Economics, 17*(3), 351–375.

Beirão, G., Patrício, L., & Fisk, R. P. (2017). Value co-creation in service ecosystems. *Journal of Service Management, 28*(2), 227–249.

Bigdeli, M., Shroff, Z. C., Godin, I., & Ghaffar, A. (2018). Health systems research on access to medicines: Unpacking challenges in implementing policies in the face of the epidemiological transition. *BMC Global Health, 2*(3), e000941. https://doi.org/10.1136/bmjgh-2018-000941

Borgonovi, E., Adinolfi, P., Palumbo, R., & Piscopo, G. (2018). Framing the shades of sustainability in health care: Pitfalls and perspectives from Western EU countries. *Sustainability, 10*(12), 4439.

Boutin-Foster, C., & Charlson, M. E. (2001). Problematic resident-patient relationships. *Journal of General Internal Medicine, 16*(11), 750–754.

Braveman, P. (2006). Health disparities and health equity: concepts and measurement. *Annual Review of Public Health, 27*, 167–194.

Callander, E. J., Corscadden, L., & Levesque, J. F. (2017). Out-of-pocket healthcare expenditure and chronic disease–do Australians forgo care because of the cost? *Australian Journal of Primary Health, 23*(1), 15–22.

Carnero, M. C. (2015). Assessment of environmental sustainability in health care organizations. *Sustainability, 7*(7), 8270–8291.

Chang, A. Y., Robinson, L. A., Hammitt, J. K., & Resch, S. C. (2017). Economics in "Global Health 2035": A sensitivity analysis of the value of a life year estimates. *Journal of Global Health, 7*(1), 010801. https://doi.org/10.7189/jogh.07.010401

Coiera, E., & Hovenga, E. S. (2007). Building a sustainable health system. *Yearbook of Medical Informatics, 16*(01), 11–18.

Conference Board of Canada. (2012). Health Spending. Do Countries Get What They Pay for When It Comes to Healthcare? Retrieved on July 10, 2020 from http://www.conferenceboard.ca/hcp/hot-topics/health-spending.aspx

Cummins, R., Mpofu, E., & Machina, M. (2015). Quality of community life indicators. In E. Mpofu (Ed.), *Community oriented health services: Practices across disciplines* (pp. 165–181). New York, NY: Springer.

Cushing, E. (2013). *Amazon Mechanical Turk: The digital sweatshop.* UTNE. Retrieved on January 26, 2020 from http://utne.com/science-and-technology/amazon-mechanical-turk-zm0z13jfzlin.aspx.

Dean, K. R., Kraurer, F., & Schmid, B. V. (2019). The epidemiology of plague in Europe: Inferring transmission dynamics from historical data. *Royal Society Open Science, 6*, 181695. https://doi.org/10.1098/rsos.181695

Denny, K., & Davidson, M. J. (2012). Area-based socio-economic measures as tools for health disparities research, policy and planning. *Canadian Journal of Public Health/Revue Canadienne De Sante'e Publique, 103*, S4–S6.

Di Nardo, F., Saulle, R., & La Torre, G. (2010). Green areas and health outcomes: A systematic review of the scientific literature. *Italian Journal of Public Health, 7*(4), 402–413.

Downes, E., Horigan, A., & Teixeira, P. (2019). The transformation of health care for patients: Information and communication technology, digiceuticals, and digitally enabled care. *Journal of the American Association of Nurse Practitioners, 31*(3), 156–161.

Faezipour, M., & Ferreira, S. (2013). A system dynamics perspective of patient satisfaction in healthcare. *Procedia Computer Science, 16*, 148–156.

Fineberg, H. V. (2012). A successful and sustainable health system—How to get there from here. *New England Journal of Medicine, 366*(11), 1020–1027.

Fischer, M. (2015). Fit for the future? A new approach in the debate about what makes healthcare systems really sustainable. *Sustainability, 7*(1), 294–312.

Frow, P., McColl-Kennedy, J. R., & Payne, A. (2016). Co-creation practices: Their role in shaping a health care ecosystem. *Industrial Marketing Management, 56*, 24–39.

Garcés, J., Ródenas, F., & Sanjosé, V. (2003). Towards a new welfare state: The social sustainability principle and health care strategies. *Health Policy, 65*(3), 201–215.

Glance, L. G., Stone, P. W., Mukamel, D. B., & Dick, A. W. (2011). Increases in mortality, length of stay, and cost associated with hospital-acquired infections in trauma patients. *Archives of Surgery, 146*(7), 794–801.

Glasgow, R. E., Goldstein, M. G., Ockene, J. K., & Pronk, N. P. (2004). Translating what we have learned into practice: Principles and hypotheses for interventions addressing multiple behaviors in primary care. *American Journal of Preventive Medicine, 27*(2), 88–101.

Gluyas, H. (2015). Effective communication and teamwork promotes patient safety. *Nursing Standard (2014+), 29*(49), 50.

Gottlieb, L., Fichtenberg, C., Alderwick, H., & Adler, N. (2019). Social determinants of health: What's a healthcare system to do? *Journal of Healthcare Management, 64*(4), 243–257.

Graves, N. (2004). Economics and preventing hospital-acquired infection. *Emerging Infectious Diseases, 10*(4), 561.

Guidotti, T. L. (2018). Sustainability and health: Notes toward a convergence of agendas. *Journal of Environmental Studies and Sciences, 8*(3), 357–361.

Harris, C., Green, S., & Elshaug, A. G. (2017). Sustainability in Health care by Allocating Resources Effectively (SHARE) 10: Operationalising disinvestment in a conceptual framework for resource allocation. *BMC Health Services Research, 17*(1), 632.

Himmelstein, D. U., Thorne, D., Warren, E., & Woolhandler, S. (2009). Medical bankruptcy in the United States, 2007: Results of a national study. *The American Journal of Medicine, 122*(8), 741–746.

Hobijn, B., & Lagakos, D. (2005). Inflation inequality in the United States. *Review of Income and Wealth, 51*(4), 581–606.

Iandolo, F., Vito, P., Fulco, I., & Loia, F. (2018). From health technology assessment to health technology sustainability. *Sustainability, 10*(12), 4748.

Jameton, A., & Pierce, J. (2001). Environment and health: 8. Sustainable health care and emerging ethical responsibilities. *Canadian Medical Associating Journal, 164*(3), 365–369.

Jameton, A., & McGuire, C. (2002). Toward sustainable health-care services: Principles, challenges, and a process. *International Journal of Sustainability in Higher Education, 3*(2), 113–127. https://doi.org/10.1108/14676370210422348

Jamison, D. T., Summers, L. H., Alleyne, G., Arrow, K. J., Berkley, S., Binagwaho, A., … Ghosh, G. (2013). Global health 2035: A world converging within a generation. *The Lancet, 382*(9908), 1898–1955.

Jaswal, A. (2013). Canada 2020 Analytical Commentary No. 3. Valuing Health in Canada. Who, How and How Much? Canada 2020: Ottawa, ON, Canada.

John, T. J., & Samuel, R. (2000). Herd immunity and herd effect: New insights and definitions. *European Journal of Epidemiology, 16*(7), 601–606.

Karanikolos, M., Mladovsky, P., Cylus, J., Thomson, S., Basu, S., Stuckler, D., … McKee, M. (2013). Financial crisis, austerity, and health in Europe. *The Lancet, 381*(9874), 1323–1331.

Keat, R., Whiteley, N., & Abercrombie, N. (Eds.). (1994). *The authority of the consumer* (pp. 1–19). London, UK: Routledge.

Khirasaria, R., Singh, V., & Batta, A. (2020). Exploring digital therapeutics: The next paradigm of modern health-care industry. *Perspectives in Clinical Research, 11*(2), 54.

Kickbusch, I., Allen, L., & Franz, C. (2016). The commercial determinants of health. *The Lancet Global Health, 4*(12), e895–e896.

Kindig, D., & Stoddart, G. (2003). What is population health? *American Journal of Public Health, 93*(3), 380–383.

Kontis, V., Bennett, J. E., Mathers, C. D., Li, G., Foreman, K., & Ezzati, M. (2017). Future life expectancy in 35 industrialised countries: Projections with a Bayesian model ensemble. *The Lancet, 389*(10076), 1323–1335.

Leetaru, K. (2016). Are research ethics obsolete in the era of big data? *Forbes*. Retrieved on January 27 from https://www.forbes.com/sites/kalevleetaru/2016/06/17/are-research-ethics-obsolete-in-the-era-of-big-data/#2353d8897aa3

Leff, B., & Montalto, M. (2004). Home hospital—Toward a tighter definition. *Journal of the American Geriatrics Society, 52*(12), 2141–2141.

Lennox, L., Maher, L., & Reed, J. (2018). Navigating the sustainability landscape: A systematic review of sustainability approaches in healthcare. *Implementation Science, 13*(1), 27.

Liaropoulos, L., & Goranitis, I. (2015). Health care financing and the sustainability of health systems. *International Journal for Equity in Health, 14*(1), 80.

Link, B. G., & Phelan, J. C. (2005). Fundamental sources of health inequalities. In D. Mechanic, L. B. Rogut, D. C. Colby & J. R., Knickman (Eds). *Policy challenges in modern health care* (pp. 71–84). New Brunswick, NJ: Rutgers University Press.

Lo Presti, L., Testa, M., Marino, V., & Singer, P. (2019). Engagement in healthcare systems: Adopting digital tools for a sustainable approach. *Sustainability, 11*(1), 220.

Lu, J. F. R., & Hsiao, W. C. (2003). Does universal health insurance make health care unaffordable? Lessons from Taiwan. *Health Affairs, 22*(3), 77–88.

Marcus, J., Kurucz, E. C., & Colbert, B. A. (2010). Conceptions of the business-society-nature interface: Implications for management scholarship. *Business & Society, 49*(3), 402–438.

Marimuthu, M., & Paulose, H. (2016). Emergence of sustainability based approaches in healthcare: Expanding research and practice. *Procedia - Social and Behavioral Sciences, 224*, 554–561.

Marmot, M. (2017). Social justice, epidemiology and health inequalities. *European Journal of Epidemiology, 32*(7), 537–546.

Marmot, M., & Commission on Social Determinants of Health. (2007). Achieving health equity: From root causes to fair outcomes. *The Lancet, 370*(9593), 1153–1163.

Marrone, S. (2007). Understanding barriers to health care: A review of disparities in health care services among indigenous populations. *International Journal of Circumpolar Health, 66*(3), 188–198.

McGrail, M. R., & Humphreys, J. S. (2015). Spatial access disparities to primary health care in rural and remote Australia. *Geospatial Health, 10*(2), 138–143.

McKenzie, S. (2004). *Social sustainability: towards some definitions.* Magill, Australia: University of South Australia, Hawke Research Institute.

Meara, J. G., & Greenberg, S. L. (2015). The Lancet Commission on global surgery global surgery 2030: Evidence and solutions for achieving health, welfare and economic development. *Surgery, 157*(5), 834–835.

Mills, A. (2014). Health care systems in low-and middle-income countries. *New England Journal of Medicine, 370*(6), 552–557.

Mintzberg, H. (2015). *Rebalancing society: Radical renewal beyond left, right, and center.* Berrett-Koehler Publishers, San Franscisco, CA: Berrett-Koehler Publishers.

Mitchell, R., & Popham, F. (2008). Effect of exposure to natural environment on health inequalities: An observational population study. *The Lancet, 372*(9650), 1655–1660.

Mohrman, S. A., Shani, A. B. R., & McCracken, A. (2012). Organizing for sustainable health care: The emerging global challenge. In *Organizing for sustainable health care.* Emerald Group Publishing Limited.

Momete, D. C. (2016). Building a sustainable healthcare model: A cross-country analysis. *Sustainability, 8*(9), 836.

Mossialos, E., Wenzl, M., Osborn, R., & Sarnak, D. (2016). *2015 international profiles of health care systems.* Canadian Agency for Drugs and Technologies in Health.

Mpofu, E. (Ed.). (2015). *Community oriented health services: Practices across disciplines.* New York, NY: Springer.

Muldoon, K. A., Galway, L. P., Nakajima, M., Kanters, S., Hogg, R. S., Bendavid, E., & Mills, E. J. (2011). Health system determinants of infant, child and maternal mortality: A cross-sectional study of UN member countries. *Globalization and Health, 7*(1), 42.

Muntaner, C., Borrell, C., Ng, E., Chung, H., Espelt, A., Rodriguez-Sanz, M., … O'Campo, P. (2011). Politics, welfare regimes, and population health: Controversies and evidence. Sociology of Health & Illness, 33(6), 946–964.

Muzyka, D., Hodgson, G., & Prada, G. (2012). The inconvenient truths about Canadian health care. In *The Conference Board of Canada.* Retrieved on May 4, 2020, from http://www.conferenceboard.ca/cashc/research/2012/inconvenient_truths.aspx

National Health Expenditure Accounts (NHEA). (2019). Retried on July 10, 2020 from https://www.cms.gov/research-statisticsdata-and-systems/statistics-trends-and-reports/nationalhealthexpenddata/nationalhealthaccountshistorical

OECD. Publishing, & Organization for Economic Cooperation and Development (OECD) Staff. (2013). *Health at a glance 2013: OECD indicators*. OECD Publishing.

Palozzi, G., Brunelli, S., & Falivena, C. (2018). Higher sustainability and lower opportunistic behaviour in healthcare: A new framework for performing hospital-based health technology assessment. *Sustainability, 10*(10), 3550.

Palumbo, R., & Manna, R. (2018). What if things go wrong in co-producing health services? Exploring the implementation problems of health care co-production. *Policy and Society, 37*(3), 368–385.

Paris, V., Slawomirski, L., Colbert, A., Delaunay, N., & Oderkirk, J. (2017). *New health technologies managing access, value and sustainability*. Paris, France: OECD Publishing.

Pencheon, D. (2015). Making health care more sustainable: The case of the English NHS. *Public Health, 129*(10), 1335–1343.

Peter, F. (2001). Health equity and social justice. *Journal of Applied Philosophy, 18*(2), 159–170.

Pinzón-Flórez, C. E., Fernández-Niño, J. A., Ruiz-Rodriguez, M., Idrovo, A. J., & Lopez, A. A. A. (2015). Determinants of performance of health systems concerning maternal and child health: A global approach. *PLoS One, 10*(3), e0120747.

Popa, I., & Ştefan, S. C. (2019). Modeling the pathways of knowledge management towards social and economic outcomes of health organizations. *International Journal of Environmental Research and Public Health, 16*(7), 1114.

Popescu, M. E., Militaru, E., Cristescu, A., Vasilescu, M. D., & Maer Matei, M. M. (2018). Investigating health systems in the European Union: Outcomes and fiscal sustainability. *Sustainability, 10*(9), 3186.

Prahalad, C. K. (2009). *The fortune at the bottom of the pyramid, revised and updated 5th anniversary edition: Eradicating poverty through profits*. FT Press.

Previtali, P., & Cerchiello, P. (2018). The prevention of corruption as an unavoidable way to ensure healthcare system sustainability. *Sustainability, 10*(9), 3071.

Ridic, G., Gleason, S., & Ridic, O. (2012). Comparisons of health care systems in the United States, Germany and Canada. *Materia Socio-Medica, 24*(2), 112.

Roberts, J. L., & World Health Organization. (1998). *Terminology: A glossary of technical terms on the economics and finance of health services* (No. EUR/ICP/

CARE 94 01/CN01). Copenhagen, Denmark: WHO Regional Office for Europe.

Ruger, J. P. (2004). Health and social justice. *The Lancet, 364*(9439), 1075–1080.

Russo, G., Moretta Tartaglione, A., & Cavacece, Y. (2019). Empowering patients to co-create a sustainable healthcare value. *Sustainability, 11*(5), 1315.

Ryan, J., Riley, P., Abrams, M., & Nocon, R. (2015). *How strong is the primary care safety net? Assessing the ability of federally qualified health centers to serve as patient-centered medical homes.* New York, NY: Commonwealth Fund.

Saksena, P., Xu, K., Elovainio, R., & Perrot, J. (2010). Health services utilization and out-of-pocket expenditure at public and private facilities in low-income countries. *World Health Report, 20*, 20.

Savedoff, W. D., & Hussmann, K. (2006). The causes of corruption in the health sector: A focus on health care systems. *Transparency International. Global Corruption Report.* London, UK: Pluto Press.

Saviano, M., Bassano, C., & Calabrese, M. (2010). A VSA-SS approach to healthcare service systems the triple target of efficiency, effectiveness and sustainability. *Service Science, 2*(1–2), 41–61.

Saviano, M., Bassano, C., Piciocchi, P., Di Nauta, P., & Lettieri, M. (2018). Monitoring viability and sustainability in healthcare organizations. *Sustainability, 10*(10), 3548.

Schroeder, K., Thompson, T., Frith, K., & Pencheon, D. (2012). *Sustainable healthcare.* New York, NY: Wiley.

Schroeder, S. A. (2013). Rethinking health: Healthy or healthier than? *The British Journal for the Philosophy of Science, 64*(1), 131–159.

Selhub, E. M., & Logan, A. C. (2012). *Your brain on nature: The science of nature's influence on your health, happiness and vitality.* John Wiley & Sons.

Sen, A. (2002). Why health equity? *Health Economics, 11*(8), 659–666.

Shostak, S. (2013). Exposed science: Genes, the environment, and the politics of population health. Univ of California Press.

Solomon, R., & Orridge, C. (2014). Defining health equity. *Healthcare Papers, 14*(2), 62–65.

Stefan, S. C., Popa, I., & Dobrin, C. O. (2016). Towards a model of sustainable competitiveness of health organizations. *Sustainability, 8*(5), 464.

Stuart, N., & Adams, J. (2007). The sustainability of Canada's healthcare system: A framework for advancing the debate. *Healthcare Quarterly (Toronto, Ont.), 10*(2), 96–103.

Summers, L. H. (2015). Economists' declaration on universal health coverage. *The Lancet, 386*(10008), 2112–2113.

Torrey, E. F., & Yolken, R. H. (2005). *Beasts of the earth: Animals, humans, and disease.* Rutgers University Press.

Tsasis, P., Agrawal, N., & Guriel, N. (2019). An embedded systems perspective in conceptualizing Canada's healthcare sustainability. *Sustainability, 11*(2), 531.

United Nations Development Program (UNDP) Istanbul International Center for Private Sector Development (IICPSD). (2014). *Barriers and opportunities at the base of the pyramid: The role of the private sector in inclusive development.* New York, NY: Author.

Verjan, C. R., Augusto, V., Xie, X., & Buthion, V. (2013). Economic comparison between Hospital at Home and traditional hospitalization using a simulation-based approach. *Journal of Enterprise Information Management, 26*(1–2), 135–153.

Voudris, K. V., & Silver, M. A. (2018). Home hospitalization for acute decompensated heart failure: Opportunities and strategies for improved health outcomes. In *Healthcare* (Vol. 6(2), p. 31). Multidisciplinary Digital Publishing Institute.

Wagstaff, A. (2002). Poverty and health sector inequalities. *Bulletin of the World Health Organization, 80*, 97–105.

Woolf, S. H., & Braveman, P. (2011). Where health disparities begin: The role of social and economic determinants—And why current policies may make matters worse. *Health Affairs, 30*(10), 1852–1859.

World Health Organization. (2007). *Everybody's business – Strengthening health systems to improve health outcomes: WHO's framework for action.* Geneva, Switzerland: Author.

World Health Organization (WHO). (2013a). *Climate change and health: A tool to estimate health and adaptation costs.* Geneva, Switzerland: Author.

World Health Organization (WHO). (2013b). *Global burden of disease.* Geneva, Switzerland: Author.

World Health Organization (WHO). (2016). *Global strategy on human resources for health: Workforce 2030.* Geneva, Switzerland: Author.

World Health Organization (WHO). (2020). *The world health report 2000. Health systems: Improving performance; WHO.* Geneva, Switzerland: Author.

World Health Organization, & Research for International Tobacco Control. (2008). WHO report on the global tobacco epidemic, 2008: the MPOWER package. World Health Organization.

Zaidi, B., & Morgan, S. P. (2017). The second demographic transition theory: A review and appraisal. *Annual Review of Sociology, 43*, 473–492.

2

Health Disparities and Their Impact on Community Health

Errol D. Crook, Carrie E. Crook, Martha I. Arrieta, and Roma Stovall Hanks

Introduction

The most important resource in a sustainable community is the people of that community. Therefore, the health of those community members is critical in creating a sustainable environment and maintaining it over generations. Observations that certain populations have poorer health and health outcomes when compared to other populations have been noted for centuries. In a speech focusing on the plight of poor people in the US, Dr. Martin Luther King, Jr. (1966) stated: "Of all the forms of inequality, injustice in health care is the most shocking and most inhumane because it often results in physical death." During the 1960s, scholars and community activists

E. D. Crook (✉) • M. I. Arrieta
University of South Alabama College of Medicine, Mobile, AL, USA
e-mail: ecrook@health.southalabama.edu; marrieta@southalabama.edu

C. E. Crook
Tulane University, New Orleans, LA, USA
e-mail: ccrook@tulane.edu

E. Mpofu (ed.), *Sustainable Community Health*,
https://doi.org/10.1007/978-3-030-59687-3_2

recorded several observations and facilitated many discussions regarding the health of poor people, especially African Americans, who comprised a disproportionate amount of the poor in the US, compared to whites. In 1985, the scholarly work surrounding the health of poor and Black people intensified with the Report of the US Department of Health and Human Services (HHS) Secretary's Task Force on Black and Minority Health (1985). The Heckler Report, as it was called, was named for the Secretary of HHS at the time, Margaret Heckler. This report examined disparities in health in ethnic minority groups in the US when compared to white Americans. Following its release, the attention to health disparities increased, and led to several other reports over the decades highlighting the impact of these disparities and their effect on the health of communities (DHHS 2004; Smedley, Stith, & Nelson, 2003; Walker, Mays, & Warren, 2004).

Learning Objectives

By the end of the chapter, the reader should be able to:

1. Define health disparity and health equity in community settings.
2. Discuss the factors that contribute to disparities in health (upstream vs. downstream).
3. Outline community populations at highest risk for health disparities.
4. Discuss the impact of health disparities on overall community health.
5. Discuss community-engaged scholarship/community-based participatory research as a tool to address community health disparities.
6. Discuss the importance of multidisciplinary efforts to achieve sustainable health equity.

The definition of health disparities has evolved since the Heckler Report. As reviewed by Braveman et al. (2011), early definitions by The National Institutes of Health (NIH) were quite literal as differences in incidence (new cases each year), prevalence (existing cases at any time), mortality, and burden of a disease or set of diseases and their related

R. S. Hanks
University of South Alabama, Mobile, AL, USA
e-mail: rhanks@southalabama.edu

syndromes that exist among specific populations. A thorough understanding of the factors leading to an observed health disparity is essential in creating a sustainable community health system. Indeed, when a community is stressed, early signals of that stress may be a negative impact on the health of the individuals in the community; increases in emergency room visits and hospital admissions are often the first observations with premature deaths and higher rates of disability following closely behind (Ford, Mokdad, & Link, 2006; Spruill et al., 2019; Stoner, Haley, Golin, Adimora, & Pettifor, 2019; Wallace, Wallace, & Rauh, 2003).

Determinants of Health Disparities The reasons for disparities in health are usually complex. In a few cases, a disparity may be related to a single gene mutation, such as in sickle cell anemia, or to a variety of genetic variations, such as in cardiovascular disease and its risk factors. However, it has become clear that health disparities are the result of many factors **upstream** of the health care system (in the community) as well as factors within the health care system itself (**downstream**). The Institute of Medicine recognized that differences in access to health services, quality of health care, health literacy, economic and educational attainment, culture, language, and location all contribute to health disparities (Christopher, 2008; Smedley et al., 2003). Disadvantage, defined by Merriam-Webster as "an unfavorable, inferior, or prejudicial condition," has been identified as a major mediator in health disparities. This recognition led to Healthy People 2020 defining health disparities as a *"particular type of health difference that is closely linked with economic, social, or environmental disadvantage. Health disparities adversely affect groups of people who have systematically experienced greater obstacles to health based on their racial or ethnic group; religion; socioeconomic status; gender; age; mental health; cognitive, sensory, or physical disability; sexual orientation or gender identity; geographic location; or other characteristics historically linked to discrimination or exclusion."*

Below are some examples of health disparities that exist in the US using the characteristics listed in the Healthy People 2020 definition:

- Individuals living in the Southeastern US have higher rates of stroke than individuals living in other parts of the US (geography) (region of country).

- Adult African Americans in the US have the highest rates of hypertension seen in the world (race/ethnicity and geography).
- Individuals from low-income groups are at higher risk for obesity than those from higher-income groups (socioeconomic status).
- Children living in poor communities have higher rates of childhood obesity when compared to children not living in poor communities (socioeconomic status, geographic location (zip code)).

Health disparities are manifest in a variety of ways, mostly in the community or upstream to the health care system. Below we list those that are commonly described.

- Earlier onset of disease
- Later time of diagnosis
- Unequal care or lower-quality care
  - Unable to see a health care provider
  - Providers not prescribing or providing the standard of care
  - The standard of care is not available to the patient
- Earlier death/lower life expectancy
- Poorer survival post-diagnosis
- Higher death rate
- Higher rates of disability/loss of quality years of life
- Increased rates of complications due to a disease

Since we are referring to conditions of dissimilarity among community populations when we discuss health disparities, our descriptions of disparities ultimately come down to comparisons. There are several groupings used to differentiate health disparities as outlined in the Healthy People 2020 definition above. In addition to that list, we think that it is important to specifically mention occupation with regard to both income and to the higher risks for injury and harmful exposures inherent within certain jobs. For example, the occupational risk of coal mining carries for lung disease. Another comparator we would like to add is access to health care.

Access to Health Care Access to health care by communities comprises many factors and overlaps with several of the differentiating factors in the Healthy People 2020 definition for health disparity. Access to health care depends on physical access to providers (e.g., physicians, dentists, nurses) and the physical spaces in which they work (e.g., hospitals, clinics, pharmacies). In rural areas, the access to these physical resources may be limited—the interplay between geography and access. Access is also dependent on an individual having the financial resources necessary to fund the interaction with a provider—the interplay between socioeconomic status and access. This is dependent upon whether one has health insurance, if the insurance is accepted by the provider, and if an individual has sufficient income, with or without health insurance, to pay for the encounter (Zabel & Stevens, 2006). Moreover, access may depend on whether one has transportation to a provider even when the individual is in relatively close proximity to a health care institution. Finally, even if physical access is available, the quality of care to which one has access has to be considered (see also Chap. 9, this volume). It is important to know if one is able to get the standard of care for that condition, as lower-quality care contributes to worsening health disparities (IOM, 2000; Zabel & Stevens, 2006; US Department of Health and Human Services, 2004; Crook & Peters, 2008).

Health Conditions There are health disparities in almost all health conditions. We direct the reader to the Center for Disease Control [CDC] website (www.cdc.gov) for data on disparities in specific health conditions. Chronic diseases such as hypertension, diabetes mellitus, and heart disease are the leading cause of death and disability in developed countries. In the US they account for 7 of every 10 deaths and affect the quality of life of at least half of the adults in the country (Strong, Mathers, Leeder, & Beaglehole, 2005). Therefore, even small disparities in life expectancy or rates of complications due to these chronic diseases may have a significant impact on the makeup, and therefore the sustainability of the health of all communities. Some disparities that have a great impact on communities and their ability to be sustainable are listed in Table 2.1 (Crook & Peters, 2008).

Table 2.1 Health conditions with significant health disparities impacting communities in the US

Breast cancer	Colon cancer
Prostate cancer	Lung cancer
Cardiovascular disease	Hepatitis C
Asthma and COPD	Maternal infant mortality
Hypertension	Obesity
Chronic kidney disease	Substance abuse
Diabetes mellitus	HIV/AIDS
Obesity	Violent trauma/homicide

See CDC website, American Public Health Association website, Levine, Schneid, Zoorob, and Hennekens (2019)

A detailed discussion outlining the characteristics of the disparities in these conditions is beyond the scope of this chapter. We will give some insight into some of them in subsequent parts of this chapter so that the reader is able to better understand how disparities in health may impact the sustainability of a community. However, in the US and many other developed nations, the disparities in health for these conditions are greatly impacted by income, wealth, educational attainment, access to health care, and other factors dependent on decades of policy decisions (Smedley et al., 2003; Walker et al., 2004; Christopher, 2008; Crook & Peters, 2008). The learner is directed to Discussion Box 2.1 and Case Study 2.1 for an exercise to help with understanding the reason a disparity may exist and its impact on a community.

Discussion Box 2.1: Racial Disparities in Breast Cancer Mortality

African American women have a lower incidence rate of breast cancer than white American women. However, African American women have a higher mortality rate from breast cancer than white American women (CDC; DeSantis, Miller, Goding Sauer, Jemal, & Siegel, 2019).

Discuss the potential reasons for this disparity in mortality between African American and white American women.

What Do You Think?

1. Might there be differences in the biology/genetics of the cancer between the two groups?
2. How might disparities in access to health care lead to this disparity in breast cancer mortality? (Consider all aspects of access to health care mentioned above.)

Self-Check Questions

1. How would you define health disparities?
2. Children in poor families have higher rates of asthma (Hill, Graham, & Divgi, 2011; Hughes, Matsui, Tschudy, Pollack, & Keet, 2017). This often leads to higher rates of absenteeism from school. Is the lower educational attainment of children from poor families a health disparities issue? Why or why not?

Case Study 2.1: Disparities Among Chronic Diseases and Life Expectancy

Consider Table 2.1. HIV/AIDS, substance abuse, and violent trauma are present at higher levels in poor, African American communities in the US (Richardson & Hemenway, 2011; Levine et al., 2019; Stoner et al., 2019; CDC). The disparities among the chronic diseases in the table also shorten life expectancy (Crook & Peters, 2008).

What Do You Think?

1. What will be the impact of these disparities on the demographic makeup of a poor community in the US compared to a wealthy community? Consider average age of community, educational attainment, number in the household, head of household, employment, percentage of community's population that is < 18 years of age.
2. What would that community look like 25 years from now?

Professional Definitions

Addressing health disparities in the context of creating and maintaining sustainable communities requires contributions from policymakers, the health care industrial complex, the business community, charitable organizations, and a broad array of scientists. As such there is a need for a common language and commonly agreed-upon terms. We will focus on four terms in this discussion: Health, Health Disparity, Health Equity, and Population Health.

Health extends beyond health care; therefore, sustaining a healthy community requires attention to the many factors that make up a healthy community. Indeed, almost all subsequent chapters in this book consider those factors in detail as community indicators or outcomes (Part Two), as well as their impact on the health of vulnerable populations (Part Four). Sustainable communities require healthy community members and, therefore, must promote the conditions that promote health and healthy habits at home, school, community, and work (Christopher, 2008; Braveman, 2010, 2014, 2017; Crook, Pierre, & Arrieta, 2019; see also Chap. 3, this volume).

The relationship between health and poverty is intertwined, and it has been suggested that policymakers and social scientists should amend the definition of poverty to include health. Sen (1999) outlined the complexity of how health impacted the overall condition of communities and that even rich countries had sectors within them that had poor health outcomes, stating "deprivation for particular groups in very rich countries can be comparable to that in the so-called third-world" (p. 620). As an example is the well-known fact that the health outcomes for African Americans, American Indians/Alaskan Natives, and Latinos in the US, a wealthy nation, are similar to or worse than individuals in poor nations. Gwartkin (2000) advocated for the consideration of health conditions in poor populations to be a primary factor in the development of health policy goals. More specifically, Gwartkin (2000) further suggested that the relative conditions of wealthy and poor be expressed to highlight the discrepancies between them so that a clearer picture was available and better solutions could be developed.

Having a clear understanding of health disparities and the sociocultural underpinnings of their existence is critical. To that end, we now have a definition for health disparity that gives emphasis to the economic and social disadvantage of groups that suffer these inequities in health (Braveman, 2014). As previously mentioned, Healthy People 2020 defined health disparities as to be influenced by socio-demographics linked with economic, social, or environmental pillars of health and well-being based on their racial or ethnic group religion; socioeconomic status; gender disability; sexual orientation *"or other characteristics historically linked to discrimination or exclusion."*

This definition does not just focus on the differences in health and health outcomes observed between groups, but rather it places emphasis on those differences that are influenced by *"someone's relative position in the social pecking order…* "(Braveman, 2014, p. 6). Braveman (2014) further observes that the characteristics listed in the Healthy People 2020 definition "can influence how people are treated in a society." (p. 6). Braveman (Braveman, 2017; Braveman et al., 2011) strongly articulated the concern about health disparities not just being rooted in the fact there are differences, but rather the reasons for the differences and concerns about **social justice.**

Health equity is about fairness and providing the opportunity for all individuals and communities to have the best health possible. Healthy People 2020 defines *health equity* as the "attainment of the highest level of health for all people." A commitment to achieving health equity is a commitment to addressing the factors that impact health disparities and doing so in a way that is appropriate for the community on which one is focused. Equality and health equity are not the same. Equality refers to having equal resources but it does not necessarily mean that those resources can be used effectively by all. In pursuit of health equity, sustainable communities match specific resources to the specific needs of the population in an effort to achieve the highest standard of health for all people (Braveman, 2010, 2014, 2017; Braveman et al., 2011; Crook et al., 2019; Hanks et al., 2018).

The science of **population health** is critical in understanding health disparities and achieving health equity. Population health science examines how health-related risks and health outcomes are distributed within and across populations (Szreter, 2003). Historically, these efforts were centered within academic institutions, the National Institutes of Health, the Centers for Disease Control, local and state departments of public health, and the counterparts to these groups in other nations. Population health often examines the role of those social factors listed in the definition of health disparities, also known as the **social determinants of health**, in the context of understanding the drivers to a community's health. Thus, the sustainable health of communities requires an ongoing assessment of their health status, needs, and trends. In the US, many large health care systems increasingly put more resources into understanding and addressing the social determinants of health of the communities they serve (See Discussion Box 2.2) (Lantz, 2018; Lantz, Lichtenstein, & Pollack, 2007).

Discussion Box 2.2: Health Insurer Preferences

You are the administrator of a health care system that serves a community with high rates of poverty, poor educational attainment, housing and food insecurity, and high unemployment. The largest employer in your community wants to support the community by designating your health care system as the preferred health care provider for its employees. The health insurance administrator for the employer is hesitant to give that designation because you do not perform well on certain quality measures that insurers follow (Lantz, 2018; Loria, 2020).

What Do You Think?

1. Consider why your health care system performs poorly on these quality measures?
2. Assuming you have the resources, what are some interventions you might implement in the community to improve your quality performance and the community's health?
3. What are challenges to addressing the social determinants of health at the community level?

History of Research and Practice

Research in health disparities has evolved since the Heckler Report shone a light on the magnitude of the issue. Health disparities research evolves over four generations (Chin, Walters, Cook, & Huang, 2007; Crook et al., 2009; Thomas, Crouse Quinn, Butler, Fryer, & Garza, 2011).

- First-generation research details the scope of the problem.
- Second-generation research details the factors contributing to and mediating the problem.
- Third-generation research designs and evaluates the impact of interventions focused on solving the problem.
- Fourth-generation research addresses structural determinants of health through comprehensive multilevel interventions and includes key bioethics principles of justice to facilitate action to eliminate health disparities.

There has been considerable progress made in first-and second-generation research. Disparities in several health conditions, overall health, and life expectancy are well documented for the US (CDC) and for many other nations (WHO 2008; The Bill & Melinda Gates Foundation 2020). We now better understand the importance of the social determinants of health in achieving health equity. In fact, the progress made in first- and second-generation research has contributed to the evolution of the aspiration goals for Healthy People, where the goal went from reducing health disparities among Americans (2000) to eliminating health disparities (2010), to the goal of achieving health equity (2020). This aspirational goal remains a work in progress.

Perhaps the biggest challenge in health disparities research is to implement third-generation/interventional studies. These studies are mostly focused on a specific disease, such as diabetes mellitus or a specific type of cancer, and the projects often cannot be implemented across multiple sites. But, medical care only accounts for 10–15% of the premature deaths we see in communities suffering inequities in health in the US and the lion's share of disparity is dependent upon genetic and social determinants (Isaacs & Schroeder, 2004; McGinnis, Williams-Russo, & Knickman, 2002; Schroeder, 2007). Combined, these limitations make it difficult to develop a simplified blueprint for sustainable health that can be applied to all communities.

In acknowledging the challenge of addressing the broader structural drivers of health, Thomas et al. (2011) have proposed the development of comprehensive interventions that operate across multiple levels of the socio-ecological model of health and address multiple outcomes, composing fourth-generation health disparities research. Inherent in fourth-generation research is the adoption of public health critical race praxis as a framework (Ford & Airhihenbuwa, 2010), which allows for the incorporation of racism and structural determinism (the forces that perpetuate focused disadvantage) as *bonafide* elements for scrutiny within a research process that is oriented toward action capable of promoting community health equity.

We and others believe that the elimination of health disparities requires an approach that is fundamentally community-based and community-focused. This type of research is called **community-based participatory research (CBPR)** and/or **community-engaged research (CEnR)**. CBPR/CEnR is not easy and requires that investigators truly invest real-time with their communities of focus. Research partners in community health must learn the community's problems and work on solutions to those problems (which may not be the primary problem of interest to the investigator), before implementing any research in that community. True CBPR/CEnR engages the community at every stage of the research, and, as such, empowers the community, contributing to the creation of a healthy sustainable community (Alvidrez, Castille, Laude-Sharp, Rosario, & Tabor, 2019; Arrieta, Wells, Parker, Hudson, & Crook, 2018; Crook et al., 2019; Graham et al., 2016; Hanks et al., 2018; Thomas et al., 2011).

Two other areas that are proving to be critical in health disparities research are mentioned below:

- **Dissemination science** studies and evaluates the methodology by which valuable information learned in third-generation studies and budding fourth-generation studies is disseminated more broadly. Interventions that are successful in addressing health disparities must be shared. Dissemination science goes beyond academic-to-academic communication but examines best practices for communication to and within the community of focus and other communities (Baumann & Cabassa, 2020).
- **Health policy research** evaluates the potential impact of a proposed policy, evaluates the impact of implemented policies, and provides data to policymakers to guide the development of health policies. This work is critical in all aspects of health disparities research and the maintenance of sustainable healthy communities.

Pertinent Sustainable Community Health-Oriented Approaches

As mentioned earlier, there is no simple path to achieving community health equity. If we think of an individual in a community who has a health issue and seeks medical care, there is a tendency to think of the starting point for that health issue at the time the individual presents to the health care establishment. However, at the time of presentation, that individual arrives with a lifetime of experiences that influence their health and the way that person interacts with the health care system. Those prior community life experiences are referred to as **upstream** (macro-level) social determinants of health (Lantz, 2018; Szreter, 2003). For example, upstream determinants could include the air quality restrictions, or lack thereof, in a particular neighborhood, or the housing quality standards governing the condition of an individual's home. Every individual will carry the sum total of these upstream determinants to each interaction with a health care provider.

The **downstream** (micro-level) determinants are those events that occur when one is negotiating the health care system. The downstream events may be quite complex, particularly when one has to manage a chronic disease or suffers a life-altering event such as a stroke. In the US and other developed nations, there have been impressive advances in medical science discoveries. While it is said that a rising tide lifts all boats, whether an individual benefits from these medical science advances depends largely on the upstream or community determinants in their life. In essence, was that person's boat in the water when the tide came in? It is beyond the scope of this chapter to discuss those downstream determinants, particularly clinical care and basic medical science, in detail. However, it is clear that all members of a community will benefit from these advances if the research or clinical program design considers the upstream determinants as they develop their program (Crook et al., 2019; Graham et al., 2016). Discussion Box 2.3 provides opportunities for learners to consider how upstream and downstream factors impact medical research.

Discussion Box 2.3: Upstream and Downstream Determinants

State University School of Medicine, located outside of a large city, is beginning a clinical trial for Drug X in the treatment of asthma. Researchers have enlisted physicians in the State University Health System to inform their asthma patients about the study in their clinics.

What Do You Think?

1. Which upstream or community determinants will determine if a patient is recruited for the study?
2. Which downstream or health care determinants will determine if a patient is recruited for the study?
3. What are some potential effects of inequities in clinical/translational research?

Community-Engaged Approaches

Community-engaged strategies have proven successful in addressing health disparities. As mentioned earlier, community engagement is not easy and takes time (see also Chap. 8, this volume). The underlying foundation for successful community engagement is **trust**. Many communities that suffer from inequities in health have been conditioned not to trust individuals and agencies that are not part of their community. They have had promises broken many times over, and often feel as if they are not heard. Ultimately, trust is built over time. Time allows for the development of relationships, the exchange of knowledge for all parties to gain more understanding regarding the needs of the community, and the time to plan a strategy (Hanks et al., 2018).

The elements of a successful community-engaged approach are outlined in Table 2.2. Successful community-engaged approaches contribute to the sustainability of communities by empowering the community. Communities are empowered by:

- Increased knowledge and self-awareness
- Enhanced voice to articulate their problems
- A solution to a problem or increased knowledge about a problem
- Connection to advocates
- The newly acquired skills remain in the community

Table 2.2 Elements for a successful community-engaged approach to a health disparity issue

Trust building	Listen to problems Make an effort to use academic resources to address some of their problems Understand their way of life Be forthcoming about who you are
Sincere investment of time	Meet regularly but be respectful of their time Recognize that their schedules may not coincide with traditional work hours
Identification of community leaders and relationship building with them	May not be "traditional leaders" Have the trust of their fellow community members
Shared recognition of the problem	The initial problem to tackle may not be what was originally intended.
Shared planning of the intervention	The community knows best what will work for them The community must believe the intervention is for its benefit
Community members as part of implementation team	Highly suggested if project requires face-to-face interactions with community members Find task appropriate for community members' skills Train community members in necessary skills Involve community members in data analysis where appropriate
Timely sharing of results with community	Let them know what you found (e.g., effectiveness of the intervention)
Maintain relationship post completion of project	Maintains friendships Adds to trust, you are present even when you do not want something Makes implementation of next project easier

See: Eng and Parker (1994); Wallerstein and Duran (2006, 2017; Hinton, Rausa, Lingafelter, and Lingafelter (1992); Hanks et al. (2018); Arrieta et al. (2018)

Some examples of groups or individuals that might implement the community-engaged approach to address health disparity issues at the community level are shown in Table 2.3.

Table 2.3 Examples of groups or individuals who may use a community-engaged approach to address a health disparity issue

Entity	Example
Local or state department of health	Address a decrease in adherence with childhood vaccination recommendations
	Address an increase in sexually transmitted infections
Health care system	Wants to understand why there is an increase in emergency department visits for conditions that should not require a visit to the emergency department (such as diabetes, hypertension, or asthma)
HIV/AIDS investigator	Wants to determine the most effective messaging for prevention of HIV, early testing of HIV, and/or treatment for those with HIV
Cancer investigator	Desires to improve cancer screening to improve cancer outcomes
Criminal justice system	Needs to alert community to and understand reasons for increase in illicit drug use
Health policy researcher	Wants to understand the impact of a proposed new policy for Medicaid health insurance
Nonprofit organization	Wants to know how best to invest their funds to address food insecurity
	Wants to combat childhood obesity by building playgrounds and needs to know where best to place them
Social worker	Wants to determine best messaging strategy to inform senior adults of new exercise classes for them

The Community Member as an Integral Part of the Team in the Community-Engaged Approach to Health Equity

As mentioned in Table 2.2, it is important to have a community member(s) as part of the team in a community-engaged approach (Wallerstein & Duran, 2006, 2017). It is important that those community members are empowered with the skills to perform their job well. Many will have natural skills, so-called natural helpers, and are individuals to whom others go to for advice and assistance (Eng & Parker, 1994).

These individuals are referred to by several titles, many of which have overlapping jobs. Some of the titles given to these team members are listed below:

- Community Health Advisor (Hinton et al., 1992; Lisovicz, Wynn, Fouad, & Partridge, 2008)
- Lay Health Advisor
- Community Health Advocate (Hanks et al., 2018)
- Navigator (Lisovicz et al., 2008)

In general, Community Health Advisors, Lay Health Advisors, and Navigators will have specific tasks. They are often tasked to focus on a specific disease or set of diseases such as cancer or diabetes mellitus. They are familiar with the upstream or community determinants for individuals in their communities. Therefore, these community liaisons are able to assist their fellow community members in managing those issues to ensure they receive the health care they deserve, are able to lead healthier lives, and are able to better manage any chronic disease(s) that may be present.

Community Health Advocates Our team works with Community Health Advocates (Hanks et al., 2018). Community Health Advocates are those natural leaders who are committed to advocate for health equity within their communities. Our approach was not to limit them to specific conditions but to allow them to choose their areas of focus. We learned that they chose areas that are impacting their communities. We partner with them to empower them with knowledge in their area of interest, resources to implement small interventions, and connections to others with similar interests to form a sustainable network. We call our community leaders Community Health Advocates in recognition of the vital link between individual health awareness and behavior change and the need for advocacy and policy change in order to advance health equity. Dean and Gilbert (2009) noted, "Social capital in the African American community has been leveraged to address health disparities directly while building political advocacy around activism on the social causes of health disparities like racial residential segregation." An example of Community Health Advocate success has been the progress they have made in identifying food deserts and food insecurities. Their advocacy has resulted in the establishment of a community-sponsored market that provides a resource for affordable fresh produce in the heart of a community with historically high health disparities.

Research Apprenticeship as a Model of Community-Engaged Research Our group has sought to enhance the impact of including community members as part of the team, and hence, increase the likelihood that communities that suffer inequities in health become more sustainable, more resilient, and closer to achieving health equity (Arrieta et al., 2018). We have included members of our partnering communities as part of our research team as research apprentices. Research apprentices are adult members of our partnering communities who are often underemployed due to the lack of formal skills and opportunity. They were trained in the principles and practice of research and were engaged in the development of protocols, the collection of data in the field, the entry of data into a statistical database, and the analysis and dissemination of data. Their inclusion enhanced the collection of data from communities that are often distrustful of research efforts and are often not included in such work. Research apprentices acquired valuable employment skills in the process that was instrumental in improving their job status (Arrieta et al., 2018).

Effective Interventions in Health Disparities Are Usually Targeted and Community-Based Perhaps the greatest challenge in the march to achieve health equity is that all communities are unique and a one-size-fits-all approach does not work. Interventions are typically designed to be implemented at the community level, with customization to the local needs and assets (see also Chap. 7, this volume). The reader is also directed to the CDC's The Health Disparities and Strategies Reports for examples of interventions targeting a variety of health disparity issues (www.cdc.gov/minorityhealth/). Disseminating the lessons learned from all attempted interventions is necessary. There is a general agreement that successful interventions should be examined for use in other communities. Those interventions will have to be modified to respect the unique characteristics of each community whether it be considerations for age, language, household size and makeup, employment, educational attainment, food and housing insecurity, and/or community trust (Crook et al., 2019). It is just as important that those interventions that were not successful are also shared. There are valuable lessons to be learned in analyzing why an intervention did not have its expected outcome. Understanding those lessons will help direct the development of future endeavors.

Cultural, Legislative, and Professional Issues

Political landscape, cultural values, and legal precedent all play a significant role in creating and alleviating health disparities. Ensuring health equity in a sustainable community health system does not mean focusing only on the interactions between health care organizations and individuals in communities, but rather taking a holistic view of a community ecosystem (see also Chap. 6, this volume). Patients carry dense, rich histories with them in every health care setting because upstream factors mold every patient's geopolitical and social landscape. Furthermore, these upstream or community factors endured over a lifetime account for the majority of health disparities (Lantz, 2018; Szreter, 2003). This means policies at the local, national, and international levels will affect the ability to create an equitable health environment, and thus a sustainable community.

In the US, health care and public health are deeply politicized topics. The political debate around health care, in particular, has intensified over the last two decades. From this debate has come some legislation that improved health care access. Examples include Part D of Medicare that provides coverage for medicines for Medicare recipients enacted under President George W. Bush and the Affordable Care Act enacted under President Barack Obama. However, there are times that the winds of political debate may limit or silence some issues. An example is the restriction of language in the 2017 Centers for Disease Control and

Prevention budget report. Words like "transgender" and "diversity" were removed from the CDC budget appropriation request. Lesbian, gay, bisexual, and transgender questions were removed from community needs assessments of older US populations (Gostin, 2018). But what does this erasure mean for health equity in individual communities? In this case, political values determined the scope of public health practice and determined who would not benefit from targeted public health measures. Such restriction could result in reduction of research and public health project funding focusing on the most vulnerable communities. Federal and state resource allocation determines which communities will have better access to health care, safer places to live, more food security, and better educational opportunities. All of these upstream factors influence the health and sustainability of communities and lack of attention to them will increase inequities in health.

It should be noted, however, that politicization of public health has also sparked national and international movements to address health disparities. These macro-level standards can increase the efficacy of community-based health approaches. Within the US, the Department of Health and Human Services has developed the Healthy People initiative to increase the prevention of disease and create an environment for Americans to live healthier lives (US Department of Health and Human Services, 2014). Similarly, the United Nations has developed worldwide Sustainable Development Goals "to end poverty, protect the planet and improve the lives and prospects of everyone, everywhere." In order to achieve this ideal, the United Nations created a 15-year agenda with 17 different goals, including specific objectives to decreased maternal mortality, child mortality, malarial rates, and HIV/AIDS rates (United Nations, n.d.). Thus, researchers and community stakeholders have the opportunity to use national and multinational support and funding to tailor health interventions to a specific community.

Ultimately, to have sustainable community health system, all members of that community must have a fair, equitable opportunity to be in the best health possible—health equity. In health care and public health, this opportunity is supported by a "safety net," an infrastructure and resources that support those members with the greatest challenges (See Discussion Box 2.4) (Arrieta, Foreman, Crook, & Icenogle, 2008; Gwartkin, 2000;

Discussion Box 2.4: Supplemental Assistance Programs

The Supplemental Nutrition Assistance Program (SNAP) and the Children's Health Insurance Program (CHIP) are examples of the public health and health care safety net in the US.

What Do You Think?

1. Think of some other programs that impact the upstream or health care system factors impacting a community's health. Think of those in the US and internationally.
2. Think of other programs that support an individual's access to health care, per se, in the US and internationally.

Keep these questions in mind as you go to subsequent chapters in this book, particularly for question 1. These are the types of programs that drive a community to sustainability and health equity.

Sen, 1999). In the health care debate, the fundamental issue is whether health care is a right or a privilege (Maruthappu, Ologunde, & Gunarajasingam, 2013). If it is the latter, then a safety net may not be necessary theoretically. If health care is a right, then the safety net is the foundation from which all other health care begins. When it comes to public health, in the US there is general agreement that a safety net needs to exist for the greater health of the nation and its communities (Institute of Medicine, 2000). For policymakers, it is the hope that when the safety net is constructed or adjusted that it is placed high enough that the momentum of those falling into it does not allow them to hit the hard ground below resulting in death or severe disability. It should be the goal that the net propels those persons "upward" into a position of sustainability and comfort in their community.

Related Disciplines Influencing Community-Oriented Health Aspects

Addressing health disparities requires a multidisciplinary, community-engaged approach. The most effective projects are implemented through true team science. Consider Case Study 2.2 below.

Case Study 2.2: Developing and Disseminating New Treatments in Underserved Communities

A basic scientist discovers a gene mutation that may explain why some women have a more aggressive form of breast cancer. A drug targeting the product of that mutation is developed. That investigator now needs to move that discovery to clinical trials and that will require a team that knows how to conduct trials and has the wherewithal to recruit the appropriate patients. To be research that is geared to foster health equity there needs to be a diverse group of participants. With the history of studies such as the US Public Health Service Study of Untreated Syphilis in the Negro Male (conducted in Tuskegee, Alabama), the investigator team may have considerable difficulty with participant recruitment. The team will need trusted community members as part of the team (navigators, community health advisors, research apprentices) to achieve recruitment and will need social scientists to help them understand how to address this mistrust.

Ultimately, the clinical trials are performed and the drug is approved for treatment of breast cancer in women with this specific gene mutation. The women with this mutation get their disease at a relatively young age before screening is recommended and the mutation occurs at higher rates in African Americans and certain Latinx populations.

What Do You Think?

1. You need to determine how to get the drug covered by multiple insurances. You need to be able to estimate the number of women who may have this mutation. What type of investigators will you bring onto your team?
2. You need to determine how to make patients from underserved communities aware of this advancement. What type of scientist(s) have the expertise for this work?
3. With this effective treatment, do you work to amend the preventive health guidelines?
4. Later it is learned that this mutation is observed in higher frequencies in women who work in and live close to certain industries. With whom do you partner to get environmental justice, social justice, and health equity for these women?

While the case study is a hypothetical, there are true stories that mirror it. It illustrates that community health disparities research and advocacy requires partnerships among any and all disciplines and individuals that have a goal of developing and maintaining sustainable communities (Arrieta et al., 2017, 2018; Crook et al., 2019). Unfortunately, communities with high levels of health disparities are often the last to benefit from advances in science, health care, and health policy. The likelihood of these communities benefitting from these advances is greatly enhanced if the implementation and research team is a multidisciplinary team with health disparities investigators involved from the beginning (Arrieta et al., 2017; Graham et al., 2016).

Research Critical to Issues Discussed in the Chapter

Often a sign of a stressed community is a downward change in the health of the community members (Spruill et al., 2019; Stoner et al., 2019; Wallace et al., 2003). In addition, when communities are stressed by natural disasters or economic downturns, those members of the community that suffer from inequities in health suffer the most (Christopher, 2008; Ford et al., 2006). Two examples are Hurricane Katrina's impact on the city of New Orleans, LA and the Northern Gulf Coast of the US in 2005 and the COVID-19 pandemic of 2020.

The images after Hurricane Katrina in New Orleans were vivid and demonstrated that those members of the community with fewer resources (upstream determinants) were the ones that suffered the greatest losses, including death. Those already suffering from health disparities had worsening of their chronic conditions and there were great challenges in mental health (Arrieta et al., 2008; Crook, Arrieta, & Foreman, 2010; Ford et al., 2006; Ridenour, Cummings, Sinclair, & Bixler, 2007). The environmental impact was felt most by those less equipped to handle it. The lack of equity and justice was apparent (Krousel-Wood, 2008; Arrieta et al., 2008).

We are in the midst of the COVID-19 pandemic at the time of the writing of this chapter. Its presence and impact on disparities in health in the US is clear and are summarized by Vickers et al. (2020). Poor communities with a large percentage of ethnic minorities suffer disparate rates of death. Moreover, members from these communities are more likely to have chronic health conditions that further increase the risk of serious infection and death. Members of these communities find themselves greatly impacted by the upstream determinants as they are more likely to be in jobs that are essential, but also low paying, requiring them to go to work where they often have difficulty practicing social distancing. Finally, many feel they do not have access to health care due to a lack of insurance or other financial resources (Vickers et al., 2020).

These examples illustrate the importance of focusing on the **social determinants of health (upstream factors)** as we work toward achieving health equity. Given the impact of poverty on health, a multidisciplinary focus on communities with concentrated poverty is warranted (Smith et al., 2017). In the intersection of poverty and health, there appears to be a threshold above which the disparities in health related to poverty lessen significantly (Chetty et al., 2016; Makaroun et al., 2017). Elevating overwhelmingly poor communities up to that threshold would be just, would move us much closer to health equity, and would make those communities more sustainable. It will be critical to assure that quality education is available in those communities, along with food and housing security, and job training and opportunity. Clearly these issues are health issues (Braveman et al., 2011).

These upstream or community determinants lead to biological disease and conditions. The lack of safe, walkable communities results in lower physical activity and higher rates of obesity. The lack of healthy food does the same and brings on earlier heart disease and stroke (see also Chap. 3). More recently, it has been observed that the stress of living in these impoverished neighborhoods is associated with the onset or worsening of several health conditions such as hypertension and HIV/AIDS (Bruce et al., 2010; Ibragimov et al., 2019; Spruill et al., 2019; Stoner et al., 2019). Of interest, while life expectancy in communities of concentrated poverty is lower than in non-impoverished communities (Chetty et al., 2016; Makaroun et al., 2017), there are many survivors in these impoverished areas. Understanding the determinants of that resilience to survive is important and will accelerate the transformation of these communities to healthy, sustainable communities.

Summary and Conclusion

As Rev. Dr. Martin Luther King, Jr. stated in 1966, "injustice in health care is the most shocking and most inhumane." Understanding and addressing health disparities is critical if a community is to be sustainable. The disparities in health in the US and internationally are largely due to disparities in the social determinants of health. Thus, the following chapters in this book all address an area impacting a community's ability to achieving sustainable health equity. Achieving health equity—a state in which all communities and all members within those communities have the best chance to have good health—is a **health social justice cause** and will be a signal, perhaps stronger than any other, that there is equity.

Self-Check Questions

1. Define health disparity and health equity in community settings.
2. What upstream and downstream factors contribute to community health disparities and how?
3. Identify community populations at highest risk for specific health disparities and why?
4. Discuss some evidence-based best practices to address community health disparities.
5. What are the strengths and limitations of community-engaged scholarship/community-based participatory research as a tool to address community health disparities?
6. Discuss the importance of multidisciplinary efforts to achieve sustainable health equity.

Discussion Questions

1. Children in poor families have higher rates of asthma (Hill et al., 2011; Hughes et al., 2017). This often leads to higher rates of absenteeism from school. Is the lower educational attainment of children from poor families a health disparities issue? Why or why not?
2. Discuss how the principles of social justice and fairness are critical to achieving health equity. Refer to examples from a community you are familiar.

Field-Based Experiential Exercise

Put yourself in the place of the Nonprofit Organization in Table 2.3. Considering your own community:

1. How would you begin to determine where a playground may have its greatest impact on childhood obesity?
2. Who would be the community leaders that would be the best partners?
3. Once you select a place, what will you do to make sure it is used?
4. What health conditions and social determinants of health will be impacted by this project?

Online Learning Resources

https://www.HealthyPeople.gov

http://www.cdc.gov/minorityhealth/ (For data on health disparities. There are several resources under the Health Disparities and Strategies Reports).

https://www.un.org/sustainabledevelopment/sustainable-development-goals/

https://www.apha.org

References

Alvidrez, J., Castille, D., Laude-Sharp, M., Rosario, A., & Tabor, D. (2019). The national institute on minority health and health disparities research framework. *American Journal of Public Health, 109*(S1), S16–S20.

Arrieta, M. I., Fisher, L., Shaw, T., Bryan, V., Hudson, A., Hansberry, S., … Washington-Lewis, C. (2017). Consolidating the academic end of a community-based participatory research venture to address health disparities. *Journal of Higher Education Outreach and Engagement, 21*(3), 113.

Arrieta, M. I., Foreman, R. D., Crook, E. D., & Icenogle, M. L. (2008). Insuring continuity of care for chronic disease patients after a disaster: Key preparedness elements. *The American Journal of the Medical Sciences, 336*(2), 128–133.

Arrieta, M. I., Wells, N. K., Parker, L. L., Hudson, A. L., & Crook, E. D. (2018). Research apprenticeship and its potential as a distinct model of peer research practice. *Progress in Community Health Partnerships: Research, Education, and Action, 12*(2), 199.

Baumann, A. A., & Cabassa, L. J. (2020). Reframing implementation science to address inequities in healthcare delivery. *BMC Health Services Research, 20*, 190.

Braveman, P. (2010). Social conditions, health equity, and human rights. *Health and Human Rights, 12*(2), 31–48.

Braveman, P. (2014). What are health disparities and health equity? We need to be clear. *Public health reports, 129*(1_suppl2), 5–8.

Braveman, P., Arkin, E., Orleans, T., Proctor, D., & Plough, A. (2017). *What is health equity?: And what difference does a definition make?.* National Collaborating Centre for Determinants of Health. Robert Wood Johnson Foundation, Princeton, NJ.

Braveman, P. A., Kumanyika, S., Fielding, J., LaVeist, T., Borrell, L. N., Manderscheid, R., & Troutman, A. (2011). Health disparities and health equity: The issue is justice. *American Journal of Public Health, 101*(S1), S149–S155.

Bruce, M. A., Beech, B. M., Crook, E. D., Sims, M., Wyatt, S. B., Flessner, M. F., … Ikizler, T. A. (2010). Association of socioeconomic status and CKD among African Americans: The Jackson heart study. *American Journal of Kidney Diseases, 55*(6), 1001–1008.

Centers for Disease Control and Prevention. (2016, October 3). *Strategies for reducing health disparities 2016 – Minority health – CDC.* Centers for Disease Control and Prevention. https://www.cdc.gov/minorityhealth/strategies2016/index.html.

Chetty, R., Stepner, M., Abraham, S., Lin, S., Scuderi, B., Turner, N., … Cutler, D. (2016). The association between income and life expectancy in the United States, 2001–2014. *JAMA, 315*(16), 1750–1766.

Chin, M. H., Walters, A. E., Cook, S. C., & Huang, E. S. (2007). Interventions to reduce racial and ethnic disparities in health care. *Medical Care Research and Review, 64*(suppl), 7S–28S.

Christopher, G. (2008). Social justice and health outcomes in vulnerable communities. *The American Journal of the Medical Sciences, 335*(4), 251–253.

Committee on Quality of Health Care in America, & Institute of Medicine Staff. (2001). *Crossing the quality chasm: A new health system for the 21st century.* National Academies Press. Washington, D.C.

Crook, E. D., Arrieta, M. I., & Foreman, R. D. (2010). Management of hypertension following hurricane Katrina: A review of issues in management of chronic health conditions following a disaster. *Current Cardiovascular Risk Reports, 4*(3), 195–201.

Crook, E. D., Bryan, N. B., Hanks, R., Slagle, M. L., Morris, C. G., Ross, M. C., … Arrieta, M. I. (2009). A review of interventions to reduce health disparities in cardiovascular disease in African Americans. *Ethnicity & Disease, 19*(2), 204.

Crook, E. D., & Peters, M. (2008). Health disparities in chronic diseases: Where the money is. *The American Journal of the Medical Sciences, 335*(4), 266–270.

Crook, E. D., Pierre, K., & Arrieta, M. A. (2019). Identifying and overcoming roadblocks that limit the translation of research findings to the achievement of health equity. *Journal of Health Care for the Poor and Underserved, 30*(5), 43–51.

Dean, L., & Gilbert, K. L. (2009). Social capital and political advocacy for African American health. *Harvard Journal of African American Public Policy, 16*, 85–95.

DeSantis, C. E., Miller, K. D., Goding Sauer, A., Jemal, A., & Siegel, R. L. (2019). Cancer statistics for African Americans, 2019. *CA: A Cancer Journal for Clinicians, 69*(3), 211–233.

Eng, E., & Parker, E. (1994). Measuring community competence in the Mississippi Delta: The interface between program evaluation and empowerment. *Health Education Quarterly, 21*(2), 199–220.

Ford, C. L., & Airhihenbuwa, C. (2010). The public health critical race methodology: Praxis for antiracism research. *Social Science & Medicine, 71*(80), 1390–1398.

Ford, E. S., Mokdad, A. H., Link, M. W., Garwin W. S., McGuire L. C. Jiles R. A., Bulluz L. S. (2006). Chronic disease in health emergencies: In the eye of the hurricane. *Preventing Chronic Disease*, 3, 1–8.

Gostin, L. O. (2018). Language, science, and politics: The politicization of public health. *JAMA, 319*(6), 541–542.

Graham, P. W., Kim, M. M., Clinton-Sherrod, A. M., Yaros, A., Richmond, A. N., Jackson, M., & Corbie-Smith, G. (2016). What is the role of culture, diversity, and community engagement in transdisciplinary translational science? *Translational Behavioral Medicine, 6*(1), 115–124.

Gwartkin, D. R. (2000). Health inequality and the health of the poor: What do we know? What can we do? *Bull World Health Org, 78*, 3–18.

Hanks, R., Myles, H., Wraight, S., Williams, M., Patterson, C., Hodnett, B., … Crook, E. (2018). A multi-generational strategy to transform health education into community action. *Progress in community health partnerships: Research, education, and action, 12*(1 Suppl), 121.

Hill, T. D., Graham, L. M., & Divgi, V. (2011). Racial disparities in pediatric asthma: A review of the literature. *Current Allergy and Asthma Reports, 11*(1), 85–90.

Hinton, A. W., Rausa, A., Lingafelter, T., & Lingafelter, R. (1992). Partners for improved nutrition and health: An innovative collaborative project. *Journal of Nutrition Education (USA).*, *24*, 67S–70S.

Hughes, H. K., Matsui, E. C., Tschudy, M. M., Pollack, C. E., & Keet, C. A. (2017). Pediatric asthma health disparities: Race, hardship, housing, and asthma in a national survey. *Academic Pediatrics, 17*(2), 127–134.

Ibragimov, U., Beane, S., Adimora, A. A., Friedman, S. R., Williams, L., Tempalski, B., … Cooper, H. L. (2019). Relationship of racial residential segregation to newly diagnosed cases of HIV among black heterosexuals in US metropolitan areas, 2008–2015. *Journal of Urban Health, 96*(6), 856–867.

Institute of Medicine. (2000). America's health care safety net: Intact but endangered. *Committee on the changing market, managed care, and the future viability of safety net providers. Washington, D.C., The National Academies Press.*

Isaacs, S. L., & Schroeder, S. A. (2004). Class – The ignored determinant of the nation's health. *The New England Journal of Medicine, 351*(11), 1137–1142.

King, M. L. (1966, March, 25). *Presentation at the second national convention of the medical committee for human rights*. Chicago, IL.

Krousel-Wood, M. A. (2008). Beyond Katrina: From crisis to opportunity. The *American Journal of the Medical Sciences, 336*, 92–93.

Lantz, P. M. (2018). The medicalization of population health: Who will stay upstream? *The Milbank Quarterly, 97*(1), 36–39.

Lantz, P. M., Lichtenstein, R. L., & Pollack, H. A. (2007). Health policy approaches to population health: The limits of medicalization. *Health Affairs, 26*(5), 1253–1257.

Levine, R. S., Schneid, R., Zoorob, R. J., & Hennekens, C. H. (2019). A tale of two cities: Persistently high homicide rates in Baltimore City compared to significant declines in New York City. *The American Journal of Medicine, 132*(1), 3–5.

Lisovicz, N., Wynn, T., Fouad, M., & Partridge, E. E. (2008). Cancer health disparities: What we have done. *The American Journal of the Medical Sciences, 335*(4), 254–259.

Loria, K. (2020). Population health management: Medicare advantage plans covering more SDOH benefits. *Managed Healthcare Executive, 30*(3), 30–31.

Makaroun, L. K., Brown, R. T., Diaz-Ramirez, L. G., Ahalt, C., Boscardin, W. J., Lang-Brown, S., & Lee, S. (2017). Wealth-associated disparities in death and disability in the United States and England. *JAMA Internal Medicine, 177*(12), 1745–1753.

Maruthappu, M., Ologunde, R., & Gunarajasingam, A. (2013). Is health care a right? Health reforms in the USA and their impact upon the concept of care. *Annals of Medicine and Surgery, 2*(1), 15–17.

McGinnis, J. M., Williams-Russo, P., & Knickman, J. R. (2002). The case for more active policy attention to health promotion. *Health Affairs, 21*(2), 78–93.

Office of Minority Health and Health Equity. (2013, November 26). *CDC Health Disparities & Inequalities Report (CHDIR) – Minority health – CDC.* Centers for Disease Control and Prevention. https://www.cdc.gov/minorityhealth/CHDIReport.html

Richardson, E. G., & Hemenway, D. (2011). Homicide, suicide, and unintentional firearm fatality: Comparing the United States with other high-income countries. *The Journal of Trauma, 70*(1), 238–243.

Ridenour, M. L., Cummings, K. J., Sinclair, J., & Bixler, D. (2007). Displacement of the underserved: Medical needs of hurricane Katrina evacuees in West Virginia. *Journal of Health Care for the Poor and Underserved, 18*(2), 369–381.

Schroeder, S. A. (2007). We can do better – Improving the health of the American people. *New England Journal of Medicine, 357*(12), 1221–1228.

Sen, A. K. (1999, May, 18). *Health in development:* Keynote address by professor Amartya Sen, Master of Trinity College, Cambridge, Nobel laureate in economics, to the fifty-second world health assembly. Geneva, Tuesday, 18 May.

Smedley, B. D., Stith, A. Y., & Nelson, A. R. (2003). *Unequal treatment: Confronting racial and ethnic disparities in healthcare.* Institute of Medicine, Committee on Understanding and Eliminating Racial and Ethnic Disparities in Health Care. Washington, DC: National Academies Press

Smith, D. A., Akira, A., Hudson, K., Hudson, A., Hudson, M., Mitchell, M., & Crook, E. (2017). The effect of health insurance coverage and the doctor-patient relationship on health care utilization in high poverty neighborhoods. *Preventive Medicine Reports, 7*, 158–161.

Spruill, T. M., Butler, M. J., Thomas, S. J., Tajeu, G. S., Kalinowski, J., Castañeda, S. F., … Ogedegbe, G. (2019). Association between high perceived stress over time and incident hypertension in black adults: Findings from the Jackson heart study. *Journal of the American Heart Association, 8*(21), e012139.

Stoner, M. C., Haley, D. F., Golin, C. E., Adimora, A. A., & Pettifor, A. (2019). The relationship between economic deprivation, housing instability and transactional sex among women in North Carolina (HPTN 064). *AIDS and Behavior, 23*(11), 2946–2955.

Strong, K., Mathers, C., Leeder, S., & Beaglehole, R. (2005). Preventing chronic diseases: How many lives can we save? *The Lancet, 366*(9496), 1578–1582.

Szreter, S. (2003). The population health approach in historical perspective. *American Journal of Public Health, 93*(3), 421–431.

The Bill & Melinda Gates Foundation. 2020. *What we do.* The Bill & Melinda Gates Foundation. https://www.gatesfoundation.org/what-we-do (accessed April 2020).

Thomas, S. B., Crouse Quinn, S., Butler, J., Fryer, C. S., & Garza, M. A. (2011). Toward a fourth generation of disparities research to achieve health equity. *Annual Review of Public Health, 32*, 399–416.

United Nations. (n.d.). *About the sustainable development goals – United Nations sustainable development.* https://www.un.org/sustainabledevelopment/sustainable-development-goals/

US Department of Health and Human Services (DHHS). (1985). *Secretary's Task Force on Black and Minority Health.* Black and Minority Health Report. Washington, D. C.

US Department of Health and Human Services. (2004). *National health care disparities report.* Rockville: Agency for Healthcare Research and Quality.

US Department of Health and Human Services. (2014). *Healthy people 2020.* https://www.healthypeople.gov/

US Department of Health and Human Services. (2020, March 2). *U.S. public health service Syphilis Study at Tuskegee.* Centers for Disease Control and Prevention. https://www.cdc.gov/tuskegee/index.html

Vickers, S. M., Britt, L. D., Hildreth, J. E. K., Johnson, R. L., King, Jr., T. E., Love, T. W., Mouton, C. P. (2020, April 13). Black medical leaders: Coronavirus magnifies racial inequities, with deadly consequences. *USA Today.* https://www.usatoday.com/story/opinion/2020/04/10/coronavirus-health-inequities-deadly-african-americans-column/5124088002/

Walker, B., Mays, V. M., & Warren, R. (2004). The changing landscape for the elimination of racial/ethnic health status disparities. *Journal of Health Care for the Poor and Underserved, 15*(4), 506.

Wallace, D., Wallace, R., & Rauh, V. (2003). Community stress, demoralization, and body mass index: Evidence for social signal transduction. *Social Science & Medicine, 56*(12), 2467–2478.

Wallerstein, N., & Duran, B. (2017). The theoretical, historical and practice roots of CBPR. In *Community-based participatory research for health: Advancing social and health equity* (pp. 17–29). Jossey Bass. Hoboken, NJ.

Wallerstein, N. B., & Duran, B. (2006). Using community-based participatory research to address health disparities. *Health Promotion Practice, 7*(3), 312–323.

White, H. L., Crook, E. D., & Arrieta, M. (2008). Using zip code-level mortality data as a local health status indicator in Mobile, Alabama. *The American Journal of the Medical Sciences, 335*(4), 271–274.

World Health Organization. (2008). *Social determinants of health: Key concepts.* Geneva: World Health Organization.

Zabel, M. R., & Stevens, D. P. (2006). What happens to health care quality when the patient pays? *Quality & Safety in Health Care, 15*(3), 146–147.

3

Creating Livable Communities

Laura M. Keyes and Abraham David Benavides

Introduction

Livable communities assure health sustainability. *Sustainability* is an ideal in which people strive to minimize the impact of social, economic, and ecological damage from the current generation for future generations (Tretter, 2013). Once we have achieved a place that is safe, secure, and affordable, every effort should be made to maintain it. Simon Dresner (2012) noted that sustainability is a relatively new term that invokes our ability to take care of our current needs without compromising future generations. As we look to improve our communities, we should consider the three components of sustainability—environment, economy, and social equity (Campbell, 1996).

L. M. Keyes (✉) • A. D. Benavides
Department of Public Administration, University of North Texas, Denton, TX, USA
e-mail: Laura.Keyes@unt.edu; Abraham.Benavides@unt.edu

E. Mpofu (ed.), *Sustainable Community Health*,
https://doi.org/10.1007/978-3-030-59687-3_3

Learning Objectives

By the end of this chapter, the reader should be able to:

1. Define livable communities and livability.
2. Outline the historical evolution of planning models and strategies toward the development of livability planning concepts and community health outcomes.
3. Analyze public health practices for livable communities encompassing housing, walkability, and community access.
4. Propose strategies to create systematic, cultural, and behavioral changes to advance livability efforts.
5. Discuss the interdisciplinary nature of research and practice in livability and capacity of a community to be livable.

Community health is typically measuring a population's rate of death and disability, but also includes broader health considerations related to mental health, social health, happiness, and life satisfaction (Christian et al., 2013). The neighborhood level allows us to examine the relationship between the livability of a place and public health outcomes. The varying trends in positive and negative health outcomes across communities suggest variations in urban form and individual access to neighborhood services, walkable environments, parks and open space, and safe communities across zip codes. Ultimately, where individuals live determines their health outcomes, although the health and wellbeing of our people's families (and ancestors) can also determine where individuals and their families live. However, the historical influences on people's lives while important are modifiable with opportunity and resources, which unfortunately cannot be assumed of and for all (see also Chap. 2, this volume).

Specific environmental amenities are important to ensure opportunities for health. Factors shaping livable environments include the overall density of residential development, street connectivity, and land-use mix (Christian et al., 2013). Livability indicators of housing, transportation, access to open space, and ability of the urban environment to support overall community vitality and viability are critical for planners measuring progress toward building a livable community and enhancing quality of life (Balsas, 2004). A review of international academic and policy documents synthesized livability indicators into 11 domains, "Crime and safety; education; employment and income; health and social services; housing; leisure and culture; local food and goods; natural environment; public open space; transportation; and social cohesion and local democracy" (Badland et al., 2014, p. 1). Urban planners promote health through livability goals addressing food access, physical activity, and access to clean water and air through design and function of community space.

Livable housing and neighborhoods matter for all individuals. The *neighborhood effect*[1] influences overall human development and achievement. For instance, poor physical housing is a strong predictor for emotional and behavioral problems as well as poor health outcomes such as chronic disease from limited physical activity. The research evidence suggests that decades of planning policies and community design patterns have resulted in many communities being less livable over time due to rapid suburbanization, displacement of lower income and minority populations, dispersed housing located away from daily needs and employment opportunities, and loss of economy due to disinvestment (Balsas, 2004). As a method for urban development, a sustainability framework helps planners navigate and the tensions between the *three Es* of economy, ecology, and equity (Campbell, 1996). This balance resolved through comprehensive plans sought to develop healthy ecosystems (Chapin, 2012). Cities found a usefulness for the application of sustainability principles and strategies such as revitalization tools, densification efforts, mixed-use development, and a focus on creating walkable communities as drivers for economic development (Long & Rice, 2019).

Research suggests a connection between individual perceptions of their *built environment* including access, walking, and social interaction to general overall life satisfaction and feelings of good health (Cao, 2016).[2] For instance, urban planning attention focusses on creating areas that achieve livability goals and improve general quality of life. As city leaders invest in the public infrastructure, housing, and commercial

---

[1] https://www.huduser.gov/portal/periodicals/em/fall14/highlight1.html

[2] Cao, X. J. (2016). How does neighborhood design affect life satisfaction? Evidence from Twin Cities. *Travel Behaviour and Society, 5*, 68–76.

development to support the principles of livability, unintended consequences of displacement and segregation may occur. Therefore, the components of sustainability help to assure that public policy is responsive to the needs and abilities of all community members especially the most vulnerable. As we begin to consider the various aspects and components of livable communities, we are confronted with several professional and legal definitions that blur a clear understanding of the concept. In the next few paragraphs, we attempt to highlight some of these definitions to bring clarity to the topic.

Professional and Legal Definitions

A livable community is composed of various items such as affordability, safety, accessibility, mobility, housing, and technology just to name a few. Therefore, it is no surprise that it can have various definitions depending on the perspective of the organization, discipline, or individual. For instance, the AARP (2019), formerly American Association of Retired People, defines livable communities as: "A livable community is one that is safe and secure, has affordable and appropriate housing and transportation options, and offers supportive community features and services. Once in place, those resources: enhance personal independence; allow residents to remain in their homes and communities as they age; and foster residents' engagement in the community's civic, economic, and social life."

Ray LaHood, former U.S. Department of Transportation Secretary, defined livability as "a community where you can take kids to school, go to work, see a doctor, go to the grocery store, have dinner and a movie, and play with your kids in a park, all without having to get into a car" (Environmental Protection Agency [EPA], 2014, n.p). Similarly, Dwira Aulia (2016) defined livability as "the sum of the factors that add up to a community's quality of life. The factors are including the built and natural environments, economic prosperity, social stability and equity, educational opportunity, and cultural, entertainment and recreation opportunities" (p. 336).

Mohamad Kashef (2016, p. 240) defined livability to refer to "*physical and social well-being parameters to sustain a productive and meaningful human existence; to participate in forming successful and self-sustaining social systems.*" Livable communities "facilitate personal independence and the engagement of residents in civic and social life" (Kochera & Bright, 2006, p. 32). From these definitions, it is apparent that livability is much more than the sum-total of the items described above from the interconnectedness of how these items work together to form and create a livable space for the people within and around a location—especially for older adults.

As we can see, a livable community is not necessarily something new or anything out of the ordinary. It appears to be a quest to establish an idyllic neighborhood that will benefit all residents.

History of Research and Practice in Livable Communities

Livable communities as a concept have had a place in the development of planning visions for over 50 years (Chapin, 2012; Ewing, 1997; Kaiser & Godschalk, 1995; Krizek & Power, 1996). National and international planning models have influenced the emergence of livable communities as a planning vision evolving from a narrow focus on land management to considering the relationship between individuals and community and their surrounding social and physical environments (Gordon & Richardson, 1997; Hodge & Robinson, 2001; Meck, 2002).

State Growth Management Acts A short review of historical planning approaches to guide and manage land and development patterns offers some perspective. During the 1960s through the 1990s, states led the debate on land use adopting *state growth management acts*. The use of land development regulations, development caps, and urban growth boundaries became prominent tools in efforts to manage development patterns (Chapin, 2012). For example, in the United States, the state of Maryland set standards for incentivizing compact development as a counter to the pressures for suburbanization and land consumption of greenspace. Portland, Oregon achieved goals for compact development and developing transit as a reasonable form of transportation through a designated urban growth boundary densifying development around transit-oriented locations. Furthermore, infrastructure issues over ecosystems primarily pushed the decisions over land management (Chapin, 2012).

Human and Natural Environments Multiple sources influenced the beginning of the sustainability movement in the late 1980s until present day including the Brundtland Commission Report which was the first to make a connection between the human and natural environments (Keeble, 1988). The report explored the balance of population growth and the protection of natural resources. Further, it advanced international and national efforts to advance policies designed to reduce both human poverty and environmental damage in developed and developing countries (Goodland, 1995). For example, the U.S. federal government under the Clinton administration (1996–2000) led national efforts to advance sustainability as criteria in planning through the Environmental Protection Agency's (EPA) office on Sustainable Development operating for four years. These efforts led to federally funded grant initiatives to support environmental protection projects at the state and local community levels. Future administrations continue federal legislative support for sustainable policy actions funding various federal agencies to fund local community projects in the areas of housing, transportation, and smart growth (Mazmanian & Kraft, 1999).

Smart Growth The Smart Growth movement borne out of the sustainability era is a precursor to Livable Communities. The Smart Growth paradigm emerged to offer a planning vision and solutions to stop urban sprawl and protect greenspace through planning incentives such as higher density bonuses and flexible design standards (Godschalk, 2004). Smart Growth principles support goals for sustainable land development by reducing individual demand for infrastructure by co-locating live, work, and play options in a community (APA, 2012). The walkability of a community is an important feature of a community designed to these standards. The smart growth model, driven mainly by special interests, created pressure for the local adoption of Smart Growth principles (Chapin, 2012). The movement centered on a creative class of workers with possible dislocation of others (Long & Rice, 2019) and did not strive to advance health-related goals.

Professional associations for planners and architects such the *American Planning Association* and *Congress for New Urbanism* developed policy guides to facilitate needed local zoning code changes to support *resilient*, compact, and walkable places (American Planning Association, 2012). The associations rely on the academic and professional membership to inform the declarations presented in a policy guide. The policy guides provide community leaders and practitioners with policy objectives to achieve livability goals both in form, function, and feature of communities; but they also inform on expected economic, land use, infrastructure, and other societal outcomes.

Livable Communities Livable communities planning expands on related movements to focus on land use and the built environment to address the individual's experience in their social and physical environment. Livability guides the local discourse of the community and policymakers on framing urban goals at a human scale (McArthur & Robin, 2019). The public health community has prominent participation in this model presenting positive health outcomes when communities foster active travel such as walking and biking (Handy & McCann, 2010).

The Centers for Disease Control and Prevention (CDC) Healthy Communities Checklist empowers the public to make connections between their *built environment* and health outcomes and advocate for housing access, employment, learning, safety, and amenities that improve their overall wellbeing. The global initiative between the *World Health Organization's (WHO) Age-Friendly Cities Initiative* and its U.S. affiliate, AARP, is actively engaging older people in the conversation about livability. Livability for older people is the connection between housing and access to community services (Hwang, Glass, Gutzmann, & Shin, 2008).

Research on aging and local government leadership finds positive results for older adult independence and health when communities facilitate housing options, develop accessible public transportation, improve the built environment, and expand access to services (Benavides & Keyes, 2015; Keyes & Benavides, 2017; Lehning, 2014 and Lui, Everingham, Warburton, Cuthill, & Bartlett, 2009).

Pertinent Sustainable Community Health Approaches

In the following section, we focus on housing, walkability, and community access approaches and their connection to creating livable communities and healthy community outcomes for community populations. In doing so, we explore housing in the context of community design, housing options across type and size, and affordability. We go on to discuss the importance of walkability as a mobility option to connect individuals to their community to promote health and physical activity. Finally, we consider the role of community design to support individual access to community supports and basic daily needs.

Housing

Where we live can influence one's ability to walk, socialize, and preserve health choices. We explore the role of housing in the context of community design and the relationship between the home, commercial activity, recreational options, and infrastructure network that connects these activities. Access to housing options across price, size, and type influences the ability for individuals to remain in their communities across their lifespan typically a result of local zoning codes. An individual's ability to enjoy a livable community is all relative to their ability to afford to reside there. Federal policies have proliferated suburban development patterns separating residential living from basic daily needs. Certain programs, grants, and nonprofits are breaking down barriers as an inclusive approach to livable communities across all ages, abilities, and income levels.

Community Design

An important factor in the discussion about the role of housing and general overall health relates to the location of the residence. Housing is a central component to the formula for creating livable places:

- Housing density associated with street connectivity patterns supports higher levels of physical activity.
- Positive perceptions of the built environment influence overall life satisfaction (including physical, social, and mental health).
- Older people that reported high satisfaction with their residence and general access to the community also reported high quality of life (Feng et al., 2018).

Livable guidelines focus attention on the condition and location of housing in the community relative to a variety of access to daily living needs and transportation options.

Planning and building with a goal of *active design* is an approach to residential development that engages architects and urban planners in neighborhood development that attempts to co-locate housing with options of safe biking, well-lit residential units, and community gardens to increase physical activity (Tannis et al., 2019). As further discussed in Discussion Box 3.1, residential location to greenway facilities and natural areas contributes to quality of life (Shafer, Lee, & Turner, 2000). Evidence finds that individuals residing in active design communities in the Bronx, New York, were likely to report increased stair usage and overall better health behaviors (Tannis et al., 2019). Residents that self-selected to live in housing developed under Australia's Livable Neighborhood's (http://www.livablehousingaustralia.org.au/) guidelines (i.e. mixed-use, interconnected streets, public transit, and access to open space and parkland) participated in higher levels of recreational walking and cycling leading to better health outcomes (Christian et al., 2013). Visualization and mapping tools support efforts to empower residents in community improvement and affordable housing location decisions with a focus on social and economic wellbeing outcomes (Klassen et al., 2013).

Discussion Box 3.1: Active Design for Healthy Lifestyles

Planning and building with a goal of *active design* is an approach to residential development that engages architects and urban planners in neighborhood development that attempts to co-locate housing with options of safe biking, well-lit residential units, and community gardens to increase physical activity (Tannis et al., 2019). According to the American Planning Association (2012), certain design features promote physical activity such as the location of housing near public transit and the connectedness of a transportation system that makes it possible to walk and cycle safely. Urban planning can foster changes in the built environment such as the construction of trails, parks, and recreational spaces to incentivize physical activity (Christian et al., 2013).

What Do You Think?

1. If you are a junior planner for a city, what key components of active living would you include in your plan to build an inclusive community?
2. Which design elements are the most important for producing positive health outcomes?
3. What steps would you take to engage the residents in a discussion about active living?
4. Are the concepts of active design and livable communities realistic?

Livability from the perspective of housing for older people relates to structure and suitability, but also the ability to integrate community-based services to support long-term options for aging in place (Hwang et al., 2008). The World Health Organization (WHO) Age-Friendly Cities Initiative and its U.S. affiliate, AARP's Age-Friendly City Network suggest that general livability requires sustainability of the current home community building practices that support the changing needs of individuals across the life span (Hwang et al., 2008). Housing is a central component to creating livability for people of all ages and abilities because an individual's access to quality housing determines the extent one can reside in that neighborhood. The AARP's Livability Index examines the opportunities for housing across accessibility, affordability, and choice across neighborhoods to determine the reality that a community can support individuals of all ages and abilities (AARP, 2019).

In the United States, local governments are leading on policy changes to expand housing options for older people to support rising demands among those aged 60 and older to age in the community (Keyes & Benavides, 2017). South Korea provides an international perspective illustrating a national response to providing low-income housing for older people by developing affordable housing operated by the state government (Hwang et al., 2008). *Universal design* guidelines direct construction standards for housing in the City of Seoul to provide non-slip bathrooms, wider doors, and convenient doorknobs as a response to the changing psychological and social needs of older people (Hwang et al., 2008).

Choices and Affordability

Housing affordability has economic meaning suggesting that individuals have access to residential units that do not exceed more than 30% of overall income. In the context of community development, Green and Haines (2015) suggest that an examination of affordability also considers the condition of the housing stock, the distribution of housing type, and overall access of housing to community amenities. Housing affordability impacts individuals in different capacities. Minorities and immigrants are less likely to ever be homeowners and have reasonable access to housing (Sirmans & Macpherson, 2003).

Housing is fundamental to health including pride associated with homeownership, and the desire to maintain an acceptable standard of living and maintain a place in the community. Housing markets, relative to the basic economics of supply and demand, should create the right mix of housing options for the buyer (Green & Haines, 2015). Regulations at the federal, state, and local levels impact the cost of housing, and zoning policies give local control over the allowable types of housing products making housing options out of reach for many individuals.

In the United States, suburban development patterns have led to the proliferation of single-family detached housing leaving few options for smaller home options, limited types, and lower price points in most communities in the United States. Federal housing policy, as further explored in Research Box 3.1, is central to the debate about choice and affordability at the local level. The *Housing Act of 1949* played a critical role in the

Research Box 3.1: The Transformation of Public Housing Policy, 1985–2011 (Goetz, 2012)

Background—Smart growth and new urbanist ideals have led local housing authorities to demolish public housing residences and replace them with subsidized units in mixed-use rental units. During the 1980s through the 1990s, the number of public housing sites reduced throughout the country through teardowns. A national commission established criteria to identify the most severely distressed properties. Policy shifts such as the federal HOPE VI program to reduce concentrated poverty led to new efforts of redevelopment. Section 8 vouchers and new subsidized units served as replacements.

Method—This research examines policy, political, and economic shifts in federal housing policy from 1985 to 2011.

Results—Federal policy shifts mainly under the U.S Department of Housing and Urban Development to reduce concentrated poverty in American cities led to new practices and housing options for low or very low-income residents. HOPE VI reforms in the 1990s resulted in both an approach to decentralize poverty through mixed-use and income residents and through the use of Section 8 voucher to expand affordable housing options in nontraditionally underserved neighborhoods. The 1990s saw an increase in pressure for redevelopment of downtown communities and neighborhoods. New Urbanists and Smart Growth planning principles no longer supported the concept of public housing. Many remaining public housing sites were demolished due to pressures of gentrification.

Conclusion and Implications—Local housing authorities have replaced the traditional public housing residence with public–private and fully private redevelopment projects that offer subsidized units mixed with market-rate rental units. Many experts conclude that while residents of the existing public housing faced displacement and separation from friends and family from the HOPE VI approach of demolition and disperse, the general sentiment suggested that public housing was in such despair that individuals were ultimately better off relocating. The practical implication for the most vulnerable individuals was the provision of fewer subsidized units being made available after the completion of the new mixed-use project.

What Do You Think?

1. What was the main goal of the federal housing policy shifts relative to public housing demolition?
2. What concepts of new urbanism and smart growth influenced the demolition of public housing?
3. What is a practical implication for the overall availability of available subsidized units?

creation of suburban communities through increased access to home mortgages. It also enabled local governments to use eminent domain for economic redevelopment purposes and some scholars attribute this with urban blight and reduction in the supply of affordable housing units across the country (Von Hoffman, 2000).

In the early 2000s, local housing authorities across the country integrated the HOPE VI federal funding program through the U.S. Department of Housing Urban Development to redevelop public housing with mixed-income communities in higher income communities (Goetz, 2011). The goal of the Hope VI projects was to integrate lower income individuals into higher income communities, increasing access to employment, better housing, and educational opportunities. Scholars reported fewer positive results of children benefiting from Hope VI housing developments; showing no overall health improvements or advances in educational attainment (Goetz, 2011). Findings suggested negative impacts on the original residents of public housing displaced during the redevelopment process. This supports the notion that maintenance of social connections and friendships matters to the general wellbeing of individuals across all income levels.

Legal barriers for housing options in suburban communities typically stem from large minimum square footage requirements for house and lot size. Trends in the development of tiny homes and smaller cottage residential units are producing positive health and environmental outcomes. Practical experience provides information on the substantial energy savings to owners of smaller units, emergency shelter after a natural disaster, and longer-term sheltering options for homeless members of the community (Kaufmann, 2015). In the United States, local *zoning* laws, as explored in Case Study 3.1, typically need revisions to allow for the construction of units with smaller square footage and permanent placement of a tiny home. Through an international lens, changes to China's national housing policy in the 1990s resulted in a fundamental shift in ideology about aging in place. Older Chinese citizens residing in urban areas tend to prefer living arrangements outside the family unit resulting in new attention to quality of life and the mixture of housing structured around daily activities (Feng et al., 2018).

Case Study 3.1: Making Communities Livable, The Case of Lifelong Mableton

The Atlanta Regional Commission in partnership with Cobb County, Georgia worked directly with the residents of the community of Mableton, Georgia to create a livable and lifelong community. The livable community planning effort started in 2010 driven by three goals: promote housing and transportation options, create healthy places, and increase access to information. The Mableton Lifelong Communities Partnership included representation from residents, community stakeholders, and local and regional government officials. The Partnership empowered local residents to have initial and ongoing input into the planning process. The planning effort led to the adoption of a new zoning code called the Mableton Form-Based Code which legalized the development of denser housing and diversity of type in closer proximity of retail. Form-based code focuses more specifically on design and scale rather than the type of use allowing creative development options to support walkability. Attention to the goal of creating healthy places led to the donation of land for the purpose of developing a working community garden. The collaborative process led to the participation of the local AARP chapter in the garden to support their "Plant a Row for the Hungry" program. Participants of the garden were instrumental in providing food from the garden to local food pantries. Cobb County Government also approved the use of public land to support the operations of a weekly farmer's market. The regional Area Agency on Aging and Cobb Senior Services provided the necessary leadership to register farmers with SNAP allowing older residents to use their vouchers to purchase fresh produce at the market. Finally, the development of a Mental Health Collaborative and partnership with Cobb County Public Health led to new efforts to address gaps in referral processes among providers; concentrating efforts on reducing the need for arrests and hospitalization, as well as reducing issues that may lead to homelessness. The efforts of the collaboration provide evidence of the need for strong leadership from the beginning and empowerment of the community residents to guide and provide input in the planning process.[3]

What Do You Think?

1. What were the main goals of the Lifelong Mableton Project and some example programs developed to achieve a livable community?
2. What was the purpose and outcomes of the community-led partnership?
3. What were the results from the efforts to make systematic changes to the areas zoning code?

[3]Keyes, L., Phillips, D. R., Sterling, E., Manegdeg, T., Kelly, M., Trimble, G., & Mayerik, C. (2014). Transforming the way we live together: A model to move communities from policy to implementation. *Journal of aging & social policy*, *26*(1–2), 117–130. http://www.ciaip.org/grantees/atlanta/

Community Resilience and Sustainability and the Nonprofit Response

The U.S. Department of Housing and Urban Development (HUD) makes available two important tax credit programs, Low Income Housing Tax Credit and the New Market Tax credit for state and local tax allocating agencies to issue credits for new and rehabilitated housing for the benefit of lower income householders.[4] Local governments may also receive and allocate Community Development Block Grant Funds for the purposes of home repair and rehabilitation with a majority focus on the benefit to low-income households. The Green House Project, a national nonprofit, uses New Market Tax credits developing homes for older people in communities matched with a transformative care model that fosters relationships and socialization with the benefit of private rooms and personal freedom as individual's age.[5] Another community-based organization such as CAAS Community Social Services is leading in the provision of tiny home developments in Detroit, Michigan for immediate shelter for individuals experiencing homelessness.[6] The CAAS program revenue portfolio includes HUD funds matched with private gifts from foundations, donors, corporations, and religious organizations.[7]

Walkability

The extent our community supports or constrains walking has a direct effect on quality of life (see also Chap. 16, this volume). Certain tools exist to help residents measure their community's walkability. Urban planners enhance walkability through attention to the design of the built environment and placement of wayfinding signage to ease navigation. One of the tenets of the livable community movement is the concept of mobility. Whether packaged as an effort to reduce carbon emissions and

---

[4] www.huduser.gov

[5] https://www.thegreenhouseproject.org/solutions/finance-green-house-home

[6] https://casscommunity.org/

[7] www.casscommunity.org/about/annual-report

reduce our dependence on oil or to design safe reliable transportation options or even to promote public health, the underlying theme is mobility options and health. The Centers for Disease Control and Prevention (2019, n.p.) notes that "Physical activity is one of the best things people can do to improve their health. It is vital for healthy aging and can reduce the burden of chronic diseases and prevent early death."

Among the topics within this broad category of mobility is walkability. Walking is the most basic form of transportation for humans. From our first steps as a baby to our last steps as an older adult, we walk to move ourselves from place to place. Walkability can also be seen as a measure that looks at the built environment to gauge how various factors lend themselves to a friendly walking experience. For instance, Ewing and Handy (2009) look at the street environment to measure how it affects walking behavior. They find that urban design and physical characteristics do contribute to a positive or negative walking experience.

Similarly, Christian et al. (2013) found in their survey of new housing developments in Perth, Australia that designed livable neighborhoods—those that supported walking—showed an increase in walkability for its residents as opposed to other types of developments. They concluded that residents living in neighborhoods with intentional livable community design patters had a higher quality of life than those that resided in suburban developments.

Walkability as a subfield of study has grown in recent years because of the importance of how we build and design our neighborhoods. We have realized that our physical structures, the places we live, the environment, all do affect our quality of life, as do our personal relationships. Therefore, designing our communities with our eyes on reducing stressors and accommodating physical health has increased in priority. For instance, several organizations as well as academics have created walkability checklists. Research Box 3.2 explores these concepts in more detail. The Neighborhood Environment Walkability Scale (NEWS) was developed to measure residents' perceptions of their environment (Saelens, Sallis, Black, & Chen, 2003). Others have added to this scale to provide environmentally related information (Brownson, Chang, Eyler, et al., 2004). Cerin, Saelens, Sallis, and Frank (2006) created a short form, NEWS-A, and recommended it be used when participant burden was a concern.

Research Box 3.2: Many Pathways from Land Use to Health: Associations Between Neighborhood Walkability and Active Transportation, Bud Mass Index, and Air Quality (Frank et al., 2006)

Background—The design of the surrounding environment has direct implications on individual health outcomes. Suburban and lower density design of communities greatly reduced the ability for residents to walk and increases their reliance on the automobile to access their daily needs. The typical design pattern for suburban communities is based on a separation of uses and a disconnected street system. For example, many suburban residential communities are built with a series of cul-de-sacs generally offering only one access point to the neighborhood. Previous research finds greater levels of obesity in less connected and less walkable areas (Ewing, Schmid, Killingsworth, Zlot, & Raudenbush, 2003).

Method—This research uses a Neighborhood Quality of Life Study to examine an association between land-use patterns and physical activity in neighborhoods throughout King County, Seattle. The research ranked block groups by walkability and then administered the survey to sample adult populations comprising 1228 adults within these block groups.

Results—Findings show a strong association between the level of walkability in a block group and the overall physical activity and levels of obesity of the residents. The findings also suggest that areas with higher levels of walkability produce fewer air pollution emissions.

Conclusion and Implications—The findings provide support for local policy changes in land use to develop walkable communities. Minor changes in walkability show support for improved public health. Investment in infrastructure to create a connected transportation system creates opportunities for residents to choose walking as an active transportation option leading to potential improvements in health outcomes. However, those already inclined to use an automobile for travel may not switch to walking.

What Do You Think?

1. What characteristics of suburban form result in autodependent lifestyles?
2. What is meant by the term active transportation?
3. In what ways do the findings support consideration of policy actions to support investments in walkability?

The U.S. Department of Transportation's National Highway Traffic Safety Administration and the Federal Highway Administration (2019) have also created a type of survey for citizens to rate how walkable their communities are. They also provide tips and resources for improving their community score.

Walkability enhances livability when the community is easy to navigate. Larger street signage, wayfinding information in the community, and information in multiple languages accommodates the diverse needs and ages of residents (Benavides, Nukpezah, Keyes, & Soujaa, 2020). Wayfinding signs and other landmarks provide helpful cues informing pedestrians and drivers with directional information to remove mobility barriers throughout the community (Walford, Samarasundera, Phillips, Hockey, & Foreman, 2011). These types of markers reduce stress and travel anxiety, and highlight potentially dangerous areas (Hwang & Ziebarth, 2015).

Community Resilience and Nonprofit Response

The City of Los Angeles Department of City Planning has also created a list intended more for developers and others that plan on making changes to sidewalks. It provides guidance and essential direction on assuring that walkability is considered and is safe. A nonprofit organization, Champions for Change, has also created a walkability checklist. Their list is intended for people to use, walk the street, and rate its walkability. Champions for Change then provides ideas for improving the walkability in neighborhoods. Organizations like America Walks, Partnership for a Walkable America, Kids Walk to School, National Center of Bicycling & Walking, Pedestrian and Bicycle Information Center, and many more are geared to making the walking experience not only healthy but fun.

Community Access

A livable community promotes access to jobs, economic opportunities, and social connections for individuals of all ages and abilities. Further, it promotes access to basic daily needs such as fresh food, health and supportive services, and mobility options. Urban planners provide examples on how to create livability for residents through leveraging technology and the capacity of the existing community assets. Having the ability to access the community with little to no constraints is an essential aspect of a livable community. The concept of community access is realized as an individual's ability to interact with the community based on their ease of connection to public services, daily needs, health and supportive services, food and nutrition, social interaction opportunities, and information. Development features supporting a community's accessibility typically include a well-connected street network, compact development that locates housing near-daily needs for living, and public transportation options. A community's overall ability to support an individual's accessibility of daily life needs is important to general public health based on the benefits received through increased opportunities for making and maintaining social connections with neighbors, friends, and family (AARP, 2019). This section examines livability through a lens of accessibility relative to the community's role in supporting daily living (e.g. access food, open space, health and supportive services, building social inclusion of community members through technology and information, and increasing community resilience through a shared sense of purpose among community stakeholders and nonprofits).

Daily Living

A livable community depends on features that help support a mix of commercial activity, retail, shopping, and other leisure activities important to accommodating the basic daily needs of its residents (Balsas, 2004). A framework for developing a healthy town center proposes basic qualities such as diversity of uses, connectivity between uses, safety, and capacity to support future growth (Department of the Environment, 1994). Perhaps the most important factor related to resident access to a diversity of uses is the overall availability and access to food (see also Chap. 5, this volume). Typically, food access becomes a determining factor for the overall livability of a community. A *food desert* is a basic planning concept that refers to a resident's lack of reasonable proximity to healthy and nutritious food options. Discussion Box 3.2 further explores the concepts in terms of water access and the importance the relationship access to clean water and livability. Researchers find that lower income individuals face higher levels of *food insecurity* due to lack of land to grow

Discussion Box 3.2: Livable Communities and the Importance of Natural Resources

An essential component of any livable community is water quality. The majority of Americans receive their drinking water from a community water system most often provided by a municipality either directly or contracted out to a water authority. The Environmental Protection Agency at the federal level and state agencies at the local level monitor water quality. In most cases and in most communities, the tap water is safe to drink and rarely is a second thought given to the quality of water. Nevertheless, contamination of drinking water can occur and in some cases our streams, lakes, and even oceans have become contaminated. The Flint Michigan water crisis was an example of local drinking water being contaminated by lead. On the west coast, the city of Imperial Beach, California just south of San Diego is currently experiencing contamination of its ocean beaches. The Tijuana River which flows from Tijuana, Mexico into the United States and empties into the Pacific Ocean just north of the U.S. border often flows untreated and is contaminated. It affects the city of Imperial Beach, the Tijuana River National Estuarine Research Reserve, Border Field State Park, and small neighborhoods necessitating the closure of beaches on the Pacific Ocean. The raw sewage and associated garbage and contaminants flowing into the United States are a public health emergency. Federal, state, and local officials have been working with their Mexican counterparts to resolve the problem for years; however, little progress has been made. If the City of Imperial Beach aspired to be a livable community, this public health emergency would need to be addressed. Although the Pacific Ocean does function like a big filter and eventually some contaminants are removed, this natural process is not sustainable, and the beaches have had to be closed more often and for longer periods in recent years. This scenario with international and multijurisdictional oversight is complicated at best and unresolvable at worst. Each community has its difficulties to overcome in its transformation to become a livable community. Some are complicated like the situation above. Others are less so but still require a number of organizations and disciplines to be involved to achieve a safe, secure, and affordable place to live.

What Do You Think?

1. Whose fault is the water contamination flow from Tijuana, Mexico? What can be done to stop the contamination flow from Tijuana, Mexico?
2. If you were the mayor of Imperial Beach, California what would you do about the water contamination problem?
3. What role do federal, state, and local agencies play in helping to resolve water quality issues?

Through our understanding of the conflict toward livable community outcomes presented in the drinking water case between Mexico and the United States, explain the importance of collaboration in achieving livability goals.

their own food and lack of access to healthy food choices (Horst, McClintock, & Hoey, 2017). Fast food options are prevalent in lower income neighborhoods' increasing access to cheap food but do not offer healthy options as a primary menu item.

The *food security* threat to lower income individuals is their displacement from livable communities when land values and housing increase due to the attractiveness of the accessibility and walkable places. Planners have multiple tools to foster livability and food access for lower income individuals through:

- Comprehensive planning that includes policies that recognize the importance of food (for livability).
- Zoning and regulations that remove barriers for urban agriculture (e.g. farms, chicken coops, beehives).
- Dedicated spaces on public lands for community gardens.
- Land trust programs to help purchase property for urban farms (Horst et al., 2017).

Cities, such as Philadelphia, Pennsylvania, promote programs to reclaim vacant land to support urban agriculture initiatives and create green public gardens. Reclaiming these unused spaces as assets in redevelopment initiatives supports sustainability efforts and fosters opportunities for residential recreation and social interaction (Gough & Accordino, 2013). For older residents, the meaning of livability relates to the realistic ability to age in place as their needs change (see also Chap. 13, this volume). Reducing the distance between where older people live and their health and medical providers reduces the need for many different mobility options, which is especially important for older people (Neal et al., 2006). It is common for an individual over 65 years of age to have up to three different doctors on average. Easy access to needed medical appointments helps ensure older people follow through and keep these appointments (Philadelphia Corporation on Aging, 2006). Access in the context of creating closer proximity to health care options is an important factor in long-term care planning for older people, typically supported through a network of community-based providers (World Health Organization, 2007).

Technology and Information

Increasing access through technology and information platforms is about strengthening an individual's and the community's linkages to resources (see also Chap. 11, this volume). Cities are integrating technology into many different planning platforms, from *interactive mapping* features on their websites to the use of apps that gauge public opinion through the smartphone. Interactive mapping tools visually illustrate the planning policies of the city and their contribution to livability goals. Technology allows the government to share information broadly across the public domain to garner more public opinion on plans for development such as parks, trails, housing, new retail, schools, and other features of the community. Tools such as online polling or *text polling* via the smartphone also allow more residents to participate in the community planning process. Technology and digital planning make everyone aware of what is going on in their community, building social cohesion and connectedness around issues.

Improved access to community information may help support the goals of older residents to remain in the home and community as their community living needs change (see also Chap. 13, this volume). For instance, social services information such as financial information to help with tax reporting and housing, legal aid, support for caregivers, and home and community-based services made available through a central hotline or website may empower older people with knowledge to remain independent and healthy. The National Association of Area Agencies on Aging Eldercare Locator provides social services information at a local level.[8]

Community Resilience and Nonprofit Response

Community partnerships are fundamental to resiliency and sustainability (Gough & Accordino, 2013). To increase local engagement and build capacity to expand services and programs for the community, planners

[8] https://eldercare.acl.gov/Public/Index.aspx

can enhance access through better design of redevelopment projects. Planners can identify and leverage existing community amenities as assets for quality of life. A case study on creating a lifelong livable community in Mableton, Georgia provides insight on the positive impacts when planning agencies partner with the business community and public health to improve quality of life (Keyes et al., 2014). The collaborative team purposefully created new access points in the community including a community garden and weekly farmer's market to locate healthy food options closer to lower income residents and to stimulate social cohesion, especially among older adults.

Inclusion and social justice are important factors for measuring a community's overall access. Some communities call on public health to help with this effort. The City of San Francisco provides insight on fostering engagement through the lens of public health promoting celebration and intergenerational events that highlight the role of individuals with disabilities as assets to the community (Yeh, Walsh, & Wallhagen, 2016). In this case, the municipal transit agency was recognized as a champion for providing free transit for older people and individuals with disabilities.

Cultural, Legislative, and Professional Issues That Impact the Sustainable Community Health Approaches

An important question related to the sustainability of health approaches through livability planning asks: how has the culture integrated the meaning of livability into language, systems, and behavior? The concept of livability as presented requires systematic changes to planning and zoning in communities that allow different types of housing across size and price point, streets, and sidewalks constructed to promote walkability and mobility options, and ease of access to the community resources that support daily living.

In order to change the existing culture to one that supports livability, we need to adjust ideology and the current know-how. The change may need to happen across many levels and is supported by the overall community's willingness to change. A process for change requires community leaders to decide to change, implement the change, and institutionalize the change (Trice & Beyer, 1993). Exploring culture, as illustrated in Case Study 3.2, helps us to know the meaning of the term, its impact on systems, and the tools and strategies needed for trying to achieve design patterns that foster livability for all individuals.

Case Study 3.2: Livable Communities in China: Housing Options for Older People

The Chinese government is now considering livability in the context of supporting older people. Cultural changes are resulting in older people seeking housing options outside of the family unity and under the roof of their children. Due to housing reforms, older people are seeking independent options in urban areas. A form of Chinese development called danwei tends to have well-connected infrastructure to support accessibility and walkability to daily needs of living. Residents living in the boundaries of a danwei tend to report high satisfaction with their quality of life. Employment locations are the driving force of the geographic location of the danwei. In addition to jobs, the community includes other important institutions such as housing, childcare, health care, and other health and supportive services. Research finds the job housing balance of the danweis results in high levels of quality of life for residents and overall positive health outcomes. However, when measured over long periods, the self-contained design of the danwei leads to a decrease in social connectedness beyond the boundaries of the community leading to higher levels of social isolation especially among older people. The danwei design has similarities to the gated communities found in suburban areas of the United States.[9]

What Do You Think?

1. What factors are influencing housing demand among older people in China?
2. What are the key components of a danwei in the context of livability?
3. What is a key finding of the long-term impact on older people living in Chinese danweis?

[9] Feng, J., Tang, S., & Chuai, X. (2018). The impact of neighbourhood environments on quality of life of elderly people: Evidence from Nanjing, China. *Urban Studies, 55*(9), 2020–2039.

Systems Change

In 2009, the U.S. Department of Transportation, the Environmental Protection Agency, and the Department of Housing and Urban Development created a partnership to promote sustainable communities. This partnership is sometimes referred to as the Partnership for Sustainable Communities (PSC) and was a first of its kind with three federal government agencies aligning policies across their various disciplines. It marked a significant way in which federal departments worked on one policy issue. The PSC based its efforts on the following six livability principles.

The U.S. HUD–DOT–EPA Partnership for Sustainable Communities Guiding Livability Principles

The livability guiding principles are grounded in community wellness for all. They work to provide people with basic needs like transportation, housing, economic means, community revitalization, and neighborhood safety.

Provide More Transportation Choices Develop safe, reliable, and economical transportation choices to decrease household transportation costs, reduce our nation's dependence on foreign oil, improve air quality, reduce greenhouse gas emissions, and promote public health.

Promote Equitable, Affordable Housing Expand location and energy-efficient housing choices for people of all ages, incomes, races, and ethnicities to increase mobility and lower the combined cost of housing and transportation.

Enhance Economic Competitiveness Improve economic competitiveness through reliable and timely access to employment centers, educational opportunities, services, and other basic needs by workers, as well as expanded business access to markets.

Support Existing Communities Target federal funding toward existing communities—through strategies like transit-oriented, mixed-use development, and land recycling—to increase community revitalization and the efficiency of public works investments and safeguard rural landscapes.

Coordinate and Leverage Federal Policies and Investment Align federal policies and funding to remove barriers to collaboration, leverage funding, and increase the accountability and effectiveness of all levels of government to plan for future growth, including making smart energy choices such as locally generated renewable energy.

Value Communities and Neighborhoods Enhance the unique characteristics of all communities by investing in healthy, safe, and walkable neighborhoods—rural, urban, or suburban.

Research has explored the use and association of the term livability across academic disciplines. A review of the literature from 1999 to 2017 shows the use of livability as a variable explored relative to community features, development, environmental features, federal and national initiatives, health and safety, housing, measurement indices, social justice, and transportation (Herrman & Lewis, 2017). The frequency of use of the term livability was highest in the literature related to transportation research, followed by published articles on development practices, with the third most commonly used term being community features. Findings show a trend in the prevalence in the use of the term and its definition but not necessarily the tools and strategies to implement in a planning and health context.

Some community strategies inform on systematic change institutionalizing livability into local community development plans. The Atlanta Regional Commission in Atlanta, Georgia developed a new strategy to combat pressures of sprawling development, environmental threats to its water resources, and health consequences due to an autocentric lifestyle. The agency, responsible for the allocation of federal transportation funding in the metro area, redirected over $350 million in federal transportation funds over a 10-year period to funding planning and implementation projects for communities to retrofit communities into walkable, denser, mixed-use, and livable places (Dobbins, 2005). The program offered multiple financial incentives for communities to develop collaborative partnerships with community stakeholders to galvanize support for zoning codes and design regulations necessary to develop livable places. For example, for many communities, outdoor al fresco dining was illegal due to zoning code restrictions.

Casc Antic of Barcelona provides an international example of systematic change through the mobilization of lower income residents in a poorer neighborhood seeking revitalization of the community to support better health, improved environmental quality, and livability (Anguelovski, 2015). In this community, over 31% of the residents are foreigners from various regions of Africa, Asia, and the Middle East. The area has experienced massive city block redevelopment projects for three decades as well as tenant harassment in efforts to force out the lowest income renters. The residents banded together and worked to raise their voice in the planning processes to ensure that the environmental conditions of the community were protected and that new spaces reflected the desires of a culturally diverse community including soccer fields, playgrounds, trees, parks, and community inspired gardens. Livability with regard to social justice can create inequities if individuals displaced from communities during redevelopment efforts and infrastructure investments price low-income individuals out of the community in redevelopment efforts.

Behavioral Change

Livability is a design and system response but also a needed change in individual behavior. Herrman and Lewis (2017) provide insight on identifying opportunities to change individual behavior to support necessary development and redevelopment changes to programs and infrastructure. Governments may need to identify the benefits for sustainable development. For instance, Herrman and Lewis (2017) highlight the strategic connection between livability and transportation in the context of reducing automobile use. In their research, they found that new communication strategies are necessary to increase transit ridership. The focus is on moving non-riders from their automobiles to transit. Most non-riders surveyed indicated they did not use transit due to a perceived inconvenience. Dialogue with the community should articulate credible benefits (Morrison et al., 2018). The authors suggest there are opportunities to influence community behavior change toward twenty-first-century problems such as climate change by playing off people's individual motivations to be heroes and a part of the solution to these problems.

The U.S. Federal Funding Initiatives to Support Community Goals Toward Livability

According to Herrman's and Lewis's (Herrman & Lewis, 2017) literature review on the concept of livability, the federal government was instrumental in bringing livability into the urban planning agenda through the leadership of the Clinton/Gore administration. The authors traced the integration of the term livability into federal guidance on advancing smart growth principles through funding initiatives for greenspace projects, traffic congestion reduction, and urban redevelopment.

The *HOPE VI Program*, a funding initiative of HUD, integrated the philosophy of livability through a targeted grant program to help cities demolish public housing and redevelop communities with mixed-use communities providing market and affordable rate units. The planning approach focused on helping lower income communities redevelop areas with the intent of attracting outside investment. Opportunities were also sought to leverage connections between the school systems, efforts to rehab housing, commercial investment, and public infrastructure (Goetz, 2011). HUD continues to support mixed financed housing programs. Helping local communities increases access to affordable units for their residents.

In 2009, the U.S. federal government, through Federal Partnership for Sustainability between HUD, EPA, and DOT, helped to institutionalize the federal goals and policies through livability by working directly with communities across the country. The program fostered the sharing of ideologies across different levels of government and among the various stakeholders, leaders, and residents of the communities. The federal funding helped incentivize local communities to make systematic changes to the planning process (Young & Hermanson, 2013). The first round of the program provided $1.5 billion in funding to local communities. The federal DOT provided over $35 million in Tiger grants to fund the second year. Additionally, communities could match these funds with other grant resources from HUD (Partnership for Sustainable Communities Brief, 2014).

The U.S. Federal Transit Administration (FTA) continues to have a prominent role in the continued efforts to institutionalize livability via the development of public infrastructure. FTA's Bus livability discretionary grants target communities that seek to invest in transportation options that increase access to work, housing, and play for lower income individuals, persons with disabilities, and older people. The agency makes this and other grant programs available to communities developing a mobility management planning approach for their vulnerable populations (Federal Transit Administration, 2016). The design and function of public infrastructure is an essential component of livability and may help to incentivize planning consideration for walkable environments that connect individuals with the things they need in the community for daily living and independence.

Related Disciplines and Comparisons

Creating livable communities is not the domain of one discipline or the combination of just a few. Designing sustainable livable communities requires the efforts of a number of disciplines. A word that perhaps unites these disciplines is professionalism. The Merriam-Webster dictionary defined professionalism as the "the conduct, aims, or qualities that characterize or mark a profession or a professional person;" and it defines a profession as "a calling requiring specialized knowledge and often long and intensive academic preparation." Various professional organizations help to create and contribute to the subject of livable communities. For instance, the International City-County Management Association (ICMA) promotes a holistic strategy including active and healthy living, aging in place, small and rural communities, smart growth, and transportation and mobility. Their website includes resources for its users. https://icma.org/

The AARP is an organization dedicated to enhancing the quality of life for people 50 and older. Their livable communities' efforts support "towns, cities, and rural areas to be great places for people of all ages. We believe that communities should provide safe, walkable streets; age-friendly housing and transportation options; access to needed services; and opportunities for residents of all ages to participate in community life." (AARP, 2019). Through AARP *Livability Index*, the general public has an awareness about how their city ranks in comparison to other places in the nation and provides them with information to advocate for local change. The Centers for Disease Control and Prevention (CDC) is another organization that has information specifically related to health and public health systems that are pertinent to livable communities. For instance, The CDC (2016, n.p.) has a Community Health Improvement (CHI) program that "brings together healthcare, public health, and other stakeholders to identify and address the health needs of communities."

Finally, we note the importance of the United Nations 2015 Sustainable Development Goals (SDGs). These goals were adopted by all United Nations members' as a "shared blueprint for peace and prosperity for people and the planet, now and into the future," (United Nations, 2019, np). These goals as presented by the United Nations and listed in Table 3.1 are consistent with livable communities and cities. Communities and all relevant stakeholders can work on all 17 SDGs or concentrate on one or just a few. The U.N. goals are summarized for important reference as follows:

Often when we think of sustainability, it is tied to climate or environmental issues. The United Nations Sustainable Development Goals, however, are comprehensive in nature and specific in purpose to address vital issues that affect all communities around the world. The SDG framework provides a reasonable outline for governments to address issues that are important to their communities. The SDG target for addressing these goals is 2030. Various metrics for tracking each goal have been put in place to monitor the progress each country has made. It is not expected that countries/cities will address all issues all at once but

Table 3.1 17 Sustainable Development Goals (SDGs)

UN sustainable development goals	Description
Goal 1—No Poverty	End poverty in all its forms everywhere
Goal 2—Zero Hunger	End hunger, achieve food security and improved nutrition, and promote sustainable agriculture
Goal 3—Good health and Wellbeing	Ensure healthy lives and promote wellbeing for all at all ages
Goal 4—Quality Education	Ensure inclusive and equitable quality education and promote lifelong learning opportunities for all
Goal 5—Gender Equality	Achieve gender equality and empower all women and girls
Goal 6—Clean Water and Sanitation	Ensure availability and sustainable management of water and sanitation for all
Goal 7—Affordable and Clean Energy	Ensure access to affordable, reliable, sustainable, and modern energy for all
Goal 8—Decent Work and Economic Growth	Promote sustained, inclusive, and sustainable economic growth, full and productive employment, and decent work for all
Goal 9—Industry, Innovation, and Infrastructure	Build resilient infrastructure, promote inclusive and sustainable industrialization, and foster innovation
Goal 10—Reduced Inequalities	Reduce inequality within and among countries
Goal 11—Sustainable Cities and Communities	Make cities and human settlements inclusive, safe, resilient, and sustainable
Goal 12—Responsible Consumption and Production	Ensure sustainable consumption and production patterns
Goal 13—Climate Action	Take urgent action to combat climate change and its impacts
Goal 14—Life Below Water	Conserve and sustainably use the oceans, seas, and marine resources for sustainable development
Goal 15—Life on Land	Protect, restore, and promote sustainable use of terrestrial ecosystems, sustainably manage forests, combat desertification, and halt and reverse land degradation and halt biodiversity loss
Goal 16—Peace, Justice, and Strong Institutions	Promote peaceful and inclusive societies for sustainable development, provide access to justice for all, and build effective, accountable, and inclusive institutions at all levels
Goal 17—Partnerships for the Goals	Strengthen the means of implementation and revitalize the global partnership for sustainable development

Source: Adapted from United Nations Sustainable Development Goals (2019)

that they become aware of the various multiple concerns facing many of the communities around the world.

Issues Related to Research and Practice

The main theoretical implications of creating livable communities are that there is not a single theory providing the assumptions for planning and development decisions to drive the movement forward. One could say there is a knowledge gap in integrative theory and methods' studies on development decisions. Several different disciplines research livability to help explain different phenomena related to quality of life and general wellbeing including urban planning, city management, public health, and gerontology. The theories for these disciplines reside within their own academic paradigm but are not mutually exclusive considering the context relates to the overall capacity of a community to be livable.

The extant of theories explaining livable communities spans from creating sustainable places through compact development (Echenique, Hargreaves, Mitchell, & Namdeo, 2012), the need for higher density to support efficient public transit and transit-oriented development (Calthorpe, 1993; Duany, Plater-Zyberk, Krieger, & Lennertz, 1991), and incentives for denser development and redevelopment toward the creation of mixed-use environments to protect rural and greenspace areas (Moeckel & Lewis, 2017). These theories tend to emphasize the planning tensions between environment, equity, and ecology in sustainability, where the livability paradigm incorporates the necessary land-use design and planning to support sustainable outcomes (Godschalk, 2004). Additionally, engagement theories grounded in collective action assumptions help to explain the mobilization of residents demanding a response to environmental threats and to provide the political, economic, and social responses to ensure a quality of life for residents, an outcome of livability planning and policies (Friedmann, 2000). Residential activists are guided by engagement theories to come together for the support of good policies to support environmental quality (Fainstein, 2014). Campbell (1996) highlights the engagement tensions between equity, equality, and environment in sustainability planning.

The public health community examines livability from the lens of health trends and community health outcomes relative to the proximity of residents to community features that support healthy lifestyles such as parks and open space (Jacobs, Wilson, Dixon, Smith, & Evens, 2009). Gerontology research theorizes livability relative to an older person's ability to make housing, transportation, and informed choices that support goals for personal health and independence (Lui et al., 2009).

The gap in a comprehensive research agenda on livability relates to the lack of interconnectedness between these disciplines, all of which are attempting to inform on the health implications from the micro (individual) to the macro (community and public policy systems). Table 3.2 illustrates the common interest toward livability policies across disciplines including planning and public administration, public health, and gerontology and the varying scale of health-related outcomes.

Table 3.2 Livability policy matrix across professional disciplines

Policy area	Action	Planning and public administration	Public health	Gerontology
Promotion of walkability and development of pedestrian facilities	Inputs	Allocation of resources for capital improvements	Resources and information for health data creation	Data documentation and communication
	Outputs	Infrastructure development	Health mapping and assessment strategies	Individual access to housing choice, mobility options, and information
	Outcomes	Features that enhance overall quality of life	Community health outcomes to reduce morbidity and mortality	Improvements to personal wellbeing

For enhanced effectiveness, future research on livability should consider the interdisciplinary nature of creating livable communities. First, research should work toward developing a theoretical model of livability applicable across disciplines. As in Table 3.2, the model should develop a clearer picture of the relationship between the livable policy area and the health impact. Second, future research should identify opportunities to build capacity across the disciplines to collaborate on empirical research that draws richer connections between livability policy decisions, equity, and health outcomes. Finally, the findings from interdisciplinary livability research would provide solutions to improve the quality of life for all residents while mitigating conflicts such as residential displacement and other impacts on vulnerable populations.

Conclusion

In this chapter, we looked at livability through physical design across community elements including streets, commercial and residential locations, and public spaces. We explained that a livable community is composed of various items such as affordability, safety, accessibility, mobility, housing, and technology. We demonstrated that a livable community was the summation of a number of metrics that composed the quality of life for individuals and families. We reviewed the history and origins of the concept and also addressed sustainability issues concerning livable communities and pertinent community health approaches. We identified the importance of housing, walkability, community access, and various cultural approaches to livability. We shared that many disciplines share a role in advancing the concept of livable communities and that their connection was professionalism. We introduced the United Nations Sustainable Development Goals and the Livability Policy Matrix that compares Planning and Public Administration, Public Health, and Gerontology across policy areas showing inputs and outputs and outcomes. Various stakeholders play a role in creating livable communities. As individuals that reside in groups and neighborhoods, we should strive to assure that the places we call home are places of refuge.

Self-Check Questions

1. Define livable communities and livability.
2. Outline the key historical events and developments in livability planning concepts and community health outcomes.
3. What are the key factors comprising a comprehensive definition of a livable community?
4. Identify and explain strategies and tools available in community planning efforts to bring about systematic change needed to create livable places.
5. What role does individual behavior play in producing positive health outcomes? Provide examples.
6. What are the primary public health practices for livable communities encompassing housing, walkability, and community access.
7. What systematic, cultural, and behavioral changes strategies would advance livability efforts for sustainable living?
8. How is livability an interdisciplinary science and practice?

Discussion Questions

1. Identify the three U.S. federal agencies advancing livability policies and explain how the principles align livability with the disciplines of the agencies. What are their specific goals for developing healthy communities?
2. What are the limitations of pre-livability planning models in addressing health outcomes and how does the livability planning vision overcome these constraints?
3. Describe the term food desert and provide examples of tools and strategies available to planners to help alleviate food insecurity. Provide an example to support your discussion.
4. What are two different technology platforms available to help increase local participation in planning decisions?
5. In your opinion, which two professional disciplines are leaders in the study of creating livable communities and why.

Field-Based Experiential Exercise

Using the resources provided for students below and the walkability assessment website resource. Take a walk in your neighborhood and explore the key components of the walkability assessment and determine the overall walkability of your community. http://www.pedbikeinfo.org/cms/downloads/walkability_checklist.pdf

Online Resources for Students

https://www.un.org/sustainabledevelopment/sustainable-development-goals/
https://www.aarp.org/livable-communities/about/
https://www.epa.gov/sites/production/files/2014-06/documents/partnership_year1.pdf
https://www.cdc.gov/physicalactivity/about-physical-activity/index.html
http://www.pedbikeinfo.org/cms/downloads/walkability_checklist.pdf
https://www.planning.org

References

AARP. (2019). *AARP livability index*. As retrieved from https://livabilityindex.aarp.org/livability-defined on October, 14, 2019.

American Planning Association. (2012). *APA policy guide on smart growth*. As retrieved from https://www.planning.org/policy/guides/adopted/smart-growth.htm on September 23, 2019.

Anguelovski, I. (2015). Healthy food stores, greenlining and food gentrification: Contesting new forms of privilege, displacement and locally unwanted land uses in racially mixed neighborhoods. *International Journal of Urban and Regional Research, 39*(6), 1209–1230.

Aulia, D. N. (2016). A framework for exploring livable community in residential environment. case study: Public housing in Medan, Indonesia. *Procedia-Social and Behavioral Sciences, 234*, 336–343.

Badland, H., Whitzman, C., Lowe, M., Davern, M., Aye, L., Butterworth, I., … Giles-Corti, B. (2014). Urban liveability: Emerging lessons from Australia for exploring the potential for indicators to measure the social determinants of health. *Social Science & Medicine, 111*, 64–73.

Balsas, C. J. (2004). Measuring the livability of an urban centre: An exploratory study of key performance indicators. *Planning Practice and Research, 19*(1), 101–110.

Benavides, A. D., & Keyes, L. (2015). A local government response to the adoption of age friendly policies. *The Journal of Aging in Emerging Economies, 5*(1), 1–27.

Benavides, A. D., Nukpezah, J., Keyes, L. M., & Soujaa, I. (2020). Adoption of multilingual state emergency management websites: Responsiveness to the risk communication needs of a multilingual society. *International Journal of Public Administration*, 1–11.

Brownson, R. C., Chang, J. J., Eyler, A. A., et al. (2004). Measuring the environment for friendliness towards physical activity: A comparison of the reliability of 3 questionnaires. *American Journal of Public Health, 94*, 474–483.

Calthorpe, P. (1993). *The next American metropolis: Ecology, community, and the American dream*. Princeton architectural press. Princton, NJ.

Campbell, S. (1996). Green cities, growing cities, just cities?: Urban planning and the contradictions of sustainable development. *Journal of the American Planning Association, 62*(3), 296–312.

Cao, X. J. (2016). How does neighborhood design affect life satisfaction? Evidence from twin cities. *Travel Behaviour and Society, 5*, 68–76.

Centers for Disease Control and Prevention. (2016). *Making the case for collaborative CHI (Community Health Improvement)*. As retrieved from https://www.cdc.gov/chinav/case/ on October 25, 2019.

Centers for Disease Control and Prevention. (2019). *About physical activity*. As retrieved from https://www.cdc.gov/physicalactivity/about-physical-activity/ on October 25, 2019.

Cerin, E., Saelens, B. E., Sallis, J. F., & Frank, L. D. (2006). Neighborhood environment walkability scale: Validity and development of a short form. *Medicine & Science in Sports & Exercise, 38*(9), 1682–1691.

Chapin, T. S. (2012). Introduction: From growth controls, to comprehensive planning, to smart growth: Planning's emerging fourth wave. *Journal of the American Planning Association, 78*(1), 5–15.

Christian, H., Knuiman, M., Bull, F., Timperio, A., Foster, S., Divitini, M., … Giles-Corti, B. (2013). A new urban planning code's impact on walking: The residential environments project. *American Journal of Public Health, 103*(7), 1219–1228.

Department of the Environment. (1994). *Vital and viable town centres, meeting the challenge*. London, UK: HMSO.

Dobbins, M. (2005). Focusing growth amid Sprawl: Atlanta's livable centers initiative [Speaking of places]. *Places, 17*, 2.

Dresner, S. (2012). *The principles of sustainability*. Routledge. New York, NY.

Duany, A., Plater-Zyberk, E., Krieger, A., & Lennertz, W. R. (1991). *Towns and town-making principles*. New York, NY: Rizzoli.

Echenique, M. H., Hargreaves, A. J., Mitchell, G., & Namdeo, A. (2012). Growing cities sustainably: Does urban form really matter? *Journal of the American Planning Association, 78*(2), 121–137.

Environmental Protection Agency. (2014). *Partnership for sustainable communities report: A year of progress for American communities.* As retrieved from https://www.epa.gov/sites/production/files/2014-06/documents/partnership_year1.pdf on October 25, 2019.

Ewing, R. (1997). Is Los Angeles–style sprawl desirable? *Journal of the American Planning Association, 63*, 107–126.

Ewing, R., Schmid, T., Killingsworth, R., Zlot, A., & Raudenbush, S. (2003). Relationship between urban sprawl and physical activity, obesity, and morbidity. *American Journal of Health Promotion, 18*(1), 47–57.

Ewing, R., & Handy, S. H. (2009). Measuring the unmeasurable: Urban design qualities related to walkability. *Journal of Urban Design, 14*(1), 65–84.

Fainstein, S. S. (2014). The just city. *International Journal of Urban Sciences, 18*(1), 1–18.

Federal Transit Administration. (2016). *Livability grant programs.* As retrieved from https://www.hud.gov/program_offices/public_indian_housing/programs/ph/hope6/mfph on October 25, 3019.

Feng, J., Tang, S., & Chuai, X. (2018). The impact of neighbourhood environments on quality of life of elderly people: Evidence from Nanjing, China. *Urban Studies, 55*(9), 2020–2039.

Frank, L. D., Sallis, J. F., Conway, T. L., Chapman, J. E., Saelens, B. E., & Bachman, W. (2006). Many pathways from land use to health: Associations between neighborhood walkability and active transportation, body mass index, and air quality. *Journal of the American Planning Association, 72*(1), 75–87.

Friedmann, J. (2000). The good city: In defense of utopian thinking. *International Journal of Urban and Regional Research, 24*(2), 460–472.

Godschalk, D. R. (2004). Land use planning challenges: Coping with conflicts in visions of sustainable development and livable communities. *Journal of the American Planning Association, 70*(1), 5–13.

Goetz, E. (2011). Gentrification in black and white: The racial impact of public housing demolition in American cities. *Urban Studies, 48*(8), 1581–1604.

Goetz, E. (2012). The transformation of public housing policy, 1985–2011. *Journal of the American Planning Association, 78*(4), 452–463.

Goodland, R. (1995). The concept of environmental sustainability. *Annual Review of Ecology and Systematics, 26*(1), 1–24.

Gordon, P., & Richardson, H. W. (1997). Are compact cities a desirable planning goal? *Journal of the American Planning Association, 63*, 95–106.

Gough, M. Z., & Accordino, J. (2013). Public gardens as sustainable community development partners: Motivations, perceived benefits, and challenges. *Urban Affairs Review, 49*(6), 851–887.

Green, G. P., & Haines, A. (2015). *Asset building & community development*. Los Angeles, CL: Sage publications.

Handy, S., & McCann, B. (2010). The regional response to federal funding for bicycle and pedestrian projects: An exploratory study. *Journal of the American Planning Association, 77*(1), 23–38.

Herrman, T., & Lewis, R. (2017). *Research initiative 2015–2017 framing livability: What is livability?* As retrieved from https://sci.uoregon.edu/sites/sci1.uoregon.edu/files/sub_1_-_what_is_livability_lit_review.pdf on April 4. 2020.

Hodge, G., & Robinson, I. M. (2001). *Planning Canadian regions*. Vancouver, Canada: UBC Press.

Horst, M., McClintock, N., & Hoey, L. (2017). The intersection of planning, urban agriculture, and food justice: A review of the literature. *Journal of the American Planning Association, 83*(3), 277–295.

Hwang, E., Glass, A. P., Gutzmann, J., & Shin, K. J. (2008). The meaning of a livable community for older adults in the United States and Korea. *Journal of Housing for the Elderly, 22*(3), 216–239.

Hwang, E., & Ziebarth, A. (2015). Walkability features for seniors in two livable communities: A case study. *Housing and Society, 42*(3), 207–221.

Jacobs, D. E., Wilson, J., Dixon, S. L., Smith, J., & Evens, A. (2009). The relationship of housing and population health: A 30-year retrospective analysis. *Environmental Health Perspectives, 117*(4), 597–604.

Kaiser, E. J., & Godschalk, D. R. (1995). Twentieth century land use planning: A stalwart family tree. *Journal of the American Planning Association, 61*, 365–385.

Kashef, M. (2016). Urban livability across disciplinary and professional boundaries. *Frontiers of Architectural Research, 5*(2), 239–253.

Kaufmann, C. (2015). *Tiny houses are becoming a big deal.* AARP Livable Communities. As retrieved from https://www.aarp.org/livable-communities/housing/info-2015/tiny-houses-are-becoming-a-big-deal.html on October 14, 2019.

Keeble, B. R. (1988). The Brundtland report: 'Our common future'. *Medicine and War, 4*(1), 17–25.

Keyes, L., & Benavides, A. (2017). Local government adoption of age friendly policies: An integrated model of responsiveness, multi-level governance and public entrepreneurship theories. *Public Administration Quarterly, 41*(1), 149.

Keyes, L., Phillips, D. R., Sterling, E., Manegdeg, T., Kelly, M., Trimble, G., & Mayerik, C. (2014). Transforming the way we live together: A model to move communities from policy to implementation. *Journal of Aging & Social Policy, 26*(1–2), 117–130.

Klassen, A. C., Michael, Y. L., Confair, A. R., Turchi, R. M., Vaughn, N. A., & Harrington, J. (2013). What does it take to sustain livable public housing communities? *Progress in Community Health Partnerships: Research, Education, and Action, 7*(1), 1–3.

Kochera, A., & Bright, K. (2006). Livable communities for older people. *Generations, 29*(4), 32–36.

Krizek, K., & Power, J. (1996). *A planner's guide to sustainable development.* Chicago, IL: American Planning Association. (PAS Report 467).

Lehning, A. J. (2014). Local and regional governments and age-friendly communities: A case study of the San Francisco Bay Area. *Journal of Aging & Social Policy, 26*(1–2), 102–116.

Long, J., & Rice, J. L. (2019). From sustainable urbanism to climate urbanism. *Urban Studies, 56*(5), 992–1008.

Lui, C. W., Everingham, J. A., Warburton, J., Cuthill, M., & Bartlett, H. (2009). What makes a community age-friendly: A review of international literature. *Australasian Journal on Ageing, 28*(3), 116–121.

Mazmanian, D. A., & Kraft, M. E. (1999). The three epochs of the environmental movement. In *Toward sustainable communities: Transition and transformation in environmental policy.* Cambridge, MA: MIT Press.

McArthur, J., & Robin, E. (2019). Victims of their own (definition of) success: Urban discourse and expert knowledge production in the Liveable City. *Urban Studies, 56*(9), 1711–1728.

Meck, S. (2002). *Growing smart legislative guidebook: Model statutes for planning and the management of change.* Chicago, IL: American Planning Association.

Moeckel, R., & Lewis, R. (2017). Two decades of smart growth in Maryland (USA): Impact assessment and future directions of a national leader. *Urban, Planning and Transport Research, 5*(1), 22–37.

Morrison, D., Shaffer, A., Lewis, R., & Lewman, H. (2018). *Framing livability: A strategic and creative communication approach to improving support for public transportation in Oregon.* Portland, OR: Transportation Research and Education Center.

Neal, M. B., Chapman, N. J., Dill, J., Sharkova, I. V., DeLaTorre, A. K., Sullivan, K. A., ... Martin, S. A. (2006). *Age-related shifts in housing and transportation demand*: A Multidisciplinary study conducted for metro by Portland State University's College of urban and public affairs, final report. As retrieved from http://www.pdx.edu/ims on October 25, 2019.

Partnership for Sustainable Communities Brief. (2014). As retrieved from https://obamawhitehouse.archives.gov/sites/default/files/uploads/SCP-Fact-Sheet.pdf on October 25, 2019.

Philadelphia Corporation for Aging. (2006). *Looking ahead: Philadelphia's aging population in 2015*. Retrieved from http://www.pcacares.org/Files/2015_report.pdf .

Saelens, B. E., Sallis, J. F., Black, J. B., & Chen, D. (2003). Neighborhood-based differences in physical activity: An environment scale evaluation. *American Journal of Public Health, 93*(9), 1552–1558.

Shafer, C. S., Lee, B. K., & Turner, S. (2000). A tale of three greenway trails: User perceptions related to quality of life. *Landscape and Urban Planning, 49*(3–4), 163–178.

Sirmans, S., & Macpherson, D. (2003). The state of affordable housing. *Journal of Real Estate Literature, 11*(2), 131–156.

Tannis, C., Senerat, A., Garg, M., Peters, D., Rajupet, S., & Garland, E. (2019). Improving physical activity among residents of affordable housing: Is active design enough? *International Journal of Environmental Research and Public Health, 16*(1), 151.

Tretter, E. M. (2013). Contesting sustainability: 'SMART Growth' and the redevelopment of Austin's Eastside. *International Journal of Urban and Regional Research, 37*(1), 297–310.

Trice, H. M., & Beyer, J. M. (1993). *The cultures of work organizations*. Englewood Cliffs, NJ: Prentice-Hall, Inc.

United Nations. (2019). *Sustainable development goals*. As retrieved from https://www.un.org/sustainabledevelopment/sustainable-development-goals/ on October 25, 2019.

United States Federal Highway Administration. (2019). *Walkability checklist*. As retrieved from http://www.pedbikeinfo.org/cms/downloads/walkability_checklist.pdf on October 25, 2019.

Von Hoffman, A. (2000). A study in contradictions: The origins and legacy of the Housing Act of 1949. *Housing Policy Debate, 11*(2), 299–326.

Walford, N., Samarasundera, E., Phillips, J., Hockey, A., & Foreman, N. (2011). Older people's navigation of urban areas as pedestrians: Measuring quality of the built environment using oral narratives and virtual routes. *Landscape and Urban Planning, 100*(1–2), 163–168.

World Health Organization. (2007). *Global age-friendly cities: A guide*. Retrieved from http://www.who.int/ageing/publications/Global_age_friendly_cities_Guide_English.pdf

Yeh, J. C., Walsh, J. S. C., & Wallhagen, M. (2016). Building inclusion: Toward an aging- and disability- friendly city. *American Journal of Public Health, 106*(11), 1947–1949.

Young, E., & Hermanson, V. (2013). *Livability literature review: Synthesis of current practice* (No. 13-2940). Transportation Research Board. Washington, DC.

4

Economics of Community Health

M. Harvey Brenner

Introduction

We live in an era where population health inequities and social justice considerations, coupled with the understanding economic factors on community health, are critical to the design and implementation of futuristic health systems (Klein and Huang, 2010). First, national income per capita presupposes that the micro-foundations of economic growth exist at both the community and national level. This is because national-level economic growth is linked to supporting community-level health as a result of tax revenues that can be distributed through government policies which include allocations for health insurance, public welfare, education, and infrastructure, all of which improve community health (Keehan et al., 2012). Second, economic factors such as Gross Domestic Product (GDP), inflation rates, and unemployment rates have been linked to population health outcomes such as overall mortality, health condition mortality, suicide, homicide, and behavioral risk factors such as the

M. H. Brenner (✉)
Johns Hopkins University, Baltimore, MD, USA
e-mail: mharveybrenner@my.unthsc.edu

E. Mpofu (ed.), *Sustainable Community Health*,
https://doi.org/10.1007/978-3-030-59687-3_4

prevalence of alcohol and smoking consumption (Shiller, 1973, 1975). Increased GDP per capita has been shown to systematically decrease mortality rates in all previously listed diagnostic categories, while increased unemployment elevated mortality and criminal justice indices (Shiller, 1975). As an example, the current COVID-19 pandemic has resulted in a radical decline in economic activity worldwide arising from efforts to contain the infection such as social distancing, stay-at-home lockdowns, and the closure of many businesses. These ameliorating public health measures have unintentionally increased the unemployment rate to levels not seen since the Great Depression of the 1930s, in conjunction with sharp declines to community income and wealth, underscoring the indissoluble link between health and economics.

Learning Objectives

By the end of the chapter, the reader should be able to:

1. Define econometric approaches to sustainable community health.
2. Describe the econometric approaches to studying the impact of economic inequality on community health outcomes.
3. Evaluate the relative merits of econometric methods to the design and implementation of sustainable population health systems.
4. Discuss how the econometric approaches to community health compare with clinical approaches.
5. Examine the research and ethics of economic health expenditure allocations on community health populations.

Community and individual socioeconomic status are major predictors of life expectancy across countries (Majer, Nusselder, Mackenbach, & Kunst, 2011). The econometric paradigm highlights that more highly educated populations have higher income and investment potential, which enable them to cope more readily with economic recessions, natural disasters, and health conditions that may arise from natural environmental degradations. Thus, the level of economic resourcing of a community enables its sustainable health in several ways:

- Higher educational levels and investment are fundamentally related to economic growth and development and to increasing longevity.
- Higher GDP growth is a basis for increased investment in education (a critical element in human capital enhancement), and human capital investment, in turn, is a source of increased economic productivity and thus increased GDP per capita.
- Investment in scientific development is a source of economic growth and development, and it is fundamental to improvements in health care and pharmaceutical technology, and thus to population life expectancy and sustainable community health.
- Increased health care expenditures, as a proportion of GDP, tend to be related to long-term upward trends in national income and health care expenditures, which in theory, would elevate population longevity.
- Increased population longevity, in turn, may theoretically represent a source of motivation for investment in education and training, which would in the long term lead to higher quality of employment and higher income.
- Employment at higher income, in turn, would lead to further population capacity to obtain health care from more advanced technology.

Economic growth has multiplier health benefits. First, while economic growth initially produces greater income inequality as societies develop, in theory, this income inequality would be reduced over time. Second, initial increases in environmental threats associated with rapid industrialization would, in theory, decline among wealthier societies as they invest more heavily in the technological means to reduce environmental sources of damage to community health. Third, the physical stress of work decreases with economic growth and development by reducing the exposure of workers to industrial chemicals, environmental sources of pollution, and contact with potentially carcinogenic substances (Stern 2018).

Potential threats to community health through economic growth relate to the following terms: westernization, industrialization, urbanization, high consumption levels, technological development, and higher income. These depictions of economic growth-related "modernization" have been thought to cause the "diseases of affluence"—that is, heart disease, malignancies, stroke, chronic obstructive pulmonary disease (COPD), diabetes, and several other chronic diseases (Ezzati et al., 2005; Novotny, 2005).

This notion that economic development "produces" elements of chronic disease incidence is partly explained by the "epidemiologic transition" concept (Dye, 2014; Mooney, 2007). The epidemiologic transition concept predicts that chronic diseases will replace infectious diseases as the major source of mortality in all countries and communities due to the industrialization processes. Modern econometric processes now incorporate into their models economic development process factors in order to determine more precisely the beneficial effects of new health care technologies and community health policies on poverty reduction and community development.

Understanding the economics of community health also takes into consideration the fact that expenditures allocated to education, nutrition, or sanitation are just as essential to population health as expenditures on medical health care (Bradley, Elkins, Herrin, & Elbel, 2011). Economic status differentials among communities are associated with community health outcomes (see Chap. 2, this volume). As such, public welfare-oriented health expenditure policies should ameliorate, but not eliminate, economic gradient effects on community health outcomes. Applied to community health, econometric approaches seek to understand the health risks and health benefits that, in combination, impinge on illness or mortality rates in community populations. Econometric approaches are essential to understanding the relative contributions that different policies can make to improve a population's health (Hidalgo & Goodman, 2013).

Sustainable community health is the ultimate positive outcome of social policies that prioritize health expenditure policies for population health. In effect, all social policies, even those involving subsidies to industry for technological development, have population health implications through wages, purchase of goods, or "weightless" items such as products of thought processes (Ståhl et al., 2006). The issue of societal benefit is more complex and often does not involve monetary enhancement, but rather, as in the case of health, mortality reduction or longevity expansion at the level of communities. In the case of community health, policies typically involve factors such as health education, construction of infrastructure, hospital development, the supply of medical or nursing personnel, and so on. The outcome may then be the rate of incidence, prevalence, or mortality due to a particular illness (e.g., heart disease). Econometric approaches involve analysis of health risks or benefits that impact on community health outcomes. For instance, econometric approaches are helpful to estimating the costs of expending welfare resources to reduce poverty rates in a general population by comparing the effects on heart disease mortality among different communities (or regions).

Definitions and Theories of the Econometric Approach to Community Health

Econometric approaches to community health (also popularly known as population health) involve statistical analysis of multiple risk and benefit factors that, in combination, influence the health of populations (Coughlin et al., 2009; Stock & Watson, 2007). Econometric approaches focus on analyzing health communities (settings, populations, disease conditions) and serve to compare the health outcomes of different populations groups to those of other groups within the same general population. They also consider "opportunity costs" (Payne et al., 1996) or how resource expenditure policies may be biased toward one type of health benefit (e.g., health insurance) rather than toward expenditure policies to improve community health of specific disease populations. For example, if the objective is to reduce heart disease mortality for an entire population, or a specific health population, econometric analysis would seek to identify the variety of health risk and benefit factors that epidemiologically influence the heart disease mortality rate in the health communities or populations of interest (Jones, 2000; Last, 2001).

Econometric approaches are epidemiologic techniques typically used at population or overall community levels to assess how population health may be influenced by particular policies (Aron et al., 2015; Lantz, Lichtenstein & Pollack, 2007; Thomson et al., 2016; Warnecke et al., 2008). In order to do this, typically, a specific policy regarding how resources are to be expended (usually monetary expenditures) (e.g., health care in the case of a particular disease), is analyzed in terms of its potential impact on a population's illness or mortality rate. However, since many different health risk and benefit factors epidemiologically influence illness or mortality rates (see also Chap. 8, this volume), estimates of health economics would have to include the most important health risk and benefit factors that influence population health outcomes, while taking into account confounding factors (VanderWeele & Shpitser, 2013). Further to this, econometric analyses also take into account the effect of health expenditure on health communities (Discussion Box 4.1),

Discussion Box 4.1: Econometric Approach to Diarrheal Disease Reduction in a Low-Income Developing Country

Consider the case in which a low-income developing country wishes to construct a policy to reduce diarrheal disease (a major source of child mortality) by the use of an expanded program of oral rehydration therapy (Keusch, Walker, Das, Horton, & Habte, 2016). The utility of the oral rehydration therapy will depend on access to potable water, which may be in short supply and require expensive efforts to increase its availability. Equally important, the effectiveness of oral rehydration therapy in reducing diarrheal disease might also depend on there being existing expenditure to reduce childhood malnutrition and increased resources being put into sanitary engineering. Thus, we have a situation where the mitigation of diarrheal disease is dependent on effective action taken simultaneously in all three areas. Moreover, the allocation of resources between these three different factors needs to be balanced in order to understand their relative contributions to the reduction of diarrheal disease. Applying the econometric approach would take into account the need for all three factors to be addressed simultaneously within a comprehensive policy, while also analyzing the expenditure cost estimations of these three factors to provide a sense of what will be required to reduce diarrheal disease among specific population groups.

Once an econometric procedure lays out the quantitative impact of the policy and its economic costs, it becomes possible for government health planners to assess how their budgets can be used to tackle this endemic problem. It then becomes clear that the politics of health expenditure premised on the local government's competing priorities would influence the health economic solutions to this problem.

It may be the case that the reduction of diarrheal disease does not actually occupy a very high priority in the current government's approach to enhancing its political position. It may well be that expenditure on education or investment in infrastructure (transportation, housing), industrial technology, and so on may be more significant priorities.

What Do You Think?

Government priorities of investment in education, infrastructure, and economic development may benefit diarrheal disease reduction in the long term more so than the typical medical-oriented approaches by a department of health. What econometric considerations would be pertinent to address the diarrheal disease in this developing country setting and how?

especially since some of the health risk or benefit factors may be intrinsically associated with health care expenditure and/or illness or mortality outcomes (National Academies of Sciences, Engineering, and Medicine, 2016; Stringhini et al., 2017; Tencza, Stokes, & Preston, 2014).

With econometric analysis of health care spending, for example, the goal may be to profile health expenditure-related factors that predict illness or mortality rates, while controlling for any factors that would risk overestimating or underestimating the actual effects of health expenditure (Jo, 2014; Murthy & Okunade, 2009). Opportunity cost is one of the most fundamental concepts in all of economic theory. The opportunity cost of pursuing a given policy, or any human action, may require the use of a significant amount of resources (i.e., finances, time, or effort); as such, pursuing an alternate course of action requiring a similar expenditure of resources may be indicated (Investopedia, 2020; Shafritz, 2019).

Gross Domestic Product (GDP) per capita is a key variable in econometric analyses at the national (or macro) level, generally pertaining to a nation or a geographical location or administratively identified population (Kuznets, 1955; Thomas, 1968). **GDP** is the total market value of a country's economy during a specified period of time as measured by the goods and services it produced. Econometric models of population health that factor in the GDP per capita typically utilize log transformation to enhance accuracy predictions.

History of Research and Practice Pertinent to Econometrics: Economic Development and Community Health

Health econometrics were developed based on the need to understand the importance of economic growth on mortality decline and increased longevity. For instance, Engel (1857) proposed that nutrition indicators predicted world population longevity. Subsequently, demographers factored in the effect of economic GDP growth on mortality (Preston, 1975). Mortality was predicted to decline with GDP growth as nations become wealthier and developed comprehensive health systems. Similarly,

McKeown (1976) argued that industrialization and economic growth since the 1850s were the primary sources of rapid decline in world mortality rates. Linking back to the hypothesis by Engel (1857) and Malthus (1809), McKeown asserted that mortality decline arose from greater availability of nutrition through agricultural revolution; but equally important, the increase in personal incomes allowing lower socioeconomic population groups to partake in the nutritional outcomes of higher agricultural productivity (see also Chap. 5, this volume on the significance on nutritional health). Thomas Malthus observed that there were cycles of agricultural productivity that related, in turn, to cycles of mortality (largely infant mortality) and fertility. According to Malthus (1809), increased agricultural yield brings about both an increased fertility rate and a declining mortality rate. This is followed by declines in agricultural productivity (in the later part of the agricultural cycle), in which fertility declines and mortality increases. As societies develop, mortality continues to decline because of higher agricultural productivity and industrial-based increases in income. Moreover, fertility declines would lead to a new equilibrium in which more industrially mature societies are found to have high industrial production, low fertility, low mortality, and increasingly aging populations in the epidemiological transition (Wilkinson, 1994).

McKeown's thesis was vigorously disputed on the claim that sanitary engineering was by far the principal source of increasing longevity—and most especially the decline in infant and child mortality, and major infectious diseases (Colgrove, 2002). However, Kass (1971) suggested that three elements constituted the fundamental basis of the relationship between economic development and increased longevity, namely, nutrition, sanitation, and architectural-residential developments that served to decrease the number of persons-per-room living in homes or apartments (Kass, 1971). The growing use of enlarged living quarters was also made possible by increasing the income and wealth of the population during periods of rapid economic development, enabling them to purchase or rent more expensive living quarters.

Presently, econometric approaches begin primarily with the behavior of the economy and its cyclical periodicities and smaller fluctuations in predicting population health. For instance, employment patterns relating to mental health hospitalization over the course of a century and a half in New York State (Brenner, 1973a) and the cyclical waves of heart disease mortality were shown to be inversely related by similar waves in employment (Brenner, 1971). Moreover, changes in employment and income were shown to be inversely (i.e., negatively) related to fetal and infant mortality (Brenner, 1973b). However, somewhat higher unemployment may also result in mortality decline associated with lower accident rates and lower atmospheric pollution due to a reduction of economic activity (Brenner, 2016; Brenner, 2017a, 2017b; Ruhm, 2015).

Precursors to the current epidemiological approaches to community health date back to the work in the seventeenth century and the first statistical tabulations for the British Registrar General Report of England and Wales in the 1840s (Graunt, 2018; Marmot & Wilkinson, 2005). This early work heralded the rise to the worldwide findings of the "health gradient" or "social gradient" in which higher socioeconomic levels of a population have increasingly lower mortality rates (Fotso & Kuate-Defo; 2005; Marmot, 2003).

A major consideration in the use of econometric techniques to understanding community health is dependent on whether findings from the historical past are sufficiently rigorous and similar to those of the potential future to allow accurate forecasting for sustainable health development. Moreover, ethical and political considerations would influence the policy options for sustainable community health based on the evidentiary data from econometric approaches (Case Study 4.1). This is an important caveat to keep in mind the strengths and limitations of econometric approaches for guiding health expenditure decision choices.

Current Econometric Approaches to Community Health

Econometrics of community health uses statistical modeling to account for health disparities in relation to the existence and extent of specific health issues. The components of an econometric model can also be used to take into account changes in population health trends overtime and fluctuations in health systems. Thus, econometric approaches to understanding community health are inclusive in applying epidemiologic techniques to test assumptions about the economics of health services and applying cross-sectional estimation and comparative health differences techniques (Case Study 4.1). These analyses would also make use of economic inequality measures and indicators of lifestyle behavioral habits.

Cross-Sectional Estimation and Comparative Health Differences Cross-section estimation can occur at a moment in time or at different points in time in order to observe how health policy changes could explain a population health indicator like heart disease mortality rates (Gerdtham, Søgaard, Jönsson, & Andersson, 1992). In this case, in order to avoid over- or under-attributing of the effect of welfare on heart disease mortality, the multivariable statistical model would include as many factors that might influence community health, apart from welfare expenditures. These would include other generally well-known factors such as measures of socioeconomic status, behavioral life habits, and risks inherent in the physical environment.

Case Study 4.1: Health Care Allocation and Econometric Approaches

When we try to make a policy choice as to how much of societal resources (especially government resources) should be allocated to health care, especially for the elderly, disabled, low educated, and otherwise vulnerable populations, the question is frequently raised as to whether this monetary allocation is appropriate. The subsequent question then becomes whether financial resources alone are effective in actually improving the health of the elderly and disabled populations. If, as some theory would suggest, the expenditure of money will do little to improve the health of the elderly, chronically ill, and low educated populations (because their high illness rates will continue despite such expenditures), it would seem that there is less moral rationale to expend monies on those populations because their health levels will remain compromised.

What Do You Think?

How would we determine whether financial resources actually reduce the illness and mortality rates of the elderly and least healthy?

Comparative health differences approaches compare population health outcomes across groupings of countries at different levels of economic development (Ho and Hendi, 2018; Woolf and Aron, 2013). Presently, the wealthiest countries (i.e., industrialized countries of the Organization for Economic Cooperation and Development) and members of the European Union show the lowest global age-adjusted mortality rates. At the lowest socioeconomic development level, the countries of Sub-Saharan Africa taken together have usually shown the highest rates of mortality. Roughly middle-income countries—many of which are rapidly developing societies—including those of Asia, Central and South America, show a middle-range level of age-adjusted mortality rates (Alkire et al., 2018; Deaton, 2002; WHO, 2020). These rates appear not to have changed substantially within the last century. The interesting discrepancies from these findings include the United States, which, belonging to the OECD, tends to have higher longevity than countries in the lowest or middle-income geographic groups. On the other hand, when one compares the United States to other OECD countries within the last decade, the United States has a comparatively lower longevity rate despite the fact that it has the highest per capita expenditure on health among OECD countries (Ho and Hendi, 2018; Woolf and Aron, 2013). Nonetheless, population-based relationships between income and lower age-adjusted mortality, as indicated earlier, are often replicated at the individual and community levels in the majority of country studies. They are a part of the fundamental inverse relationship between socioeconomic status and health, referred to as the health gradient (Marmot, 2003; Adler et al., 1994).

Economic Inequality Measures Economic inequality is often measured by the Gini index as it assesses the magnitude of differences between socioeconomic groups (Piketty, 2015) that are at risk of increasing their overall community mortality rate (Wilkinson, 1990, 1992). While the amount of income and resulting access to resources by community members is crucial for their health, economic inequalities can also be identified by comparing the health outcomes of individuals from the same community with exposure to the same risks and protections from ill-health and mortality. Necessarily, the evidence on the health of communities based on the Gini index is mixed (Subramanian and Kawachi,

2004; Wilkinson, 1990, 1992). The issue seems to be that there are certain communities and geographic areas in which economic inequality prominently influences mortality rates and other areas in which it does not. This difference may be related to unknown subcultural and ethnic or racial disparities. This brings us to the next general topic which has had a lengthy history in epidemiology but has only sporadically entered econometric research.

Table 4.1 below provides an example of a full statistical model for US states, using a pooled cross-sectional time-series analysis (with random effects) from 2000 to 2014 for individuals in the 75–84 year-old age group. This model indicates that higher national health care expenditures (as percent of GDP) decreased mortality rates of the population aged over 65, controlling for the effects of the Great Recession and its aftermath (GDP per capita and the unemployment rate), smoking,

Table 4.1 Influence of health care expenditures on older population mortality during and following the Great Recession

Age-specific all cause of death mortality in total population age 75–84, 2000–2014			
Predictor	Coef.		95% CI
Five-year lag of GDP per capita, thousand US$ constant 1997	−36.18	***	(−41.02,−31.34)
Health care expenditures as percentage of total GDP	−23.86	***	(−28.95,−18.76)
One-year lag of unemployment rate in total population 65+	20.27	***	(10.55,29.98)
Five-year lag of prevalence (%) of daily smokers in total population 65+	70.22	***	(60.21,80.23)
One-year lag of CO_2 emissions from fossil fuel combustion in gram/capita	7.49	***	(4.24,10.73)
Adolescent fertility (birth rate per 100,000 girls 10–14)	43.55	***	(35.70,51.40)
Constant	5435.00	***	(5083.31,5786.69)

Notes: *, **, *** denote statistical significance of the estimated coefficient at 5%, 1%, and 0.1% confidence level respectively

Additional dummy variable adjustments were made for the following regions: (i) Great Lakes (ii) Rocky Mountains (iii) Texas, North Dakota, and Wyoming (iv) Iowa and West Virginia (v) Utah and Idaho (vi) Arizona and New Mexico (vii) Florida and Hawaii

Pooled cross-sectional analysis for the years 2000–2014 and 50 US states

atmospheric pollution, early climate change (via CO_2 emissions), as well as intergenerational poverty (via the adolescent birth rate). GDP per capita was lagged five years to allow for the effects of health care technology investment that would lead to pharmaceutical and other innovations. Smoking prevalence for those over age 65 was lagged five years in order to permit the beneficial effects of smoking cessation (or curtailment) to result in cardiovascular mortality decline. This evidence makes the clear case that age discrimination in allocation of health care resources routinely increases death rates for large numbers of older persons in the United States.

Life Behavioral Habits Population health is influenced by the prevalence of alcohol and tobacco consumption, legal and illicit drug consumption, body mass index, as well as environmental sources of damage to health, such as CO_2 and fine particulate matter $PM_{2.5}$, that are threats to atmospheric pollution and climate change (see also Chaps. 6, 12, and 13, this volume). These additional health risk and benefit factors are not only sources of increase or decrease in illness rates, but may also influence the extent of health care expenditures. Applying a multivariable econometric model, it is possible to reliably isolate the effects of lifestyle variables on health outcomes with health expenditures.

Cultural, Legislative, and Professional Issues Related to Economics of Community Health

In Western countries, racial minority communities have increased health vulnerabilities from their lower socioeconomic status and generally show higher mortality rates, although there are exceptions (Lariscy et al., 2015; Markides and Eschbach, 2011). While epidemiological studies have focused research on individual experiences of members of ethnic populations, the community-level analysis would be more revealing (see also Chap. 8). Questions remain as to whether ethnic differences in health outcomes actually are reflections of economic status, rather than idiosyncrasies of cultural, environmental, dietary, or genetic basis. Econometric research tends to incorporate ethnic population differences into broader models in order to determine whether ethnic groups themselves contribute to the explanation of observed health outcomes across ethnic groups, while accounting for specific effects of (say) income, education, or economic inequality—which would have already been accounted for in the econometric model. A more traditional econometric approach would identify specific ethnic groups in terms of their health-related outcomes, such as mortality, and utilize ethnic-specific mortality as a dependent variable (a dependent variable is the outcome we wish to predict, or understand, in relation to a particular explanatory factor) or independent variable. For instance, a goal might be to compare the influence of ethnic-specific unemployment rates separately on each of the mortality outcomes of various ethnic groups, looking to examine the relative strength of unemployment on mortality. The standard social gradient hypothesis is that higher income ethnic groups would show the weakest impact of their unemployment rate among members of that ethnic group—and thus on mortality. The procedure of comparing the impact of each of the predictive variables on each ethnic groups' morality rate would then be undertaken. Similarly, comparative studies could be implemented to examine effects by sex differences since historic health differences between the sexes are related to economic differences in their social roles. Studies have reported a "gender paradox," (using gender to mean sex) in which women manifest higher rates of illness (morbidity) while they also show consistently lower rates of mortality as compared to men from infancy to old age (Bird and Rieker, 2008; Cockerham, 2017; Springer and Mouzon, 2011). After the fact, explanations for the gender paradox include the claim that occupational situations of men expose them to greater ergonomic and environmental risks as well as greater incidence of stress, placing them at higher risk for mortality. Other explanations include that excessive alcohol use by men and other stress reducing and coping mechanisms to manage industrial and occupational tension to explain sex differences in mortality. The complexities to the health of vulnerable population (e.g., those with disabilities, older adults) are perplexing and econometrics analysis may provide solutions for policy implementation (see Table 4.1 above).

COVID-19 is politicized in the United States, with the more conservative politicians insisting on taking economic amelioration as a first priority, whereas the scientific and public health community and more "liberal" politicians have emphasized a stricter public health approach to ending the COVID-19 pandemic. Both sides of the political divide, and many voices in between, will continue to attempt to utilize econometric approaches to estimating the impact of the pandemic and separately, the COVID-19-induced recession, on subsequent mortality and damaged mental health. Nonetheless, there have been some studies to suggest that in the very short run, decreased unemployment is associated with increased overall mortality (Ruhm, 2000) and also from accidents or air pollution following economic recessions (Miller et al. 2009). Other studies covering the period of the Great Recession 2007–2009 do not report the same relationships (Ruhm, 2015), although some studies reported strong relations between increased unemployment and declining GDP during recessions in the short run and for a lengthy period thereafter (Brenner, 1976, 1979, 1984, 2016, 2017a, 2017b; Pool, Burgard, Needham, Elliott, Langa, & De Leon, 2018).

The advantage of use of econometric approaches to understanding population health is in the quantification of the solutions in both monetary and non-monetary terms. Econometric approaches also help to make clear some of the other non-monetary policy options that could also benefit (or harm) the existing state of community health, including behavioral and physical environmental factors that influence health outcomes.

Related Disciplines Influencing Community-Oriented Health Aspects

Epidemiology is at present perhaps the most prominent discipline that has both a traditional function in the health sciences and econometric modeling. The epidemiologic objective is to understand the distribution of health, illness, and mortality patterns among individuals and populations (see also chap. 8 on community epidemiology, this volume).

Econometric methods increasingly apply epidemiological approaches to study the health impact of macroeconomic phenomena such as rates of median income, unemployment rates, labor force participation rates, and educational investments among communities or larger populations.

Applied to mental health and wellbeing, econometric approaches study the extent or intensity of adverse life events (e.g., loss of income or employment or severe disability) and psychological stressors associated with suicide, homicide, unintended accidents, pathological alcohol abuse (Brenner, 1987), opioid use, and tobacco addiction. Econometric analysis also analyzes the coping mechanisms of individuals in the use of psychotropic drugs or medications, intensified by use of alcohol, opioids, or tobacco. These coping mechanisms are, in principle, typically influenced by existing social ties, marriage and divorce patterns, and other communal relations. Variables representing both the stressor component and the coping component can be entered into an econometric model to isolate the potential impact of a stressor or coping mechanism on a mental health outcome, such as the suicide rate.

Econometric approaches are also relevant to studying the influence of particular surgical procedures, medicines, and rehabilitation techniques on health outcomes involving improved health, adverse events, and survival in the in-patient or out-patient setting. Applying multivariable regression techniques, the intention is to focus on a particular outcome, such as congestive heart failure mortality in the hospital setting, and try to predict which particular diagnoses, comorbidities, and/or medical or surgical procedures serve to improve mortality outcomes. In such models, major sets comprised of variables that contain records of demographic backgrounds (age, gender, race/ethnicity), and socioeconomic and health services that are available in communities or neighborhoods of patients. These data can be analyzed at the postal code or zip code levels which indicate a patient's general residential location as identified in the census. From such census data, it is possible to characterize patients according to socioeconomic status, the level of existing environmental community development, and availability of access to health care resources (see Lichtenberg, 1998, 2014, for example).

This case scenario in Discussion Box 4.2 is contrary to the egalitarian and population-based moral position that the populations that are most

Discussion Box 4.2: Age Discrimination and the Years-of-Life-Lost (YLL) Approach

Table 4.1 shows data from an econometric study that examines age bias implicit in population health measures, such as YLL, and their implications for community health. YLL is a measure of assessing the health of a community or population that can be used at virtually any geographic or political level. This approach is frequently utilized in health economics as compared to the more standard epidemiological or clinical measures such as mortality or diagnosis-specific illness rates, symptoms, signs, or other health outcomes such as survival following treatment (Kaplan-Meier curves). This YLL method, however, is founded on the grounds that longer lived (i.e., older) populations will inevitably have poorer health and they will then lose fewer years of life under conditions of medical treatment compared to younger populations. According to YLL, it seems logical to devote more health resources (as well as other community resources) to the younger population rather than the older population (subgroups).

What Do You Think?

1. What are the "community" policies, as well as moral and ethical questions, of whether age should be a principal criterion of the distribution of health care and other community resources where improved health and survival are the potential outcomes?
2. What are the arguments against providing targeted health care and allocating increased community resources to the elderly, as well as lower-educated and lower income populations, and why?

in need are the ones that are most "deserving" of the allocation of public resources (Strandberg et al., 2015). This will be true even though allocation to those most needy, least "productive," and most dependent will provide the lowest potential of advancing the wellbeing of the population that is not elderly, ill, disabled, or unemployed (Maestas et al., 2016). On that view, loses of general societal benefit, and even wellbeing, can be experienced by community members because monetary resources can potentially be allocated to education, infrastructure, other human capital, and technological and business investment instead. This is the central issue of opportunity cost, where funds that would benefit the most needy and dependent members of society are allocated to support wellbeing of the actively employed and healthy population.

Research Issues Critical to the Econometric Approach

A major challenge confronting econometric research at the community population level is conflicting interpretation of study findings. For example, there is extensive epidemiological evidence which indicates that, with rare exceptions, individual persons subject to unemployment are at a significantly higher risk of illness and mortality due to their increased susceptibility to a wide range of diseases (Roelfs et al., 2011). The problem at the community or macro-population level is that unemployment patterns are related to changes in many other economic indicators (GDP per capita, median income, automobile sales, housing purchases, welfare expenditures, health care expenditures, etc.). Thus, at the population level, it is occasionally very difficult to isolate the impact of unemployment as distinguished from other factors that may be related to more general economic loss. In this regard, studies have found that, at the community level, it is precisely those subpopulations that have lost employment that are at higher risk for illness and mortality as compared to persons who retain employment even at times when the general population indicator of heightened unemployment is on the increase. Therefore, it remains a challenge in econometric analysis to minimize the "ecological fallacy" (Schwartz, 1994; McNamee, 2005; Kahlert et al., 2017) in which one immediately interprets macro phenomena such as the unemployment rate in a community as precisely reflecting the employment status of individuals within that community.

Another major challenge is in the forecasting capacity of econometric models. The assumption is that when either cross-sectional or time-series analyses are used, findings are true to the recent past. This is the basis of the forecasting character of such statistical models. However, due to political, environmental, major economic, or behavioral changes, not encompassed in past econometric models, extrapolations to scenarios in the future are challenging. An important and recent example is the unprecedented increase in the US unemployment rate secondary to the COVID-19 pandemic. The level of unemployment has not been witnessed since the Great Depression in the 1930s and is substantially larger and occurring at a more rapid pace compared to the Great Recession of 2007–2009. An important policy question arises as to whether one can make inferences of how economic function, mental health, and the physical health of populations are affected by major events (i.e., COVID-19) that contribute to economic recessions, as compared to the specific effects of the pandemics themselves (including cardiovascular effects and perhaps suicide). The current measurable effects of COVID-19 on mortality include increased fatality rates among minority populations, especially African Americans, Hispanics, and Native Americans. It is assumed that the high mortality rates among these minority ethnic groups is largely due to their relatively low-income status and the fact that they have a higher exposure rate to COVID-19 due to working in close contact with other members of the population.

The question of whether we can develop forecasts on illness and mortality rates based on the economic decline of the Great Recession is daunting, especially because the period of time over which the COVID-related recession has manifested is extremely short in comparison to a standard recession. Second, we have limited capacity to forecast the precarious position of low-income minority groups depending on past recessional experiences as demonstrated by the effects of population interaction during the COVID-19 pandemic, which has shown how extremely vulnerable minority groups really are. Nevertheless, it is important to try and have a near-term sense of the impact of the COVID-19-related economic recession on the health of specific communities, and in particular ethnic minorities (Brenner 2020). The reason for the crucial importance of having a forecast of the short- and long-term future with respect to COVID-19 relates to both Congressional rescue packages that may be required to sustain the economy and public health policies to minimize the effect of COVID-19 itself.

Summary and Conclusions

Econometric approaches to community health, or population health, are statistical approaches that incorporate multiple risk and benefit factors into models that attempt to explain, and sometimes forecast, the impact of economic, political, behavioral, and environmental changes on community health outcomes, such as illness or mortality rates. Typically, the econometric statistician is interested in determining the impact of a particular risk or benefit factor on the occurrence of a unique community health outcome. However, in order to accomplish this, two notions must be kept in mind. The first is that health in itself is a multivariate process, including the effects of multiple events and trends on the occurrence of illness and mortality. Therefore, in order to isolate the importance and impact of any one (policy-related) factor, one must take into account several other important risk or benefit factors, so as to be able to control for their presence. If one does not proceed in such a manner, then the implicit interrelations among different risk and benefit factors, and especially with the key factor being hypothesis-tested, then the key factor may be overestimated, underestimated, or confused with the effect of another predictive factor.

Secondly, if the key predictive factor under consideration and its impact on sustainable community health represent part of a health-improvement policy, then alternatives to that policy to improve health intrinsically compete with other community (or perhaps national) policies which could also be used to improve health. Given this view, it can be stated that health is a by-product of many policies—as stated in phrase "health in all policies" (Ståhl et al., 2006). This viewpoint emanates from the general policy orientation of public health specialists that community health should be seen as reflecting a full range of social determinants, rather than representing a singular policy orientation. For example, programs that benefit infant mortality reduction may benefit from coexisting with policies that support reduction in community poverty rates, increase overall health expenditures, involve specific medical or surgical technologies and public health approaches to hygiene, furnish greater access to potable water, and improve general education or enhanced health education.

"Health in all policies" is a phrase that denotes that community policies which intended to be of benefit to a population are likely to have either a direct or an indirect impact on some component of sustainable community health (see Chap. 18, this volume). Therefore, both the health risk and benefit elements that are known to influence a particular community health problem, and alternative policy approaches to mitigate that problem, usually need to be considered. This policy approach represents the "opportunity cost" econometric approach of utilizing a specific policy rather than one among many others to alleviate a health problem. In considering the different policy approaches to health betterment, one needs to take into account the political, economic, and effectiveness of a chosen policy in reducing community rates of illness or mortality. Practically, in considering policy options, one must undertake intensive cost and benefit analyses of the key policy choice as compared to the presumably next most advantageous option.

Self-Check Questions

1. Define econometric approaches as they apply to sustainable community health.
2. What are the origins of econometrics? What were the issues that drove its emergence? Who were its main early representatives?
3. How would you describe the econometric approaches to studying the impact of economic inequality on community health outcomes?
4. What are the relative merits of econometric methods to the design and implementation of sustainable population health systems?
5. What are the key matters for research and practice implementation of econometric approaches to community health?
6. What are the research and ethics economic health expenditure allocation issues that apply to use of econometric approaches to community health populations?

Discussion Questions

1. What do you perceive as the potential role of econometrics in assessing the influences of atmospheric pollution and climate change on economic productivity, and thus on community health?
2. How would you assess the impact of government health care expenditures on mortality, considering opportunity costs, in an econometric evaluation?
3. Consider econometric approaches to the aging population health and what benefits would follow from the related findings.
4. How would you differentiate between the economic status and racial/ethnic effects on differences in community health outcomes?

Field-Based Experiential Exercises

Identify a community site (e.g., non-profit organization, community organization) in your area that works with vulnerable populations affected by health disparities, unemployment, poverty, disability, or challenges to physical and mental health. Spend at least two hours per week volunteering over the course of the semester (your school may already have community partnerships in place). Engage in ethical and respectful ways with the individuals served by the organization and keep a weekly journal in which you reflect critically on the following questions.

1. What are the community needs the organization serves and what intervention strategies does the organization offer? How would these needs and strategies relate to one another?
2. Which community groups are being served? What are the demographics of these community groups? What are the roles of multiple socioeconomic factors that impact community health and wellbeing?
3. How does the intervention offered by the community organization integrate with other local resources available to the community? What are the comparative benefits and costs of this intervention? How might a multivariable econometric approach be used to evaluate the effectiveness of this intervention?
4. In what ways can you, as a volunteer, help intervene and make a positive difference in community health issues? In what ways could econometric analyses contribute to a better understanding of the specific community health issues at hand and ameliorate or mitigate their impact?

Online Learning Resources

International Monetary Fund: What is Econometrics? https://www.imf.org/external/pubs/ft/fandd/2011/12/basics.htm

WHO Key facts. https://www.who.int/news-room/fact-sheets/detail/diarrhoeal-disease

CDC Hygiene in lower income countries. https://www.cdc.gov/healthywater/hygiene/ldc/index.html

CDC Definition of Epidemiology. https://www.cdc.gov/csels/dsepd/ss1978/lesson1/section1.html

Gunasekara, F. I., Carter, K., & Blakely, T. (2008). Glossary for econometrics and epidemiology. *Journal of Epidemiology & Community Health, 62*(10), 858–861.

https://pdfs.semanticscholar.org/5bb4/53f7579225eb908fc8c727272e1f71bbe5d5.pdf

The World Bank, GDP per capita trends. https://data.worldbank.org/indicator/NY.GDP.PCAP.CD

Logarithmic GDP per capita. https://stats.stackexchange.com/questions/118373/what-are-the-advantages-of-using-log-gdp-per-capita-versus-simple-gdp-per-capita

GDP per capita and country comparisons. https://www.thebalance.com/gdp-per-capita-formula-u-s-compared-to-highest-and-lowest-3305848

References

Adler, N. E., Boyce, T., Chesney, M. A., Cohen, S., Folkman, S., Kahn, R. L., & Syme, S. L. (1994). Socioeconomic status and health: The challenge of the gradient. *American Psychologist, 49*(1), 15.

Alkire, B. C., Peters, A. W., Shrime, M. G., & Meara, J. G. (2018). The economic consequences of mortality amenable to high-quality health care in low-and middle-income countries. *Health Affairs, 37*(6), 988–996.

Aron, L. Y, Dubay L., Zimmerman, E., Simon, S. M., Chapman, D., & Woolf, S. H. (2015). *Can income-related policies improve population health?* Research report The Urban Institute. https://www.urban.org/research/publication/can-income-related-policies-improve-population-health/view/full_report. Accessed 16 May 2020.

Bird, C. E., & Rieker, P. P. (2008). *Gender and health: The effects of constrained choices and social policies.* New York, NY: Cambridge University Press.

Bradley, E. H., Elkins, B. R., Herrin, J., & Elbel, B. (2011). Health and social services expenditures: Associations with health outcomes. *BMJ Quality and Safety, 20*(10), 826–831.

Brenner, M. H. (1971). Economic changes and heart disease mortality. *American Journal of Public Health, 61*(3), 606–611. https://doi.org/10.2105/ajph.61.3.606

Brenner, M. H. (1973a). *Mental illness and the economy.* Cambridge, MA: Harvard University Press.

Brenner, M. H. (1973b). Fetal, infant and maternal mortality during periods of economic instability. *International Journal of Health Services, 3*(2), 145–159. https://doi.org/10.2190/UM5L-TVN7-VDFR-UU0B

Brenner M. H. (1976). *Achieving the goals of the employment act of 1946—Thirtieth anniversary review.* Estimating the social costs of national economic policy: Implications for mental and physical health, and criminal aggression. U.S. Congress, Joint Economic Committee, U.S. Government Printing Office, Washington, DC. https://www.jec.senate.gov/reports/94th%20Congress/Other%20Reports/Estimating%20the%20Social%20Costs%20of%20National%20Economic%20Policy.%20Implications%20for%20Mental%20and%20Physical%20Health%20and%20Criminal%20Aggression%20(791).pdf. Accessed May 15, 2020.

Brenner, M. H. (1979). Mortality and the national economy. A review, and the experience of England and Wales, 1936–76. *Lancet, 2*(8142), 568–573.

Brenner M. H. (1984). *Estimating the effects of economic change on National Health and social well-being,* Joint Economic Committee, US Congress, US Government Printing Office, Washington, DC. https://www.jec.senate.gov/reports/98th%20Congress/Estimating%20the%20Effects%20of%20Economic%20Change%20on%20National%20Health%20and%20Social%20Well-Being%20(1262).pdf. Accessed May 15, 2020.

Brenner, M. H. (1987). Economic change, alcohol consumption and heart disease mortality in nine industrialized countries. *Social Science & Medicine, 25*(2), 119–132.

Brenner M. H. (2016). *The impact of unemployment on heart disease and stroke mortality in European Union countries.* Available at: http://ec.europa.eu/social/main.jsp?catId=738&langId=en&pubId=7909&furtherPubs=yes. Accessed 15 May 2020.

Brenner, M. H. (2017a). Years of life lost, age discrimination, and the myth of productivity. *American Journal of Public Health, 107*(10), 1535–1537. https://doi.org/10.2105/AJPH.2017.30402

Brenner, M. H. (2017b). Small networks, evolution of knowledge and species longevity: Theoretical integration and empirical test. *Chaos, Solitons & Fractals, 104*, 314–322.

Brenner M. H. (2020). Will there be an epidemic of corollary illnesses linked to a COVID-19 – Related recession? *American Journal of Public Health.* Published online ahead of print May 14, 2020: e1–e2. https://doi.org/10.2105/AJPH.2020.305724)

Cockerham, W. C. (2017). *Medical sociology* (4th ed.). New York, NY: Routledge.

Colgrove, J. (2002). The McKeown thesis: A historical controversy and its enduring influence. *American Journal of Public Health, 92*(5), 725–729. https://doi.org/10.2105/ajph.92.5.725

Coughlin, S. S., Beauchamp, T. L., & Weed, D. L. (Eds.). (2009). *Ethics and epidemiology.* Oxford, UK: Oxford University Press.

Deaton, A. (2002). Policy implications of the gradient of health and wealth. *Health Affairs, 21*(2), 13–30.

Dye, C. (2014). After 2015: Infectious diseases in a new era of health and development. *Philosophical Transactions of the Royal Society, B: Biological Sciences, 369*(1645), 20130426.

Engel, E. (1857). *Die productions- und Consumptionsverhaltnisse des Konigreichs Sachsen. [The production and consumption conditions of the kingdom of Saxony].* Reprinted in Engel's *Die Lebenskosten belgischer Arbeiter-Familien* [Cost of living in Belgian working-class families]. Dresden: Germany, 1895.

Ezzati, M., Vander Hoorn, S., Lawes, C. M., Leach, R., James, W. P. T., Lopez, A. D., … Murray, C. J. (2005). Rethinking the "diseases of affluence" paradigm: Global patterns of nutritional risks in relation to economic development. *PLoS Medicine, 2*(5), e133.

Fotso, J. C., & Kuate-Defo, B. (2005). Measuring socioeconomic status in health research in developing countries: Should we be focusing on households, communities or both? *Social Indicators Research, 72*(2), 189–237.

Gerdtham, U. G., Søgaard, J., Jönsson, B., & Andersson, F. (1992). A pooled cross-section analysis of the health care expenditures of the OECD countries. In *Health economics worldwide* (pp. 287–310). Dordrecht, The Netherlands: Springer.

Graunt, J. (2018). Natural and political observations. In *The economics of population* (pp. 17–28). London, UK: Routledge.

Hidalgo, B., & Goodman, M. (2013). Multivariate or multivariable regression? *American Journal of Public Health, 103*(1), 39–40. https://doi.org/10.2105/AJPH.2012.300897

Ho, J. Y., & Hendi, A. S. (2018). Recent trends in life expectancy across high income countries: Retrospective observational study. *BMJ, 362*, k2562.

Investopedia. *Opportunity cost.* https://www.investopedia.com/terms/o/opportunitycost.asp. Accessed 16 May 2020

Jo, C. (2014). Cost-of-illness studies: Concepts, scopes, and methods. *Clinical and Molecular Hepatology, 20*(4), 327.

Jones, A. M. (2000). Health econometrics. In *Handbook of health economics* (Vol. 1, pp. 265–344). Amsterdam, The Netherlands: Elsevier.

Kahlert, J., Gribsholt, S. B., Gammelager, H., Dekkers, O. M., & Luta, G. (2017). Control of confounding in the analysis phase – An overview for clinicians. *Clinical Epidemiology, 9*, 195.

Kass, E. H. (1971). Infectious diseases and social change. *The Journal of Infectious Diseases, 123*(1), 110–114.

Keehan, S. P., Cuckler, G. A., Sisko, A. M., Madison, A. J., Smith, S. D., Lizonitz, J. M., … Wolfe, C. J. (2012). National health expenditure projections: Modest annual growth until coverage expands and economic growth accelerates. *Health Affairs, 31*(7), 1600–1612.

Keusch, G. T., Walker, C. F., Das, J. K., Horton, S., & Habte, D. (2016). Diarrheal diseases. In D. Control (Ed.), *Priorities: Reproductive, maternal, newborn, and child health* (3rd ed., pp. 163–185). Washington, DC: World Bank.

Klein, R., Huang, D. (2010). *Defining and measuring disparities, inequities, and inequalities in the healthy people initiative.* National Center for Health Statistics Centers for Disease Control and Prevention. https://www.cdc.gov/nchs/ppt/nchs2010/41_klein.pdf

Kuznets, S. (1955). Economic growth and income inequality. *The American Economic Review, 45*(1), 1–28.

Lantz, P. M., Lichtenstein, R. L., & Pollack, H. A. (2007). Health policy approaches to population health: The limits of medicalization. *Health Affairs, 26*(5), 1253–1257.

Lariscy, J. T., Hummer, R. A., & Hayward, M. D. (2015). Hispanic older adult mortality in the United States: New estimates and an assessment of factors shaping the Hispanic paradox. *Demography, 52*(1), 1–14.

Last, J. M. (2001). *Dictionary of epidemiology* (4th ed., p. 61). New York, NY: Oxford University Press.

Lichtenberg F. R. (1998). *Pharmaceutical innovation, mortality reduction, and economic growth.* NBER Working Paper No. 6569. https://www.nber.org/papers/w6569

Lichtenberg, F. R. (2014). Pharmaceutical innovation and longevity growth in 30 developing and high-income countries, 2000–2009. *Health Policy and Technology, 3*(1), 36–58.

Maestas, N., Mullen, K. J., & Powell, D. (2016). *The effect of population aging on economic growth, the labor force and productivity* (No. w22452). Cambridge, MA: National Bureau of Economic Research.

Majer, I. M., Nusselder, W. J., Mackenbach, J. P., & Kunst, A. E. (2011). Socioeconomic inequalities in life and health expectancies around official retirement age in 10 Western-European countries. *Journal of Epidemiology and Community Health, 65*(11), 972–979.

Malthus, T. R. (1809). *An essay on the principle of population, as it affects the future improvement of society* (Vol. Vol. 2). London, UK: J. Johnson.

Markides, K. S., & Eschbach, K. (2011). Hispanic paradox in adult mortality in the United States. In *International handbook of adult mortality* (pp. 227–240). Dordrecht, The Netherlands: Springer.

Marmot, M., & Wilkinson, R. (Eds.). (2005). *Social determinants of health.* Oxford, UK: Oxford University Press.

Marmot, M. G. (2003). Understanding social inequalities in health. *Perspectives in Biology and Medicine, 46*(3), S9–S23.

McKeown, T. (1976). *The modern rise of population.* New York, NY: Acad. Press.

McNamee, R. (2005). Regression modelling and other methods to control confounding. *Occupational and Environmental Medicine, 62*(7), 500–506.

Miller, D. L., Page, M. E., Stevens, A. H., & Filipski, M. (2009). Why are recessions good for your health? *American Economic Review, 99*(2), 122–127.

Mooney, G. (2007). Infectious diseases and epidemiologic transition in Victorian Britain? Definitely. *Social History of Medicine, 20*(3), 595–606.

Murthy, V. N., & Okunade, A. A. (2009). The core determinants of health expenditure in the African context: Some econometric evidence for policy. *Health Policy, 91*(1), 57–62.

National Academies of Sciences, Engineering, and Medicine. (2016). Chapter 3: Socioeconomic and behavioral factors that influence differences in morbidity and mortality. In *Improving the health of women in the United States: Workshop summary*. Washington, DC: The National Academies Press. https://doi.org/10.17226/23441. https://www.nap.edu/read/23441/chapter/4. Accessed 16 May 2020

Novotny, T. E. (2005). Why we need to rethink the diseases of affluence. *PLoS Medicine, 2*(5), e104.

Payne, J. W., Bettman, J. R., & Luce, M. F. (1996). When time is money: Decision behavior under opportunity-cost time pressure. *Organizational Behavior and Human Decision Processes, 66*(2), 131–152.

Piketty, T. (2015). *The economics of inequality*. Cambridge, MA: Harvard University Press.

Pool, L. R., Burgard, S. A., Needham, B. L., Elliott, M. R., Langa, K. M., & De Leon, C. F. M. (2018). Association of a negative wealth shock with all-cause mortality in middle-aged and older adults in the United States. *JAMA, 319*(13), 1341–1350.

Preston, S. H. (1975). The changing relation between mortality and level of economic development. *Population Studies, 29*, 231–248. https://doi.org/10.2307/2173509

Roelfs, D. J., Shor, E., Davidson, K. W., & Schwartz, J. E. (2011). Losing life and livelihood: A systematic review and meta-analysis of unemployment and all-cause mortality. *Social Science & Medicine, 72*(6), 840–854.

Ruhm, C. J. (2000). Are recessions good for your health? *The Quarterly Journal of Economics, 115*(2), 617–650.

Ruhm, C. J. (2015). Recessions, healthy no more? *Journal of Health Economics, 42*(C), 17–28.

Schwartz, S. (1994). The fallacy of the ecological fallacy: The potential misuse of a concept and the consequences. *American Journal of Public Health, 84*(5), 819–824.

Shafritz, J. (2019). *International encyclopedia of public policy and administration, Volume 3*. Google books, https://books.google.com/books?id=3CuNDwAAQBAJ&pg=PT665&lpg=PT665&dq=opportunity+cost+Int

ernational+Encyclopedia+of+Public+Policy&source=bl&ots=1Gh0bEynNp&sig=ACfU3U3Oo8r4EdfP_ixRH4ymawy2XcGbAg&hl=en&sa=X&ved=2ahUKEwjXqZze17npAhUGHqwKHeB_DVQQ6AEwAXoECAsQAQ#v=onepage&q=opportunity%20cost%20International%20Encyclopedia%20of%20Public%20Policy&f=false

Shiller, R. J. (1973). A distributed lag estimator derived from smoothness priors. *Econometrica: Journal of the Econometric Society, 41*, 775–788.

Shiller, R. J. (1975). *Alternative prior representations of smoothness for distributed lag estimation* (No. w0089). National Bureau of Economic Research.

Springer, K. W., & Mouzon, D. M. (2011). "Macho men" and preventive health care: Implications for older men in different social classes. *Journal of Health and Social Behavior, 52*(2), 212–227.

Ståhl, T., Wismar, M., Ollila, E., Lahtinen, E., & Leppo, K. (Eds.). (2006). *Health in all policies: Prospects and potentials*. Helinski, Finland: Ministry of Social Affairs and Health STM Finnish Institute of Occupational Health National Public Health Institute Stakes European Observatory on Health Systems and Policies.

Stern, D. I. (2018). The environmental Kuznets curve. In *Companion to Environmental Studies* (Vol. 49,54, pp. 49-54). London, UK: Routledge in Association with GSE Research.

Stock, J. H., & Watson, M. W. (2007). *Introduction to econometrics* (Addison-Wesley series in economics) (2nd ed.). Boston, MA: Pearson Addison Wesley.

Strandberg, T., Pietikäinen, S., Maggi, S., Harkin, M., & Petermans, J. (2015). Against age discrimination. *The Lancet, 386*(9991), 337–338.

Stringhini, S., Carmeli, C., Jokela, M., Avendaño, M., Muennig, P., Guida, F., … Chadeau-Hyam, M. (2017). Socioeconomic status and the 25× 25 risk factors as determinants of premature mortality: A multicohort study and meta-analysis of 1· 7 million men and women. *The Lancet, 389*(10075), 1229–1237.

Subramanian, S. V., & Kawachi, I. (2004). Income inequality and health: What have we learned so far? *Epidemiologic Reviews, 26*(1), 78–91.

Tencza, C., Stokes, A., & Preston, S. (2014). Factors responsible for mortality variation in the United States: A latent variable analysis. *Demographic Research, 21*(2), 27–70. https://doi.org/10.4054/DemRes.2014.31.2

Thomas, D. S. T. (1968). Social aspects of the business cycle (Vol. 1). Gordon and Breach., Dinda, Soumyananda. Environmental Kuznets curve hypothesis: A survey. *Ecological Economics, 49* (4) (2004), 431–455.

Thomson, K., Bambra, C., McNamara, C., Huijts, T., & Todd, A. (2016). The effects of public health policies on population health and health inequalities in European welfare states: Protocol for an umbrella review. *Systematic Reviews, 5*(1), 57.

VanderWeele, T. J., & Shpitser, I. (2013). On the definition of a confounder. *Annals of Statistics, 41*(1), 196–220. https://doi.org/10.1214/12-aos1058

Warnecke, R. B., Oh, A., Breen, N., Gehlert, S., Paskett, E., Tucker, K. L., … Hiatt, R. A. (2008). Approaching health disparities from a population perspective: The National Institutes of Health centers for population health and health disparities. *American Journal of Public Health, 98*(9), 1608–1615. https://doi.org/10.2105/AJPH.2006.102525

WHO. *Global burden of disease.* https://www.who.int/healthinfo/global_burden_disease/GlobalHealthRisks_report_part2.pdf. Accessed 16 May 2020.

Wilkinson, R. G. (1990). Income distribution and mortality: A 'natural'experiment. *Sociology of Health & Illness, 12*(4), 391–412.

Wilkinson, R. G. (1992). Income distribution and life expectancy. *BMJ, 304*, 165–168.

Wilkinson, R. G. (1994). The epidemiological transition: From material scarcity to social disadvantage? *Daedalus, 23*, 61–77.

Woolf, S. H., & Aron, L. (Eds.). (2013). *"Poorer Health Throughout Life" U.S. Health in international perspective: Shorter lives, poorer health.* Washington, DC: National Academies Press. https://www.nap.edu/read/13497/chapter/6. Accessed 16 May 2020

[illegible]

Part II

Policies and Practices in Sustainable Community Health

5

Nutrition Practices to Grow Healthy Communities

Kathleen Davis and Clare Brock

Introduction

Diet and nutrition-related disparities within the US and other developed nations are often related to population characteristics, including socioeconomics and demographics. These disparities may be even more dramatic in developing nations. Within developed nations, social programs and geographic factors may influence whether disadvantaged persons have similar access to healthy and culturally appropriate food as more advantaged members of society. Individuals with limited resources, such as those of low educational attainment, low socioeconomic status, those with children, and the elderly, are often particularly disadvantaged as are marginalized populations.

K. Davis (✉) • C. Brock
Texas Woman's University, Denton, TX, USA
e-mail: kdavis10@twu.edu; cbrock1@twu.edu

E. Mpofu (ed.), *Sustainable Community Health*,
https://doi.org/10.1007/978-3-030-59687-3_5

Learning Outcomes

By the end of the chapter, the reader should be able to:

1. Define the terms food environment and community food environment, and list factors influencing these.
2. Compare and contrast hunger, food security, food sovereignty, and food justice.
3. Discuss the history of traditional approaches to reducing rates of food insecurity and improving diet quality in the US and globally, including governmental minimum calorie and educational strategies, non-governmental organization (NGO) educational efforts, governmental, private, and NGO agricultural programs, and other approaches.
4. Outline reasons why traditional approaches to reducing food insecurity and diet-related disparities have had limited success.
5. Describe current key approaches to reducing food insecurity and improving diet quality, including research-based approaches using the social-ecological-based models and policy diffusion models.
6. Outline cultural, legal, professional, and ethical issues that promote or inhibit adoption of policies and implementation of programs that would improve food environments around the globe.
7. Discuss innovations in community nutrition research and practice that have been important for reducing food insecurity and improving diet quality, both in the US and globally.

The Right to Food is recognized by the United Nations (UN), but not by the US (UN Food and Agriculture Organization (FAO) Committee on World Food Insecurity, 2017). As would be expected, approaches to ensure that individuals have access to food vary a great deal across the US, and also around the world. A variety of factors affect individual diet and nutrition, including but not limited to those proximal to the person: taste and enjoyment, convenience, price, education, access to healthcare, cultural and religious beliefs, wealth, and income inequality. The more distal, contextual factors that affect individual diet and nutrition include political systems, geographic characteristics, city planning, governmental policy at both the state and national levels, tradition, the openness of a society, the tolerance for corruption within government and business, climate change, as well as agricultural practices and characteristics, and industry concentration. Individuals within a society may have access to resources to improve their diet, depending on their own understanding of what makes a healthy diet, their access to food and income, and the actual structure of the society. But in some circumstances, individuals are powerless to change or improve their diet quality due to inequities in systemic factors.

Definitions and Theories of Nutrition and Diet-Related Disparities and Healthy Food Environments

Nutrition and **diet-related disparities** can be defined as differences in dietary intake, dietary behaviors, and dietary patterns in different segments of the population, resulting in poorer dietary quality and inferior health outcomes for certain groups and an unequal burden in terms of disease incidence, morbidity, mortality, survival, and quality of life (Beydoun et al., 2016; Satia, 2009). In the US, racial and ethnic minorities (e.g., Black, Hispanic, American Indian/Alaska natives) often experience diet-related disparities, including poorer nutrient intake and dietary patterns (Diez Roux, Mujahid, Hirsch, Moore, & Moore, 2016; Lee-Kwan, Moore, Blanck, Harris, & Galuska, 2017; Satia, 2009; Wang et al., 2014).

Diet-related disparities are evident beginning in infancy, as breastfeeding rates show disparities according to socioeconomic status (SES) and race/ethnicity (Anstey, Chen, Elam-Evans, & Perrine, 2017; Davis, Li, Adams-Huet, & Sandon, 2017). Specifically, those in the lowest SES groups have the lowest breastfeeding initiation and persistence rates, and non-Hispanic, Black infants have consistently lower breastfeeding persistence rates compared to non-Hispanic Whites, even in the face of overall increases in breastfeeding rates (Anstey et al., 2017; Davis et al., 2017). Unsurprisingly, poverty is the leading contributor to nutrition/diet-related disparities, and it plays a major role in reducing access to safe and nutritious foods. These disparities are present on a global scale as well. For instance, lower income countries experience lower rates of breastfeeding rates overall, as well as higher rates of iron-deficiency anemia, malnutrition, and stunting compared to high-income countries (Fanzo, Hawkes, & et al., 2018), although duration of breastfeeding is generally shorter in the developed countries (Victora et al., 2016).

There are several terms that are used in discussing community nutrition issues: hunger, food security, and food insecurity (Food and Agriculture Organization of the United Nations, 2003). **Hunger** is a broad term that describes the want or scarcity of food available to individuals or within a country. **Food security** is a more specific condition in which "all people, at all times, have physical, social, and economic access to sufficient, safe, and nutritious food that meets their dietary needs and food preferences for an active and healthy life;" and "**food insecurity** exists when people do not have adequate physical, social, or economic access to food" (Food and Agriculture Organization of the United Nations, 2003) (See Discussion Box 5.1).

Discussion Box 5.1: Food Insecurity in the US and Global Food Insecurity

In the US, 11.1% of households were food insecure at some time in 2018, down from a peak of 14.9% in 2011 (United States Department of Agriculture Economic Research Service, 2019b). Food insecurity in the US is usually measured by using one of three versions of the US Household Food Security Module (USDA ERS Survey Tools, 2018).

Globally, food insecurity is highest in low-income countries and lowest in high-income countries (USDA Economic Research Service, 2017) (see Figure below). The UNFAO Voices of the Hungry project developed a validated tool to measure global food insecurity modeled on the US module, called the Food Insecurity Experience Scale (FIES) (United States Department of Agriculture, 2017). The development of the FIES has allowed for comparisons across countries using the same tool, increasing the depth of understanding of the problem globally. Sub-Saharan Africa has almost double the food insecurity of any region. Latin America and the Caribbean and South Asia are the regions that are the second and third most food insecure, respectively (United States Department of Agriculture, 2017).

(continued)

(continued)

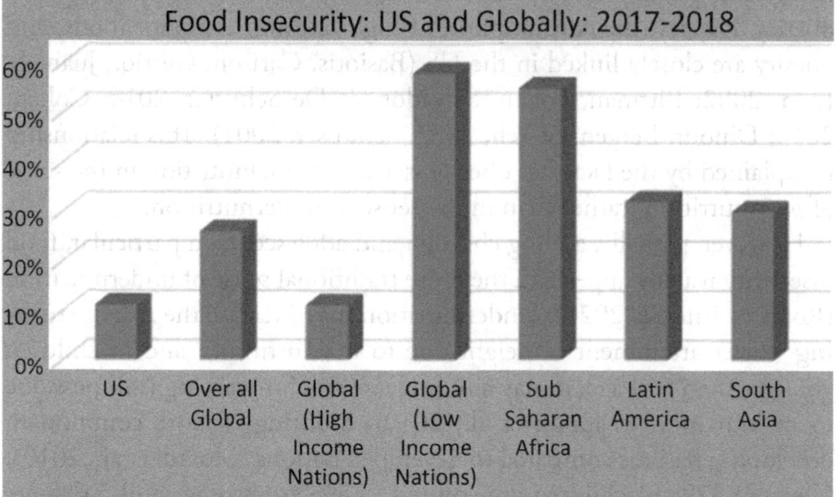

What Do You Think?

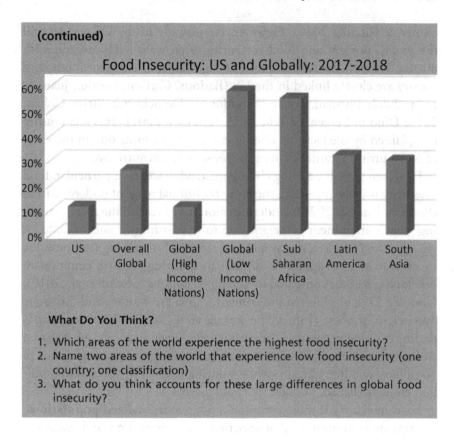

1. Which areas of the world experience the highest food insecurity?
2. Name two areas of the world that experience low food insecurity (one country; one classification)
3. What do you think accounts for these large differences in global food insecurity?

One might expect food insecurity to be associated with low weight status and **undernutrition**. However, this is not always the case. Food insecurity is associated with lower dietary quality, at least among some races/ethnicities in the US (Leung & Tester, 2019), and food insecurity is generally associated with higher obesity rates (Hernandez, Reesor, & Murillo, 2017). Nonetheless, whether food insecurity is associated with obesity or low weight status, it is almost always accompanied by malnutrition.

Malnutrition is a general term that refers to poor nutrition, which could include energy and protein over or undernutrition, as well as deficiencies or excesses of particular nutrients (World Health Organization (WHO), 2016). Historically, the type of malnutrition defined as **undernutrition** has been the primary concern both in the US and globally

(Roser & Ritchie, 2020). However, increasingly in the US and around the world, poverty and food insecurity often walk hand-in-hand with obesity. Poverty, the consumption of high-fat, nutrient-poor foods, and obesity are closely linked in the US (Basiotis, Carlson, Gerrior, Juan, & Lino, 2002; Bittman, Pollan, Salvador, & De Schutter, 2014; Cohen, 2013; Dinour, Bergen, & Yeh, 2007; Schlosser, 2001). This relationship is explained by the fact that obesity is a sign of malnutrition in the sense of *poor* nutrition, rather than in the sense of undernutrition.

However, globally, among children and adolescents in particular, food insecurity usually appears in the more traditional sense of undernutrition (Roser & Ritchie, 2020). Undernutrition may increase the risk of **stunting** (lower attainment of height due to malnutrition), and in children younger than 5 of age, it may also increase risk for **wasting** (thinness due to malnutrition) (Moradi et al., 2019). Stunting is more common in developing nations compared to developed nations (Moradi et al., 2019). Globally 22% of children are stunted, 7.5% are wasted, and 5.6% are overweight or obese (Fanzo, Hawkes, & et al., 2018). While stunting is declining in many countries, it has increased in countries in sub-Saharan Africa (Akombi et al., 2017); meanwhile, stunting and wasting coexist in 16 million children globally, while stunting and obesity coexist in 8 million globally (Fanzo, Hawkes, & et al., 2018).

Food insecurity is more common among **marginalized populations**, "groups or communities that experience discrimination and exclusion (social, political, and economic) because of unequal power relationships across economic, political, social, and cultural dimensions" (National Collaborating Centre for Determinants of Health, 2019). For example, in the US, low-income populations are *most* likely to be food insecure, followed by single women with children; however, elderly populations are *less* likely compared to the overall population to be food insecure (USDA ERS, 2019a, 2019b, 2019c). Families with young adults with Intellectual and Developmental Disabilities (IDD) have more than double the odds of living in food-insecure households compared to adults without limitations (36.7% food insecure versus 14.9%) (Brucker & Nord, 2016). In another marginalized population, transgender adults, higher rates of "food-related stress" have not been found; however, this evaluation was not based on more standard measures of food insecurity (Henderson, Jabson, Russomanno, Paglisotti, & Blosnich, 2019).

An emerging area of nutrition and health research explores how people access food. A **food environment** can be defined as "the physical presence of food that affects a person's diet; a person's proximity to food store locations; the distribution of food stores, food service, and any physical entity by which food may be obtained; a connected system that allows access to food" (Centers for Disease Control, 2010). Food environments can exist on a macro scale, involving government, policy, industry, and media food environments as well as on a more intermediate scale, involving school, worksite, community, and home food environments (Glanz, 2009; see also Chap. 16, this volume). One definition of community food environment (at least in developed countries) refers to the "number, type, location, and accessibility of food outlets" (Glanz, 2009). More recent definitions of food environment have sought to build a model that involves three domains: the physical, social, and person-centered environments, all of which work together and interrelate to influence food choices, dietary consumption, and diet-related disease risk (Lytle & Myers, 2017).

There has been a recent focus on measuring the physical environment domain of community food environments, especially within the US, by using current technology to map availability. One example of this is the Food Environment Atlas (USDA, 2019), which has a searchable map of the US with visual representation of grocery store availability, fast food availability, food insecurity, and much more. In 2019, researchers made the first attempt at mapping the global food environment and global food sustainability (Bene et al., 2019). These efforts at measuring and mapping various aspects of the food environment can help community health experts better evaluate the likelihood that their communities have food environments supportive of a good diet and nutritional health. That is, healthier community food environments are more likely to result in communities that have lower food insecurity, less hunger, and fewer diet-related disparities.

As discussed later in this chapter, numerous approaches have been used to try to address food insecurity and diet-related disparities. Many of these have been top-down approaches, but a concept gaining popularity is **food sovereignty** ("the right of peoples, communities, and countries to define their own agricultural, labor, fishing, food and land policies which are ecologically, socially, economically and culturally appropriate to their unique circumstances") (Rome Non-Governmental Organization (NGO)/CSO Forum for Food Sovereignty, 2002, n.d.). Food sovereignty includes "the right to have food and to *produce food*, which means that all people have the right to safe, nutritious and culturally appropriate food and to food-producing resources and the ability to sustain themselves and societies" (Rome NGO/CSO Forum for Food Sovereignty, 2002, n.d.). Whereas both food security and food sovereignty are concerned with the ability of persons to meet their nutritional needs as well as cultural needs for appropriate foods, food sovereignty activists concerned with **food justice** also consider how food is produced with respect to *justice* in the fields and workplaces, as well as with respect to the land and environment (Gottlieb & Joshi, 2010).

History of Research and Practice in Reducing Diet-Related Disparities

History in the US

Historically, several approaches have been taken to addressing food insecurity and diet-related disparities. A timeline of important milestones can be seen in Fig. 5.1. For instance, the earliest versions of food and nutrition programs run by the US government began during World War I

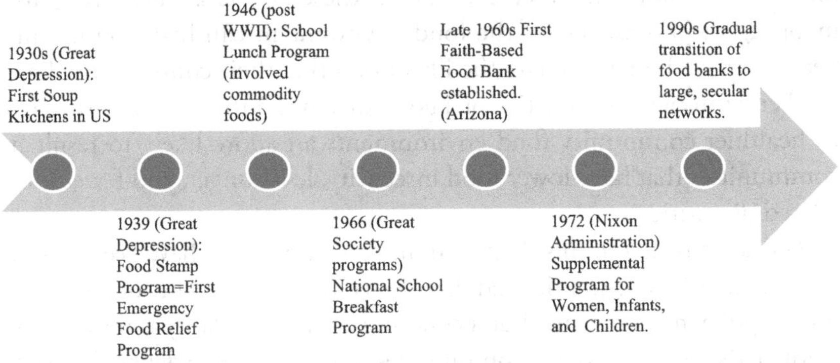

Fig. 5.1 History of nutrition and diet research and practice

under the Herbert Hoover administration. The program provided food relief to Americans and other people starving and trapped behind German lines (Levine, 2010). But it was not until the 1930s, in response to the Great Depression (1928–1933), that the US government began to embrace the task of shaping American food consumption. Early attempts at anti-hunger programs began with soup kitchens and bread lines, which were emergency food relief efforts, usually provided by faith-based organizations as part of their missions (O'Brien, Staley, Uchima, Thompson, & Torres Aldeen, 2004; Poppendieck, 1999). These early programs were largely, in their "goals, structure, and administration, more a subsidy for agriculture than a nutrition program" (Levine, 2010, p. 39).

Early American food programs grew from the notion that America "had an interest not only in agricultural productivity but in human productivity as well" (Levine, 2010, p. 34). The Supplemental Nutrition Assistance Program (SNAP) began as an emergency food relief effort in 1939 called the Food Stamp Program (Committee on Examination of the Adequacy of Food Resources and SNAP Allotments, et al., 2013). The National School Lunch Program (NSLP) was established in 1946 under the Truman administration to provide free and reduced-cost lunches to needy children, while serving as an outlet for surplus **commodity foods** (USDA Food and Nutrition Service (FNS), 2019a, 2019b). The National School Breakfast Program (NSBP) was established 20 years later in 1966 with a similar goal (United States Department of Agriculture Food and Nutrition Research Service, 2019a). Finally, the Supplemental Program for Women, Infants, and Children (WIC) began in 1972 as a pilot project to provide nutrition assistance and education to pregnant women, infants, and young children (National WIC Association, 2019).

These programs evolved over the years, slowly moving away from their primary role as **commodity** distribution programs toward the aim of providing nutritious food for Americans facing food insecurity. Nonetheless, many of these programs still function as commodity programs to some extent. **Commodities** in the US refer to foods in their raw form or processed via canning, drying, cheese-making, and so on and which are available in amounts greater than the food markets can sell. The US government buys some of these foods from farmers or processors to stabilize domestic food prices, and **food banks** and other programs help to distribute these foods to needy families. This shift in the aim of US government food programs toward providing more nutritious food occurred, in part, as a response to public criticism. In the 1960s, for instance, research showed that there were very few free lunches being provided through the NSLP, and that two out of three eligible children were not being served (Levine, 2010). It was not until the 1970s, in the face of direct criticism by anti-hunger and civil rights activists, including the Black Panther Party, that the USDA reluctantly issued standards and guidelines to make local food programs more accessible (Levine, 2010). Slowly, the USDA improved standards and made a greater effort to provide more nutritious food, set standards for what foods should be provided, and educate participants on making good food choices.

Since the establishment of government-run food programs such as the NSLP in 1946, we commonly conceive of **food policy** (governmental laws and regulations affecting production and distribution of food) as occurring on the federal level. However, state and local governments, food banks, individual communities, community organizations, healthcare organizations, and NGOs have a powerful role in shaping community health and nutrition. Each may take different approaches.

Food banks are a fairly recent phenomenon and represent a local-level attempt to alleviate poverty and its associated ills. John Van Hengel established the first food bank in Phoenix, Arizona, US, in the late 1960s, as a way of more systematically distributing food that would otherwise be wasted, in a kind of modernized **gleaning** initiative (United States Food Bank Network, 2019). Since their inception, food banks have evolved from single faith-based organizations (e.g., Catholic, Baptist, Methodist, etc.) to ecumenical (multi-faith based) organizations to fully secular organizations (O'Brien et al., 2004). This transition occurred during the 1980s during a major recession. The Temporary Emergency Food Assistance Program was launched in response to the recession, with food commodities being distributed en masse via local governments (O'Brien et al., 2004). As commodity stocks began to run low, the federal government mandated a pilot program to trial distributing a greater variety of commodity foods through food banks while local governments simultaneously began to extract themselves from this role (O'Brien et al., 2004). Today, most US food banks are large networks that serve large regions through smaller **food pantries**. In addition, they are hubs of advocacy for policy change and may alert the federal government of major economic changes occurring on the community level (O'Brien et al., 2004). Beyond merely providing food, many of these pantries have made efforts to include nutrition education, culinary programs, and information on shopping and budgeting to participants in an effort to help lower income people stretch their food dollar and purchase nutritious foods.

Another type of intervention began with research by geographers and economists. Beginning in the 1960s, researchers delved into mapping the location and availability of food, in addition to considering its relative affordability. In the early 1990s, researchers in the UK coined the term **food deserts** to describe areas with low availability of high-quality, affordable foods and lack of safe or accessible transportation to obtain food (Cummins & Macintyre, 2002; National Research Service, 2009). Since the early 2000s, research in the US into food deserts as a barrier to food access for SNAP participants or other low-income individuals has been ongoing. Interest in **community gardening**, farmer's markets, mobile farmer's markets, and community partnerships to provide grocery stores in urban, low-income, and rural areas has been increasing. Yet there is mixed evidence regarding the efficacy of these efforts at addressing food deserts and improving access to and intake of healthy foods among lower income people (Mack, Tong, & Credit, 2017; Smith & Morton, 2009).

Global History

Mirroring the development of food assistance programs in the US, systematic global attempts to quantify and address diet-related disparities across nations and continents did not begin until after the end of World War II with the establishment of the UNFAO. In 1951, a Rockefeller Center Mexican Agricultural Program paper detailing the "World Food Problem" described a potential problem of food scarcity worldwide caused by overpopulation, which could lead to instability. This line of thinking helped to usher in the Green Revolution, an effort to export crop science to various countries and using fertilizers, irrigation, and pesticides to increase crop production (Otter, 2010).

Since the mid-2000s, efforts to increase global food security have gone far beyond the 1960s efforts to improve crop yields and provide emergency hunger relief in nations in crisis. Current efforts focused on achieving "Zero Hunger" were launched by the UN Secretary General in 2012. This Zero Hunger Challenge represents an effort to end hunger and essentially provide food security to all people. The program goals include: "ending all forms of malnutrition; doubling the agricultural productivity and incomes of small-scale food producers; ensuring sustainable food production systems; increasing investment in agriculture; correcting and preventing trade restrictions and distortions in world agricultural markets; and adopting measures to ensure the proper functioning of commodity markets" (UN, 2019). These efforts are supported by four UN agencies: The World Food Programme, the World Bank, the FAO, and the International Fund for Agricultural Development (UN, 2019). As the UN and others try to ensure sustainable food production systems, climate change contributes to an environment of uncertainty about how this will be ensured in the future. (See Discussion Box 5.2 Climate Change.) Essentially, because of climate change and other factors, global efforts have shifted their focus toward both worldwide food security and toward food justice, discussed below.

Discussion Box 5.2: Climate Change

Globally, even as incomes have increased and access to healthy food has generally expanded (particularly since the global food crisis of the 1970s), new challenges to local food production have come to light. Climate change has had considerable impact on the kinds and quantities of foods grown in many regions of the world. New estimates suggest that there may be as many as 200 million climate change migrants by 2050, in large part due to disruption of food and water supplies in parts of the world (Brown, 2008). This will impact food security for both the migrants and for the countries to which they migrate. Additionally, climate change will also impact the types of foods that can be grown in various regions of the world, and the availability of food and water supplies around the world.

See the **online resource**, then answer the following questions: https://www.nytimes.com/2019/06/05/opinion/guatemala-migrants-climate-change.html?smid=nytcore-ios-share

What Do You Think?

1. How do global factors such as food security and climate change interact with policy making and decisions within the US?
2. Do you think the US has a role in fighting climate change and ensuring global food security? If so, what do you think this role should be?

Sustainable Community Health-Oriented Approaches to Reduce Diet-Related Disparities

In this section, we first briefly describe more traditional approaches to address diet- and nutrition-related disparities in the US and globally that use the following strategies: minimum calorie-level standards, global approaches, food justice promotion, genetically modified foods, and poverty reduction. We then describe two more recent approaches that are being applied somewhat more successfully to address these complex problems: the **social-ecological model** (McElroy, Bibeau, Steckler, & Glanz, 1988) and the **policy diffusion models** (Berry & Berry, 2007). Social-ecological approaches seek to address food insecurity by addressing multiple influences on health behaviors, including intrapersonal, interpersonal, organizational, community, and public policy levels (Sallis, Owen, & Fisher, 2008). **Policy diffusion models** (Berry & Berry, 2007) have sought to apply evidence-based policies deemed effective on a smaller scale, scoping to larger levels of policy implementation (Mosier & Thilmany, 2016; Shipan & Volden, 2008).

Minimum Calorie-Level Approaches

In the US, government-level approaches to address both food security and diet-related disparities have included the NSLP, NSBP, WIC, SNAP, and others. These programs provide food at free or reduced cost to those who cannot afford sufficient food. In the past, the NSLP and NSBP tended to focus on achieving minimum calorie levels at meals. However, following the passage of the Healthy, Hunger-Free Kids Act (HHFKA) in 2010, efforts to provide more whole fruits and vegetables, require more whole grains, and limit both sugary beverages and high-sodium foods were strengthened. Research shows HHFKA has increased the dietary quality of meals, yet there have been efforts to relax these guidelines (Campbell, Crulcich, & Folliard, 2019).

Historically, the NSLP and NSBP have been popular among legislators because they have been shown to reduce food insecurity modestly (by about 4%), reduce the likelihood of nutritional inadequacies among children, and are associated with less obesity among participants (Food Research and Action Center, 2017). School breakfast and lunch programs have also been shown to improve educational outcomes, including improving academic achievement and rates of tardiness in schools among low-income children both in the US (Kennedy & Davis, 1998; Millichap, 1989) and worldwide (Afridi, Barooah, & Somanathan, 2019).

SNAP has also had some success, with a 2015 White House report showing that households that receive SNAP benefits experienced 30% less food insecurity than would otherwise be the case (Fisher, 2017, 194). Yet, paradoxically, SNAP participants have been found in some studies to have lower quality diets compared to SNAP-eligible non-participants (Zhang et al., 2018). One possible explanation for the lower quality diets of SNAP participants may be in part that they are likely to be younger, female, and less likely to be non-Hispanic white, meaning that they are more likely to be especially vulnerable to poverty and of lower educational status, factors which have been independently linked to low diet quality (Zhang et al., 2018). Another explanation for this phenomenon is the cyclical nature of food access under the SNAP program. SNAP benefits, which are intended to be supplemental rather than the primary source of food, often run out after just two or three weeks, leaving recipients to depend on the cheapest available foods (often junk foods) until the next SNAP deposit arrives on their electronic benefits card, a phenomenon called a "food stamp cycle" (Dinour et al., 2007; Nestle, 2013). Yet evidence suggests that SNAP has largely been unsuccessful at permanently relieving food insecurity with levels staying unchanged in the population (Gundersen & Oliveira, 2001; Olson & Holben, 2002; Taren, Clark, Chernesky, & Quirk, 1990).

Programs such as US Women, Infants and Children (WIC) have also been popular because they are effective, to a degree. Women who participate in WIC are more likely to give birth to infants who survive infancy and who have higher mental development scores at age 2 (Center for Budget and Policy Priorities, 2017). Children whose mothers participate in WIC have higher iron intake, lower consumption of added sugars, and reduced fat intake as a percentage of energy intake (Center for Budget and Policy Priorities, 2017). However, while WIC promotes breastfeeding and employs breastfeeding counselors, participants in WIC are somewhat less likely to breastfeed compared to non-participants (Center for Budget and Policy Priorities, 2017). There is also some evidence that WIC participation improves food security of children (Center for Budget and Policy Priorities, 2017).

NGOs such as food banks and food pantries also play a role at alleviating short-term food insecurity in the US by being a stop gap to people who either do not want to enroll in or who do not receive enough assistance from SNAP, while promoting enrollment in SNAP. In addition to providing food, food banks supplement programs such as WIC by providing additional nutrition education and information on how to create food budgets. In addition, faith-based organizations in the US that operate independent food pantries help fill gaps in services in smaller, rural communities, and other areas not served by food banks (Rural Health Information Hub, 2019).

Food Justice-Oriented Approaches

Food justice approaches aim to confront the world's growing food needs while recognizing the peoples' right to produce and feed themselves in ways that encourage sustainable production, respect local and indigenous cultures, and allow women to have a fundamental role in the production, marketing, and management of food (UK Agricultural Biodiversity Compendium, 2002). To address issues of food justice, numerous agricultural researchers have tried to promote research into methods of farming that rely less on industrial support, are environmentally sustainable, and promote food justice (Gleissman, 2015). They have favored techniques that do not require intensive tillage or irrigation, that avoid factory farming of animals, and that promote agrobiodiversity (Gleissman, 2015). In India, for example, where intensive irrigation threatens to create a water crisis, farmers have greatly increased their interest in organic farming (Zwerdling, 2009a, 2009b). Likewise, in the US there has been more of an effort to focus on the environmental costs of growing food (Charles, 2016), with greater interest among farmers in the potential greater return on the dollar that growing food organically may yield, which may positively impact farmer's lives but may simultaneously make produce and healthier foods less affordable. These agricultural approaches are interesting but also influence just one aspect of complex nutrition-related disparities. They have not, so far, become widespread enough to significantly affect poverty or food security worldwide.

Global Approaches

The United Nations is arguably the world's organization with the most resources dedicated to improving global food insecurity and diet-related disparities (World Food Programme, 2020a, 2020b). The World Food Programme, just one of the UN's four organizations that addresses nutrition issues, spends more than $2 billion annually in the countries it aids (World Food Programme, 2020a, 2020b). Three additional UN organizations mediate issues related to food insecurity and diet-related disparities, including the World Bank, the UNFAO, and the International Fund for Agricultural Development (UN, 2019), and these organizations also spend significant money each year on these issues. The UNFAO (which has an additional budget of about $2.6 billion annually) (United Nations Food and Agriculture Organization, 2020) has set forth guidelines for individual countries to ensure the food security of its citizens in the Rome Principle for Sustainable Food Security, which focuses on country-owned and country-led plans (UNFAO, 2017). It makes six key recommendations meant to promote food justice and food sovereignty, which are summarized and paraphrased below (UNFAO, 2017):

1. Countries should set up and/or strengthen interagency mechanisms and communication between agencies responsible for food security and nutrition strategies, policies, and programs.
2. Those mechanisms should be coordinated at a high level of government and be supported by law, including diverse agencies from agriculture, education, health, and more.

3. National food security and nutrition strategies should address availability and access to food, utilization, and stability of the food environment and strengthen local and national food systems.
4. Countries should create or strengthen mechanisms for coordinating strategies and actions with local governments, and they should set up multi-stakeholder platforms at local and national levels to help design, implement, and monitor food security and nutrition strategies, legislation, policies, and programs. Stakeholders should include local governments, civil society, the private sector, farmer's organizations, small-scale and traditional food producers, women's and children's associations, and representatives of groups most affected by food insecurity.
5. Countries should develop and/or strengthen mapping and monitoring to allow for better coordination and accountability.
6. In designing national food security and nutrition strategies and programs, countries should consider unintentional consequences of programs on other countries.

This set of recommendations is intended to promote food justice among member countries of the UN. It is also intended to encourage cross-sector communication and coordination. This UNFAO guidance document (2017) goes on to offer seven steps countries should take to implement the Right to Food Guidelines, beginning with step 1 ("identifying who the food insecure are, where they live, and why they are hungry") and ending with step 7 ("establish accountability and claims mechanisms, which may be judicial, extrajudicial or administrative, to enable rights-holders to hold governments accountable and to ensure that corrective action can be taken without delay when policies or programs are not implemented or delivering the expected services"). However, problematically, these recommendations do not carry the force of national law in the member countries. UN member countries vary widely in the degree to which they try to live up to these recommendations. In fact, food security and nutrition programs are still fragmented in many countries, including the US, as described above. Thus, these global strategies have been in many ways solely aspirational.

NGOs have been integral to the UN's role in addressing food insecurity since its inception. At least for the past 30 years, NGOs focused on alleviating hunger and improving poverty have been using a rights-based approach to advocating for adequate food in support of the UN's Right to Food (Windfuhr, 1997). They are advocating for food sovereignty and food justice. Many organizations have become innovators in trying to change outcomes and create healthier food environments (Foodtank, 2017). Among the organizations advocating for change include A Growing Culture, which is a global NGO that promotes agroecology and connects farmers across the globe (http://www.agrowingculture.org/). Heifer International has been working since 1944 to give animals to needy families, teach them principles of agroecology so that they can raise them responsibly, and then encourage sharing of the animal offspring with other families (https://www.heifer.org/). Many other organizations have done hunger relief in various countries globally throughout the twentieth and into the twenty-first centuries. However, notably, none of these efforts have decreased the need for the relief in any sustainable way.

Genetically Modified Foods Approaches

Given the limited success in alleviating hunger through various governmental and NGO means, there has been some interest in recent years regarding the possible role of agriculture in reducing diet-related disparities. **Genetically modified or genetically engineered foods (GM, GMOs, or GE foods)** refer to foods which have been produced via the manipulation of plant or animal genes via biotechnology (Gleissman, 2015). Current GM plant varieties available in the US and worldwide have tried to reduce pesticide and herbicide use, increase drought tolerance, and improve plant hardiness with the aim of increasing crop yields and food security and reducing environmental impact, but with mixed results (Gleissman, 2015; Hakim, 2016). In spite of the trending scientific consensus that GM foods are generally safe to consume, GM foods have been controversial in the US (Funk & Kennedy, 2016), and vehemently resisted in many other countries for safety concerns (Freedman, 2013; Gleissman, 2015).

The grounds for objection to GM foods have varied. Whereas in the US, most people seem suspicious of GM foods on health and safety grounds, most US Americans have been less concerned about the justice of such products (See Research Box 5.1). However, worldwide, and even among some US farmers, there is concern about how the use of GM seeds impacts farmers. GM seeds must be purchased from seed companies and cannot be collected and replanted, adding cost to farmers (Gleissman, 2015); further, legal battles have erupted as a result of cross-pollination between farms using convention seeds and those using GM seeds. (See Research Box 5.1: The New Food Fights.)

Research Box 5.1 Funk, C., & Kennedy, B. (2016, December 1). The New Food Fights: US Public Divides Over Food Science. Pew Research Center

The scientific evidence on the safety of GM foods is fairly strong. According to the Pew Research Center, "88% of members of the American Association for the Advancement of Science (AAAS) and 92% of working Ph.D. biomedical scientists said it is safe to eat genetically modified foods" (Funk & Kennedy, 2016). Yet, the American public is generally still uncomfortable with and uninformed about GM foods.

Research on public opinion shows that many US citizens have "soft" views on GM Foods. These feelings tend to be based on limited knowledge, and they may change over time and be sensitive to question wording. These soft feelings reflect a general uncertainty about what GM foods are, whether they are safe to eat, and how they were produced. One-quarter of US adults say that they are not sure if GM foods are safe to eat, and another one-third of Americans say that GM foods are worse for their health, in spite of the scientific consensus in favor of these foods (Funk & Kennedy, 2016). Looking at this table on US attitudes toward GM foods, consider the following questions.

What Do You Think?

1. Given the scientific consensus on the safety of GM foods, how could scientists and practitioners better inform communities about the potential benefits and costs of GM foods?
2. What are the possible implications of GM foods for the food justice movement, given the questions of seed ownership, and how can this be balanced with the potential health benefits?
3. What role should public opinion have in the development of US and worldwide food and health policy?

(continued)

(continued)

Beliefs about GM foods includes some with "soft" views

% of U.S. adults who say foods with genetically modified ingredients are generally ___ than foods with no genetically modified ingredients

If given an option of saying "not sure"	
Worse for health	33%
Neither better nor worse for health	32
Better for health	7
Not sure	26
No answer (vol.)	1
Among those saying "not sure"	
Worse for health	22
Neither better nor worse for health	58
Better for health	11
No answer (vol.)	10
Views about GM foods when both questions are combined	
Worse for health	39
Neither better nor worse for health	48
Better for health	10
No answer or refused to lean (vol.)	3

Note: Respondents were first given the option of answering "not sure" when asked about the health impacts of genetically modified foods. Those respondents (and any who gave no answer) were then asked which option they were "leaning" toward if they had to choose.
Source: Survey conducted May 10-June 6, 2016.
"The New Food Fights: U.S. Public Divides Over Food Science"
PEW RESEARCH CENTER

Online Resource: See the full Pew Research report on public opinion and food science here: https://www.pewresearch.org/science/2016/12/01/the-new-food-fights/

In spite of the legal battles and controversy, proponents of GM foods maintain that they have considerable potential to address malnutrition worldwide. Golden Rice, for instance, has been widely advertised as a possible panacea to vitamin A deficiencies in developing countries. Yet, Golden Rice and similar GM foods have been criticized because they decrease environmental seed diversity and may place undue cost burdens on low-income farmers. In short, GM foods, while generally considered safe, still pose significant questions regarding their role in the food system, both in terms of reducing nutrition disparities and in terms of achieving food justice. (See Discussion Box 5.3).

Discussion Box 5.3: Golden Rice

Politicians, activists, and community leaders have often struggled to balance access to nutritious food with respect for traditional, cultural practices and farmers rights. One example of the complexity of this balance is Golden Rice. Golden Rice is a genetically modified rice that contains beta-carotene, a source of vitamin A. Vitamin A deficiency is common throughout parts of Asia and Africa and can lead to blindness and sometimes fatal infection. A single bowl of this rice can provide 60% of a child's daily requirement for vitamin A (Charles, 2013). Golden Rice has been approved by multiple government agencies, including Food Standards Australia, Health Canada, and the U.S. Food and Drug Administration. The developers of the rice are also working on iron and zinc varieties to address other micronutrient deficiencies among impoverished communities (Lynas, 2018). However, GM crops, and especially Golden Rice have been met with considerable suspicion.

One controversy around Golden Rice emerged when a Tufts researcher, in partnership with three Chinese collaborators, conducted a study regarding the safety and benefit of Golden Rice for children ages 6 to 8 years old. The researchers published the findings of study in *The American Journal of Clinical Nutrition*, showing that golden rice is a promising source of Vitamin A in Chinese children. However, soon after publication, Greenpeace issued a statement that the Chinese children were used as "guinea pigs." Researchers had not followed university or international protocol for research on human subjects. The Tufts researcher has been barred from doing human research for two years and will undergo training in research on human subjects, and the Chinese collaborators were punished for their participation by the Chinese government, which then financially compensated the parents of the child study subjects (Enserink, 2013).

What Do You Think?

1. What do you think are the drawbacks of promoting the use of a product like Golden Rice among lower income, local farmers? What are the potential benefits of such a product for a community with endemic vitamin A deficiency? What are the alternatives?
2. If you were part of an NGO or a local public health official addressing endemic vitamin A deficiency in the population you served, what steps could you take to decide what would be best received and most effective in your population?

Although Golden Rice has been approved by multiple governing agencies, opponents argue that there are other cheaper and more culturally sensitive ways to address vitamin deficiencies, and many people remain highly suspicious of the safety of these crops (Charles, 2013; Lynas, 2018). Golden rice has not been free to develop, and the approval cost has been quite large. Opponents argue that children would have to eat enormous quantities to receive benefits. They also suggest that vitamin A supplements would be just as effective, are proven to be safe, and are cheaper. However, many in the science and nutrition community point out that dietary supplements have been available for years and have not solved the problem, and Golden Rice is not intended to be the *only* dietary source of vitamin A. Golden Rice is intended to be another tool in the toolbox for fighting against malnutrition (Wesseler & Zilberman, 2014). In fact, research suggests that the opposition to Golden Rice, and the resulting delays in its approval and use may have cost 1.4 million life years in India alone during the last decade due to malnutrition and associated complications (Wesseler & Zilberman, 2014). In spite of its commercial origins (originally developed by Syngenta) and lingering opposition, Golden Rice may soon be made available, for free, to farmers in parts of Asia and India (Haq, 2018).

Poverty Reduction-Oriented Approaches

There have also been various novel attempts to address the stubbornly high numbers of the population who continue to suffer from malnutrition due to poverty. **Microfinance** has been one effort to allow those in poverty to have small loans to start small businesses in an effort to improve the status of women and other marginalized populations (Njiraini, 2015). There is research to support the idea that if women are given more ability to be independent financially or to have independent income, the family is more likely to consume a nutritionally complete diet and to have overall improvements in education and poverty (McGuire, Popkin, & Vittas,

1990). While not popular in many countries, there have also been efforts to give money without strings attached to low-income families in efforts to improve nutrition status and alleviate poverty. One recent experience in this realm was disappointing in its results, which, in addition to attitudes toward the poor, may limit this approach's use in the future (Aizenman, 2018).

Social Ecological Model (SEM) Approaches

As previously noted, social-ecological approaches to reducing food insecurity and diet-related disparities aim to influence five, interrelated determinants of these outcomes. These include the following:

1. **Intrapersonal factors** (factors related to the individual such as knowledge, attitudes, behaviors, etc.)
2. **Interpersonal factors and primary groups** (informal and formal relationships such as family and friend relationships)
3. **Institutional factors** (social organizations such as churches with their rules for interaction)
4. **Community factors** (relationships among various organizations, institutions, and informal networks)
5. **Public policy** (national, state, and local laws and policies)

According to this model, these five levels reflect the areas available for health promotion programming, as well as the levels that could be analyzed to understand causes of community health-related problems and then design appropriate interventions (McElroy et al., 1988). See also UN Rome Principle for Sustainable Food Security (UNFAO Committee on World Food Security, 2017).

Applying a social-ecological approach to addressing obesity in the US, the CDC, in partnership with the National Center for Chronic Disease Prevention and Health Promotion, issued a "Health Equity

Resource Toolkit for State Practitioners Addressing Obesity Disparities" (CDC Toolkit, 2017). This toolkit sought to address all five levels of the SEM. Other examples of how the SEM has been used include studies that have tried to use the SEM to determine better ways to address **social determinants of health** such as fruit and vegetable intake (Robinson, 2008), access to enough food (Andress, 2017) and to estimate the contribution of environmental factors to childhood obesity (Ohri-Vachaspati et al., 2015).

Increasingly, food insecurity status is documented in electronic health records, and community partnerships are being developed among hospitals and healthcare organizations, schools, food pantries, food banks, and more to address food insecurity (Health Research & Educational Trust, 2017). Family-based approaches have also been tried in healthcare settings in order to address levels 1 and 2 (inter and intrapersonal factors) of the SEM, as well as integrating various community and policy changes to influence levels 3 through 5.

Globally, there have also been several efforts to use the SEM to analyze the extent to which nutrition programming works to improve nutrition-related outcomes and to inform interventions to improve the overall nutritional health of communities. For example, DeLorme et al., (2018) used the SEM to analyze the success of the Kanyakla Nutrition Program, which used social networks to promote increased nutrition knowledge and support changes in infant- and child-feeding practices. This analysis revealed that while intrapersonal factors were influenced, participants recognized barriers emanating from community, institutional, and policy levels (DeLorme et al., 2018). Caperone et al., (2019) used qualitative methods to create their own ecological model of the determinants of dietary behavior in Nepalese adults with diabetes and high blood glucose, whereas, Scott, Ejikeme, Clottey, & Thomas used the Social Continuum model to build their own ecological model of obesity determinants in sub-Saharan Africa (2012). (See Research Box 5.2 for an example of how the SEM has been used to the factors that affect food choices in public school children.)

Research Box 5.2 Developing and Applying a Socio-Ecological Model to the Promotion of Healthy Eating in the School

Townsend, N., Foster, C. (2013). *Publ. Health. Nutr.* 16(6): 1101–1108.

Background: Schools are a popular venue for healthy diet-promoting interventions because of the potential to reach such a broad audience and have adequate time to deliver interventions.

Method: The authors collected data from 6693 students aged 11 to 16 in Wales, U.K., from questionnaires evaluating self-reported dietary choices and correlates of choices as well as data from teachers on school approaches to providing food. They used multilevel analysis to study the association of each level of the SEM on student diet choices.

Results: Interpersonal factors such as the student's individual social environment influenced food choices more than interpersonal factors such as knowledge and attitudes. School factors, such as food policies, were more strongly associated with food choice compared to community aspects of the school.

Conclusion and Implications: These results can be used to target interventions at the aspect of the SEM which most strongly relate to student food choice.

What Do You Think?

1. How did the authors use the SEM in this study?
2. Were you surprised to find out that social/interpersonal factors impacted food choice more strongly than individual/intrapersonal factors like knowledge?
3. If you were designing an intervention to improve student diet choices, how might these results influence you? What type of intervention would you design?

Policy Diffusion Model Approaches

As noted above, policy diffusion approaches to addressing dietary and nutrition disparities aim to understand and create change in these important nutrition outcomes by influencing only level 6 of the SEM (policy). These approaches seek to implement effective and innovative policies in local communities, which may then spread to other areas where the policy is likely to be effective. Berry and Berry developed the model in 1990 to account for how the internal characteristics of a community or region (finances, political system, regional determinants) and regional effects, such as the number of neighboring communities that adopt a policy, may together influence policy adoption. As they state, "the probability of state innovation is directly related to the motivation to innovate, inversely related to the strength of obstacles to innovation, and directly related to the availability of resources for overcoming those obstacles" (Berry & Berry, 1990, p. 410).

It is well known that low-income neighborhoods have poorer quality food environments, including less access to fresh foods, and greater access to food sources that promote unhealthy eating, such as gas stations and convenience stores or "dollar stores" (Hilmers, Hilmers, & Dave, 2012). Such areas, in which there is very limited access to quality, fresh food, are known as poor-quality retail food environments (Ford & Dzewaltowski, 2008, 216; Walker, Keane, & Burke, 2010). Minorities in low-income communities have particularly poor access to healthful and affordable food sources, and their communities are often the targets of "junk food" and fast food advertising (Brock & Sparrow, 2016). Because such food environments lead to poor health outcomes and higher health costs, local governments have taken an active role in shaping food and nutrition choices available to their citizens. Such policy initiatives have included the following:

- **Sugar-sweetened beverage (SSB) taxes:** Several cities in the US have begun taxing SSBs, in hopes of encouraging healthier consumer behavior. While SSB taxes are typically seen as regressive, and may disproportionately burden lower income respondents, revenues from these taxes may be used for community health programs to offset potential harms (Oddo et al., 2019).
- **Formula business restrictions:** Cities that have struggled with an overabundance of dollar stores and a dearth of grocery stores have begun creating "formula business restrictions." These restrictions on dollar stores, paired with efforts to encourage grocery stores to move into low-income communities, have had some success in bringing more grocery stores to areas that were previously food deserts (Gonzalez, 2019).
- **Food desert mitigation:** Philadelphia's "healthy corner store initiative" focused on promoting healthy and affordable food choices in a profitable way for store owners. This program is aimed at providing access to healthy, affordable food and promoting healthy practices in the community by educating youth via school programs and connecting certain customers to additional health services such as blood pressure checks (The Food Trust, 2012).

- **Urban Food Forests:** Dozens of food forests have sprung up around the US. In 2015, there were over 80 of these food forests, and more have been planted since (Bukowski, 2018). Atlanta Mayor, Keisha Lance Bottom, intends to end food insecurity in the city by increasing access to fresh food and developing a greater diversity of food systems, with the goal of putting 85% of Atlanta residents within a half mile of fresh food by 2021. The city has developed an urban food forest that is expected to be one of the largest in the country (Renner, 2019).
- **Mobile food programs:** Other cities have relied on non-profit initiatives, which often use mobile food programs to distribute food to low-income and food-barren areas. Two examples are the Fresh Express in downtown Phoenix, AZ, and the Twin Cities Mobile Market. Both buses are funded through non-profits and grants and are intended to combat food deserts through providing healthful and affordable options in under-resourced areas (Tulane School of Social Work, 2018). Garden on the Go, created by Indiana University Health by its Department of Community Outreach and Engagement, is another such program. This mobile food program also offers healthy cooking demonstrations and certain health services such as flu shots (Gura, 2015).
- **Food Sovereignty Ordinances:** Some communities have taken an even more active approach to local food control. In the spring of 2011, four towns on Maine's Down East coast passed ordinances declaring food sovereignty. Local farmers who wanted exemptions from various state regulations, which were particularly onerous for small farmers, advocated for the declaration (Clark & Teachout, 2012). Similar food sovereignty resolutions have been considered in towns in Vermont, Wyoming, Arizona, Massachusetts, and California (Clark & Teachout, 2012). However, these declarations of local food sovereignty have generally had little to do with fighting food insecurity or hunger. Rather, many of these towns have been concerned with preventing the use of genetically modified seeds, avoiding onerous regulations, or fighting other perceived corporate or "big-government" takeovers.

Each of these examples of local policy initiatives illustrates how local ideas have developed because of locally perceived needs but have then diffused into other areas with similar needs or characteristics. Some have been found lacking locally or have been limited by perceived difficulty of or opposition to implementation (e.g., SSB taxes), thus they have been limited in their spread. Other initiatives that have been viewed more positively and that have required fewer resources to implement have moved into many communities in the US and other nations (e.g., mobile food programs, urban forests). Research Box 5.3 describes how national-level policy action may influence policy adoption locally or within states/provinces.

Research Box 5.3 Diffusion of Food Policy in the US: The Case of Organic Certification

Mosier, Samantha L. & Dawn Thilmany, (2016). *Food Policy* v61: 80–91.

Background: The US represents the largest organic market globally, and it is regulated both federally and at the state level, allowing researchers to study the dynamic between state and local governance, as well as policy diffusion across states.

Method: Researchers rely on a policy diffusion approach, using event history analysis as a basis for modeling diffusion to examine the adoption of organic food and agriculture legislation in the US states from 1976 to 2010. They test the impact of the adoption of federal organic legislation (National Organic Program) as a potential motivator for state adoptions and bureaucratic rulemaking as potential positive effects on state adoptions.

Results: Results suggest that federal intervention into the marketplace has some effect on state policy decisions. Specifically, authors found that when the federal government passed legislation but there was delayed implementation of that legislation in food markets, state adoption was spurred. States were found to adopt policy on a "separate but parallel track" to the federal government, indicating an interdependent relationship, particularly under conditions of uncertainty.

Conclusion and Implications: This test of the Policy Diffusion model indicates that the federal government can motivate states to adopt and update laws in order to comply with new legislation, as well as serve as a catalyst for states to pursue their own regulations, should national efforts fail.

What Do You Think?

1. What can community health experts learn about policy adoption from studying the above example regarding the National Organic Program adoption and implementation?
2. For communities hoping to implement community health policies, such as soda taxes, through governmental regulation, is federal action necessary? How can local governments impact public health?
3. What can policy diffusion teach us about the spread of ideas? How can community health practitioners leverage this for their benefit?

These local policy approaches are not without shortfalls. While urban agriculture movements, small farms, and Community Support Agriculture (CAS) are local initiatives billed as sustainable community solutions to hunger, research shows that these approaches are highly associated with both whiteness and economic privilege, producing food that is generally cost prohibitive for lower income individuals (Alkon & McCullen, 2011; Farmer, Chancellor, Robinson, West, & Weddell, 2014; Reynolds, 2015). The above approaches are also largely concerned with food security in urban environments, which only accounts for a portion of the population struggling with food insecurity in the US and around the world.

One emerging solution is the creation of "**food policy councils**" to assist with the development of new policies and to facilitate change on a local level (Gottlieb & Joshi, 2010). These councils can take different forms, from subcommittees on city councils, to community-based organizations, to public-non-profit hybrids. But in all their forms, these organizations are instrumental in promoting food justice and food sovereignty by taking advantage of local communities and states as "laboratories of democracy" and models for policy that may then spread to other areas. For instance, Washington's 2008 Local Farms-Healthy Kids Act, legislation pushed by local farm to school and public health advocates, created a variety of programs to support local food production, as well as establishing a farmer's market technology project that allowed farmers' markets throughout the state to increase access by SNAP recipients to food sold at these markets (Gottlieb & Joshi, 2010). This legislation, in particular, was considered a breakthrough and seen as a model for other states to follow.

Cultural, Legislative, and Professional Issues That Impact the Specific Sustainable Community Health Approaches to Reducing Food Insecurity and Diet-Related Disparities

There are a variety of cultural, legislative, and professional issues that may positively or negatively impact community health approaches to reducing food insecurity and diet-related disparities. These issues include food cultures and socioeconomic status (Kirkpatrick, Dodd, Reedy, &

Krebs-Smith, 2012; Wang & Chen, 2011); immigration status (Abraído-Lanza, Dohrenwend, Ng-Mak, & Turner, 1999; Hummer, Robers, Amir, & Forbes, 2000); disparities in subsidies for large versus small farmers (Nestle, 2013); access to fresh food (Gottlieb & Joshi, 2010); regulations on food safety (Schanes, Dobernig, & Gozet, 2018) and distribution (Rack, 2018); and lack of coordination between health and nutrition professionals (Popkin et al., 2019).

Individual and Social Characteristics: Food Cultures, Immigration, and SES

Communities that have strong cultural traditions supporting cooking and consumption of whole foods may be relatively easily encouraged to consume more foods associated with health such as vegetables, fruits, and whole grains as compared to communities in which highly processed foods are regarded as cultural foods. Research indicates that **Healthy Eating Index** (HEI) scores (a measure of how closely an individual's dietary pattern adheres to dietary advice in the Dietary Guidelines for Americans) tend to be lower in lower SES households (Kirkpatrick et al., 2012; Wang & Chen, 2011). However, they also differ by race and ethnicity, with non-Hispanic black households having lower HEI scores and lower consumption of healthy foods compared to non-Hispanic white households (Kirkpatrick et al., 2012; Li et al., 2017; Wang & Chen, 2011), and Hispanic households having higher HEI compared to all non-Hispanic households (Kirkpatrick et al., 2012; Wang & Chen, 2011).

The relatively higher HEI of Latino households, and particularly of first-generation households, compared to white Americans or American-born children of immigrants, is known as the "Latino Paradox" (Abraído-Lanza et al., 1999; Hummer et al., 2000). This paradox suggests that acculturation into American society tends to worsen people's health significantly (Escarce, Morales, & Rumbaut, 2001). People who have lived in the US for longer periods of time are more likely to consume higher-fat foods, including red meat, cheese, and fried foods (Neuhouser, Thompson, Coronado, & Solomon, 2004). In short, acculturation into the mainstream American lifestyle is likely a cause of poorer eating habits for second-generation youths, as well as other American-born populations

(Brock & Sparrow, 2016). Consistent with this conclusion is the finding that in lower income, Southern African American and lower income white, non-Hispanic, communities, obesity presents as more of a problem. This is possibly a reflection of valuing processed foods as more culturally appropriate (soda, fried chicken, ice cream) (Dodd, Briefel, Cabili, Wilson, & Crepinsek, 2013; Yang, Buys, Judd, Bower, & Locher, 2013). Encouraging the adoption of foods that are not considered to be as culturally appropriate among these populations may prove to be more difficult. (See Discussion Box 5.4 to consider how one should account for culture when planning a program to improve adoption of healthy eating patterns.)

Indigenous communities are especially vulnerable to food insecurity (Jernigan, Huyser, Valdes, & Watts, 2016; Pindus & Hafford, 2019). Between 2000 and 2010, studies showed that American Indians and Alaskan Natives were nearly twice as likely to be food-insecure when compared with whites (Jernigan et al., 2016), with another study

Discussion Box 5.4: Cultural Food Practices Versus Healthy Eating Practices

Dietitians, nutritionists, health education specialists, and other community health professionals often spend a lot of time and programming effort teaching individuals and families about healthy eating, healthier foods, and healthier ways of cooking. However, getting people to change their individual dietary intake and food practices is difficult. Despite years of educational efforts, dietary quality in Westernized nations is often poor and fails to meet standards for fruit, vegetable, and whole-grain intake (Kirkpatrick et al., 2012; Wang & Chen, 2011). Part of the reason for the resistance to change is that people may adhere strongly to their cultural foodways and usual ways of eating. In particular, as discussed earlier in this chapter, in the US, foods that hold cultural significance in Black culture may have lower nutritional value (Dodd et al., 2013; Yang et al., 2013).

What Do You Think?

1. What factors do you think are important to consider when planning a program to educate members of a community on healthier ways of eating?
2. What steps would you take to make sure you understand what types of food or cooking methods are important to or hold special significance for your community members?
3. What do you think is the best way to approach providing culturally sensitive nutrition education in a community? Would you identify community stakeholders and influencers first? How might you use their knowledge and insight to inform a program?

indicating over 90% of Native American households experience food insecurity in some areas of the US (Sowerwine, Mucioki, Sarna-Wojcicki, & Hillman, 2019). Yet, food security remains unaddressed in these communities, threatening their health and wellbeing.

Governmental programs have been primarily responsible for improving food security among Native Americans, with the USDA providing nutrition assistance to tribal communities through their Food Distribution Program on Indian Reservations (FDPIR). While the FDPIR program is intended to be a supplemental nutrition program, focused on meeting minimal nutritional requirements, in 38% of recipient households the program was the sole or primary source of food (Pindus & Hafford, 2019). FDPIR participants have considerably worse health outcomes than the general population, though the FDPIR package has been revised in recent years to provide healthier and more nutritious foods (Pindus & Hafford, 2019). Even among Native Americans with sufficient food, there is often a lack of access to desired native foods (Sowerwine et al., 2019). Some community activists, experts, and researchers have provided specific tools to help communities assess their food approaches and move in the direction of food sovereignty (see, e.g., Alicia Bell-Sheeter's Food Sovereignty Assessment tool) for indigenous and rural communities in particular (2004).

Likewise, globally when income levels rise, so do intake of protein foods and processed foods, especially SSBs and meats, which may previously have been considered out of reach and thus more desirable (Popkin, Adair, & Ng, 2012; World Health Organization, 2019). Other cultural barriers to ending food insecurity may include lack of awareness of governmental and NGO assistance available and stigma associated with needing or receiving help (Meza, Altman, Martinez, & Leung, 2019; Popkin et al., 2019).

Policy Barriers: Farming Subsidies, Food Safety, and Food Distribution

Policy barriers to adopting sustainable community approaches to reducing these disparities may include on the national scale US government policies that subsidize large farming organizations without adequately supporting smaller farmers or which may not promote the growth of healthier foods (Nestle, 2013). Food justice advocates emphasize the important role that food markets, supported by both local and federal policy, can have in establishing a collaborative relationship between local farmers and communities. Small policy changes, such as requiring WIC-approved stores to carry fruits and vegetables to remain approved as WIC vendors, can have significant impacts (Gottlieb & Joshi, 2010). Gottlieb and Joshi suggest that policy change can "make the link between the store, the shopper, and the farmer, creating the basis for fresh food access for all" (2019, 163).

Additional policy changes that could improve food and hunger-related outcomes in America have to do with food safety rules, which are important to prevent foodborne illness, local, state, and federal regulations. One much-needed improvement is clearer regulations on company practices regarding use-by and sell-by dates, which currently unregulated, may sometimes promote food waste (Schanes et al., 2018) or make sharing unused food with those who are needy burdensome (Rack, 2018). Improved policies in this area could have a considerable impact on improving food security and should be relatively easy to achieve.

Professional Barriers: Problems of Coordination

Professional barriers to these sustainable initiatives to reduce diet-related disparities and food insecurity may include doctors, dietitians, NGO leaders, schools, and so on who are used to working independently on projects or who may have little knowledge about other approaches, or who work on such initiatives in ways that do not support or recognize related initiatives (Popkin et al., 2019). In addition, company and organization policies and procedures as well as different organization goals may make deep collaborations difficult. Thus, collaborative efforts to reduce food insecurity or diet-related disparities may be plagued by collaborations that are too surface level to have long-term impact or that may not be deeply integrated into the communities they intend to serve.

In addition, in the US, and globally in many cases, politics have become more divisive and harmful to the sustainable health of communities. Some sectors of society, particularly "blue-collar workers" may express social disinterest or hostility to the poor (Lauter, 2016), and there has been an increase in the partisan divide in the US between those who are more likely to believe the US economy is fair and that poverty is more related to lack of effort (Republican-leaning) and those who believe that the US economy is not fair and that poverty is more related to circumstances beyond individual control (Dunn, 2018). Some of this may relate to compassion fatigue. Twenty-four-hour news cycles of bad news may give people the impression that large problems are insoluble and that government initiatives prop up the lazy rather than improving diet and health outcomes that may save healthcare dollars and ultimately benefit society. In addition, US Americans may misperceive that recognizing economic, social, and cultural rights, including the right to food, necessitates direct provision of food and services to everyone and requires charity (Chilton & Rose, 2009). Thus, human rights approaches to addressing food insecurity have not gained ground in the US (Chilton & Rose, 2009).

In this sense, some of the biggest obstacles to comprehensive anti-hunger programs lie in attitudes about poverty and the recipients of aid programs. The recipients of SNAP, welfare, and other government subsidy programs have long been characterized as "welfare mothers," and "deadbeat dads," and these characterizations have a considerable impact on how target populations are treated in policy design (Ingram, Schneider, & Deleon, 2007). As a result, the policies designed to aid these populations are restricted and particularistic, rather than universal (like Social Security) (Mettler, 1998). Because Americans are inherently suspicious of SNAP and welfare policy recipients, they continually prefer policies designed to be more restrictive, rather than expansive, in aiding these populations. For instance, Americans continually debate restricting the use of SNAP funds on SSBs, but rarely discuss widely expanding the nutrition safety net through increased funding for SNAP and other programs (Fisher, 2017). Instead, each year, Republican legislators, in particular, have sought to defund or massively restrict food assistance programs through the farm bill, which would have calamitous consequences to community health.

These issues are not limited to the US. People in other nations also tend to oppose programs that are perceived as giving aid to those who are not deserving of it. Interestingly, a 2010 study by Barrientos and Neff evaluated attitudes toward poverty across the world, according to region, development status, and wealth status. They found the majority of respondents in countries throughout the world believe that there is little chance of impoverished persons escaping poverty. This was a consistent view across levels of economic development (Barrientos & Neff, 2010). However, in spite of this attitude, about one-third or more of respondents across all regions and country classifications (e.g., low versus high income) regarded the causes of poverty as laziness more than unfairness (Barrientos & Neff, 2010). Respondents from South Asia and East Asia held a different point of view, indicating higher levels of the population believed laziness was a primary cause of poverty. Such views may limit attempts to attack the causes of food insecurity and poverty worldwide, threatening the health of communities.

There are other barriers to improvements in these areas as well. Research shows that in the US food insecurity and body weight are positively correlated. The more food insecure a person is, the more likely he or she is to struggle with obesity (Basiotis et al., 2002; Dinour et al., 2007; see also Chap. 16, this volume). There are several interrelated reasons for this: the first reason is access to healthy and affordable foods (consider the food deserts and poor-quality retail food environments previously mentioned); and the second has to do with inconsistent food availability (such as the food stamp cycle). Current anti-hunger policies are not designed for resolving long-term food insecurity or promoting healthful diets—they are valuable, but insufficient (Fisher, 2017). Long-term solutions will need to be aimed at achieving economic and social justice, recognizing food as a human right, generating wealth for low-income and rural communities, and business models grounded in paying employees living wages (Fisher, 2017, 267).

Research Critical to Diet and Nutrition Health Improvements

Resolving hunger and food insecurity, removing disparities in food security and improving the quality of diet for all people will require a concerted effort within the US, within individual countries, and coordination on a worldwide scale. We briefly discuss key issues for research on nutrition improvements for sustainable community health.

There are several key issues introduced in this chapter that will require further research in order to help resolve complex, change-resistant issues such as food insecurity and diet-related disparity. A basic challenge to solving these problems involves having good data so that one knows when progress has been made. Another key area of research involves identifying better ways to promote good dietary quality and sustainability of diet while respecting cultural foodways and individual physiological and psychological needs surrounding food. The trend toward global diets becoming more animal-based and less sustainable with increased wealth and industrialization has created another area to be addressed. How to integrate sustainable agricultural practices into an industrialized agriculture system without reducing production or cost and availability of food is another major issue. In addition, food waste affects both access to food and climate change. Attitudes toward poverty and food insecurity are other major issues to be researched so that one can better understand how to influence them in order to expand the policy tools available to fight them. Finally, gaining a better understanding of how to promote equity across cultures and within a culture, including in healthcare, is another major area for research.

Addressing such issues will require cooperation from disciplines such as agricultural science, environmental science, public health, sociology and social work, nutrition, business, government, and other entities. Poverty and lack of adequate educational systems for all, inequitable criminal justice systems, racism, sexism, and economic injustice all contribute to food insecurity disparities (Odoms-Young, 2018). Thus, answers to such issues are not easy, and solutions cannot be achieved without ensuring that the people most at risk for hunger, such as women and people of color have a seat at the table and greater agency in developing policy and community health-based solutions (Fisher, 2017).

While developed nations and open nations tend to do a better job of tracking indicators of food insecurity and diet-related disparity compared to less-developed nations or closed societies, this data is often inconsistent within and between nations and may not rely on the same measures. As mentioned earlier in the chapter, the US recently launched an interactive Food Access Research Atlas (https://www.ers.usda.gov/data-products/food-access-research-atlas/go-to-the-atlas/) that allows better identification at a granular level of specific areas where low-income, low vehicle access, and low grocery store access exist, thereby reducing access to affordable, nutritious food. Indonesia has a similar project to map and track food insecurity

(World Food Programme, 2015). However, many countries lack such good quality data and certainly do not have it mapped. Bene and Prager recently published a paper in which they modeled worldwide food system sustainability (2019). However, they lacked data to inform their map in many areas (Bene et al., 2019). Developing an interactive atlas and map for the world similar to the one the US has developed would assist NGOs and governments in identifying the areas that need the greatest assistance and help them actively track progress. Thus, collaborations among geographers, community and public health professionals, and experts in technology will be needed to make such data tools a reality.

Discovering better ways to promote dietary quality while respecting cultural foodways and combating the trend toward less sustainable, more animal-based diets is another major area for research. A large group of scholars recently collaborated on a large-scale project to determine what type of diet could be environmentally sustainable and better promote food justice. Willett et al. recently published "Food in the Anthropocene: the EAT-Lancet Commission on Healthy Diets from Sustainable Food Systems" (2019). In this document, they brought together 19 commission members from 16 countries, including experts from human health, agriculture, political science, and environmental science to look at the best evidence for sustainable diets and sustainable food systems. The diet they proposed was radical in its recommendation for very low consumption of meats, poultry, eggs, and fish, and higher consumption of certain grains and legumes. Rather than being applauded for being daring in its approach, in many quarters it was criticized for being unrealistic and unfair to agriculture (particularly toward ranchers). The report acknowledged that the diet could not be a one-size-fits-all approach but would need to be culturally relevant; however, it was still far from well-received. Even advice such as "eat less" has faced criticism and objection. Thus, it is clear that in addition to needing better modeling and scientific knowledge to tackle hunger, food insecurity, and diet-related disparities, having political will and agreement on how to address these issues will be an ongoing effort requiring collaboration and also conversation to build consensus.

Agricultural methods that reduce reliance on carbon-based energy, decrease use of intensive tillage or irrigation, avoid factory farming of animals, promote agrobiodiversity, and use local resources wisely will increase the biological and environmental health of the land (Gleissman, 2015). However, such methods that employ principles of agroecology may be difficult to integrate into already industrialized agricultural systems and countries seeking to advance and become more modern. In addition, these methods may sometimes increase the cost of food, which if individuals do not experience similar gains in income can reduce access to food. Thus, research into ways of building support for and use of these methods among family farmers, industrialized farmers, and farming companies will be important. Policy changes requiring changes in farming methods may be best received when they are locally sensitive and involve farmer input.

The UN Zero Hunger Challenge (2015) seeks to end world hunger through ongoing work to alleviate poverty and encourage responsible governments among member countries, by promoting environmentally responsible agriculture, and by eliminating food loss and food waste (United Nations Food and Agriculture Organization, 2017). The USDA has also recently begun to emphasize waste reduction as part of a strategy to improve environmental sustainability and reduce hunger (United States Department of Agriculture, 2019a). About $1 trillion worth of food is wasted worldwide each year (about one-third of the world's food supply) (World Food Program USA, 2020). Essentially, the amount of food wasted annually could feed double the number of people in the world who are currently hungry (World Food Program USA, 2020). Thus, research into and practice innovations that find ways to store those foods, create long-lasting foods from perishables, make locally sourced food available to the people who need it, and quickly recover and redistribute wasted food is essential. (World Food Program USA, 2020).

Modifying attitudes toward poverty will also require ongoing research and innovation. So long as one-third of the population sees poverty as primarily an issue of laziness (Barrientos & Neff, 2010), there will be resistance, often among the most powerful members of society, to efforts to help the poor improve their condition. Thus, continued research into how attitudes toward poverty influence policy and how these attitudes may be changed is needed in order to advance innovation in policy initiatives to bring equity to nutrition and health outcomes.

A number of centers are advancing research into the determinants of the quality and equity of healthcare (United States Department of Veterans Affairs, 2020). This research is important for nutrition-related equity issues as well. Are there assumptions among healthcare workers about Black women in the US, for example, or indigenous women that may make them less likely to promote breastfeeding? Do the ways physicians, dietitians, or community health workers counsel people of certain races/ethnicities show lack of care and thus decrease the likelihood that the recommended practices will be adopted? Ongoing research in these areas is needed as well.

Other issues of interest to community health and nutrition professionals in the future may include how technology can be used to influence food choices. In other words, the use of mobile health (mHealth) [text messaging, apps, telemedicine, and more] is a quickly growing area of innovation, which may be used to promote healthy behaviors, increase access to nutrition programs, and more. In addition to existing research on how social relationships may affect nutrition choices and health practices, future research should evaluate the effect of social media use on food choice and better determine ways to use it to improve dietary quality.

Regarding the thorny, complex issues of hunger, food insecurity, and diet-related disparities, individuals and organizations often want to help by providing service, advice, or other help with strings attached. This is highly related to the social construction of the population being served, and those in poverty are nearly always socially constructed in such a way that policy becomes restrictive and proscriptive (Ingram et al., 2007). Systems can become stubbornly averse to change and unaware of their own bias, making change slow and difficult, even with infusions of money and goodwill. There is a temptation to offer simple solutions (panaceas) to complex problems, but research shows that communities and governments can work together to overcome complex **collective-action problems** (Ostrom, 2007), including, quite possibly, the problem of hunger and nutrition disparities.

Summary and Conclusion

The long history of disparities in food security and diet-related disparities both within the US and throughout the world may make one feel that these are insoluble problems. Lower SES individuals and marginalized groups have long suffered not only higher rates of hunger and malnutrition, but also a number of other (often related) ills, including increased risk of early death and poor health outcomes. However, because diet and nutrition status are strong contributors to health outcomes, healthcare costs, and mortality, these are problems worth solving. Some may argue that it is too expensive or unfair to expect that all, including those perceived as social loafers, would have enough to eat or that they would have healthy foods available to choose to eat. However, it is clear that with one-third of food supplies being lost or wasted, attaining food security is not impossible. It is merely difficult.

Whereas many have tried and failed before, the ability to communicate quickly on a global scale means that experts in these various areas have the potential to be able to communicate about possible solutions, and that stakeholders may be able to harness the power of social media in an era of the almost ubiquitous mobile phone to make those in need aware of the resources available. We are in a new age and should continue to strive for removing food insecurity disparities and improving dietary quality.

Self-Check Questions

1. What is the difference between the terms hunger and food insecurity?
2. What are some similarities and differences among the terms food security, food justice, and food sovereignty?
3. Outline the history of nutrition and food security programs highlighting the key development for the sustainable health of communities?
4. What are the relative merits and demerits of leading nutrition and food insecurity programs considered in this chapter?
5. How are culture, law, and policy barriers, and also enablers, of nutrition and food security? Discuss with reference to specific programs and settings.
6. What are the key issues of research to guide the science and practice of nutrition and food security around the world?

Discussion Questions

1. How are the terms malnutrition, undernutrition, obesity, stunting, and wasting similar or different in a setting you know? Explain your answer.
2. How might some common diet-related disparities that tend to be present in lower income groups and in certain ages, races/ethnicities be addressed ensuring health sustainability?
3. What are some UN efforts to reduce food insecurity and improve diet quality? How successfully may these implement around the world and why?

Experimental Fieldwork Exercises

1. Find a place to volunteer that involves working with marginalized populations who are food insecure. This could be a local food pantry, soup kitchen, or other feeding program, community center, or other community agency. Commit to spending at least 5 hours per week (65 hours over a 16-week semester). During your time volunteering, try to interact with those served by the agency in appropriate, informal ways. Practice showing care, concern, and interest in others. You could do this by asking how someone's day is going, asking, remembering, and calling him/her by name (if appropriate), inviting someone to sit and talk with you, or other small acts. **At the end of each volunteer period, reflect on the following:**

    - What types of individuals are seeking food assistance? (age, gender, race/ethnicity, income status, etc.)
    - Are the people you see those you would expect to see seeking assistance? Why or why not?
    - Where do you think you have gotten your expectations regarding who in society needs more help?
    - Do you think that those seeking help would have been able to predict that they would need help someday?
    - Is the experience one that you would be comfortable going through in order to seek help yourself?
    - Was the dignity of each person preserved by the experience?
    - Were the foods offered to participants foods that they found acceptable in terms of culture, nutrition, taste, enjoyment, and so on?
    - What are the best aspects of the program for which you volunteered?
    - What aspects would you improve or change if you could?
    - Do you think this program is sustainable? Why or why not?
    - What do you think could make the program one that local people no longer needed?

2. Visit a non-governmental organizations (NGO) and local government or community-based organization to learn of their approaches to reducing food insecurity and improving diet quality. Which ones do you think have been effective? Which do you think has the most promise?

Online Resources

Disrupting Food Insecurity: Tapping Data for Strategies That Tackle the Root Causes

From the Urban Institute. A county-by-county map of food insecurity in the US, together with resources to disrupt food insecurity, including identifying community barriers, addressing transportation issues, recognizing and addressing disparities, and more. https://apps.urban.org/features/disrupting-food-insecurity/

Community Food Forest Map

Interactive map and research on community food forests throughout the US. This website also provides resources and identifies best practices for urban food forests and community gardening. https://communityfoodforests.com/community-food-forests-map/

Open Secrets: Lobbying Disclosure Profile for Agribusiness

Open Secrets provides data on lobbying activity by sector, allowing for research into political participation and influence. This page provides data on lobbying activity in the Agribusiness sector, with particular information about the amount of money spent, the number of lobbyists, and the percent of lobbyists who are "revolvers," or former government employees. https://www.opensecrets.org/federal-lobbying/sectors/summary?cycle=2019&id=A

Global Nutrition Report Resources

An annual report from a group comprised high-level members of government, donor organizations, civil society, multilateral organizations, and the business sector, which describes the state of global nutrition. This page includes resources to learn more about malnutrition overall, nutrition profiles of individual countries, and case studies illustrating areas in which progress have been made: https://globalnutritionreport.org/resources/

References

Abraído-Lanza, A., Dohrenwend, B., Ng-Mak, D. S., & Turner, J. B. (1999). The Latino mortality paradox: A test of the 'Salmon Bias' and healthy migrant hypotheses. *American Journal of Public Health, 89*(10), 1543–1548.

Afridi, F., Barooah, B., & Somanathan, R. (2019). *Hunger and performance in the classroom* (Discussion paper series). I A Z Institute of Labor Economics.

Aizenman, N. (2018, September 14). *Which foreign aid programs work? The US runs a test—But won't talk about it.* National Public Radio, Goats and Soda. Retrieved November 8, 2019, from https://www.npr.org/sections/goatsandsoda/2018/09/14/647212387/which-foreign-aid-programs-work-the-u-s-runs-a-test-but-wont-talk-about-it

Akombi, B. J., Agho, K. E., Hall, J. J., Wali, N., Renzaho, A., & Merom, D. (2017). Stunting, wasting and underweight in sub-Saharan Africa: A systematic review. *International Journal of Environmental Research and Public Health, 14*(8), 863.

Alkon, A. H., & McCullen, C. G. (2011). Whiteness and farmers markets: Performances, perpetuations ... contestations? *Antipode, 43*(4), 937–959.

Andress, L. (2017). Using a social ecological model to explore upstream and downstream solutions to food access for the elderly. *Cogent Medicine, 4*, 1–18.

Anstey, E. H., Chen, J, Elam-Evans, L. D., & Perrine, C. G. (2017). Racial and geographic differences in breastfeeding—United States, 2011–2015. *Morbidity and Mortality Weekly Report, 66*(27), 723–727. https://www.cdc.gov/mmwr/volumes/66/wr/mm6627a3.htm. Accessed 11 May 2019.

Barrientos, A., & Neff, D. (2010). Attitudes to chronic poverty in the 'global village'. *Social Indicators Research, 100*, 101–114.

Basiotis, P. P., Carlson, A., Gerrior, S. A., Juan, W. Y., & Lino, M. (2002). *The healthy eating index: 1999–2000*. Washington, DC: U.S. Department of Agriculture, Center for Nutrition Policy and Promotion.

Bell-Sheeter, A. (2004). *Food sovereignty assessment tool*. Fredericksburg, VA: First Nations Development Institute. Retrieved April 22, from https://www.indigenousfoodsystems.org/sites/default/files/tools/FNDIFSATFinal.pdf

Bene, C., Prager, S. D., Achicanoy, H. A. E., Toro, P. A., Lamotte, L., Bonilla, C., ... Mapes, C. R. (2019). Global map and indicators of food system sustainability. *Scientific Data, 6*(279). Retrieved April 14, 2020, from https://www.nature.com/articles/s41597-019-0301-5.pdf

Berry, F. S., & Berry, W. D. (1990). State lottery adoptions as policy innovations: An event history analysis. *The American Political Science Review, 84*(2), 396–415.

Berry, F. S., & Berry, W. D. (2007). Innovation and diffusion models in policy research. In P.A. Sabatier (Ed.), *Theories of the policy process* (pp: 223–260). Boulder, CO: Westview Press. Retrieved April 17, 2020, from http://edwardwimberley.com/courses/IntroEnvPol/theorypolprocess.pdf#page=229

Beydoun, M. A., Beydoun, H. A., Mode, N., Dore, G. A., Canas, J. A., Eid, S. M., & Zonderman, A. B. (2016). Racial disparities in adult all-cause and cause-specific mortality among us adults: Mediating and moderating factors. *BMC Health, 16*(1113), 1–13.

Bittman, M., Pollan, M., Salvador, R., & De Schutter, O. (2014, November 7). How a national food policy could save millions of lives. *The Washington Post*. Retrieved September 21, 2015, from https://www.washingtonpost.com/opinions/how-a-national-food-policy-could-save-millions-of-american-lives/2014/11/07/89c55e16-637f-11e4-836c-83bc4f26eb67_story.html

Brock, C. R., & Sparrow, B. H. (2016). Race, ethnicity, and the politics of food. In D. L. Leal, T. Lee, & M. Sawyer (Eds.), *The Oxford handbook of racial and ethnic politics in the United States*. New York, NY: Oxford University Press.

Brown, O. (2008). *Migration and climate change* (Report no. 31). Geneva, Switzerland: International Organization for Migration. https://www.iom.cz/files/Migration_and_Climate_Change_-_IOM_Migration_Research_Series_No_31.pdf

Brucker, D. L., & Nord, D. (2016). Food insecurity among young adults with intellectual and developmental disabilities in the United States: Evidence from the National Health Interview Survey. *American Journal on Intellectual and Developmental Disabilities, 121*(6), 520–532.

Bukowski, C. (2018). *Community food forests: Collaboratively growing and harvesting food in forest-like ecosystems*. Retrieved November 11, 2019, from https://communityfoodforests.com/community-food-forests-map/

Campbell, E., Crulcich, S., & Folliard, J. (2019). A call for action to address food insecurity by strengthening child nutrition programs. *Journal of the Academy of Nutrition and Dietetics, 119*(10), 1747–1749.

Caperone, L., Arjyal, A., Puja, K. C., Kuikel, J., Newell, J., Peters, R., … King, R. (2019). Developing a socio-ecological model of dietary behaviour for people living with diabetes or high blood glucose levels in urban Nepal: A qualitative investigation. *PLoS One, 14*(3), 1–27.

Center on Budget and Policy Priorities. (2017, March 29). *WIC works: Addressing the nutrition and health of low-income families for 40 years*. Policy Futures. Retrieved November 6, 2019, from https://www.cbpp.org/research/food-assistance/wic-works-addressing-the-nutrition-and-health-needs-of-low-income-families

Centers for Disease Control. (2010, June 3) *General food environment resources*. Retrieved April 14, 2020, from https://www.cdc.gov/healthyplaces/healthtopics/healthyfood/general.htm

Centers for Disease Control. (2017). *Health equity resource toolkit for state practitioners addressing obesity disparities*. https://www.cdc.gov/nccdphp/dnpao/state-local-programs/health-equity/pdf/toolkit.pdf

Charles, D. (2013, March 7). In a grain of golden rice, GMO foods. *NPR*. Retrieved September 27, 2019, from https://www.npr.org/sections/thesalt/2013/03/07/173611461/in-a-grain-of-golden-rice-a-world-of-controversy-over-gmo-foods

Charles, D. (2016, May 5). The environmental cost of growing food. *NPR*, All Things Considered. Retrieved November 5, 2019, from https://www.npr.

org/sections/thesalt/2016/05/05/476600965/the-environmental-cost-of-growing-food

Chilton, M., & Rose, D. (2009). A rights-based approach to food insecurity in the United States. *American Journal of Public Health, 99*(7), 1203–1211.

Clark, S., & Teachout, W. (2012). *Slow Democracy: Rediscovering Community, Bringing Decision Making Back Home.* Junction, VT: Chelsea Green Publishing.

Cohen, D. (2013). *A big fat crisis: The hidden forces behind the obesity epidemic —And how we can end it.* New York, NY: Nation Books.

Committee on Examination of the Adequacy of Food Resources and SNAP Allotments; Food and Nutrition Board; Committee on National Statistics; Institute of Medicine; National Research Council. (2013). History, background, and goals of the supplemental nutrition assistance program. In J. A. Caswell & A. L. Yaktine (Eds.), *Supplemental nutrition assistance program: Examining the evidence to define benefit adequacy.* Washington, DC: National Academies Press (US). https://www.ncbi.nlm.nih.gov/books/NBK206907/

Cummins, S., & Macintyre, S. (2002). "Food deserts:" evidence and assumptions in health policy making. *BMJ, 325*, 436–438.

Davis, K. E., Li, X., Adams-Huet, B., & Sandon, L. (2017). Infant feeding practices and dietary consumption of US infants and toddlers: National Health and Nutrition Examination Survey (NHANES) 2003–2012. *Public Health Nutrition, 21*(4), 711–720.

DeLorme, A. L., Gavenus, E. R., Salmen, C. R., Benard, G. O., Mattah, B., Bukusi, E., & Fiorella, K. J. (2018). Nourishing networks: A social-ecological analysis of a network intervention for improving household nutrition in Western Kenya. *Social Science & Medicine, 197*, 95–103.

Diez Roux, A. V., Mujahid, M. S., Hirsch, J. A., Moore, K., & Moore, L. V. (2016). The impact of neighborhoods on cardiovascular risk: The MESA neighborhood study. *Global Heart, 11*(3), 353–363.

Dinour, L. M., Bergen, D., & Yeh, M. (2007). The food insecurity–obesity paradox: A review of the literature and the role food stamps may play. *Journal of the American Dietetic Association, 107*(11), 1952–1961.

Dodd, A. H., Briefel, R., Cabili, C., Wilson, A., & Crepinsek, M. K. (2013). Disparities in consumption of sugar-sweetened beverages by race/ethnicity and obesity status among U.S. schoolchildren. *Journal of Nutrition Education and Behavior, 45*(3), 240–249.

Dunn, A. (2018, October 4). *Partisans are divided over the fairness of the US economy and why people are rich or poor.* Pew Research Center. Retrieved November 8, 2019, from https://www.pewresearch.org/fact-tank/2018/10/04/partisans-are-divided-over-the-fairness-of-the-u-s-economy-and-why-people-are-rich-or-poor/

Enserink, M. (2013, September 18). Golden rice not so golden for tufts. *Science Magazine*. Retrieved September 27, 2019, from https://www.sciencemag.org/news/2013/09/golden-rice-not-so-golden-tufts

Escarce, J. J., Morales, L. S., & Rumbaut, R. G. (2001). The health status and health behaviors of Hispanics. In M. Tienda & F. Mitchell (Eds.), *Hispanics and the future of America* (pp. 362–409). Washington, DC: National Academies Press.

Fanzo, J., & Hawkes, C., et al. (2018). *Global nutrition report 2018*. Retrieved September 13, 2019, from https://globalnutritionreport.org/reports/global-nutrition-report-2018/

Farmer, J. R., Chancellor, C., Robinson, J. M., West, S., & Weddell, M. (2014). Agrileisure: Farmers markets, CSAs, and the privilege in eating local. *Journal of Leisure Research, 46*(3), 313.

Fisher, A. (2017). *Big hunger: The unholy alliance between corporate America and anti-hunger groups*. Cambridge, MA: MIT Press.

Food and Agriculture Organization of the United Nations. (2003). *Trade reforms and food security*. Retrieved April 22, 2020, from fao.org/3/y4671e06.htm

Food Research and Action Center. (2017). *Benefits of school lunch*. Retrieved November 5, 2019, from https://www.frac.org/programs/national-school-lunch-program/benefits-school-lunch

Foodtank: The Thinktank for Food. (2017). *After a global hunger increase, These 17 organizations innovate for change*. Retrieved from https://foodtank.com/news/2017/09/17-organizations-fighting-hunger/

Ford, P. B., & Dzewaltowski, D. A. (2008). Disparities in obesity prevalence due to variation in the retail food environment: Three testable hypotheses. *Nutrition Reviews, 66*(4), 216–228.

Freedman, D. H. (2013). The truth about genetically modified food. *Scientific American, 309*(3), 107–112.

Funk, C., & Kennedy, B. (2016, December 1). *The new food fights: US public divides over food science*. Pew Research Center. Retrieved November 5, 2019, from https://www.pewresearch.org/internet/wp-content/uploads/sites/9/2016/11/PS_2016.12.01_Food-Science_FINAL.pdf

Glanz, K. (2009). Measuring food environments: A historical perspective. *American Journal of Preventive Medicine, 6*(4S), S93–S98.

Gleissman, S. (2015). *Agroecology: The ecology of sustainable food systems* (3rd ed.). Boca Raton, FL: CRC Press.

Gonzalez, S. (2019, May 9). Planet money: Dollar stores' effects on communities. *National Public Radio*. Retrieved November 11, 2019, from https://www.npr.org/2019/05/09/721685190/planet-money-dollar-stores-effects-on-communities

Gottlieb, R., & Joshi, A. (2010). *Food justice*. Cambridge, MA: MIT Press.

Gundersen, C., & Oliveira, V. (2001). The food stamp program and food insufficiency. *American Journal of Agricultural Economics, 83*(4), 875–887.

Gura, L. (2015, June 29). *Garden on the go: Indiana University Health*. Stakeholder health: Transforming health through community partnership. Retrieved November 11, 2019, from https://stakeholderhealth.org/garden-on-the-go-indiana-university-health/

Hakim, D. (2016, October 29). Doubts about the promised bounty of genetically modified crops. *New York Times*. Retrieved November 5, 2019, from https://www.nytimes.com/2016/10/30/business/gmo-promise-falls-short.html?_r=2

Haq, S. (2018, November 28). A golden solution. *Dhaka Tribune*. Retrieved September 27, 2019, from https://www.dhakatribune.com/opinion/op-ed/2018/11/28/a-golden-solution

Health Research and Educational Trust. (2017). *Social determinants of health: Food insecurity and the role of hospitals*. Chicago, IL: Health Research and Educational Trust. Retrieved November 6, 2019, from http://www.hpoe.org/Reports-HPOE/2017/determinants-health-food-insecurity-role-of-hospitals.pdf

Henderson, E. R., Jabson, J., Russomanno, J., Paglisotti, T., & Blosnich, J. R. (2019). Housing and food stress among transgender adults in the United States. *Annals of Epidemiology, 38*, 42–47.

Hernandez, D. C., Reesor, L., & Murillo, R. (2017). Gender differences in the food insecurity-overweight and food insecurity-obese paradox among low-income adults. *Journal of the Academy of Nutrition and Dietetics, 117*, 1087–1096.

Hilmers, A., Hilmers, D. C., & Dave, J. (2012). Neighborhood disparities in access to health foods and their effects on environmental justice. *American Journal of Public Health, 102*(9), 1644–1654.

Hummer, R. A., Robers, R. G., Amir, S. H., & Forbes, D. (2000). Adult mortality differentials among Hispanic subgroups and non-Hispanic whites. *Social Science Quarterly, 81*(1), 459–476.

Ingram, H., Schneider, A. L., & Deleon, P. (2007). Social construction and policy design. In P. Sabatier (Ed.), *Theories of the policy process*. Boulder, CO: Westview Press.

Jernigan, V. B. B., Huyser, K. R., Valdes, J., & Watts, S. V. (2016). Food insecurity among American Indians and Alaska natives: A National profile using the current population survey-food security supplement. *Journal of Hunger & Environmental Nutrition, 12*, 1–10.

Kennedy, E., & Davis, C. (1998). US Department of Agriculture School breakfast program. *The American Journal of Clinical Nutrition, 67*(4), 798S–803S.

Kirkpatrick, S. I., Dodd, K. W., Reedy, J., & Krebs-Smith, S. M. (2012). Income and race/ethnicity are associated with adherence to food-based dietary guidance among US adults and children. *Journal of the Academy of Nutrition and Dietetics, 112*(5), 645–635.

Lauter, D. (2016). How do Americans view poverty? *LA Times*. Retrieved November 8, 2019, from https://www.latimes.com/projects/la-na-pol-poverty-poll/

Lee-Kwan, S. H., Moore, L. V., Blanck, H. M., Harris, D. M., & Galuska, D. (2017). Disparities in state-specific adult fruit and vegetable consumption—United States, 2015. *Centers for Disease Control Weekly Morbidity and Mortality Report, 66*(45), 1241–1247.

Leung, C. W., & Tester, J. M. (2019). The association between food insecurity and diet quality varies by race and ethnicity: An analysis of National Health and Nutrition Examination Survey 2011–2014 results. *Journal of the Academy of Nutrition and Dietetics, 119*(10), 1676–1686.

Levine, S. (2010). *School lunch politics: The surprising history of america's favorite welfare program (politics and society in modern America)*. Princeton, NJ: Princeton University Press.

Li, W., Youssef, G., Procter-Gray, E., Olendzki, B., Cornish, T., Hayes, R., … Magee, M. F. (2017). Racial differences in eating patterns and food purchasing behaviors among urban older women. *The Journal of Nutrition, Health & Aging, 21*(10), 1190–1199.

Lynas, M. (2018, May 25). *US FDA approves golden rice*. Cornell Alliance for Science. Retrieved September 27, 2019, from https://allianceforscience.cornell.edu/blog/2018/05/us-fda-approves-golden-rice/

Lytle, L., & Myers, A. (2017). *Measures registry user guide: Food environment.* National Collaborative on Childhood Obesity Research. Retrieved April 14, 2020, from http://nccororgms.wpengine.com/tools-mruserguides/wp-content/uploads/sites/2/2017/NCCOR_MR_User_Guide_Food_Environment-FINAL.pdf

Mack, E. A., Tong, D., & Credit, K. (2017). Gardening in the desert: A spatial optimization approach to locating gardens in rapidly expanding urban environments. *International Journal of Health Geographics, 16*(1), 37.

McElroy, K. R., Bibeau, D., Steckler, A., & Glanz, K. (1988). An ecological perspective on health promotion programs. *Health Education Quarterly, 15*(4), 351–377.

McGuire, J.S., Popkin, B., & Vittas, D. (1990). *Helping women improve nutrition in the developing world: Beating the zero sum game.* World Bank library. https://elibrary.worldbank.org/doi/pdf/10.1596/0-8213-1415-7

Mettler, S. (1998). *Dividing citizens: Gender and federalism in new deal public policy.* Ithaca, NY: Cornell University Press.

Meza, A., Altman, E., Martinez, S., & Leung, C. W. (2019). 'It's a feeling that one is not worth food': A qualitative study exploring the psychosocial experience and academic consequences of food insecurity among college students. *Journal of the Academy of Nutrition and Dietetics, 119*(10), 1713–1721.

Millichap, J. G. (1989). School breakfast program and school performance. *Pediatric Neurology Briefs, 3*(10), 80.

Moradi, S., Mirzababaei, A., Mohammadi, H., ParisaMoosavian, S., Arab, A., Jannat, B., & Mirzaei, K. (2019). Food insecurity and the risk of undernutrition complications in children and adolescents. *Nutrition, 62*, 52–60.

Mosier, S., & Thilmany, D. (2016). Diffusion of food policy in the U.S.: The case of organic certification. *Food Policy, 61*, 80–91.

National Collaborating Centre for Determinants of Health. *Glossary: Marginalized populations.* Retrieved November 8, 2019, from. http://nccdh.ca/glossary/entry/marginalized-populations

National Research Council (US). (2009). *The public health effects of food deserts: Workshop summary.* Washington, DC: National Academies Press (US). 1, Introduction. Available from: https://www.ncbi.nlm.nih.gov/books/NBK208016/

National WIC Association. (2019). *WIC program overview and history.* Retrieved November 8, 2019, from. https://www.nwica.org/overview-and-history

Nestle, M. (2013). *Eat drink vote.* New York, NY: Rodale.

Neuhouser, M. L., Thompson, B., Coronado, G. D., & Solomon, C. C. (2004). Higher fat intake and lower fruit and vegetables intakes are associated with greater acculturation among Mexicans living in Washington state. *Journal of the American Dietetic Association, 104*(1), 51–57.

Njiraini, J. (2015). *Microfinance: Good for the poor?* New York, NY: Africa Renewal. Retrieved November 8, 2019, from https://www.un.org/africarenewal/magazine/august-2015/microfinance-good-poor

O'Brien, D., Staley, E., Uchima, S., Thompson, E., & Torres Aldeen, H. (2004). *The charitable food assistance system: The sector's role in ending hunger in America.* The UPS Foundation and the Congressional Hunger Center 2004 hunger forum discussion paper. Retrieved October 28, 2019, from https://www.hungercenter.org/wp-content/uploads/2012/10/The-Charitable-Food-Assistance-System-Americas-Second-Harvest.pdf

Oddo, V. M., Krieger, J., Knox, M., Saelens, B. E., Chan, N., Pinero Walkinshaw, L., … Joens-Smith, J. C. (2019). Perceptions of the possible health and economic impacts of Seattle's sugary beverage tax. *BMC Public Health, 19*(910), 1–13.

Odoms-Young, A. (2018). Examining the impact of structural racism on food insecurity: Implications for addressing racial/ethnic disparities. *Family & Community Health, 41*(2), 1–5.

Ohri-Vachaspati, P., DeLia, D., DeWeese, R. S., Crespo, N. C., Todd, M., & Yedidia, M. J. (2015). The relative contribution of layers of the social ecological model to childhood obesity. *Public Health Nutrition, 18*(11), 2055–2066.

Olson, C. M., & Holben, D. H. (2002). Position of the American dietetic association: Domestic food and nutrition security. *Journal of the American Dietetic Association, 102*(12), 1840–1847.

Ostrom, E. (2007). A diagnostic approach for going beyond panaceas. *PNAS, 104*(39), 15181–15187.

Otter, C. (2010). Feast and famine: The global food crisis. *Origins: Current Events in Historical Perspective,* 3(6). Retrieved October 31, 2019, from https://origins.osu.edu/article/feast-and-famine-global-food-crisis

Pindus, N., & Hafford, C. (2019). Food security and access to healthy foods in Indian country: Learning from the food distribution program on Indian reservations. *Journal of Public Affairs, 19*(3), 1–8.

Popkin, B., Adair, L. S., & Ng, S. W. (2012). Now and then: The global nutrition transition—The pandemic of obesity in developing countries. *Nutrition Reviews, 70*(1), 3–21.

Popkin, S. J., Gilbert, B., Harrison, E., Arena, O., DuBois, N., & Waxman, E. (2019). *Research report: Evidence-based strategies to end childhood food inse-*

curity and hunger in Vermont. Urban Institute. Retrieved November 8, 2019, from https://www.urban.org/sites/default/files/publication/99831/evidence-based_strategies_to_end_childhood_food_insecurity_and_hunger_in_vt_1.pdf

Poppendieck, J. (1999). *Sweet charity: Emergency food and the end of entitlement*. London, UK: Penguin Books.

Rack, P. (2018). *Contributors to food waste in local restaurants and obstacles to food donation* (Honors thesis). Fort Worth, TX: Texas Christian University. Retrieved November 8, 2019, from https://repository.tcu.edu/bitstream/handle/116099117/22365/Rack__Petra-Honors_Project.pdf?sequence=1&isAllowed=y

Renner, R. (2019). *Atlanta's food forest will provide fresh fruit, nuts, and herbs to forage*. City Lab. Retrieved November 11, 2019, from https://www.citylab.com/environment/2019/06/urban-food-forest-local-agriculture-atlanta-fresh-produce/590869/

Reynolds, K. (2015). Disparity despite diversity: Social injustice in New York city's urban agriculture system. *Antipode, 47*(1), 240–259.

Robinson, T. (2008). Applying the socio-ecological model to improving fruit and vegetable intake among low-income African Americans. *Journal of Community Health, 33*, 395–406.

Rome NGO/CSO Forum for Food Sovereignty. (2002). *Food sovereignty: A right for all*. Retrieved November 9, 2019, from https://nyeleni.org/spip.php?article125

Roser, M., & Ritchie, H. (2020). *Hunger and undernourishment*. Published online at OurWorldInData.org. Retrieved from: https://ourworldindata.org/hunger-and-undernourishment

Rural Health Information Hub. (2019). *Rural hunger and access to healthy food*. Retrieved November 6, 2019, from https://www.ruralhealthinfo.org/topics/food-and-hunge

Sallis, J. F., Owen, N., & Fisher, E. B. (2008). Ecological models of health behavior. In Glanz et al. (Eds.), *Health behavior and health education*. San Francisco, CA: Wiley and Sons.

Satia, J. A. (2009). Diet-related disparities: Understanding the problem and accelerating solutions. *Journal of the American Dietetic Association, 109*(4), 610–615.

Schanes, K., Dobernig, K., & Gozet, B. (2018). Food waste matters – A systematic review of household food waste practices and their policy implications. *Journal of Cleaner Production, 182*, 978–991.

Schlosser, E. (2001). *Fast food nation: The dark side of the all-American meal*. New York, NY: Houghton Mifflin Company.

Scott, A., Ejikeme, C. S., Clottey, E. N., & Thomas, J. G. (2012). Obesity in sub-Saharan Africa: Development of an ecological theoretical framework. *Health Promotion International, 28*(1), 3–16.

Shipan, C. R., & Volden, C. (2008). The mechanisms of policy diffusion. *American Journal of Political Science, 52*(4), 840–857.

Smith, C., & Morton, L. W. (2009). Rural food deserts: Low-income perspectives on food access in Minnesota and Iowa. *The Journal of Nutrition Education and Behavior, 41*(3), 176–187.

Sowerwine, J., Mucioki, M., Sarna-Wojcicki, D., & Hillman, L. (2019). Reframing food security by and for native American communities: A case study among tribes in the Klamath River basin of Oregon and California. *Food Security, 11*, 579–607.

Taren, D. L., Clark, W., Chernesky, M., & Quirk, E. (1990). Weekly food servings and participation in social programs among low income families. *American Journal of Public Health, 80*(11), 1376–1378.

The Food Trust. (2012). *What we do: In corner stores.* Retrieved November 11, 2019, from http://thefoodtrust.org/what-we-do/corner-store

Tulane University School of Social Work. (2018). Food deserts in America (Infographic). *Tulane University School of Social Work Online Magazine.* Retrieved November 11, 2019, from. https://socialwork.tulane.edu/blog/food-deserts-in-america

U.K Agricultural Biodiversity Compendium. (2002). *FAO world food summit – 5 ½ years later.* Retrieved April 22, 2020, from ukabc.org/wfs5+.htm

United Nations. (2019). *United Nations: Global issues – Food.* Retrieved November 8, 2019, from https://www.un.org/en/sections/issues-depth/food/index.html

United Nations Food and Agriculture Organization. (2003). Chapter 2: Food security, concepts and measurement. In *Trade and food security: Conceptualizing the linkages expert consultation.* Rome, Italy. Retrieved October 10, 2019, from http://www.fao.org/3/y4671e/y4671e06.htm#TopOfPage

United Nations Food and Agriculture Organization. (2017). *Global Initiative on food loss and waste.* Retrieved November 8, 2019, from http://www.fao.org/3/a-i7657e.pdf

United Nations Food and Agriculture Organization. (2020). *Strategic planning.* Retrieved April 23, 2020, from http://www.fao.org/about/strategic-planning/en/

United Nations Food and Agriculture Organization Committee on World Food Security. (2017).*Global strategic framework for food security and nutrition.*

Retrieved April 15, 2020, from: http://www.fao.org/cfs/home/products/onlinegsf/5/en/

United Nations Zero Hunger Challenge. (2015). *United Nations zero hunger challenge: United for a sustainable world.* United Nations Food and Agriculture Organization. Retrieved November 8, 2019, from http://www.fao.org/un-expo/system/files_force/attaches/the_zero_hunger_challenge_-_united_for_a_sustainable_worlda5f3.pdf?download=1&download=1

United States Department of Agriculture (2017). *The national school lunch program.* Retrieved April 15, 2020, from https://fns-prod.azureedge.net/sites/default/files/resource-files/NSLPFactSheet.pdf

United States Department of Agriculture. (2019). *Food waste FAQs.* Retrieved November 8, 2019, from https://www.usda.gov/foodwaste/faqs

United States Department of Agriculture Economic Research Service. *Household food insecurity in the US in 2018.* Retrieved October 28, from https://www.ers.usda.gov/publications/pub-details/?pubid=94848

United States Department of Agriculture Economic Research Service. (2019a). *Who are the world's food insecure? Identifying the risk factors of food insecurity around the world.* Retrieved October 28, 2019, from https://www.ers.usda.gov/amber-waves/2019/june/who-are-the-world-s-food-insecure-identifying-the-risk-factors-of-food-insecurity-around-the-world/

United States Department of Agriculture Economic Research Service. (2019b). *Survey tools.* Retrieved October 28, 2019, from https://www.ers.usda.gov/topics/food-nutrition-assistance/food-security-in-the-us/survey-tools/

United States Department of Agriculture Economic Research Service. (2019c). *Food environment atlas.* Retrieve April 14, 2020, from https://www.ers.usda.gov/data-products/food-environment-atlas/go-to-the-atlas/

United States Department of Agriculture Food and Nutrition Research Service. (2019a). *National school lunch program: Feeding the future with healthy school lunches.* Retrieved November 8, 2019, from https://www.fns.usda.gov/nslp

United States Department of Agriculture Food and Nutrition Research Service. (2019b). *School breakfast program: Program history.* Retrieved November 8, 2019, from https://www.fns.usda.gov/sbp/program-history

United States Department of Veterans Affairs. (2020). *Center for health equity research and promotion.* Retrieved April 23, 2020, from https://www.cherp.research.va.gov/

United States Food Bank Network. (2019). *Feeding America: Our history.* Retrieved November 8, 2019, from https://www.feedingamerica.org/about-us/our-history

Victora, C. G., Bahl, R., Barros, A. J., França, G. V., Horton, S., Krasevec, J., … Group, T. L. B. S. (2016). Breastfeeding in the 21st century: Epidemiology, mechanisms, and lifelong effect. *The Lancet, 387*(10017), 475–490.

Walker, R. E., Keane, C. R., & Burke, J. G. (2010). Disparities and access to healthy food in the United States: A review of food deserts literature. *Health & Place, 16*(5), 876–884.

Wang, D. D., Leung, C. W., Li, Y., Ding, E. L., Chiuve, S. E., Hu, F. B., & Willett, W. C. (2014). Trends in dietary quality among adults in the United States, 1999 Through 2010. *JAMA Internal Medicine, 174*(10), 1587–1595.

Wang, Y., & Chen, X. (2011). How much of racial/ethnic disparities in dietary intakes, exercise, and weight status can be explained by nutrition and health-related psychosocial factors and socioeconomic status among US adults? *Journal of the American Dietetic Association, 111*(12), 1904–1911.

Wesseler, J., & Zilberman, D. (2014). The economic power of the Golden Rice opposition. *Environment and Development Economics, 19*, 724–742.

Willett, W., Rockström, J., Loken, B., Springmann, M., Lang, T., Vermeulen, S., … Jonell, M. (2019). Food in the Anthropocene: The EAT–lancet commission on healthy diets from sustainable food systems. *The Lancet, 393*(10170), 447–492.

Windfuhr, M. (1997). *NGOs and the right to adequate food.* Food First Information and Action Network (FIAN) International. Retrieved April 17, 2020, from http://www.fao.org/3/w9990e/w9990e04.htm

World Food Program USA. (2020). *Hunger and food waste.* Retrieved April 18, 2020, from https://www.wfpusa.org and https://www.wfpusa.org/stories/8-facts-to-know-about-food-waste-and-hunger/

World Food Programme. (2015). *Indonesia-food security and vulnerability atlas 2015.* Retrieved April 19, 2020, from https://www.wfp.org/publications/indonesia-food-security-and-vulnerability-atlas-2015

World Food Programme. (2020a). *Overview.* Accessed April 23, 2020, from https://www.wfp.org/overview

World Food Programme. (2020b). *Supply chain.* Accessed April 23, 2020, from https://www.wfp.org/supply-chain

World Health Organization. (2016). *What is malnutrition?* Retrieved April 22, 2020, from https://www.who.int/features/qa/malnutrition/en/

World Health Organization. (2019). *Programmers and projects: Availability and changes in consumption of animal products.* Retrieved November 8, 2019, from https://www.who.int/nutrition/topics/3_foodconsumption/en/index4.html

Yang, Y., Buys, D. R., Judd, S. E., Bower, G. A., & Locher, J. L. (2013). Favorite foods of older adults living in the black belt of the United States. *Appetite, 63*, 18–23.

Zhang, F. F., Liu, J., Rehm, C. D., Wilde, P., Mande, J. R., & Mozaffarian, D. (2018). Trends and disparities in diet quality among US adults by supplemental nutrition assistance program status. *JAMA, 1*(2), e180237.

Zwerdling, D. (2009a). 'Green revolution' trapping India's farmers in debt. *National Public Radio*, Morning edition. Retrieved November 5, 2019, from https://www.npr.org/2009/04/14/102944731/green-revolution-trapping-indias-farmers-in-debt

Zwerdling D. (2009b). In India, bucking the 'revolution' by going organic. *National Public Radio*, Morning edition. Retrieved November 5, 2019, from https://www.npr.org/templates/story/story.php?storyId=104708731

6

Community Substance Use Safety

Justin R. Watts, Bradley McDaniels, Solymar Rivera-Torres, Danielle Resiak, Elias Mpofu, and Sonia Redwine

Introduction

Substance use is typical of communities across the globe, and mostly for health promotion. Ordinarily, most community members use substances in such a way that does not result in substance use disorders (SUDs) and would not negatively impact the health, social, or vocational outcomes of community members (National Institute of Health [NIH], 2015; Substance Abuse and Mental Health Services Administration [SAMHSA], 2017). Substance use disorders are defined as a chronic disease in which individuals continue to engage in substance use despite severe life

J. R. Watts (✉) • B. McDaniels • S. Rivera-Torres • S. Redwine
University of North Texas, Denton, TX, USA
e-mail: Justin.Watts@unt.edu; Bradley.McDaniels@unt.edu; solymarrivera@my.unt.edu; Sonia.redwine@unt.edu

D. Resiak
University of Sydney, Sydney, NSW, Australia
e-mail: danielle.resiak@sydney.edu.au

E. Mpofu (ed.), *Sustainable Community Health*,
https://doi.org/10.1007/978-3-030-59687-3_6

consequences in several life domains—social, vocational, and/or health (APA, 2013; WHO, 2018). They involve using substances in ways that increase the risk of harm to the user and those around them (e.g., strained relationships, health issues, poor functioning at work, driving while intoxicated) (SAMHSA, 2017). Substance use disorders have received a great deal of consideration in the literature, overshadowing strategies to enhance safer substance use within communities for sustainable community health.

Learning Objectives

By the end of this chapter, the reader should be able to:

1. Define the concept of community safety for sustainable health.
2. Outline the history of research and practice related to community substance use safety and substance use health promotion.
3. Discuss current and prospective approaches to community substance use health and wellbeing.
4. Consider the cultural, legislative, and professional issues that impact sustainable community-oriented approaches addressing substance use and promoting healthy behaviors related to substances.
5. Identify multidisciplinary and interdisciplinary evidence-based practices for substance use literacy and control for community health.
6. Critically examine issues for research and other forms of scholarship on community substance use safety important for long-term community health.

Some communities have safer levels of substance use than others, despite variations in substance types, and purposes of typical use. Communities with substance use safety are those which demonstrate low risk consumption rates and are protective of community member's

E. Mpofu
University of North Texas, Denton, TX, USA

University of Sydney, Sydney, NSW, Australia

University of Johannesburg, Johannesburg, South Africa
e-mail: Elias.Mpofu@unt.edu

health and functioning in relation to substance use. The World Health Organization (WHO, 1998) proposed a Safe Communities model by which the wellbeing of community members is a shared collective responsibility of those who make up the community, and those who collaboratively identify and mobilize resources for the collective good (see also Svanström, 2012). To date, the Safe Communities model has been implemented in over 150 communities around the globe in the context of injury prevention (Rahim, 2005; Svanström, 2012); but these models have not been implemented to address safer substance use despite their potential. The continued lack of investigation and adoption of safe community modeling is a missed opportunity for the global community which is presently preoccupied with programs to mitigate SUDs. This chapter aims to address an apparent gap in the literature on community substance use safety research and practice.

To place into context the significance of substance use safety to community health, we note that the WHO (2004) reported 3–16% of the world population met the criteria for a SUD including alcohol use disorders. As an example, the National Survey on Drug Use and Health estimated 8.4% of Americans (20.2 million adults) meet the criteria for SUDs (both alcohol and substance use) (SAMHSA, 2017). Although most societies and communities do not demonstrate remarkable rates of substance misuse/abuse or SUDs, the social and economic toll is considerable (SAMHSA, 2017). For instance, substance misuse/abuse in the United States is a public health concern, resulting in an annual economic burden of $740 billion when considering costs related to criminal offenses, lost productivity, and health-related issues (National Institute on Drug Abuse [NIDA], 2017). Substance use disorders are distinctive in their social, economic, and environmental impacts, and their mitigation requires the involvement of many professions and disciplines (e.g., education, environmental health, foreign affairs, health financing, law enforcement, medicine, mental health, public health). In fact, the United Nations SDG goals include the need for interventions to treat addictive disorders (United Nations Development Program [UNDP], 2014). Regrettably, the historical focus on SUDs has also had the unintended effect of encouraging stigmatization by the broader society of individuals with SUDs, who are often inaccurately perceived as compromised in morals and dangerous

(Tu, Yan, Li, & Watts, 2019). Understandably, federal, state, and international agencies are invested in managing SUDs; and we assume the view that the design and implementation of community substance use safety interventions would be cost effective and sustainable long term.

Some communities have lower rates of SUDs and their SUD preventative assets for community substance use safety are understudied. Adding to the complexity of the scholarship on substance use safety is the fact that there is divergence in how communities define acceptable and problematic use, as well as risk and protective factors associated with SUDs (Gureje, Vazquez-Barquero, & Janca, 1996; Hawkins, Van Horn, & Arthur, 2004). Identifying characteristics of safer substance use practices within communities is essential to sustaining their health systems, preventing substance misuse/abuse, and fostering healthier community relationships with substances. Community substance use safety programs are by design a sustainable health system practice. Sustainable health systems address the social, economic, and environmental factors that serve as supports or barriers toward population health (Fineberg, 2012). These systems should be offered at "sufficient intensity for the sustained achievement of desirable program goals and population outcomes" (Shelton, Cooper, & Stirman, 2018, p. 56). In other words, sustainability implies the continuation of practices long after the implementation phase (Bond et al., 2014). Safe community approaches are effective in injury prevention (Spinks, Turner, Nixon, & McClure, 2005; Svanström, 2012) and hold great promise for community substance use safety.

Professional and Legal Definitions and Theories Relating to Substance Use

According to the WHO (1998), safe communities are those characterized by the presence of infrastructure, systems, and programs for collaboration with cross-sectorial governance structures for overall wellbeing. They also include long-term, sustainable programs that are responsive to the needs of community members across the life span, life situations, and events. We translate this definition to imply that communities that engage in safer substance use practices have sustainable promotion systems and practices for healthier consumption of substances, which would aim to prevent SUDs by identifying and utilizing community assets for substance use literacy. Substance use literacy refers to knowledge and practices for safer use of substances (Dermota et al., 2013).

According to the United States Department of Health and Human Services (USDHHS, 1999) and American Psychiatric Association (APA, 2013), substance use occurs on a spectrum beginning with experimentation with substances and ending with SUDs. Most individuals fall somewhere in between these two extremes, as they engage more frequently in social or non-problematic substance use that results in negligible health or social consequences. Alcohol consumption that occurs within established health guidelines is one such example in which some communities have lower risk for harmful consumption than others (Stockwell, Heale, Chikritzhs, Dietze, & Catalano, 2002). Similarly, communities vary widely in their (mis)use of prescription pain medications, which would impact the long-term health of community members (Kanouse & Compton, 2015; Shepherd, 2014). It is important to note that *how* a society views the nature of a problem influences solutions for the problem. Accordingly, public opinions largely influence perceptions of what constitutes substance misuse/abuse (Henninger & Sung, 2014). For instance, within population segments, how alcohol or substance usage is perceived is largely dependent upon the environment in which the use occurs (Bellis, Hughes, & Lowey, 2002). Indeed, recreational use of alcohol and other substances is considered normal in some communities, although it may be criminalized in environments with stricter legal guidelines (Nicholson, Duncan, & White, 2002; Stockwell et al., 2002). Unfortunately, recreational substance use may include the misuse of prescription medications (Benotsch, Koester, Luckman, Martin, & Cejka, 2011), which would be harmful in all cases.

Different substances carry varying degrees of risk that largely depend on the substance type, amount used, purity, method of procurement, and any preexisting health issues an individual may have. The use of non-prescription substances such as alcohol taken in moderation may carry the potential to benefit health in lowering the risk for cardiovascular disease for some, but not others (Durrant & Thakker, 2003). Thus, safe substance use within communities is complex in nature and influenced by individual and community-wide beliefs regarding the nature and acceptability of substance use, the environmental context, and types of substances that may be available in communities.

In addition to the dynamic nature of community substance use safety norms, the policies governing substance use/abuse in societies also change over time. For instance, in the early twentieth century, supporters of the Prohibition Movement sought to ban alcohol use largely for religious reasons in an attempt to control public alcohol consumption (Blocker, 2006). This movement initially reduced the rates of consumption, but they resulted in high levels of organized crime. Essentially, alcohol supply was perceived as *the problem*; therefore, the solution involved terminating the supply. This movement eventually gave rise to the War on Drugs in the 1970s, in which steeper criminal penalties were encouraged for substance-related crimes. In this case, a moral understanding (i.e., substance use, abuse/misuse, being perceived as a character flaw, weakness or sin) of substance use encouraged substance-using behaviors to be perceived as *the problem*. As paradigms shift in the twenty-first century, research continues to advance our understanding of substance misuse/abuse. Current research has explored the biological, psychological, and social origins of substance misuse/abuse and SUDs in order to provide more effective treatment (Henninger & Sung, 2014). Likewise, predispositions (e.g., mental health issues, previous trauma, adverse childhood experiences) influence individuals' substance abuse/misuse (Ghodse, Herrman, Maj, & Sartorius, 2011; Waite & Ryan, 2019). Substance use disorders are now widely recognized as a disease and largely a public health challenge as opposed to a judicial issue (WHO, 2009). (See Discussion Box 6.1).

Discussion Box 6.1: Are Substance Use Disorders a Disease?

There has been some debate as to whether SUDs meet the criteria of a disease. It can be difficult to distinguish between what constitutes a disease, from behaviors or characteristics that a society might merely find disturbing (Scully, 2004). The National Institute on Drug Abuse (2018) defines SUDs as a chronic health condition or disease characterized by unremitting substance use resulting in considerable changes to the brain. Chronic substance use interrupts the normal functioning of the brain, and, if not addressed through some form of treatment, can be life-threatening. Areas of the brain that involve the experience of reward, stress, and self-control are significantly affected. Moreover, these changes in the brain explain continued drug use despite serious consequences because areas in the brain that would usually recognize the risks associated with these detrimental behaviors are damaged and under-functioning.

Historically, substance-using behaviors have been perceived as marginalized behaviors in society, as many view these behaviors as a "choice" or "moral issue." A common reason why many individuals begin to develop a SUD is that they use substances to *feel better* (SAMHSA, 2017); in these cases, individuals are usually coping with an underlying mental health issue or facing significant distress (often due to trauma). Consistent with the Self-Medication Model, individuals may act as their "own pharmacist" in using substances to mitigate their symptoms (Khantzian, 1999). Research has shown that individuals who experience child maltreatment, trauma, disability, or have other mental health conditions are at significantly greater risk for developing SUDs because these conditions are associated with intense psychological, emotional, and at times physical pain. Externally, it might look like individuals are *choosing* to use a substance despite severe life consequences; however, substance use in these cases is an attempt to alleviate pain for individuals who may have relatively few coping strategies (e.g., tools to regulate emotions, social support etc.) as they may not have otherwise learned due to impoverished relational environments throughout development. Those who have child-maltreatment histories also use substances earlier, which has a more profound impact on brain-development and further incites the risk for continued use (Felitti et al., 1998). As individuals begin to rely more heavily on substances, they may begin to develop physical and/or psychological dependence on a substance (typically after long term, chronic use). Many individuals with SUDs also continue to use substances to avoid uncomfortable withdrawal symptoms, and, in addition to the significant changes to the brain that result in continued use, substance use is compounded (SAMHSA, 2017). In cases of severe SUDs, the brain's reward system is overridden, persuading the individual that substances are required for survival. This reward system functions much faster than the areas of the brain responsible for reasoning, planning, and decision-making (Inaba & Cohen, 2011), causing negative emotional states (e.g., distress) rather than objective reasoning to determine whether someone will continue substance use.

What Do You Think?

1. What is your opinion of defining SUDs as a disease?
2. Why would some individuals be conflicted about defining SUDs as a disease?
3. What would be supports that communities could provide in order to prevent the development and maintenance of SUDs and promote recovery of SUDs?

History of Research and Practice Related to Community Substance Use

There is a long history of hallucinogenic botanical substance use by communities across the globe (Parsche, Balabanova, & Pirsig, 1993). While the specific types of drugs used in ancient cultures are not well documented, there is evidence to suggest some substances could be likened to those available in contemporary societies (Baez et al., 2000). In modern times, substance use studies and practices have predominantly focused on identifying and classifying substances hierarchically by their habit-forming potential, as well as their degree of risk and acceptance for medical use in community samples (Durrant & Thakker, 2003), in addition to their capacity for performance enhancement in competitive sporting (Dimeo, 2008).

According to Durrant & Thakker (2003), over the past 100 years, classification schemes for mental disorders have shown substantial variation, with diagnostic criteria being removed, added, or significantly revised. Such variability cross-culturally throughout history is indicative of current concepts of substance abuse and dependency being a byproduct of cultural–historical contingencies (Durrant & Thakker, 2003). Suggesting, as the way in which substance use disorders are conceptualized changes, so too does the way they are addressed within the community. Humane social attitudes and environments responsive to public health are seen as preventative of disease (Mann, 1984). Services for people with a SUD are largely shaped by a country or region's philosophical approach to drug treatment (zero tolerance or harm minimization) (Resiak, Mpofu, & Athanasou, 2016). The later, places emphasis on public health, steering away from individualistic theories and pathology, and placing greater attention on the social and environmental influence on behavior (Groves, 2018). Reviewing the historical development of concepts of health, while paying particular attention to the development of problematic substance use can highlight how current concepts have been informed, and through an exploration of cross-cultural differences in the understanding of substance-related problems, innovative modes of practice can be constructed (Durrant & Thakker 2003).

In recent years, there has been a revitalization in social attitudes and environments for managing community health (Mann, 1984), prioritizing community-friendly substance consumption practices or those to minimize harm from substance use consumption (Groves, 2018; Resiak et al., 2016; see also "pertinent sustainable community substance use health-oriented approaches section below"). These approaches have been backed by initiatives that emphasize collaborative partnerships between law enforcement, health, and other support agencies (Hughes & Hughes, 2007). Examples of such collaborations include Needle and Syringe Programs (NSPs), Opioid Substitution Therapies, and Medically Supervised Injecting Centres (MSIC) aimed at minimizing the risk of harm and potential transmission of blood-borne viruses associated with injecting drug use (see Bull, Denham, Trevaskes, & Coomber, 2016; North Richmond Community Health Wulempuri-Kertheba, 2017). In Australia, law enforcement is encouraged to work collaboratively with community members, and to develop positive relationships with local NSPs which in turn can lead to positive outcomes for people who inject drugs and the wider community alike (Bull et al., 2016). Many communities are trending toward harm minimization approaches, some also advocating decriminalization to avoid minor possession charges, and better aligning to current community attitudes toward substance use within a wider political context more accepting of diversity in substance use patterns (Groves, 2018).

Pertinent Sustainable Community Substance Use Health-Oriented Approaches

Community substance use safety-oriented approaches seek to develop both individual and community capacities while giving adequate attention to expanding public policy *and* community action, and thus promoting mental health wellbeing (see also Chap. 7, this volume). As previously noted, safe community approaches prioritize engagement of key stakeholders and the community members themselves (WHO, 1998). In this section, we discuss evidence-based safer community substance use practices including, substance use education and literacy, harm reduction, zero tolerance or judicial policies, prevention, caring communities, family strengthening, case management, and integrated services.

Substance Use Literacy and Education One of the most promising yet understudied constructs relating to the prevention of substance misuse/

abuse is health literacy. Substance use literacy has been championed in the general population as essential for both the attainment and maintenance of adequate health; but among individuals with problematic substance use, limited research is available (Degan, Kelly, Robinson, & Deane, 2019). Health literacy refers to one's ability to actively engage in health-related activities designed to promote adequate health (WHO, 2009). Furthermore, health literacy encompasses one's ability to access, comprehend, and communicate one's health needs with healthcare practitioners (Degan et al., 2019). The existing literature has clearly established that health literacy enhances health-related quality of life in sustainable ways (Osborn, Paasche-Orlow, Bailey, & Wolf, 2011; Rowlands et al., 2015). The value of substance use literacy for sustainable community health is illustrated by the fact that school-based drug prevention programs have emphasized abstinence based knowledge about harms associated with habit-forming substances (Teesson, Newton, & Barrett, 2012). Examples of substance use and health literacy programs for school students include Climate Schools Online Prevention (Climate Schools, 2020) and Positive Choices (Positive Choices, 2020), which provide resources for teachers and parents designed to encourage informed decision-making.

In Australia, school-based prevention programs have proved successful in reducing substance use disorders among school learners (Teesson et al., 2012). Drug education and health literacy are incorporated into the teaching and learning curriculum in Australian schools, as it's considered to play a pivotal role in shaping a normative culture of safety, moderation, and informed decision-making (Education and Training, 2020). Such initiatives aim to intercept early initiation of drug use to reduce the risk of development of substance use disorders, comorbid mental health conditions, juvenile offending, impaired educational performance, and education early departure; all of which have a direct impact on an individual's current and future functioning (Teesson et al., 2012).

Harm Reduction Approaches Many countries (e.g., Portugal, Germany, the Netherlands, and Australia) have implemented harm reduction approaches to contend with substance misuse/abuse. In such settings, substance misuse/abuse is perceived as a public health issue, recognizing that for some, abstinence is not possible nor desired (Tsui, 2000). Rather than use of punitive measures on those who consume substances beyond safety guidelines, harm minimization promotes health and welfare support, and services are implemented to reduce the overall risks of harm to the individual and the wider community (Single, 1995). The focus of harm reduction is to reduce the overall effects of substance use from a holistic health perspective inclusive of economic and social issues, as opposed to a narrow focus that perceives drug consumption as a primary outcome measure (Canadian Paediatric Society, 2008). Largely, harm reduction approaches have evidence for promoting safer substance use consumption, improving overall health outcomes of individuals who use substances and have been found to be cost-efficient (Wilson, Donald, Shattock, Wilson, & Hurt, 2015). Harm reduction approaches also have evidence to reduce rates of opioid overdose and blood-borne virus transmission while improving access to health and welfare services (Dolan, Kimber, Fry, & Fitzgerald, 2000). See Discussion Box 6.2 for an illustration of a harm reduction program.

Community-based harm reduction efforts involve services, actions, policies, and programs that are intended to reduce substance-related harm to both the individual and the community-at-large (WHO, 2020). Examples of harm reduction models include: (a) needle and syringe exchange programs to reduce the risks associated with intravenous drug use (HIV, Hepatitis B and C transmission), (b) providing education to communities regarding the risks associated with excessive alcohol use or the hazards associated with simultaneous polysubstance use, (c) opioid substitution therapies, and (d) medically supervised injecting centers/ drug consumption rooms.

Sharing used needles and syringes poses a significant risk of blood-borne virus transmission among individuals who engaged in intravenous substance use. Despite the growing number of needle and syringe programs (NSPs) globally, stigmatizing beliefs, fear of being exposed, and limited access to sterile injecting equipment are reported barriers to safer injecting practices (Islam, Stern, Conigrave, & Wodak, 2008). In Australia, the provision of evidence-based NSPs was the foundation response in 1986 (NSW Government Health, 2017) to reducing the transmission of HIV, and subsequently Hepatitis B and C infections, among people who inject drugs (Australian Government Department of

Discussion Box 6.2: The Opioid Epidemic in the United States, Using Harm Reduction to Reduce Fatal Opioid Overdose

The "Opioid Epidemic" has been a major public health crisis in the United States for several decades. According to the U.S. Department of Health and Human Services (2019), 47,600 individuals died from opioid overdose in 2016, which equates to more than 130 deaths per day. In the late 1990s, pharmaceutical companies misrepresented the addictive potential of opioid pain relievers causing prescriptions to increase substantially. In 2012, there were 259 million prescriptions written for opioid medications, enough for each American adult to have their own prescription (American Society of Addiction Medicine, 2016). Most (63.4%) of those who reported misusing prescriptions did so to relieve physical pain, 11.7% reported using the substances to "feel good" or "get high," 14.1% reporting misusing prescriptions to "relax" or to "help with negative feelings or emotions," and 4.5% were using them to "fall asleep" (SAMHSA, 2017). As the epidemic progressed, four out of five new heroin users started out using prescription opioids and transitioned to heroin because many (94%) reported that getting prescription opioids was far more difficult and costly. This led to the prevalent misuse of prescription and non-prescription opioids, causing the U.S. Department of Health and Human Services to be forced to declare the epidemic a public health emergency in 2017. This forced communities, legislators, and physicians to begin addressing the issue on a larger scale by expanding access to treatment and recovery services for those who needed it, promoting the use of overdose-reversing drugs, enhancing public health surveillance to better understand the epidemic, providing supports for research on pain management and substance use disorders, and to continue to advance best practices for pain management. In communities globally, trained first responders can administer Naloxone (Narcan/Evzio) in the event of an opiate overdose. Naloxone is an oxymorphone derivative used to reverse the resulting respiratory depression, sedation, and euphoria of an opioid overdose (Shorthouse, 2017 #2851). It works as a competitive antagonism at all opioid receptor sites and is administered via intravenous, intramuscular, or subcutaneous methods. While Naloxone is commonly administered as a single dose, in rare cases continuous infusion may be required due to its short duration of action. For individuals with an opioid use disorder, side effects may include acute withdrawal, drowsiness at high doses, and arrhythmias or hypertension in predisposed patients (Shorthouse, 2017 #2851). These medications can provide an opportunity for those who overdose to have another possibility to find support to pursue recovery.

What Do You Think?

1. What would the prospects and limits of health literacy be in mitigating prescription drug (mis)use?
2. What does the transition from licit to illicit use of medications look like in a community you are familiar with?
3. What benefits and potential risks do you perceive from the use of Naloxone to reverse opioid use overdose in a community you are familiar with?

Health and Ageing, 2010; Commonwealth of Australia, 2010) as cited in (Treloar, McLeod, Yates, & Mao, 2014).

Australian NSPs are publicly funded, low-threshold services where individuals have access to sterile injecting equipment through distinct channels namely: (a) primary outlets (stand-alone), (b) secondary outlets (within an existing service, such as sexual health center, community health center, or hospital emergency department), and (c) vending machines and outreach programs (Treloar et al., 2014). The primary objective of an NSP is to minimize the risk of transmission of blood-borne viruses associated with injecting drug use (NSW Government Health, 2017). As described in Islam et al. (2013), service provision is non-judgmental and largely anonymous with few exceptions (e.g., referral to external service provider). Staff employed in these settings commonly have extensive experience working with people who inject drugs (PWID) and a sound knowledge of associated medical and psychiatric co-morbidities and risk exposures (Islam et al., 2013).

While a diverse client base access NSPs each year, guidelines for service provision in Australia call for particular attention to be paid to those identified as highly vulnerable (e.g., the homeless, sex workers, those who are HIV positive, young at-risk injectors, Aboriginal and Torres Strait Islander People, and those from Culturally and Linguistically Diverse communities; NSW Government Health, 2017). Rich scholarship exists pertaining to the broader implications of safer injecting practice, which include: policing practices affecting NSP access, and a combination of the great costs of illicit substances and poverty resulting in sharing resources to purchase drugs (Treloar et al., 2014).

Zero-Tolerance or Judicial Policy Approaches In many cases, state and federal governments who adopt a judicial policy approach attempt to deter substance-using behavior with more severe societal consequences (e.g., incarceration for minor drug crimes) (Heitzeg, 2009). Examples of zero-tolerance approaches include the establishment of a legal drinking age to prevent minors from using alcohol; penalties (e.g., loss of license for driving under the influence of alcohol or other drugs) for high-risk substance-related behaviors; imprisonment for the supply, transport, or distribution of illicit substances; and the establishment of dry communities (i.e., communities in which no alcohol is supplied or consumed), the latter approach could also be considered a harm reduction approach.

Judicial policy interventions focus on punishment and control more than health and empathy (Arnold, 2016). For example, in the United States, implementation of judicial policies has resulted in large racial disparities causing communities of color to be disproportionately more likely to face criminal penalties relating to substance use than Whites (Mitchell & Caudy, 2015). Unfortunately, research has shown that incarceration is not always the most effective means of deterring problematic use. Many who are detained for minor illicit substance possession are, while in prison, exposed to habitual and harmful substances and inconsistently receive treatment of any type, which results in continued use and recidivism (Chandler, Fletcher, & Volkow, 2009; Hall et al., 2012; WHO, 2009). The WHO (2009) has emphasized that substance misuse/abuse and SUDs should be treated as a health issue as opposed to a judicial issue, suggesting the viability of harm reduction approaches.

Prevention-Orientated Approaches Prevention activities are intended to support and educate communities to circumvent the use and misuse of drugs and the development of SUDs (SAMHSA, 2020). Examples of prevention efforts within communities might include (a) providing parents or caregivers with instrumental support and education to discuss substance use with their children, or (b) providing educational information to community members regarding the risks of binge drinking or underage alcohol use. In addition to these educational campaigns, some prevention efforts might focus on recognizing risk factors for problematic or disordered substance use and providing instrumental support (e.g., counseling services, psychoeducation, family supports) to address some of the underlying etiology. (See Research Box 6.1 for description of child maltreatment, which is a significant underlying factor that is linked to later substance-related issues).

One example of a prevention and health promotion program would include Life Skills Training (LST), which was developed as a classroom-based universal prevention and health promotion program for middle school students designed to educate adolescents about the risks associated with alcohol, tobacco, and cannabis in an attempt to mitigate use (Botvin, Baker, Renick, Filazzola, & Botvin, 1984). Life Skills Training was

Research Box 6.1 Adverse Child Experiences, a Risk Factor for Problematic Substance Use (Felitti et al., 1998)

Background. Felitti et al. (1998) conducted a groundbreaking and innovative study that was one of the first to examine the impact of early adversity (e.g., child-maltreatment, child abuse, or neglect) on health outcomes later in adulthood. This study exposed the relationship between adverse childhood experiences and chronic illness and disability later in life, while examining other risky health behaviors associated with child adversity (i.e., substance use) that also impact functioning.

Method. Participants in this study were $N = 13{,}494$ adults who completed a medical evaluation at a large Health Maintenance Organization. Participants were also given a questionnaire focused on adverse childhood experiences, which included seven categories of abuse that occurred before the age of 18 (sexual, physical, psychological, witnessing domestic violence, or living with caregivers who engaged in problematic substance use, had issues with mental illness or suicide ideation, or were ever in prison).

Key findings. Participants who experienced four or more adversities in childhood had a significant increase in risk for problematic alcohol use or drug use. Using the existing data to further clarify the link between early adversities and substance-using behaviors, Dube, Anda, Felitti, Edwards, and Croft (2002) established that four or more adversities were associated with a four-fold risk of alcohol use disorder, even when adjusting for maternal and paternal alcohol use. Researchers (Dube, Felitti, Dong, Chapman, & Giles, 2003) found that each adverse experience resulted in a two to four-fold increase in the probability of early substance use, while participants with five or more adversities were seven to ten times more likely to report problematic illicit drug use or substance use disorder.

Conclusion and implications: Adverse childhood experiences are a significant risk factor for chronic disease and disability later in life, in addition to placing individuals at significant risk for engaging in problematic substance use early in development. As the severity of adversity increases, so does the risk of problematic substance use.

What Do You Think?

1. How might you explain the nature of the relationship between adverse childhood experiences and substance-using behaviors later in development? Would you believe that early adversity would result in substance use later in life regardless of community living situation? If yes, why and how? If no, why and how?
2. If you were to develop a prevention program for children to reduce problematic substance use, keeping in mind the relationship between early adversity and early initiation of substance use, what types of programming might be beneficial at the individual, family, and community level?

designed to both decrease risk factors and foster protective factors at the individual, peer, and community level. The intervention utilizes a cognitive-behavioral and social learning approach to engage students to develop general social skills (e.g., effective communication, conflict resolution, assertiveness), personal self-management skills (e.g., coping with anxiety and anger, decision-making), and drug resistance skills (e.g., understanding consequences of choices, coping with media and peer pressure; Botvin et al., 1984). Several longitudinal studies have been conducted and have reported reductions in cigarette smoking (Spoth, Randall, Trudeau, Shin, & Redmond, 2008), binge drinking (Botvin, Griffin, Diaz, & Ifill-Williams, 2001a, 2001b), cannabis use (Botvin, Baker, Dusenbury, Tortu, & Botvin, 1990), and illicit drug use (Botvin et al., 2000; Crowley, Jones, Coffman, & Greenberg, 2014; Spoth, Clair, Shin, & Redmond, 2006). A significant aspect of community-wide prevention is understanding the available community assets for preventing disordered substance use in the first place, while working to find solutions for community members to address underlying issues that might increase the risk of using substances as a primary, albeit maladaptive, coping mechanism.

Caring Community Approaches Community-wide problems like SUDs require community-level solutions in addition to individual interventions (Aguirre-Molina & Gorman, 1996). Health-related professionals and researchers have sought to articulate novel ideas and new information leading to the advancement of successful programs, resulting in greater effectiveness in preventing problematic substance use and addressing substance misuse/abuse. Across all services, efforts aimed at developing individual and community capacities, giving adequate attention to beneficial public policy, and sustainable community actions have emerged (Stirman et al., 2012). The new initiatives have generally moved from prevention and risk reduction to having a larger focus on community-based treatment models that focus on inclusion and recovery. Those services fall under one or more of the categories of (a) prevention, (b) intervention, and (c) treatment and recovery support.

Communities that Care (CTC; Hawkins & Catalano Jr, 1992) is a community-based prevention system that is implemented in early adolescence to promote healthy development while reducing antisocial behavior. CTC is based on the theories of public health promotion and community competence (Butterfoss, Goodman, & Wandersman, 1993), which target the specific needs of disparate communities to decrease risk factors and increase protective factors. This may be accomplished by educating youth, who often lack the ability and information to make informed decisions about unhealthy behaviors and the risks of initiating substance use. Research findings suggest that CTC is an effective and sustainable program designed to reduce health-risk behaviors that support critical public health objectives (Oesterle et al., 2018). CTC is designed to increase the adoption of evidence-based prevention, increase the collaboration between service providers, and utilize the program to address high-risk behaviors as prioritized in the community. CTC has demonstrated effectiveness at significantly reducing problematic alcohol, tobacco, cannabis, and illicit drug use through age 21 (Hawkins et al., 2008).

Approaches to Strengthen Families According to Griffin and Botvin (2010), prevention programs focused on youth and families typically provide educational materials that provide skills training for youth in school settings or provide instruction for parents to effectively monitor and converse with children regarding substance use. Other outlets for prevention might also involve community programs that address these issues, while also utilizing media outlets (e.g., providing media campaigns to challenge or correct assumptions of normal substance use) or addressing these issues through public policy (e.g., restricting purchase of substances like nicotine or alcohol by setting the minimum age to purchase). Community-wide prevention programming typically revolves around the most commonly used substances (e.g., alcohol, nicotine, and cannabis), largely because these substances pose the greatest risk to public health due to widespread use.

The Strengthening Families Program 10–14 (SFP; Kumpfer, Molgaard, & Spoth, 1996) is an evidence-based program designed to teach skills to improve outcomes for youth in high-risk families. The program teaches parents how to create boundaries and rules for their youth by utilizing community resources to protect against substance abuse and related challenging behaviors. Contained within the SFP curriculum are topics related to understanding the consequences of problematic behavior, handling stress, and dealing with peer pressure. A number of studies and meta-analyses have provided strong evidence to support the effectiveness of the SFP in reducing problematic alcohol use and illicit drug use, while promoting psychosocial skill development (Foxcroft, Ireland, Lister-Sharp, Lowe, & Breen, 2003; Foxcroft, Ireland, Lowe, & Breen, 2002; Spoth, Redmond, & Shin, 2001).

Integrative Approaches Integrative approaches are designed to treat SUDs from a community-based perspective utilizing case management supports. Vanderplasschen, Rapp, Wolf, and Broekaert (Vanderplasschen, Rapp, De Maeyer, & Van Den Noortgate, 2019) define case management as a highly integrated and synchronized approach to service delivery and persistent supportive care, which improves access to resources for enhanced functioning in the community. That notion implies a linkage between individuals and multiple relevant services within their communities. Case management should facilitate connections with different types of services and encourage sustained retention of available resources by facilitating involvement and promoting continued participation in treatment plans. Moreover, interdisciplinary programs (e.g., community-based case management) that promote optimal health and welfare of clients can be an effective response to addressing substance abuse/misuse and SUDs (Penzenstadler, Machado, Thorens, Zullino, & Khazaal, 2017; Vanderplasschen et al., 2019).

Assertive Community Treatment Assertive Community Treatment (ACT; Drake et al., 1998; Penzenstadler, Soares, Anci, Molodynski, & Khazaal, 2019) is a case management approach with evidence of (a) higher rates of service contact (e.g., follow up to referral resources), (b) reduced hospital utilization, and (c) improved treatment engagement of community members with problematic substance use (Penzenstadler et al., 2019). Assertive Community Treatment is known for assertive outreach, direct service delivery, and the use of multidisciplinary teams to address issues related to mental health. Drake et al. (1998) conducted a study examining the effectiveness of ACT interventions compared to general case management. The treatment group (e.g., ACT) received integrative treatment from multidisciplinary teams involved in direct service provision in their living environment and access to resources 24 hours a day. Researchers (Drake et al., 1998) found that participants in the ACT group had lesser rates of attrition, higher rates of remission related to drug use, and greater reported quality of life, while other measures (hospital days, stable community days, psychiatric symptoms, and remission of alcohol use) remained relatively similar between the treatment and comparison groups.

One example of an integrative model to address co-occurring trauma and substance use is The Seeking Safety Model (Najavits, 2002). This integrative approach is tailored to enhance cognitive, behavioral, and interpersonal skills, while simultaneously addressing issues related to trauma and SUDs. This form of integrated support is designed to enhance motivation and promote insight by helping individuals to understand connections between their trauma and substance use. Seeking Safety emphasizes five fundamental principles that include (a) safety is the primary aim, (b) integrated treatment, (c) ideals as a primary focus, (d) four content areas (cognitive, behavioral, interpersonal, and case management), and (e) attention to clinician's processes (Najavits, 2002).

Though some community supports are available to address substance misuse/abuse, community members often experience barriers to engaging in treatment for SUDs (e.g., cost, lack of insurance, transportation issues), which has led to peer support groups (e.g., Alcoholics Anonymous or Self-Management and Recovery Training [SMART Recovery]) to become the primary mode of treatment sought in the United States (SAMHSA, 2013). These groups are an example of an ecological approach to treating SUDs. Though these groups are often utilized as a primary treatment approach by individuals who experience barriers, they are best seen as a supportive community resource that provides continuing care after treatment has been completed (or as an adjunct to treatment). These groups are free of cost, widely accessible, and provide an environment for participants to gain instrumental and peer support, structure, accountability, peer modeling, and an opportunity to acquire behaviors and undertakings that promote recovery (Moos, 2008; Solomon, 2004). Research has consistently demonstrated strong positive correlations between engagement in peer support groups and subsequent recovery (O'Sullivan, Watts, Xiao, & Bates-Maves, 2015).

Cultural, Legislative, and Professional Issues That Impact Community Substance Use Health

For most of the last century, substance-related issues were treated as social issues and were largely managed at the individual level and less often through existing social infrastructure (Weitzman, Folkman, Folkman, & Wechsler, 2003). However, major social movements, policies, and cultural orientations have influenced the safer use of substances in communities. Moreover, legal regulations also impact the availability and accessibility of substances and perceptions of their safety (Chomynova, Miller, & Beck, 2009; Piontek, Kraus, Bjarnason, Demetrovics, & Ramstedt, 2013). Low-resource countries typically have a different profile as it relates to safer substance use compared to medium and high resource settings (Buck, 2011; Myers, Kline, Doherty, Carney, & Wechsberg, 2014).

Influences of Social Movements and Culture In some communities, the prevalence and acceptability of substance use is explained by the significant socio-political events and cultural assertions of the historically disenfranchised. For instance, in the United States, as the rates of SUDs rose in the 1970s among college students and soldiers returning from the Vietnam War, the healthcare system experienced an influx of people with SUDs, and were ill-equipped to effectively manage the burgeoning demand for care (Read, Wood, Davidoff, McLacken, & Campbell, 2002). During this time, the socio-political views surrounding SUDs were largely disparaging and frequently met with both civil and criminal justice interventions (White, 1998). In the United States, communities of racial minorities and individuals with low socioeconomic status (SES) have lower substance use safety and often encounter poor access to SUD treatment.

Legal Regulation Each country and region have a legal precedent that explains patterns of normative or illicit use of substances. In the United States, healthcare treatments for individuals with SUDs are a controversial debate. For instance, of the heavily debated and divisive legislative changes include—the *Paul Wellstone and Pete Domenici Mental Health Parity and Addiction Equity Act* (MHPAEA, 2008) and the *Patient Protection and Affordable Care Act* ([ACA, 2010]; Congressional Budget Office, n.d.). Historically, SUDs were not provided the same coverage as other medical and surgical conditions (e.g., diabetes, hypertension, cardiovascular disease), therefore individuals seeking treatment were largely left to carry the burden independently or forgo treatment altogether. The MHPAEA and ACA, together, reshaped the landscape of the SUD prevention, diagnosis, and treatment and ushered in a new era of integrated healthcare, which reduced the gaps in both access to and quality of care (Congressional Budget Office, n.d). The goal of the MHPAEA was to require insurers and health plans to cover SUDs in a manner congruent to or at parity (e.g., comparable copays, deductibles, quantitative treatment limits) with the treatment of all other medical and surgical benefits (MHPAEA, 2008).

While the MHPAEA (2008) was instrumental in ensuring that people with SUDs had coverage and access to the necessary services, the ACA (2010) went further and established comprehensive healthcare reforms allowing people with SUDs to obtain and maintain coverage, offset the financial burdens associated with purchasing insurance, and access coverage for preexisting conditions (Beronio, Glied, & Frank, 2014). More specifically, the ACA (2010) increased the number of people who qualified for Medicaid by providing substantial federal funding to offset the costs of each state's expanded Medicaid system (Banthin, 2013) and mandated that people with preexisting conditions (e.g., SUDs) be able to keep their current coverage or obtain new coverage. Additionally, the ACA offered tax credits to individuals with SUDs to allow them to purchase insurance coverage through each state's Health Insurance Marketplace (Patient Protection and Affordable Care Act, 2010). The literature examining the effects of the ACA largely agree that Medicaid coverage increased and resulted in reductions in insurance rates for low-income adults (Buchmueller, Levinson, Levy, & Wolfe, 2016; Sommers, Blendon, & Orav, 2016).

International Policy and Legislation From a global perspective, in 2015, the United Nations member states implemented new policies to assuage the sequelae of substance-related issues internationally (see also Discussion Box 6.3). More specifically, the Millennium Development Goals (MGDs), which had been in place for 15 years, were replaced by the *Sustainable* Development Goals (SDGs) following adoption by the United Nations General Assembly (Lund et al., 2018). Included among the SDGs is the goal to improve on the prevention and treatment of SUDs through national policies, addressing economic factors (e.g., socio-

Discussion Box 6.3: Examining Decriminalization in Portugal

In the 1990s, Portugal faced a major public health crisis as it relates to substance use (United States Library of Congress, 2016). In 1998, Portugal's government created a Commission for the National Strategy for Drug Control, which was tasked with recommending a national policy to address problematic substance use. As implementations progressed in 2001, the government endorsed Decree-Law No. 183, which provided a comprehensive plan for policies related to preventing substance use, reducing risks associated with substance use, and application of harm reduction strategies. As this strategy progressed, the government generated programs and instrumental supports to enhance public health by improving awareness and creating systematic supports that would increase treatment referrals for individuals with SUDs. Later, policymakers advocated for the decriminalization of substance use and further classified previous criminal sanctions for drug use, possession, and acquisition as lower level administrative offenses. In an independent review of this legislation, authors (Hughes & Stevens, 2007) described a moderate increase in cannabis use after legislation was enacted in Portugal, while heroin use decreased significantly. Researchers also found that drug-related deaths reduced by 59% between 1999 and 2003, and, after having the highest rate of HIV among substance users in Europe in 1999, Portugal experienced a 17% decrease in reports of novel drug-associated cases of HIV.

What Do You Think?

1. What are your views on substance use decriminalization as a sustainable community health approach?
2. What community level measures would be appropriate to assess and track the effectiveness of decriminalization as a community substance use safety approach?

economic deprivation, crime; Lund et al., 2018), environmental factors (e.g., natural disasters, war, political violence; Truong & Ma, 2006), and social/cultural factors (e.g., education, social relationships, group membership; Esch et al., 2014; Santini, Koyanagi, Tyrovolas, Mason, & Haro, 2015). The realization of the SDG goals for SUDs should result in safer and healthier countries (Costanza et al., 2016).

Low-Resource Settings Although there is significant variability across counties, low-resource settings generally have significant barriers for community substance use safety (Falck et al., 2007; Smalley, Warren, & Klibert, 2012). The use of illicit drugs in rural communities is a significant and ever-increasing public health concern. Social determinants of health (e.g., water and sanitation, agriculture and access to food, access to healthcare and social services, support for employment, safe and supportive work environments, living environment and housing, education, and transportation) are strongly associated with access to resources for health and wellbeing, however these resources are often lacking in low-resource areas (Bambra et al., 2010). In the United States, rural communities have experienced an increase in the rates of opioid use disorder (Palombi, Olivarez, Bennett, & Hawthorne, 2019). However, some low-resource settings have social capital for community substance use safety (e.g., strong sense of community and support networks). Community substance use safety is likely with community participation (Patel et al., 2018), improved access to evidence-based interventions (Wainberg et al., 2017), and appropriate policy changes for healthy substance consumption.

Related Disciplines Influencing Community Substance Use Safety

Community substance use safety requires interdisciplinary approaches combining psychological, medical, and sociocultural approaches (Sdrulla & Chen, 2015) across the continuum of care: prevention, treatment, and recovery (Substance Abuse and Mental Health Services Administration, 2020). For instance, primary care providers and those employed in community medicine are likely to play a role in the prevention and treatment of substance use disorders. Case managers and social workers can provide support to community substance use safety programs utilizing collaborative team approaches to improve health and wellbeing. Other professionals who may be invested in addressing community substance use may include licensed professional counselors (LPCs) with addiction training or certification, psychologists, rehabilitation counselors, or psychiatrists.

Issues for Research and Other Forms of Scholarship

Few studies apply a community or population-based perspective to understand patterns of substance use across the life span (Griffin & Botvin, 2010). This is despite the fact that community-centric approaches enhance substance use safety. In addition, there is a paucity of research evidence on substance health literacy among the general population (Degan et al., 2019), and on the sustainability of evidence-based practices in safe substance use (Glasgow & Chambers, 2012; Proctor et al., 2015).

Despite important advancements in the prevention and treatment of SUDs over the past two decades, along with the countless research-years and billions of dollars that have been applied searching for workable solutions, the gap between research and practice (i.e., knowledge translation) remains exceedingly wide and persistent (National Institute on Drug Abuse, 2018). In spite of the robust body of evidence that supports the effectiveness of prevention programs for individuals and communities, few programs have been widely implemented or dependably executed (Dusenbury, Brannigan, Falco, & Hansen, 2003). This limits the potential for these programs to have a positive impact on the health of community members.

While patterns of substance misuse across the life span are relatively predictable (Griffin & Botvin, 2010; SAMHSA, 2017), research is needed on best practices for that *early* substance use safety and literacy, minimizing risk for substance misuse later in life. There is a need for studies on the efficacy and sustainability of harm reduction strategies across communities, and complementary early prevention strategies to reduce the risk of problematic substance use, such as substance use literacy (Adkins & Corus, 2009). Studies are also needed on best protocols to train physicians in safe prescription practices for community substance use safety (Pushpakom et al., 2019; Singh & Pushkin, 2019) as well as in coordination of care with populations with SUDs (Gale, Hansen, & Williamson, 2017).

Summary and Conclusion

Community substance use safety is under-researched, although critical to sustainable community health. It has unique strength in addressing

Self-Check Questions

1. What factors characterize healthy communities as they relate to substance use?
2. Briefly describe the history of research that has informed substance use safety and health promotion.
3. List and describe several promising approaches to community substance use health.
4. Identify and describe issues that affect sustainable approaches to promote healthier behaviors related to substances. Include cultural, legislative, and professional issues.
5. What is meant by the term multidisciplinary, and what evidence is available to support the assertion that substance use prevention and treatment requires such an approach?
6. Identify and describe areas of potential research that are needed for communities to develop healthier and safer relationships with substances.

Discussion Questions

1. Describe the social and economic impact of substance misuse/abuse.
2. Describe the impact of the Prohibition movement and the War on Drugs on substance misuse/abuse.
3. Australia has been at the forefront of innovative substance use/abuse policies. Describe some of the approaches that have been implemented and the associated outcomes.
4. What is meant by "health literacy," and what role does it play in preventing substance misuse/abuse?
5. Describe different harm reductions approaches and discuss their impact on substance misuse/abuse.
6. Describe prevention approaches and evaluate their effectiveness.
7. What is the difference between harm reduction and prevention?
8. Discuss the challenges that communities face in trying to effectively address substance use/misuse. Include healthcare, legislative, and available resources.
9. Describe the disciplines involved in establishing a sustainable substance use community. What role does each play?
10. Why is a community-centric approach needed to adequately address substance use/misuse?

Experiential/Field Exercises

1. **Abstinence Project**. Consider giving up something that you do on a daily basis for 30 days (examples might be social media, watching television, eating fast food, nicotine use, etc.), but make sure that this will not have an adverse effect on your health. You might consider telling one or two people close to you about your decision to serve as social support and also for accountability. As you progress through the next 30 days, consider times when you might be more tempted to engage in this behavior, consider factors that contribute to this (stress, boredom, social influences, etc.), and also consider the purpose behind the behavior (e.g., if giving up nicotine, consider antecedents of this behavior and really analyze when you do it and why).
   1.1. Consider the factors that contributed to your success and also factors that might have contributed to a lack of success with the project. Explain your choices.
   1.2. What are substance use safety issues from your field study? How may they be different in another setting and why.
2. **Substance use in the media.** In many cases, the media can provide both accurate and inaccurate depictions of substance use. Select a movie or other media clip that depicts substance use of any kind. Consider the following questions:
   2.1. How is substance use depicted in the media clip of your choice?
   2.2. How might this be different from a community substance use safety perspective?

Key Online Resources

Substance Abuse Mental Health Services Administration. https://www.samhsa.gov/
World Health Organization. https://www.who.int/substance_abuse/en/
National Institute on Drug Abuse. https://www.drugabuse.gov/
Centers for Disease Control and Prevention. https://www.cdc.gov/pwid/addiction.html

community assets both to promote healthy substance use as well as to prevent and reduce the harm associated with problematic substance use. Key elements in community substance use safety include substance use education and literacy, harm reduction, zero tolerance or judicial policy, prevention, caring communities, family strengthening, case management, and integrated services. Community engagement can provide valuable insight into understanding the role of substance use in distressed and diverse communities while identifying innovative solutions. Expanding access to adequate, culturally sensitive, evidence-based approaches specifically designed to foster community engagement to decrease problematic substance use is essential for sustainable community health. Substance use literacy is key to community substance use safety through prevention, intervention, and harm reduction approaches. It is important to adopt safe community policies for community substance use safety, building them into existing community health research and care programs for sustainability.

References

Adkins, N. R., & Corus, C. (2009). Health literacy for improved health outcomes: Effective capital in the marketplace. *Journal of Consumer Affairs, 43*(2), 199+.

Aguirre-Molina, M., & Gorman, D. M. (1996). Community-based approaches for the prevention of alcohol, tobacco, and other drug use. *Annual Review of Public Health, 17*(1), 337–358.

American Psychiatric Association. (2013). *Diagnostic and statistical manual of mental disorders* (5th ed.). Washington, DC: Author.

American Society of Addiction Medicine. (2016). *Opioid addiction 2016 facts and figure*. Retrieved from: https://www.asam.org/docs/default-source/advocacy/opioid-addiction-disease-facts-figures.pdf

Arnold, H. (2016). The prison officer. In H. Arnold (Ed.), *Handbook on prisons*. London, UK: Routledge.

Baez, H., Castro, M. M., Benavente, M. A., Kintz, P., Cirimele, V., Camargo, C., & Thomas, C. (2000). Drugs in prehistory: Chemical analysis of ancient human hair. *Forensic Science International, 108*(3), 173–179.

Bambra, C., Sowden, A., Wright, K., Gibson, M., Whitehead, M., & Petticrew, M. (2010). Tackling the wider social determinants of health and health inequalities: Evidence from systematic review. *Journal of Epidemiology & Community Health, 64*(4), 284.

Banthin, J. (2013). *Understanding CBO's Medicaid coverage projections under the Affordable Care Act.* In Presentation to academy Health annual research meeting, Baltimore, MD.

Bellis, M. A., Hughes, K., & Lowey, H. (2002). Healthy nightclubs and recreational substance use: From a harm minimisation to a healthy settings approach. *Addictive Behaviors, 27*(6), 1025–1036.

Benotsch, E. G., Koester, S., Luckman, D., Martin, A. M., & Cejka, A. (2011). Non-medical use of prescription drugs and sexual risk behavior in young adults. *Addictive Behaviors, 36*(1–2), 152–156.

Beronio, K., Glied, S., & Frank, R. (2014). How the affordable care act and mental Health parity and addiction equity act greatly expand coverage of behavioral health care. *Journal of Behavioral Health Services and Research, 41*, 410–428.

Blocker, J. (2006). Did prohibition really work? Alcohol prohibition as a public health innovation. *American Journal of Public Health, 96*, 233–243.

Bond, G. R., Drake, R. E., McHugo, G. J., Peterson, A. E., Jones, A. M., & Williams, J. (2014). Long-term sustainability of evidence-based practices in community mental health agencies. *Administration and Policy in Mental Health and Mental Health Services Research, 41*(2), 228–236.

Botvin, G. J., Baker, E., Dusenbury, L., Tortu, S., & Botvin, E. M. (1990). Preventing adolescent drug abuse through a multimodal cognitive-behavioral approach: Results of a 3-year study. *Journal of Consulting and Clinical Psychology, 58*(4), 437.

Botvin, G. J., Baker, E., Renick, N. L., Filazzola, A. D., & Botvin, E. M. (1984). A cognitive-behavioral approach to substance abuse prevention. *Addictive Behaviors, 9*(2), 137–147.

Botvin, G. J., Griffin, K. W., Diaz, T., & Ifill-Williams, M. (2001a). Drug abuse prevention among minority adolescents: Posttest and one-year follow-up of a school-based preventive intervention. *Prevention Science, 2*(1), 1–13.

Botvin, G. J., Griffin, K. W., Diaz, T., & Ifill-Williams, M. (2001b). Preventing binge drinking during early adolescence: One-and two-year follow-up of a school-based preventive intervention. *Psychology of Addictive Behaviors, 15*(4), 360.

Botvin, G. J., Griffin, K. W., Diaz, T., Scheier, L. M., Williams, C., & Epstein, J. A. (2000). Preventing illicit drug use in adolescents: Long-term follow-up data from a randomized control trial of a school population. *Addictive Behaviors, 25*(5), 769–774.

Buchmueller, T. C., Levinson, Z. M., Levy, H. G., & Wolfe, B. L. (2016). Effect of the affordable care act on racial and ethnic disparities in health insurance coverage. *American Journal of Public Health, 106*(8), 1416–1421.

Buck, J. A. (2011). The looming expansion and transformation of public substance abuse treatment under the Affordable Care Act. *Health Affairs, 30*(8), 1402–1410.

Bull, M., Denham, G., Trevaskes, S., & Coomber, R. (2016). From punishment to pragmatism: Sharing the burden of reducing drug-related harm. *Chinese Journal of Comparative Law, 4*(2), 300–316.

Butterfoss, F. D., Goodman, R. M., & Wandersman, A. (1993). Community coalitions for prevention and health promotion. *Health Education Research, 8*(3), 315–330.

Canadian Paediatric Society. (2008). Harm reduction: An approach to reducing risky health behaviours in adolescents. *Paediatrics & Child Health, 13*(1), 53–60.

Chandler, R., Fletcher, B., & Volkow, N. (2009). Treating drug abuse and addiction in the criminal justice system: Improving public health and safety. *Journal of the American Medical Association, 301*, 183–190.

Chomynova, P., Miller, P., & Beck, F. (2009). Perceived risks of alcohol and illicit drugs: Relation to prevalence of use on individual and country level. *Journal of Substance Use, 14*(3–4), 250–264.

Climate Schools. (2020). *Climate schools online prevention.* Retrieved from: https://www.climateschools.com.au/

Congressional Budget Office. (n.d.). *Insurance coverage provisions of the Affordable Care Act.* CBOS March 2015 baseline.

Costanza, R., Daly, L., Fioramonti, L., Giovannini, E., Kubiszewski, I., Mortensen, L. F., … Wilkinson, R. (2016). Modelling and measuring sustainable wellbeing in connection with the UN sustainable development goals. *Ecological Economics, 130*, 350–356.

Crowley, D. M., Jones, D. E., Coffman, D. L., & Greenberg, M. T. (2014). Can we build an efficient response to the prescription drug abuse epidemic? Assessing the cost effectiveness of universal prevention in the PROSPER trial. *Preventive Medicine, 62*, 71–77.

Degan, T. J., Kelly, P. J., Robinson, L. D., & Deane, F. P. (2019). Health literacy in substance use disorder treatment: A latent profile analysis. *Journal of Substance Abuse Treatment, 96*, 46–52.

Dermota, P., Wang, J., Dey, M., Gmel, G., Studer, J., & Mohler-Kuo, M. (2013). Health literacy and substance use in young Swiss men. *International Journal of Public Health, 58*(6), 939–948.

Dimeo, P. (2008). *A history of drug use in sport: 1876–1976: Beyond good and evil.* London, UK: Routledge.

Dolan, K., Kimber, J., Fry, C., & Fitzgerald, J. (2000). Drug consumption facilities in Europe and the establishment of supervised injecting centres in Australia. *Drug and Alcohol Review, 19*(3), 337.

Drake, R. E., McHugo, G. J., Clark, R. E., Teague, G. B., Xie, H., Miles, K., & Ackerson, T. H. (1998). Assertive community treatment for patients with co-occurring severe mental illness and substance use disorder: A clinical trial. *American Journal of Orthopsychiatry, 68*(2), 201–215.

Dube, S. R., Anda, R. F., Felitti, V. J., Edwards, V. J., & Croft, J. B. (2002). Adverse childhood experiences and personal alcohol abuse as an adult. *Addictive Behaviors, 27*(5), 713–725.

Dube, S. R., Felitti, V., Dong, M., Chapman, D., Giles, W., & Anda, R. (2003). Childhood abuse, neglect, and household dysfunction and the risk of illicit drug use. *The adverse childhood experiences study. Pediatrics, 111*, 564–572.

Durrant, R., & Thakker, J. (2003). *Substance use and abuse: Cultural and historical perspectives.* Oakland, CA: Sage.

Dusenbury, L., Brannigan, R., Falco, M., & Hansen, W. (2003). A review of research on fidelity of implementation: Implications for drug abuse prevention in school settings. *Health Education Research, 18*, 237–256.

Education and Training. (2020). *Drug education.* Retrieved from https://www.education.vic.gov.au/school/teachers/teachingresources/discipline/physed/Pages/drugeducation.aspx

Esch, P., Bocquet, V., Pull, C., Couffignal, S., Lehnert, T., Graas, M., … Ansseau, M. (2014). The downward spiral of mental disorders and educational attainment: A systematic review on early school leaving. *BMC Psychiatry, 14*(1), 237.

Falck, R. S., Wang, J., Carlson, R. G., Krishnan, L. L., Leukefeld, C., & Booth, B. M. (2007). Perceived need for substance abuse treatment among illicit stimulant drug users in rural areas of Ohio, Arkansas, and Kentucky. *Drug and Alcohol Dependence, 91*(2–3), 107–114.

Felitti, V. J., Anda, R. F., Nordenberg, D., Williamson, D. F., Spitz, A. M., Edwards, V., Koss, M. P., & Marks, J. S. (1998). Relationship of childhood abuse and household dysfunction to many of the leading causes of death in adults: The adverse childhood experiences (ACE) study. *American Journal of Preventive Medicine, 14*(4), 245–258.

Fineberg, H. V. (2012). A successful and sustainable health system—How to get there from here. *New England Journal of Medicine, 366*(11), 1020–1027.

Foxcroft, D., Ireland, D., Lowe, G., & Breen, R. (2002). Primary prevention for alcohol misuse in young people. *Cochrane Database of Systematic Reviews, 3.* https://doi.org/10.1002/14651858.CD003024

Foxcroft, D. R., Ireland, D., Lister-Sharp, D. J., Lowe, G., & Breen, R. (2003). Longer-term primary prevention for alcohol misuse in young people: A systematic review. *Addiction, 98*(4), 397–411. https://doi.org/10.1046/j.1360-0443.2003.00356.x

Gale, J. A., Hansen, A. Y., & Williamson, M. E. (2017). *Rural opioid prevention and treatment strategies: The experience in four states.* Portland, ME: Maine Rural Health Research Center.

Ghodse, H., Herrman, H., Maj, M., & Sartorius, N. (Eds.). (2011). *Substance abuse disorders: Evidence and experience.* Chichester, UK: John Wiley & Sons.

Glasgow, R. E., & Chambers, D. (2012). Developing robust, sustainable, implementation systems using rigorous, rapid and relevant science. *Clinical and Translational Science, 5*(1), 48–56.

Griffin, K., & Botvin, G. (2010). Evidence-based interventions for preventing substance use disorders in adolescents. *Child and Adolescent Psychiatric Clinics of North America, 19*, 505–526.

Groves, A. (2018). 'Worth the test?' pragmatism, pill testing and drug policy in Australia. *Harm Reduction Journal, 15*(1), 12.

Gureje, O., Vazquez-Barquero, L., & Janca, A. (1996). Comparisons of alcohol and other drugs: Experience from the WHO collaborative cross-cultural applicability research study. *Addiction, 91*, 1529–1538.

Hall, W., Babor, T., Edwards, G., Laranjeira, R., Marsden, J., Miller, P., … West, R. (2012). Compulsory detention, forced detoxification and enforced labour are not ethically acceptable or effective ways to treat addiction. *Addiction, 107*, 1891–1893.

Hawkins, J. D., Brown, E. C., Oesterle, S., Arthur, M. W., Abbott, R. D., & Catalano, R. F. (2008). Early effects of communities that care on targeted risks and initiation of delinquent behavior and substance use. *Journal of Adolescent Health, 43*(1), 15–22.

Hawkins, J. D., & Catalano, R. F., Jr. (1992). *Communities that care: Action for drug abuse prevention.* San Francisco, CA: Jossey-Bass.

Hawkins, J. D., Van Horn, M., & Arthur, M. (2004). Community variation in risk and protective factors and substance use outcomes. *Prevention Science, 4*, 213–220.

Heitzeg, N. A. (2009). Education or incarceration: Zero tolerance policies and the school to prison pipeline. In *Forum on public policy online* (Vol. 2009, No. 2). Urbana, IL: Oxford Round Table.

Henninger, A., & Sung, H. (2014). History of substance abuse treatment. In G. Bruinsma & D. Weisburd (Eds.), *Encyclopedia of criminology and criminal justice.* New York, NY: Springer.

Hughes, C., & Hughes, C. (2007). Evidence based policy or policy based evidence? The role of evidence in the development and implementation of the illicit drug diversion initiative. *Drug and Alcohol Review, 26*, 363–368.

Hughes, C., & Stevens, A. (2007). *The effects of decriminalization of drug use in Portugal*. London, UK: The Beckly Foundation Drug Policy Programme.

Inaba, D., & Cohen, W. (2011). *Uppers, downers, all arounders: Physical and mental effects of psychoactive drugs*. Medford, OR: CNS.

Islam, M. M., Stern, T., Conigrave, K. M., & Wodak, A. (2008). Client satisfaction and risk behaviors of the users of syringe dispensing machines: A pilot study. *Drug and Alcohol Review, 27*(1), 13–19.

Islam, M. M., Topp, L., Iversen, J., Day, C., Conigrave, C., & Maher, L. (2013). Healthcare utilization and disclosure of injecting drug use among clients of Australia's needle and syringe programs. *Australian and New Zealand Journal of Public Health, 37*, 148–154.

Kanouse, A. B., & Compton, P. (2015). The epidemic of prescription opioid abuse, the subsequent rising prevalence of heroin use, and the federal response. *Journal of Pain & Palliative Care Pharmacotherapy, 29*(2), 102–114.

Khantzian, E. J. (1999). *Treating addiction as a human process*. Northvale, NJ: Jason Aronson.

Kumpfer, K. L., Molgaard, V., & Spoth, R. (1996). The strengthening families program for the prevention of delinquency and drug use. In R. D. Peters & R. J. McMahon (Eds.), *Preventing childhood disorders, substance abuse, and delinquency* (3rd ed., pp. 241–267).

Lund, C., Brooke-Sumner, C., Baingana, F., Baron, E. C., Breuer, E., Chandra, P., … Medina-Mora, M. E. (2018). Social determinants of mental disorders and the sustainable development goals: A systematic review of reviews. *The Lancet Psychiatry, 5*(4), 357–369.

Mann, R. D. (1984). *Modern drug use: An enquiry on historical principles*. Dordrecht, Netherlands: Springer.

Mitchell, O., & Caudy, M. (2015). Examining racial disparities in drug arrests. *Justice Quarterly, 32*, 288–313.

Moos, R. H. (2008). Active ingredients of substance use–focused self-help groups. *Addiction, 103*, 387–396.

Myers, B., Kline, T. L., Doherty, I. A., Carney, T., & Wechsberg, W. M. (2014). Perceived need for substance use treatment among young women from disadvantaged communities in Cape Town, South Africa. *BMC Psychiatry, 14*(1), 100.

Najavits, L. (2002). *Seeking safety: A treatment manual for PTSD and substance abuse*. New York, NY: Guilford Publications.

National Institute of Health. (2015). *News releases*. https://www.nih.gov/news-events/news-releases/10-percent-us-adults-have-drug-use-disorder-some-point-their-lives

National Institute on Drug Abuse. (2017). *Trends and statistics*. Retrieved from: https://www.drugabuse.gov/related-topics/trends-statistics

National Institute on Drug Abuse. (2018). *The science of drug use and addiction: The basics*. Retrieved from: https://store.samhsa.gov/system/files/sma11-4648.pdf

Nicholson, T., Duncan, D. F., & White, J. B. (2002). Is recreational drug use normal? *Journal of Substance Use, 7*(3), 116–123.

North Richmond Community Health Wulempuri – Kertheba. (2017). *Medically supervised injecting room*. Retrieved from: https://nrch.com.au/services/medically-supervised-injecting-room/

NSW Health Organizations. (2017). Retrieved from: https://www.health.nsw.gov.au/annualreport/Publications/2017/nsw-health-organisations.pdf

O'Sullivan, D., Watts, J., Xiao, Y., & Bates-Maves, J. (2015). Refusal self-efficacy among SMART recovery members by affiliation length and meeting frequency. *Journal of Addictions and Offender Counseling, 37*, 87–101.

Oesterle, S., Kuklinski, M. R., Hawkins, J. D., Skinner, M. L., Guttmannova, K., & Rhew, I. C. (2018). Long-term effects of the communities that care trial on substance use, antisocial behavior, and violence through age 21 years. *American Journal of Public Health, 108*(5), 659–666.

Osborn, C., Paasche-Orlow, M., Bailey, S., & Wolf, M. (2011). The mechanisms linking health literacy to behavior and health status. *American Journal of Health Behavior, 35*, 118–128.

Palombi, L., Olivarez, M., Bennett, L., & Hawthorne, A. N. (2019). Community forums to address the opioid crisis: An effective grassroots approach to rural community engagement. *Substance Abuse: Research and Treatment, 13*, 1–7.

Parsche, F., Balabanova, S., & Pirsig, W. (1993). Drugs in ancient populations. *The Lancet, 341*(8843), 503.

Patel, V., Saxena, S., Lund, C., Thornicroft, G., Baingana, F., Bolton, P., … Herrman, H. (2018). The lancet commission on global mental health and sustainable development. *The Lancet, 392*(10157), 1553–1598.

Patient Protection and Affordable Care Act, 42 U.S.C. § 18001 et seq. (2010).

Paul Wellstone and Pete Domenici Mental Health Parity and Addiction Equity Act of 2008, H.R. 6983 (2008).

Penzenstadler, L., Machado, A., Thorens, G., Zullino, D., & Khazaal, Y. (2017). Effect of case management interventions for patients with substance use disorders: A systematic review. *Frontiers in Psychiatry, 8*, 51.

Penzenstadler, L., Soares, C., Anci, E., Molodynski, A., & Khazaal, Y. (2019). Effect of assertive community treatment for patients with substance use disorder: A systematic review. *European Addiction Research, 25*(2), 56–67.

Piontek, D., Kraus, L., Bjarnason, T., Demetrovics, Z., & Ramstedt, M. (2013). Individual and country-level effects of cannabis-related perceptions on cannabis use. A multilevel study among adolescents in 32 European countries. *Journal of Adolescent Health, 52*(4), 473–479.

Positive Choices. (2020). *Drug and alcohol information*. Retrieved from https://positivechoices.org.au/

Proctor, E., Luke, D., Calhoun, A., McMillen, C., Brownson, R., McCrary, S., & Padek, M. (2015). Sustainability of evidence-based healthcare: Research agenda, methodological advances, and infrastructure support. *Implementation Science, 10*(1), 88.

Pushpakom, S., Iorio, F., Eyers, P. A., Escott, K. J., Hopper, S., Wells, A., … Pirmohamed, M. (2019). Drug repurposing: Progress, challenges and recommendations. *Nature Reviews Drug Discovery, 18*(1), 41+.

Rahim, Y. (2005). Safe community in different settings. *International Journal of Injury Control and Safety Promotion, 12*(2), 105–112.

Read, J. P., Wood, M. D., Davidoff, O. J., McLacken, J., & Campbell, J. F. (2002). Making the transition from high school to college: The role of alcohol-related social influence factors in students' drinking. *Substance Abuse, 23*(1), 53–66.

Resiak, D., Mpofu, E., & Athanasou, J. (2016). Drug treatment policy in the criminal justice system: A scoping literature review. *American Journal of Criminal Justice, 41*(1), 3–13.

Rowlands, G., Protheroe, J., Winkley, J., Richardson, M., Seed, P., & Rudd, R. (2015). A mismatch between population health literacy and the complexity of health information: An observational study. *British Journal of General Practice, 65*, 379–386.

Santini, Z. I., Koyanagi, A., Tyrovolas, S., Mason, C., & Haro, J. M. (2015). The association between social relationships and depression: A systematic review. *Journal of Affective Disorders, 175*, 53–66.

Scully, J. (2004). What is a disease? *European Molecular Biology Organization, 5*, 650–653.

Sdrulla, A., & Chen, G. (2015). A multidisciplinary approach to the management of substance abuse. In A. Kaye, N. Vadivelu, & R. Urman (Eds.), *Substance abuse: Inpatient and outpatient management for every clinician*. New York, NY: Springer.

Shelton, R. C., Cooper, B. R., & Stirman, S. W. (2018). The sustainability of evidence-based interventions and practices in public health and health care. *Annual Review of Public Health, 39*, 55–76.

Shepherd, J. (2014). Combating the prescription painkiller epidemic: A national prescription drug reporting program. *American Journal of Law & Medicine, 40*(1), 85–112.

Shorthouse, J. (2017). *Oxford dictionary of anesthesia*. Oxford, UK: Oxford university press.

Singh, R., & Pushkin, G. (2019). How should medical education better prepare physicians for opioid prescribing? *AMA Journal of Ethics, 21*, 636–641.

Single, E. (1995). Defining harm reduction. *Drug and Alcohol Review, 14*, 287–290.

Smalley, K. B., Warren, J. C., & Klibert, J. (2012). Health risk behaviors in insured and uninsured community health center patients in the rural US south. *Rural and Remote Health, 12*, 2123.

Solomon, P. (2004). Peer support/peer provided services underlying processes, benefits, and critical ingredients. *Psychiatric Rehabilitation Journal, 27*, 392–401.

Sommers, B. D., Blendon, R. J., & Orav, E. J. (2016). Both the 'private option' and traditional Medicaid expansions improved access to care for low-income adults. *Health Affairs, 35*(1), 96–106.

Spinks, A., Turner, C., Nixon, J., & McClure, R. (2005). The 'WHO safe communities' model for the prevention of injury in whole populations. *Cochrane Database of Systematic Reviews, 18*(2), CD004445.

Spoth, R. L., Clair, S., Shin, C., & Redmond, C. (2006). Long-term effects of universal preventive interventions on methamphetamine use among adolescents. *Archives of Pediatrics & Adolescent Medicine, 160*(9), 876–882.

Spoth, R. L., Randall, G. K., Trudeau, L., Shin, C., & Redmond, C. (2008). Substance use outcomes 5½ years past baseline for partnership-based, family-school preventive interventions. *Drug and Alcohol Dependence, 96*(1–2), 57–68.

Spoth, R. L., Redmond, C., & Shin, C. (2001). Randomized trial of brief family interventions for general populations: Adolescent substance use outcomes 4 years following baseline. *Journal of Consulting and Clinical Psychology, 69*(4), 627.

Stirman, S. W., Kimberly, J., Cook, N., Calloway, A., Castro, F., & Charns, M. (2012). The sustainability of new programs and innovations: A review of the empirical literature and recommendations for future research. *Implementation Science, 7*(1), 17.

Stockwell, T. R., Heale, P., Chikritzhs, T. N., Dietze, P., & Catalano, P. (2002). How much alcohol is drunk in Australia in excess of the new Australian alcohol guidelines? *The Medical Journal of Australia, 176*(2), 91–92.

Substance Abuse and Mental Health Services Administration. (2013). *Results from the 2012 National Survey on Drug Use and Health: Summary of national findings (NSDUH Series H-46, HHS publication No. [SMA] 13-4795).* Retrieved from http://www.samhsa.gov/data/sites/ default/files/NSDUHresults2012/NSDUHresults2012.pdf

Substance Abuse and Mental Health Services Administration. (2020). *A window into addiction treatment and primary care integration.* Retrieved from: https://www.integration.samhsa.gov/about-us/esolutions-newsletter/integrating-substance-abuse-and-primary-care-services

Substance Abuse Mental Health Services Administration. (2017). *Trends in substance use disorders among adults aged 18 or older.* Retrieved from: https://www.samhsa.gov/data/sites/default/files/report_2790/ShortReport-2790.html

Svanström, L. (2012). It all started in Falköping, Sweden: Safe communities–global thinking and local action for safety. *International Journal of Injury Control and Safety Promotion, 19*(3), 202–208.

Teesson, M., Newton, N. C., & Barrett, E. L. (2012). Australian school-based prevention programs for alcohol and other drugs: A systematic review. *Drug and Alcohol Review, 6*, 731–736.

Treloar, C., McLeod, R., Yates, K., & Mao, L. (2014). What's the cost of finding the right fit? The cost of conducting NSP business in a range of modalities. *Contemporary Drug Problems, 41*(1), 41–56. https://doi.org/10.1177/009145091404100103

Truong, K. D., & Ma, S. (2006). A systematic review of relations between neighborhoods and mental health. *Journal of Mental Health Policy and Economics, 9*(3), 137–154.

Tsui, M. S. (2000). The harm reduction approach revisited: An international perspective. *International Social Work, 43*(2), 243–251.

Tu, W. M., Yan, M., Li, Q., & Watts, J. (2019). Attitudes toward disabilities among students in college settings: A multidimensional scaling analysis with biplot. *The Australian Journal of Rehabilitation Counseling, 25*, 79–96.

United Nations Development Program. (2014). *Human development report 2014: Sustaining human progress reducing vulnerabilities and building resilience.* Retrieved from: http://hdr.undp.org/sites/default/files/hdr14-report-en-1.pdf

United States Department of Health and Human Services. (1999). *Blending perspectives and building common ground. The spectrum of substance use, abuse, and addiction.* Retrieved from: https://aspe.hhs.gov/report/blending-perspectives-and-building-common-ground/spectrum-substance-use-abuse-and-addiction

United States Department of Health and Human Services. (2019). *What is the U.S. opioid epidemic*? Retrieved from: https://www.hhs.gov/opioids/about-the-epidemic/index.html

Vanderplasschen, W., Rapp, R. C., De Maeyer, J., & Van Den Noortgate, W. (2019). A meta-analysis of the efficacy of case management for substance use disorders: A recovery perspective. *Frontiers in Psychiatry, 10*, 186.

Wainberg, M. L., Scorza, P., Shultz, J. M., Helpman, L., Mootz, J. J., Johnson, K. A., … Arbuckle, M. R. (2017). Challenges and opportunities in global mental health: A research-to-practice perspective. *Current Psychiatry Reports, 19*(5), 28.

Waite, R., & Ryan, R. A. (2019). *Adverse childhood experiences: What students and Health professionals need to know*. London, UK: Routledge.

Weitzman, E. R., Folkman, A., Folkman, M. P., & Wechsler, H. (2003). The relationship of alcohol outlet density to heavy and frequent drinking and drinking-related problems among college students at eight universities. *Health & Place, 9*(1), 1–6.

White, W. L. (1998). *Slaying the dragon: The history of addiction treatment and recovery in America*. Bloomington, IL: Chestnut Health Systems/Lighthouse Institute.

Wilson, D., Donald, B., Shattock, A., Wilson, D., & Hurt, N. (2015). The cost-effectiveness of harm reduction. *International Journal of Drug Policy, 26*, 5–11.

World Health Organization. (1998). *WHO safe communities*. Retrieved from: http://www.phs.ki.se/csp/default.htm

World Health Organization. (2004). *Global health observatory data: Prevalence of drug use disorders*. Retrieved from: https://www.who.int/gho/substance_abuse/burden/drug_prevalence/en/

World Health Organization. (2009). *Assessment of compulsory treatment of people who use drugs in Cambodia, China, Malaysia and Viet Nam: An application of human rights principles*. Geneva, Switzerland: World Health Organization.

World Health Organization. (2018). *International classification of diseases for mortality and morbidity statistics* (11th Revision). Retrieved from https://icd.who.int/browse11/l-m/en

World Health Organization. (2020). *Harm reduction*. Retrieved from: http://www.euro.who.int/en/health-topics/communicable-diseases/hivaids/policy/policy-guidance-for-areas-of-intervention/harm-reduction.

7

Community Mental Health Resourcing

Justin R. Watts, Elias Mpofu, Qiwei Li, Veronica Cortez, and Ganesh Baniya

Introduction

Approximately 25–30% of the global population self-reported at least one mental health disorder in the past year, identifying mental disorders as a primary source of poor health and disability globally (Lancet Global Mental Health Group [LGMHG], 2007; World Health Organization [WHO], 2020). The United Nations (UN) emphasizes human rights for

J. R. Watts (✉) • Q. Li • V. Cortez • G. Baniya
University of North Texas, Denton, TX, USA
e-mail: Justin.Watts@unt.edu; Qiwei.Li@unt.edu; Veronica.Cortez@unt.edu; Ganesh.Baniya@unt.edu

E. Mpofu
University of North Texas, Denton, TX, USA

University of Sydney, Sydney, NSW, Australia

University of Johannesburg, Johannesburg, South Africa
e-mail: Elias.Mpofu@unt.edu

E. Mpofu (ed.), *Sustainable Community Health*,
https://doi.org/10.1007/978-3-030-59687-3_7

individuals with mental illnesses in treatment services and the importance of patient participation in treatment planning (Thornicroft, Deb, & Henderson, 2016). While the burden of mental health on communities is widely acknowledged, approaches have typically been reactive, clinic-focused, and exclusionary to specific communities. The relative neglect of a holistic community approach in treating mental health disorders fails to take into account the enormous cost of recycling patients through the health care system that often results from a narrow focus on clinical care approaches. Moreover, the fact that less than half of diagnosable adults with any form of mental illness receive inpatient or outpatient care with medication management or psychotherapy (Substance Abuse Mental Health Services Administration [SAMHSA], 2018), suggests that the community is the ultimate resource for promoting mental health wellbeing (Barry, Clarke, Petersen, & Jenkins, 2019). While factors including stigma and discrimination, transportation barriers, lack of insurance, and denial of mental health issues frequently interfere with individuals seeking clinical mental health care (Thornicroft et al., 2010); community-oriented mental health initiatives would reach far more people, providing accessible, low-cost resources in the context of community living needs (Rosen, O'Halloran, Mezzina, & Thompson, 2015).

Learning Objectives

After reviewing this chapter, the reader should be able to:

1. Define the notion of sustainable community mental health resourcing.
2. Consider professional legal frameworks that bear on sustainable community mental health practices.
3. Outline the history of research and practice in community mental health practices.
4. Distinguish between attributes of flourishing and languishing as applied to community mental health.
5. Examine the influences of culture, legislation, and policy on community mental health resourcing.
6. Evaluate the evidence base for community mental health practices and emerging or promising practices.

Clinic-based services are also impeded by their incredibly poor reach in that few people in need would have the capacity to access those services for a variety of reasons, including cost capitation systems that exclude the poor, and those with no or subminimal insurance coverage. This is particularly so in developing countries with underdeveloped formal health care systems (LGMHG, 2007), or historically disadvantaged racial minority communities in the US (McGuire & Miranda, 2008). Poor access to mental health care is exacerbated by the fact that clinic-based mental health services tend to be concentrated in affluent urban centers, which limits access to people in rural areas and small towns. The dire shortage of mental health professionals is well known around the globe (WHO, 2018), in addition to the need for cultural competence among the workforce to address mental health issues across racial/ethnic, gender, and other intersections of identity.

Only a fraction of mental health professionals have a working knowledge or lived knowledge of the neighborhoods they provide services to, which undercuts their potential to serve as community mental health resources. Community psychiatry is the major exception, as it addresses the continuum of mental health needs with an emphasis on wellness promotion rather than a focus on pathology (Caplan, 2013). Thornicroft et al. (2016) proposed the idea that, in order to achieve community mental health, approaches must be adopted to "address population mental health needs in ways that are accessible and acceptable; building on the goals and strengths of people who experience mental illnesses; promoting a wide network of supports, services and resources of adequate capacity; and by emphasizing services that are both evidence-based and recovery-oriented" (p. 276).

Professional Definitions and Theories of the Specific Sustainable Community Mental Health

Community mental health resourcing encompasses practices at the community level for improving individual mental health, premised on the assumption that healthier communities, overall, make for environments that experience flourishing rather than languishing related to mental health (Miller, Paschall, & Svendsen, 2006; Thornicroft et al.,

2016). This is a lived health approach that prioritizes people's relationship assets and other resources for improving their wellbeing. The WHO (2004) defines mental health as "a state of well-being in which the individual realizes his or her own abilities, can cope with the normal stresses of life, can work productively and fruitfully, and is able to make a contribution to his or her community." *Community* mental health resourcing is consistent with the WHO vision for mental health, which aspires for the collective improvement of the mental health of individuals within communities.

Communities that have the resources to provide *optimal* mental health services to their members are characterized by practices that are inclusive of diversity in mental health presentations and are representative of *all* community members, while providing for healthy interaction spaces for individuals and groups around mutually shared values (Rosen et al., 2015). While comprehensive community mental health would include access to (a) primary care mental health (e.g., pharmacological treatments, mental health counselors, and emphasizing assessment and case management), (b) general services to support mental health in the community (e.g., outpatient treatment, inpatient treatment, long-term residential care, support for career development), and (c) specialized services for mental health care (Thornicroft et al., 2016; Thornicroft & Tansella, 2004), community mental health resourcing, with its wellness orientation, aims to prevent community members from needing psychiatric care in the first place, and aims to provide psychiatric care in the context of the everyday community living needs of the person. The community [rather than the clinical] focus has built in sustainability in that, community members, rather than mental health agencies, are both providers and beneficiaries of each other's mental health wellbeing. Community mental health resourcing achieves this collective good by strengthening communities and neighborhoods, and providing culturally informed mental health support resources to meet the unique needs of each community in a sustained way (see Case Example 7.1).

Case Example 7.1 Examining Flourishing Mental Health in Collectivist and Individualist Societies

In collectivistic cultures (i.e., cultures that emphasize the needs of a community above individual needs), like many American Indian (AI) and Alaskan Native (AN) communities, mental health is only one component of complete health and wellness; in many cases a very large number of individuals within these communities experience flourishing mental health (Kading et al., 2015). As opposed to concentrating on pathology, many AI/AN communities focus on the connection and balance among physical health, emotional health, spiritual health, and mental health as markers of holistic health (Urban Indian Health Institute, 2012). In some cases, excessive focus on pathology and inflexible designations of wellness may perhaps over-stigmatize communities, and may instigate misinterpretations of cultural expressions of distress (Kirmayer, Dandeneau, Marshall, Phillips, & Williams, 2011). This is likely to further marginalize individuals within these communities that are experiencing issues related to mental health, and reduce the likelihood of seeking support. In some AI communities, perceptions of discrimination are associated with poorer mental health; however, individuals who more actively participate in traditional activities within the community (e.g., playing traditional games, gathering traditional medicines, and engaging in customary food gathering) are shown to experience increased mental health (Kading et al., 2015). The Urban Indian Health Institute (2012) conducted a systematic review of 79 studies regarding factors and resources that improve mental health in AI/AN communities. Results illustrated that a focus on community and family, in addition to the inclusion of cultural knowledge and practice into mental health care, both improved the mental health of individuals within the community.

In individualistic cultures (cultures that emphasize the needs of an individual above the collective needs of the group), similar to that of Western culture (i.e., the US), more emphasis is placed on independence (Scott, Ciarrochi, & Deane, 2004). In this case, behaviors are usually structured by an individual's attitudes, and there is a larger emphasis on emotional independence and self-understanding. In a study of 276 undergraduates, Scott et al., (2004) found that participants who reported higher levels of idiocentrism tended to have less satisfying social support systems, less skill in managing emotions, sought less support from family and friends regarding mental health, and reported higher levels of hopelessness and suicide ideation. Researchers (Scott et al., 2004) emphasized the importance of citizenship and enhancing social supports as a means of assisting individuals within individualistic cultures to begin to address some of the dysfunctional beliefs that might worsen symptoms, and teach individuals to enhance emotional competence.

What Do You Think?

1. How do characteristics of collectivistic cultures differ from individualistic cultures in their resourcing of community mental health?
2. From your experience, how has your culture of origin shaped your view on mental health and the resources for it?

History of Research and Practice Pertinent to Community Mental Health

The mental health of community members was historically a collective responsibility, placing the responsibility of individual care on members of the community (Rosen et al., 2015). Over time, research and practices progressed as the classification of mental functions was utilized to identify treatable needs to be served by the field of psychiatry (Palha & Marques-Teixeira, 2012; Shorter, 2008). From a community health perspective, these designations and diagnoses would be rather arbitrary, as some communities would consider many conditions to not warrant a mental illness diagnosis and treatment at care centers removed individuals from community life (American Psychiatric Association [APA], 2013; Rosenhan, 1973). In developing countries, community mental health resourcing prioritizes community participation rather than a clinical diagnosis (Canino & Alegria, 2008; Rosen et al., 2015). For instance, many of the diagnostic criteria for mental illness in Western cultures do not account for mental ill-health in developing countries (Canino & Alegria, 2008) (see Research Box 7.1). A culture's predominant beliefs regarding the development and maintenance of mental health symptoms shape how individuals within a specific culture might understand or label symptoms of mental health problems, and the manner in which individuals describe their experiences with mental illness. Moreover, how individuals understand the causes of mental health issues affect the frequency and manner in which they engage with treatment and seek support (Choudhry, Mani, Ming, & Khan, 2016).

The deinstitutionalization movement was in recognition that people would recover and better sustain their mental health wellbeing while living in the community rather than in specialist care services that risked the criminalization of individuals with atypical mental health (Dixon & Goldman, 2003). Although the deinstitutionalization movement may not have had the necessary community resourcing supports to facilitate sustainable mental health of those with severe mental illness (Teich, 2016), it provided a role for family members to be more involved in

Research Box 7.1 Psychiatric Diagnosis: Is It Universal or Relative to Culture? (Ho et al., 1996)

Background. Culture plays a considerable role in understanding how specific populations understand the development, maintenance, and treatment of mental health disorders. Examining mental health through an exclusively Western perspective may be inappropriate, and in some cases harmful to community members. In order to examine constructs related to hyperactivity and antisocial characteristics among Chinese children, researchers (Ho et al., 1996) examined teachers' and parents' responses for children's behavior on the Rutter Behavioral Questionnaire.

Method. Participants in this study were $N = 3069$ seven-year-old Chinese Boys in Hong Kong. Parents and teachers completed Rutter Behavioral Questionnaire (e.g., questions involve how often children have difficulty staying seated, how often children squirm or fidget, etc.) for each participant, providing information on hyperactivity, neurotic behaviors, and/or antisocial behaviors.

Key findings. Findings suggested that participants in this study (i.e., seven-year-old Chinese boys) demonstrated nearly *twice* the amount of hyperactivity as children of the same age and gender in the West.

Conclusion and implications: Mental health and understanding of behaviors are largely influenced by social contexts and cultural norms; a universal view of mental health conditions is largely inadequate. The cross-cultural legitimacy of mental health disorders varies considerably by culture (contingent on the disorder). It is essential to emphasize the role of *culture* and *context* in determining ways in which mental health issues might manifest. In this particular study, authors (Ho et al., 1996) concluded that results were unable to determine whether hyperactivity was more common among this sample when comparing to rates of hyperactivity in Western culture. The authors further emphasized that cross-cultural differences in Chinese adults' expectations and level of acceptance for behaviors outlined in the assessment were a very conceivable justification for the higher rates of hyperactivity among children in this sample. In this case, Chinese adults may perceive some behaviors children exhibited to be more severe due to their cultural perspective.

What Do You Think?

1. What implications do the study findings have for mental health community resourcing with cultural diversity?
2. What are the prospects and limitations of cultural beliefs on resourcing for community mental health wellbeing?

caring for their loved ones (Novella, 2010). Nonetheless, the fact that rates of homelessness and incarceration increased among individuals with mental health illness (Teich, 2016) serves as evidence of the under-resourcing of the mental health wellbeing for people who were deinstitutionalized, contributing to their neglect for successful community living.

Recent developments in community mental health resourcing are premised on a framework of cultural competence (i.e., having the knowledge, skills, and processes requisites to function appropriately in culturally diverse circumstances or communities) that focuses on developing cultural *partnerships* within communities rather than clinical diagnosis (Gopalkrishnan, 2018). The trending community mental health practices for developing partnerships for health with communities would allow for more equitable relationships between mental health wellbeing policy makers, care providers (inclusive of families), and community members for successful living in the communities people belong to (Caplan, 2013; Rosen et al., 2015). Developing collaborative partnerships also provides opportunities for inclusive community health wellbeing practices that seek to proactively meet the needs of all community members across different life domains, rather than focus on practices that are predicated on diagnosis (Murray & Skull, 2003; Rosen et al., 2015).

Globally, many countries have transitioned to community mental health programs as an alternative to psychiatric hospitalization (Barry et al., 2019; Markström, 2014). These types of programs have grown in popularity worldwide and show promise in treatment of individuals with mental illnesses (Markström, 2014). Community mental health programs are treatment focused and typically include the use of multidisciplinary teams in the context of primary health care, providing person-centered care with social support networks (Kroenke & Unutzer, 2017). *Person-centered care* emphasizes the client as the expert on themselves and encourages a collaborative emphasis between the practitioner and client to achieve positive outcomes (Gehart, 2012). Community mental health is intermediary to sustainable community mental health wellbeing that is aimed at promoting wellness for all regardless of clinical mental health statuses.

Pertinent Sustainable Community Mental Health-Oriented Approaches

Sustainable community mental health approaches are those that are designed to be accessed by typical community members, while also with scope to provide for those with atypical mental health needs resulting from mental illness. These approaches should provide a seamless connection between services that support both the physical and mental health issues of community members, are participatory in nature, while addressing environmental and social justice issues community members experience (see also Chap. 6, this volume). Sustainable community mental health resourcing includes the following qualities: access and reach, scalability, feasibility and flexibility, and integrated health systems.

Access and Reach Qualities Mental health wellbeing access and reach are sustainability pillars of community wellbeing. Access qualities include ease of reach of necessary mental health wellbeing services within the community setting, in addition to having access and reach being a typical part of resourcing for community living (Kazdin & Rabbitt, 2013; Patel et al., 2018). This requires a health and wellness approach to population health so that people have multiple layers and opportunities to thrive in normative community settings, meeting their everyday living needs (Ammon, Curry, Hardy, Lax, & Tracy, 2015). Reach and access are closely related resources for mental health wellbeing in that reach enhances inclusive access to community members who would otherwise not be served by traditional clinic-based services, including the rural and urban poor, racial/ethnic and gender minorities. As examples, minority groups in the US face a variety of disparities in accessing mental health wellbeing resources, including clinical care (Wang, Demler, & Kessler, 2002). They also live in in-hospitable community settings with disparities in housing, transportation, and neighborhood safety (Neckerman et al., 2009; Williams & Jackson, 2005), which increases their levels of mental health stress (Sternthal, Slopen, & Williams, 2011). Further, their reach of targeted services to meet their physical health, mental health, and community living needs is poor (Lo, Cheng, & Howell, 2014), and explained in part by a general mistrust of medical personnel based on intergenerational discriminatory experience and stigmatization (Raphael & Stoll, 2013; Sanchez, Ybarra, Chapa, & Martinez, 2016). There is evidence to suggest that social policies and programs designed to provide for racial minorities in the US resulted in better health outcomes in those communities (Miller, Pollack, & Williams, 2011; Sanchez et al., 2016) as would selective, targeted, and indicated interventions for enhancing reach and access to wellness resources for other mainstream community members (Mpofu, 2005). In other words, the resourcing of community mental health wellbeing should be tailored to the demographics of the community, rather than a one-size-fits-all approach (McGorry, Bates, & Birchwood, 2013). A public health mental health policy that enables access to wellness resources in everyday settings of ordinary community members would promote overall community mental health wellbeing (Slade, 2012; see Discussion Box 7.1).

Discussion Box 7.1: Mental Health Stigma

According to SAMHSA (2013), mental health is an essential element of complete health; however, stigma (a set of unfair, often realistic beliefs about a disease or condition) is a large reason why many individuals fail to talk about issues related to mental health (and is largely informed by societal and cultural values). For instance, in the US, individuals with mental illness are often perceived as more dangerous than those without mental health issues, and those with mental health issues are in many cases blamed for their illness. Often, individuals begin to internalize these publicly endorsed, unfounded beliefs, resulting in self-stigma, which can interfere with help-seeking behaviors and foster negative self-perceptions. Self-stigma can also exacerbate their mental health symptoms. Research has in many cases shown that facilitating understanding of the connection between mental health and overall health increases the likelihood that community members with mental health issues will discuss mental health issues and seek supports when needed. It is vital that community mental health practitioners and administrators seek ways in which they can help community members to boost acceptance, reduce misperceptions, and improve negative beliefs associated with mental illness.

What Do You Think?

1. In what ways does stigma likely create access and reach barriers for community members with mental health issues?
2. How may community health workers assist in increasing community mental health access and reach?

Expansion of a nonprofessional workforce or use of community health workforce allows for a more intimate connection with the community by those who know the community well, often through lived experience (Alem, Jacobsson, & Hanlon, 2008). Use of community mental health workers as a workforce optimizes reach into the most vulnerable community segments, as community mental health workers are trained in task-shifting skills, which means that they can provide a broad range of services: consultation, direct services, referral, and follow-up. These task-shifting capabilities enable both timely and targeted services to community members, with community mental health workers acting as liaisons who bridge the gap between the mental health needs of the community and available services (of which the ordinary community members may be unaware). Task shifting with community health workers has strong evidence of effectiveness in developing countries treating communicable and non-communicable diseases (WHO, 2008). Community mental health workers have been successfully trained to deliver mental health services with task shifting following brief training (Buttorff et al., 2012). Expansion of mental health service settings is when quality mental health services are provided to community members across their typical social service access points, without requiring clinic hours. This calls for innovative mental health services design in which community members can elect the supports they need, which are aimed to optimize their wellbeing, and are inclusive of their mental health needs. When health care services prioritize user setting preferences, health outcomes are optimized and at lower delivery costs (Mühlbacher, Bethge, Reed, & Schulman, 2016).

Scalability Scalability refers to the capacity of mental health services to be provided on a larger community level with integration of typical health-oriented services that the community offers (Milat, King, Bauman, & Redman, 2013). Scalability spans the social, economic, and environmental pillars of community mental health wellbeing. As an environmental sustainability pillar, enhanced scalability permits community mental health services to respond to emerging needs from unforeseen events such as natural disasters (e.g., earthquakes, or forest fires) or man-made disasters (e.g., major industrial accidents). Disadvantaged communities typically contend with a number of these environmental stressors in addition to other social and economic sustainability-related health issues, including limited physical activity, diabetes, obesity, unhealthy food choices and options, poor access to health care services, asthma, issues related to hygiene, and lack of dental and eye care facilities (Alfonso, Jackson, Jackson, Hardy, & Gupta, 2015). Sustainable community mental health services require scalability to the extent possible to address these community concerns comprehensively, which if left unattended could aggravate, if not cause, mental health problems. Mental health wellbeing resourcing scalability would involve timely collaborative adoption of proven community wellness practices, human social capital partnerships, and financing policies for sustainability (Meredith, Branstrom, Azocar, Fikes, & Ettner, 2011; Milat et al., 2013). By using a collaborative approach, community members and their health providers can easily scale up the resources with evidence to enhance community mental health wellbeing in sustainable ways (see also Gehart, 2012).

Affordability The economic pillar of sustainable community mental health wellbeing is premised on its cost-effectiveness to the community members. Affordable community mental health wellbeing resourcing includes livable communities (see Chap. 3, this volume), nutrition security (see Chap. 5 this volume), and access to quality, low-cost mental health care. Individuals with severe mental illnesses are at a higher risk of being uninsured and more likely to enroll in public insurance programs like Medicare and Medicaid (Drake & Latimer, 2012; Garfield, Zuvekas, Lave, & Donohue, 2011). As an example of initiatives primarily focused on decreasing costs of community mental health wellbeing, the US enacted the Mental Health Parity and Addiction Equity Act in 2008, providing insurance coverage for individuals seeking mental health and substance use treatment, while placing importance on treating behavioral health conditions with the same significance as other medical conditions

(Barry & Huskamp, 2011). Subsequently, the Patient Protection and Affordable Care Act (PPACA) in 2010 was enacted, increasing coverage for an estimated 27.6 million Americans with severe mental illnesses.

In countries with universal medical care services, affordability is provided for through state subsidies, with all community members having access to basic health care services, inclusive of mental health care. As an example, Germany and Sweden provide national health insurance, which gives access to mental health care (Altenstetter, 2003; Amroussia, Gustafsson, & Mosquera, 2017). A cost-effective resourcing of community mental health services is through community-preventative and health-promotive programs aimed at improving the mental health of community members (Grosios, Gahan, & Burbidge, 2010).

Feasibility and Flexibility These service qualities refer to the adaptability or customization of services to local community conditions all while considering the diverse needs of the community (Kazdin & Rabbitt, 2013). When providers and community members collaborate to co-design community mental health wellbeing programs, it enhances the feasibility for these programs to achieve their purpose (Rosen et al., 2015). Co-designed community mental health wellbeing programs have the advantage to incorporate existing community assets, contributing to their sustainability (Hackett, Mulvale, & Miatello, 2018; Robert et al., 2015). Co-designed services would also have built in flexibility as to how they are implemented, for who and for what needs, and under what community living circumstances. Feasibility and flexibility of community mental health wellbeing is premised on data gathering from community members on their lived mental health needs and the solutions they perceive to be tailored to their needs.

Community-based participatory approaches (CBPA) are a proven resource for the design and implementation of sustainable community mental health systems for feasibility and flexibility. CBPA is a collaborative approach that includes all health partners to combine knowledge and action for social change to improve community health (Stacciarini, Shattell, Coady, & Wiens, 2011). It is as much a method for developing sustainable community health systems as a practice orientation. Applying CBPA to the design of sustainable community mental health systems would include the following steps (Minkler & Wallerstein, 2008):

(a) Defining membership boundaries
(b) Identifying community strengths and resources to build or enhance services
(c) Empowering the community through power sharing to address social inequities
(d) Promoting co-learning and capacity building among all partners
(e) Focusing on action for the mutual benefit of all partners
(f) Emphasizing public health problems of local relevance
(g) Development of health care systems through a cyclical and iterative process
(h) Wide dissemination of findings and knowledge gained with and for the community
(i) Building commitment to sustainability

Integrated Physical and Mental Health Wellbeing Providing comprehensive community mental health promotion services requires interventions across multiple life domains, including physical health, social functioning, employment, and overall quality of life (Corrigan, Druss, & Perlick, 2014). As an example, the resourcing of community mental health wellbeing requires addressing physical health needs, as mental health is strongly associated with physical health and vice versa (LGMHG, 2007; Sharan et al., 2009). Failure to recognize the interdependency of physical and mental health wellbeing would harm community mental health wellbeing from mis-targeted and inefficiently framed wellness resources (Carroll, 2017). Consequently, resources for sustainable community mental health wellbeing would achieve a multiplier effect in addressing *both* the physical health needs of the community (see also Chaps. 3 and 5 this volume) and the mental health needs of the community (see also Chap. 6, this volume; see Discussion Box 7.2).

Integrating physical and mental health care is best practice in community-oriented mental health (Lake & Turner, 2017; Rosen et al., 2015)

Discussion Box 7.2: The Connection Between Physical and Mental Health

According to Thornicroft (2011), there is a significant association between physical and mental health. In fact, many groups with mental health diagnoses experience shorter life expectancy than community members without mental illness (some groups experiencing a 20% reduction). There are many factors that account for this disparity, as those with mental illness often experience added risk factors for chronic disease (e.g., heart disease, diabetes), side effects from medication intended to address mental health issues, suicide risk, and poorer access to comprehensive health care than community members without mental health issues. Rodgers et al. (2018) also noted that many community health systems fail to provide integrated care (care in which mental and physical health issues are treated simultaneously) which is intended to improve overall community health outcomes, as many experience barriers such as (a) poor communication among community health providers, (b) ambiguity as to which professionals are responsible for care that addresses physical health, and (c) failure to recognize the impact of stigma within the community in which services are provided.

What Do You Think?

1. What are the benefits to sustainable health of integrating physical and mental health care?
2. Given the adage that "there is no health without mental health," how would mental health be a pathway to sustainable community health?

and is cost effective (Dovidio & Fiske, 2012; Lee-Tauler, Eun, Corbett, & Collins, 2018; Satel & Klick, 2005). This integrated and collaborative approach to community mental health promotion is a promising method for ensuring the sustainability of community health services.

Cultural, Legislative, and Professional Issues That Impact Sustainable Community Mental Health Approaches

As previously discussed, accessibility and cost-effectiveness are key to the resourcing of community mental health wellbeing (see also Bruckner, Singh, Snowden, Yoon, & Chakravarthy, 2019). However, cultural communities have unique needs (Wang et al., 2002), as cultural beliefs influence how community members understand mental health and the necessary supports for sustainable community mental health wellbeing. For instance, people from collectivist cultures would embrace a community living and wellbeing view of mental health compared to a view framed on mental illness experience, which they would view as punitive and embarrassing (Karim, Saeed, Rana, Mubbashar, & Jenkins, 2004; Lee, Lee, Chiu, & Kleinman, 2005). These cultures may undervalue clinic-based services, preferring community engagement-oriented activities for health promotion across the spectrum of mental health statuses. Thus, it is important for practitioners in community mental health to understand the role of culture in each community member's approach to mental health and wellbeing (see Research Box 7.2).

Legal and Professional Issues Sustainable community mental health wellbeing requires multi-sectoral human services resourcing collaboration to realize its full potential. In many jurisdictions, there are fewer laws and professional codes for providing community mental health wellbeing. Rather, the prevalent laws and professional codes are for clinical mental health services, which neglect the mental health wellbeing needs of the community as a whole (Rosen et al., 2015). Moreover, although the availability of clinical mental health services is an important aspect of community mental health webbing, it remains underfunded in most jurisdictions (Tomlinson & Lund, 2012). Moreover, clinical mental health services are typically provided as a stand-alone health service and not coordinated with physical health needs (Sipe et al., 2015; Sundararaman, 2009), while the major insurance players are less invested in community wellbeing resources as in providing clinical care services. Ideally, community mental health wellbeing should be included within general health policy and community planning (Funk & Sataceno, 2009; Jenkins, 2003; Sipe et al., 2015).

Research Box 7.2 Examining Disparities in Community Mental Health Service Utilization in Minority Communities

Background. Minority communities have experienced poorer outcomes related to health and mental health, issues that are largely attributed to fundamental lack of service access. The mental health needs of minority groups continue to be unaddressed. Communities that are poor, and have high concentrations of ethnic/racial minorities frequently lack the necessary resources to maintain or improve community member's mental health at an acceptable level. In line with social selection theory, White community members may be less represented within poorer neighborhoods because of social privilege, and only in the most extreme cases of mental illness would reside in higher poverty neighborhoods. The purpose of this study was to determine whether racial/ethnic disparities existed in communities with differing poverty levels.

Method. Participants in this study were $N = 78{,}085$ residents of New York City who received services from state-funded community mental health care facilities. Researchers conducted statistical analysis examining demographic, diagnostic, and service utilization as primary variables of interest in both high-income and low-income areas.

Key findings. Findings suggested that there were significant racial and ethnic disparities in mental health service utilization within communities. Researchers concluded that minority groups may access community mental health services through different means than White counterparts. Minority groups were significantly more likely to utilize emergency services for mental health issues than Whites. Racial and ethnic disparities in mental health service utilization continued, but were actually more noticeable in lower poverty areas. Regardless of socioeconomic status, minority community members (Black and Hispanic) were less likely to be referred to community mental health services by self, family, or friends; they were more likely to be referred by law enforcement officials in lower poverty areas. Results indicated that the use of community mental health services was both more coercive and less collaborative in lower poverty areas for minority community members.

Conclusion and implications: Researchers concluded that community mental health services should be personalized to address the differing needs of minority communities in different community settings. In addition to this, investigating routes to mental health service access and utilization is essential in order to generate more suitable routes for minority community members within lower poverty areas.

What Do You Think?

1. What steps can community health agencies take to reduce stigma and discrimination among low-poverty minority communities as it relates to mental health?
2. What barriers do you believe minority community members experience when seeking adequate mental health care in both low and high poverty settings? What can be done to address these barriers?

Related Disciplines Influencing Community-Oriented Health Aspects

Community mental health is an interdisciplinary field—as a holistic understanding of many different fields and disciplines is necessary to provide quality and comprehensive care. Behavioral and social sciences have a significant influence on community mental health as they also focus on individual *and* societal factors that impact overall health and wellbeing.

Behavioral and Social Sciences Behavioral and social sciences address various issues that directly affect the overall success or failure of community mental health efforts which include supporting wellbeing, allocating cost-effective health care equitably, utilizing health care organizations, observing providers' performance, and examining the psychological and social consequences of illness and death. They also include following the social and psychological impact of treatment on recovery, transmitting resources and principles across generations, recording mechanisms of social support, authenticating the outcomes of methods of care, and generating health decisions (National Research Council Committee for Monitoring the Nation's Changing Needs for Biomedical, Behavioral, and Clinic Personnel, 2005). These issues are important to understand as an interface between these matters determines how individuals make decisions while accessing services for mental health.

Community psychology is a behavioral science discipline that focuses on challenges encountered by individuals living with serious mental illness in the community (Glenwick, 1979). The field is separated from clinical psychology, which placed a deeper influence on the individual when it came to conceptualizing and treating mental health issues. Community psychologists emerged as new innovators to address complex social problems encountered while working in community mental health settings (Townley, Brown, & Sylvestre, 2018). Community psychology has contributed to the community mental health field to address issues related to social justice, multiculturalism, collaboration, and citizen participation and empowerment because community mental health not only focuses on an individual's illness, it also focuses on an individuals' strengths, aspirations, and capabilities through a recovery perspective (Thornicroft et al., 2016). Furthermore, the field of psychology has developed and advanced itself with many subspecialties such as clinical, counseling, rehabilitation, community, pediatric psychology, health psychology (a.k.a. behavioral medicine or medical psychology), and clinical neuropsychology, with practitioners providing services in primary, secondary, and tertiary care settings.

Interdisciplinary and Multidisciplinary Teams Multidisciplinary and interdisciplinary teams that address community mental health issues commonly include psychiatrists, medical staff, psychologists, social workers, counselors, peer support specialists, case managers, and vocational rehabilitation specialists. However, these teams have often been discussed in regard to their role in treating individuals with mental illness rather than in promoting community mental health. It is important to understand the complexities of interagency teams as they coordinate services within and outside of the agency. In a multistage qualitative research study, researchers (Andvig, Syse, & Severinsson, 2014) aimed to understand the interdisciplinary collaboration related to mental health providers employed in community mental health care settings in Norway. This study found that several factors were important to consider in developing effective collaborations within and between agencies, which involved (a) developing organizational strategies and interaction styles to enhance collaboration between professionals, (b) enhancing communication skills of all stakeholders, (c) developing methods of enhancing coordination and accountability, and (d) improved expert discernment into principles and circumstances required for effective resolutions, with a larger role of leadership in terms of organizing services and providing feedback to practitioners. In addition to these factors, Bronstein (2003) emphasized the importance of clearly defining and understanding the roles of professionals within collaborations. It is also essential for professionals within collaborations to be flexible, emphasize cooperative ownership of goals, and reflect on processes that occur within and between agencies.

Research Issues and Other Forms of Scholarship on Sustainable Community Mental Health

Community mental health wellbeing is an emerging area of practice and the vast majority of current mental health systems are for clinical mental health care services. Community wellbeing tends to be provided for through primary health care or facility-based services (Chisholm, 2005), often overlooking the non-medical mental health needs of the community (Mpofu, 2015). Regardless, even the clinical mental health care services tend to be underfunded and ill-equipped to address pressing community mental health needs (Bond et al., 2014; Thornicroft & Tansella, 2004). Studies are needed that focus on the resourcing of multisectoral and sustainable holistic community mental health wellbeing systems.

Creating sustainable community mental health wellbeing programs requires integrated life-span, life-space approaches aimed to provide for the needs of specific groups within the communities (Bunting & Stacciarini, 2019; Currier, Stefurak, Carroll, & Shatto, 2017; Dossett et al., 2018; Kelley, Haas, Felber, Travis, & Davis, 2019; Rodríguez, Garcia, Blizzard, Barroso, & Bagner, 2018; Slewa-Younan, Blignault, Renzaho, & Doherty, 2018; Weng & Spaulding-Givens, 2017). Moreover, studies are needed to determine the most effective and suitable approaches in creating accessible community mental health services, peer support interventions, tailoring mental health services to diverse communities, educating the community about stigma and discrimination, and whether budget reallocation can help maintain and sustain community mental health programs (Aarons, Wells, Zagursky, Fettes, & Palinkas, 2009; Thornicroft et al., 2010).

Community mental health wellbeing research and practice could also benefit in community reach utilizing telehealth or telemedicine. These remote health support systems have tremendous potential for accessibility and reach, scalability, affordability, and providing integrated health services (see also Chaps. 10 and 11, this volume). With decreased cost of telehealth devices, distance services are now considered a cost-effective approach for community-based mental health services, particularly for low-income and rural areas (Neufeld, Case, & Serricchio, 2012), although fewer health providers have training in use of telehealth care services (McClellan, Florell, Palmer, & Kidder, 2020). Telehealth also allows mental health services to reach underrepresented or marginalized population such as those in jails (Zaylor, Whitten, & Kingsley, 2000), children and adolescents (Myers et al., 2018), and veterans (Azevedo, Weiss, Webb, Gimeno, & Cloitre, 2016). As an example, during the COVID-19 pandemic period, telehealth services have proved a critical resource for community mental health care reducing infection rates that may have been caused by office visits (Zhou et al., 2020). Studies are needed on telehealth resourcing of community mental health wellbeing beyond clinical care services by providers.

Conclusion

Community is an essential aspect of the mental health of individuals. Yet, the literature on community mental health is preponderantly biased toward clinical care services, often neglecting the mental health needs of the community at large. We discussed the notion of sustainable community mental health resourcing for wellbeing taking into account possible influences of diversity in cultural orientation (i.e., collectivistic and individualistic), prioritizing community living needs rather than pathology, and applying multidisciplinary approaches. We propose a need for research and practices on the design and implementation of community mental health wellbeing programs for accessibility and reach, affordability, scalability, and integration across life domains.

Self-Check Questions

1. Define community mental health wellbeing resourcing.
2. Outline the historical research and/or practices on community mental health wellbeing. Identify the key events that are progressing this movement.
3. Identify and characterize community mental health resourcing approaches and their merits.
4. What has been the influences of culture, law, and policy in the resourcing of community mental health wellbeing?
5. What disciplines are involved in community mental health wellbeing and in what roles?
6. Identify and discuss two to three leading issues for research and other forms of scholarship on community mental health wellbeing resourcing.

Discussion Questions

1. What are some ways in which your culture has shaped your views on community mental health wellbeing resourcing? How do you think these views would shape community member participation in their personal and collective health?
2. Think of a community you are familiar with and identify the major barriers and enablers to their community mental health wellbeing?

Experiential/Field Exercises

1. Seek an opportunity to visit with a community organization to learn of how their services contribute to community mental health wellbeing? What other ways could the community organization be serving to meet the mental health needs of the community beyond what they shared?
2. **Online Scavenger Hunt**. Work alongside classmates, and consider local resources for community mental health wellbeing. Consider questions that you might have about those resources and the answers to which would contribute to community mental health wellbeing with implementation.
3. **Perspective Taking.** Select a community of interest and write about their likely mental health wellbeing assets and liabilities. Also, consider how this community could be resourced for better mental health wellbeing.

Key Online Resources

Substance Abuse Mental Health Services Administration
SAMSA.gov
National Institute on Mental Health
https://www.nimh.nih.gov/index.shtml
American Psychiatric Association
https://www.psychiatry.org/
National Alliance on Mental Illness
https://www.nami.org/
World Health Organization
https://www.who.int/mental_health/en/

References

Aarons, G. A., Wells, R. S., Zagursky, K., Fettes, D. L., & Palinkas, L. A. (2009). Implementing evidence-based practice in community mental health agencies: A multiple stakeholder analysis. *American Journal of Public Health, 99*(11), 2087–2095.

Alem, A., Jacobsson, L., & Hanlon, C. (2008). Community-based mental health care in Africa: Mental health workers' views. *World Psychiatry, 7*(1), 54–57.

Alfonso, M. L., Jackson, G., Jackson, A., Hardy, D., & Gupta, A. (2015). The willow hill community health assessment: Assessing the needs of children in a former slave community. *Journal of Community Health, 40*(5), 855–862.

Altenstetter, C. (2003). Insights from health care in Germany. *American Journal of Public Health, 1*, 38–44.

American Psychiatric Association. (2013). *Diagnositc and statistical manual of mental disorders* (5th ed.). https://doi.org/10.1176/appi.books.9780890425596

Ammon, N., Curry, M. B., Hardy, J., Lax, G., & Tracy, H. (2015). *Health in all policies*. Health Connections Network. Retrieved on July 17, 2020, from https://www.scph.org/sites/default/files/editor/2015_HiAP_Report.pdf

Amroussia, N., Gustafsson, P., & Mosquera, P. (2017). Explaining mental health inequalities in northern Sweden: A decomposition analysis. *Global Health Action, 10*, 1–10.

Andvig, E., Syse, J., & Severinsson, E. (2014). Interprofessional collaboration in the mental health services in Norway. *Nursing Research and Practice, 1*, 1–8.

Azevedo, K. J., Weiss, B. J., Webb, K., Gimeno, J., & Cloitre, M. (2016). Piloting specialized mental health care for rural women veterans using STAIR delivered via telehealth: Implications for reducing health disparities. *Journal of Health Care for the Poor and Underserved, 27*(4), 1–7.

Barry, C. L., & Huskamp, H. A. (2011). Moving beyond parity – Mental health and addiction care under the ACA. *The New England Journal of Medicine, 365*(11), 973–975.

Barry, M. M., Clarke, A. M., Petersen, I., & Jenkins, R. (Eds.). (2019). *Implementing mental health promotion.* New York: Springer Nature.

Bond, G. R., Drake, R. E., McHugo, G. J., Peterson, A. E., Jones, A. M., & Williams, J. (2014). Long-term sustainability of evidence-based practices in community mental health agencies. *Administration and Policy in Mental Health and Mental Health Services Research, 41*(2), 228–236.

Bronstein, L. R. (2003). A model for interdisciplinary collaboration. *Social Work, 48*, 297–306.

Bruckner, T. A., Singh, P., Snowden, L. R., Yoon, J., & Chakravarthy, B. (2019). Rapid growth of mental health services at community health centers. *Administration and Policy in Mental Health and Mental Health Services Research, 46*, 670–677.

Bunting, A., & Stacciarini, J. M. (2019). CBPR pilot intervention: Understanding social health determinants of mental health in rural Latinos. *University of Florida Journal of Undergraduate Research, 20*(2). https://doi.org/10.32473/ufjur.v20i2.106248

Buttorff, C., Hock, R. S., Weiss, H. A., Naik, S., Araya, R., Kirkwood, B. R., … Patel, V. (2012). Economic evaluation of a task-shifting intervention for common mental disorders in India. *Bulletin of the World Health Organization, 90*(11), 813–821.

Canino, G., & Alegria, M. (2008). Psychiatric diagnosis: Is it universal or relative to culture. *Journal of Child Psychology and Psychiatry, 49*, 237–250.

Caplan, G. (Ed.). (2013). *An approach to community mental health* (Vol. 3). New York: Routledge.

Carroll, A. E. (2017). The high costs of unnecessary care. *JAMA, 318*(18), 1748–1749.

Chisholm, D. (2005). Choosing cost-effective interventions in psychiatry: Results from the CHOICE programme of the World Health Organization. *World Psychiatry, 4*(1), 37–44.

Choudhry, F., Mani, V., Ming, L., & Khan, T. (2016). Beliefs and perceptions about mental health: A meta-synthesis. *Neuropsychiatric Disease and Treatment, 12*, 2807–2818.

Corrigan, P. W., Druss, B. G., & Perlick, D. A. (2014). The impact of mental illness stigma on seeking and participating in mental health care. *Psychological Science in the Public Interest, 15*(2), 37–70.

Currier, J. M., Stefurak, T., Carroll, T. D., & Shatto, E. H. (2017). Applying trauma-informed care to community-based mental health services for military veterans. *Best Practices in Mental Health, 13*(1), 47–65.

Dixon, L. B., & Goldman, H. H. (2003). Forty years of progress in community mental health: The role of evidence-based practices. *Australian and New Zealand Journal of Psychiatry, 37*(6), 668–673.

Dossett, E., Kiger, H., Munevar, M. A., Garcia, N., Lane, C. J., King, P. L., … Segovia, S. (2018). Creating a culture of health for perinatal women with mental illness: A community-engaged policy and research initiative. *Progress in Community Health Partnerships: Research, Education, and Action, 12*(2), 135–144.

Dovidio, J. F., & Fiske, S. T. (2012). Under the radar: How unexamined biases in decision-making processes in clinical interactions can contribute to health care disparities. *American Journal of Public Health, 102*(5), 945–952.

Drake, R. E., & Latimer, E. (2012). Lessons learned in developing community mental health care in North America. *World Psychiatry, 11*(1), 47–51.

Funk, M., & Sataceno, B. (2009). *Improving health systems and services for mental health*. World Health Organization. https://apps.who.int/iris/bitstream/handle/10665/44219/9789241598774_eng.pdf;jsessionid=BE4B4C43B03CF965AD03F26A9C4E3D54?sequence=1

Garfield, R. L., Zuvekas, S. H., Lave, J. R., & Donohue, J. M. (2011). The impact of national health care reform on adults with severe mental disorders. *American Journal of Psychiatry, 168*(5), 486–494.

Gehart, D. R. (2012). The mental health recovery movement and family therapy, part II: A collaborative, appreciative approach for supporting mental health recovery. *Journal of Marital and Family Therapy, 38*(3), 443–457.

Glenwick, D. S. (1979). The role and definition of community psychology. *American Psychologist, 34*(6), 559.

Gopalkrishnan, N. (2018). Cultural diversity and mental health: Considerations for policy and practice. *Frontiers in Public Health, 6*, 1–7.

Grosios, K., Gahan, P., & Burbidge, J. (2010). Overview of healthcare in the UK. *The EPMA Journal, 1*, 529–534.

Hackett, C. L., Mulvale, G., & Miatello, A. (2018). Co-designing for quality: Creating a user-driven tool to improve quality in youth mental health services. *Health Expectations: An International Journal of Public Participation in Health Care and Health Policy, 21*(6), 1013–1023.

Ho, T., Leung, P., Luk, E., Taylor, E., Bacon-Shone, J., & Mak, L. (1996). Establishing the constructs of childhood behavioral disturbances in a Chinese population: A questionnaire study. *Journal of Abnormal Child Psychology, 24*, 417–431.

Jenkins, R. (2003). Supporting governments to adopt mental health policies. *World Psychiatry : Official Journal of the World Psychiatric Association (WPA), 2*(1), 14–19.

Kading, M., Hautala, D., Palombi, L., Aronson, B., Smith, R., & Walls, M. (2015). Flourishing: American Indian positive mental health. *Society and Mental Health, 5*, 203–217.

Karim, S., Saeed, K., Rana, M. H., Mubbashar, M. H., & Jenkins, R. (2004). Pakistan mental health country profile. *International Review of Psychiatry, 16*(1–2), 83–92.

Kazdin, A. E., & Rabbitt, S. M. (2013). Novel models for delivering mental health services and reducing the burdens of mental illness. *Clinical Psychological Science, 1*(2), 170–191.

Kelley, F. R., Haas, G. L., Felber, E., Travis, M. J., & Davis, E. M. (2019). Religious community partnerships: A novel approach to teaching psychiatry residents about religious and cultural factors in the mental health care of African-Americans. *Academic Psychiatry, 43*(3), 300–305.

Kirmayer, L., Dandeneau, S., Marshall, E., Phillips, M., & Williams, K. (2011). Rethinking resilience from indigenous perspectives. *Canadian Journal of Psychiatry, 56*(2), 84–91.

Kroenke, K., & Unutzer, J. (2017). Closing the false divide: Sustainable approaches to integrating mental health services into primary care. *Journal of General Internal Medicine, 32*(4), 404–410.

Lake, J., & Turner, M. S. (2017). Urgent need for improved mental health care and a more collaborative model of care. *The Permanente Journal, 21*, 17–24.

Lancet Global Mental Health Group. (2007). Scale up services for mental disorders: A call for action. *Lancet, 370*, 1241–1252.

Lee, S., Lee, M. T., Chiu, M. Y., & Kleinman, A. (2005). Experience of social stigma by people with schizophrenia in Hong Kong. *The British Journal of Psychiatry, 186*(2), 153–157.

Lee-Tauler, S. Y., Eun, J., Corbett, D., & Collins, P. Y. (2018). A systematic review of interventions to improve initiation of mental health care among racial-ethnic minority groups. *Psychiatric Services, 69*(6), 628–647.

Lo, C., Cheng, T., & Howell. (2014). Access to and utilization of health services as pathway to racial disparities in serious mental illness. *Community Mental Health Journal*, 50, 251–270.

Markström, U. (2014). Staying the course? Challenges in implementing evidence-based programs in community mental health services. *International Journal of Environmental Research and Public Health, 11*(10), 10752–10769.

McClellan, M. J., Florell, D., Palmer, J., & Kidder, C. (2020). Clinician telehealth attitudes in a rural community mental health center setting. *Journal of Rural Mental Health, 44*(1), 62.

McGorry, P., Bates, T., & Birchwood, M. (2013). Designing youth mental health services for the 21st century: Examples from Australia, Ireland and the UK. *The British Journal of Psychiatry, 202*(s54), s30–s35.

McGuire, T. G., & Miranda, J. (2008). New evidence regarding racial and ethnic disparities in mental health: Policy implications. *Health Affairs, 27*(2), 393–403.

Meredith, L. S., Branstrom, R. B., Azocar, F., Fikes, R., & Ettner, S. L. (2011). A collaborative approach to identifying effective incentive for mental health clinicians to improve depression care in large managed behavioral healthcare organization. *Administration and Policy in Mental Health and Mental Health Services Research, 38*(3), 193–202.

Milat, A. J., King, L., Bauman, A. E., & Redman, S. (2013). The concept of scalability: Increasing the scale and potential adoption of health promotion interventions into policy and practice. *Health Promotion International, 28*(3), 285–298.

Miller, B. J., Paschall, B., & Svendsen, M. (2006). Mortality and medical comorbidity among patients with serious mental illness. *Psychiatric Services, 57*, 1482–1487.

Miller, W. D., Pollack, C. E., & Williams, D. R. (2011). Healthy homes and communities: Putting the pieces together. *American Journal of Preventive Medicine, 40*(1), S48–S57.

Minkler, M., & Wallerstein, N. (Eds.). (2008). *Community-based participatory research for health: From process to outcomes*. San Francisco, CA: John Wiley & Sons.

Mpofu, E. (2005). Selective interventions in counseling African-Americans with disabilities. In D. A. Harley & J. M. Dillard (Eds.), *Contemporary mental health issues among African Americans* (pp. 235–254). Alexandria, VA: American Counseling Association.

Mpofu, E. (2015). *Community oriented health services: Practices across disciplines*. New York: Springer Publishing Company.

Mühlbacher, A. C., Bethge, S., Reed, S. D., & Schulman, K. A. (2016). Patient preferences for features of health care delivery systems: A discrete choice experiment. *Health Services Research, 51*(2), 704–727.

Murray, S., & Skull, S. (2003). Re-visioning refugee health: The Victorian immigrant health programme. *Health Services Management Research, 16*, 141–146.

Myers, K., Cummings, J. R., Zima, B., Oberleitner, R., Roth, D., Merry, S. M., … Stasiak, K. (2018). Advances in asynchronous telehealth technologies to improve access and quality of mental health care for children and adolescents. *Journal of Technology in Behavioral Science, 3*(2), 87–106.

National Research Council (US) Committee for Monitoring the Nation's Changing Needs for Biomedical, Behavioral, and Clinic Personnel. (2005). *Advancing the nation's health needs: NIH research training programs*. Washington, DC: National Academies Press (US).

Neckerman, K. M., Lovasi, G. S., Davies, S., Purciel, M., Quinn, J., Feder, E., … Rundle, A. (2009). Disparities in urban neighborhood conditions: Evidence from GIS measures and field observation in New York City. *Journal of Public Health Policy, 30*(1), S264–S285.

Neufeld, J., Case, R., & Serricchio, M. (2012). Walk-in telemedicine clinics improve access and efficiency: A program evaluation from the perspective of a rural community mental health center. *Journal of Rural Mental Health, 36*(2), 33.

Novella, E. J. (2010). Mental health care in the aftermath of deinstitutionalization: A retrospective and prospective view. *Journal of Health Philosophy and Policy, 18*(3), 222–238.

Palha, A., & Marques-Teixeira, J. (2012). The emergence of psychiatry in Portugal: From its roots to now. *International Review of Psychiatry, 24*(4), 334–340.

Patel, V., Saxena, S., Lund, C., Thornicroft, G., Baingana, F., Bolton, P., … Herrman, H. (2018). The lancet commission on global mental health and sustainable development. *The Lancet, 392*(10157), 1553–1598.

Raphael, S., & Stoll, M. A. (2013). Assessing the contribution of the deinstitutionalization of the mentally ill to growth in the U.S. incarceration rate. *The Journal of Legal Studies, 42*(1), 187–222.

Robert, G., Cornwell, J., Locock, L., Purushotham, A., Sturmey, G., & Gager, M. (2015). Patients and staff as codesigners of healthcare services. *British Medical Journal, 350*, g7714.

Rodgers, M., Dalton, J., Harden, M., Street, A., Parker, G., & Eastwood, A. (2018). Integrated care to address the physical health needs of people with severe mental illness: A maping review of the recent evidence on barriers, facilitators and evaluations. *International Journal of Integrated Care, 18*, 1–12. https://doi.org/10.5334/ijic.2605

Rodríguez, G. M., Garcia, D., Blizzard, A., Barroso, N. E., & Bagner, D. M. (2018). Characterizing intervention strategies used in community-based mental health care for infants and their families. *Administration and Policy in Mental Health and Mental Health Services Research, 45*(5), 716–730.

Rosen, A., O'Halloran, P., Mezzina, R., & Thompson, K. S. (2015). International trends in community-oriented mental health services. In E. Mpofu (Ed.), *Community oriented health services: Practices across disciplines* (pp. 315–343). New York, NY: Springer.

Rosenhan, D. (1973). On being sane in insane places. *Science, 179*(4070), 250–258.

Sanchez, K., Ybarra, R., Chapa, T., & Martinez, O. N. (2016). Eliminating behavioral health disparities and improving outcomes for racial and ethnic minority populations. *Psychiatric Services, 67*(1), 13–15.

Satel, S. L., & Klick, J. (2005). The institutes of medicine report: Too quick to diagnose bias.*Perspectives in. Biology and Medicine, 48*(1), 15–23.

Scott, G., Ciarrochi, J., & Deane, F. (2004). Disadvantages of being an individualist in an individualistic culture: Idiocentrism, emotional competence, stress and mental health. *Australian Psychologist, 39*, 143–153.

Sharan, P., Gallo, C., Gureje, O., Lamberte, E., Mari, J. J., Mazzotti, G., … Saxena, S. (2009). Mental health research priorities in low- and middle-income countries of Africa, Asia, Latin America and the Caribbean. *The British Journal of Psychiatry, 195*(4), 354–363.

Shorter, E. (2008). History of psychiatry. *Current Opinion in Psychiatry, 21*(6), 593.

Sipe, T. A., Finnie, R. K. C., Knopf, J. A., Qu, S., Reynolds, J. A., Thota, A. B., … Nease, D. E. (2015). Effects of mental health benefits legislation. *American Journal of Preventive Medicine, 48*(6), 755–766.

Slade, M. (2012). Everyday solutions for everyday problems: How mental health systems can support recovery. *Psychiatric Services, 63*(702–704), 2012. https://doi.org/10.1176/appi.ps.201100521)

Slewa-Younan, S., Blignault, I., Renzaho, A., & Doherty, M. (2018). *Community-based mental health and wellbeing support for refugees: An evidence check rapid review.* Sax Institute. https://www.saxinstitute.org.au/wp-content/uploads/Community-based-mental-health-and-wellbeing-support-for-refugees_FINAL.pdf

Stacciarini, J. R., Shattell, M. M., Coady, M., & Wiens, B. (2011). Review: Community-based participatory research approach to address mental health in minority populations. *Community Mental Health Journal, 47*(5), 489–497.

Sternthal, M. J., Slopen, N., & Williams, D. R. (2011). Racial disparities in health: How much does stress really matter? 1. *Du Bois Review: Social Science Research on Race, 8*(1), 95.

Substance Abuse and Mental Health Services Administration. (2013). *Community conversations about mental health: Information brief.* Retrieved from: https://store.samhsa.gov/system/files/sma13-4763.pdf

Substance Abuse and Mental Health Services Administration. (2018). *Key substance use and mental health indicators in the United States: Results from the 2017 National Survey on Drug Use and Health* (HHS Publication No. SMA 18-5068, NSDUH Series H-53). Rockville, MD: Center for Behavioral Health Statistics and Quality, Substance Abuse and Mental Health Services Administration. Retrieved from https://samhsa.gov/data/

Sundararaman, R. (2009). *The U.S. mental health delivery system infrastructure: A primer.* Congressional Research Service. https://fas.org/sgp/crs/misc/R40536.pdf

Teich, J. (2016). Better data for better mental health services. *Issues in Science and Technology, 32*(2), 61–67.

Thornicroft, G. (2011). Physical health disparities and mental illness: The scandal of premature mortality. *The British Journal of Psychiatry, 199*, 441–442. https://doi.org/10.1192/bjp.bp.111.092718

Thornicroft, G., Alem, A., Antunes Dos Santos, R., Barley, E., Drake, R. E., Gregorio, G., … Wondimagegn, D. (2010). WPA guidance on steps, obstacles and mistakes to avoid in the implementation of community mental health care. *World Psychiatry, 9*(2), 67–77.

Thornicroft, G., Deb, T., & Henderson, C. (2016). Community mental health care worldwide: Current status and further developments. *World Psychiatry, 15*(3), 276–286.

Thornicroft, G., & Tansella, M. (2004). Components of a modern mental health service: A pragmatic balance of community and hospital care: Overview of systematic evidence. *The British Journal of Psychiatry, 185*(4), 283–290.

Tomlinson, M., & Lund, C. (2012). Why does mental health not get the attention it deserves? An application of the Shiffman and Smith framework. *PLoS Medicine, 9*(2), e1001178. https://doi.org/10.1371/journal.pmed.1001178

Townley, G., Brown, M., & Sylvestre, J. (2018). Community psychology and community mental health: A call for reengagement. *American Journal of Community Psychology, 61*(1–2), 3–9.

Urban Indian Health Institute, Seattle Indian Health Board. (2012). *Addressing depression among American Indians and Alaska natives: A literature review.* Seattle, WA: Urban Indian Health Institute.

Wang, P. S., Demler, O., & Kessler, R. C. (2002). Adequacy of treatment for serious mental illness in the United States. *American Journal of Public Health, 92*(1), 92–98.

Weng, S. S., & Spaulding-Givens, J. (2017). Informal mental health support in the Asian American community and culturally appropriate strategies for community-based mental health organizations. *Human Service Organizations: Management, Leadership & Governance, 41*(2), 119–132.

Williams, D. R., & Jackson, P. B. (2005). Social sources of racial disparities in health. *Health Affairs, 24*(2), 325–334.

World Health Organization. (2004). *Promoting mental health: Concepts, emerging evidence, practice.* Retrieved on May 10, from https://www.who.int/mental_health/evidence/en/promoting_mhh.pdf

World Health Organization. (2008). *Task shifting: Rational redistribution of tasks among health workforce teams: Global recommendations and guidelines.* Geneva, Switzerland: World Health Press.

World Health Organization. (2018). *Mental health: Massive scale-up of resources needed if global targets are to be met.* Retrieved from: https://www.who.int/mental_health/evidence/atlas/atlas_2017_web_note/en/

World Health Organization. (2020). *Mental disorders affect one in four people.* Retrieved on May 16, 2020, from: https://www.who.int/whr/2001/media_centre/press_release/en/

Zaylor, C., Whitten, P., & Kingsley, C. (2000). Telemedicine services to a county jail. *Journal of Telemedicine and Telecare, 6*(1_suppl), 93–95.

Zhou, X., Snoswell, C. L., Harding, L. E., Bambling, M., Edirippulige, S., Bai, X., & Smith, C. (2020). The role of telehealth in reducing the mental health burden from COVID-19. *Telemedicine and e-Health, 26*(4), 377–379.

8

Community Epidemiological Approaches

Chisom Nmesoma Iwundu, Diana Kuo Stojda, Kirsteen Edereka-Great, and Heath Harllee

Introduction

The community context is an important determinant of the health outcomes of its members through its existing assets and liabilities. A community's health assets include the individuals and families comprising the community, the various organizations and associations that make up a community's "civil society" and its primary health care systems (Morgan, Ziglio, & Davies, 2010; Rotegård, Moore, Fagermoen, & Ruland, 2010). Communities are unequal in their health assets that impact sustainable health and the distribution of morbidity and mortality (Morgan et al., 2010). These types of assets would also vary depending on the demographic makeup of the community (Hornby-Turner, Peel, & Hubbard, 2017; Klemera, Brooks, Chester, Magnusson, & Spencer, 2017). Moreover,

C. N. Iwundu (✉) • K. Edereka-Great • H. Harllee
University of North Texas, Denton, TX, USA
e-mail: Chisom.Odoh@unt.edu; Heath.Harllee@unt.edu

D. K. Stojda
University of Louisville, Louisville, KY, USA

E. Mpofu (ed.), *Sustainable Community Health*,
https://doi.org/10.1007/978-3-030-59687-3_8

diseases and health-related outcomes are not randomly distributed in communities; rather, certain distinct features exist that predispose, enable, or protect communities from various diseases and health outcomes (Gordis, 2014; Hornby-Turner et al., 2017; Klemera et al., 2017). Thus, understanding key health information, as well as identifying assets for managing wellbeing and diseases nested within communities and their associated risk factors, is important for sustainable community health.

Learning Objectives

By the end of the chapter, readers will be able to:

1. Define community epidemiology in the context of sustainable community health.
2. Trace the history of theory and practice in community-oriented epidemiology.
3. Identify and discuss current community epidemiological approaches and emerging high prospect approaches.
4. Contextualize community epidemiology in culture, legislation, and professional issues on relevance and scope of practice.
5. Identify and describe the interdisciplinary nature of community epidemiology.
6. Evaluate the research and practice evidence in community epidemiology.

Traditionally, epidemiologic approaches aim to provide the expertise in identifying communities with a disproportionate burden of diseases via disease quantification, determining disease etiology, and implementing intervention measures (Friis, 2010; Coffman et al., 2017; Smith et al., 2018; Susser, 1973; Vangeepuram et al., 2018). By definition, epidemiology is the study of the frequency, distribution, determinants of diseases and health-related states, and its application to control health problems in populations (Friis, 2010). Epidemiology approaches have a long association with public health services (Susser, 1973). With the use of epidemiological approaches, it is possible to account for disparities in community health problems while identifying communities needing the most attention. Many factors that affect the health of communities are unique to the communities, and identifying these factors enables communities to address their health needs (Kjærgård, Land, & Bransholm Pedersen, 2014).

Community epidemiology is an emerging field that goes beyond individual focus and traditional epidemiology in that it involves the identification of health assets of communities and how community members are involved in their own health promotion (Alvarez-Dardet, Morgan, Cantero, & Hernán, 2015; Mckenzie, 2018; Smith, 1998). Community epidemiology considers the cultural comprehension of the social assets of community health including social, economic, and environmental indicators of health (Alvarez-Dardet et al., 2015). Further, community epidemiology focuses on formal, informal, personal, and symbolic community health assets for wellness by incorporating community knowledge and community involvement (Botello et al., 2013). By extension, community epidemiology refers to research and practices aimed to explore, discover, implement, and evaluate community wellbeing interventions for sustainability (Mckenzie, 2018; Smith, 1998).

Community health assets mapping is a community-based participatory research (CBPR) approach that utilizes epidemiologic methods to understand the socioeconomic context of health conditions, and the relevance of mitigation approaches to the community members (Collins et al., 2018; Israel, Schulz, Parker, & Becker, 1998; Morgan et al., 2010; Salimi et al., 2012). Resolving health issues at the community level can aid community members in their decision of how best to utilize community resources, prevent and control emergent diseases, as well as evaluate program effectiveness. Understanding of community epidemiology can assist in the ability to identify, access, and interpret relevant community data for the development of programs that enhance the lives of community members.

Professional and/or Legal Definitions and Theories

As previously noted, epidemiology is the study of the frequency and distribution of determinants of diseases and health-related states, and its application to control health problems in populations (Friis, 2010; John, 2001). The three main types of epidemiological approaches consist of interventional epidemiology, descriptive epidemiology, and analytic epidemiology (John, 2001).

Interventional epidemiology uses the results from the descriptive and analytic approaches to implement public health actions and health interventions in communities. Informing interventional disease processes are screening tests which provide early detection to reduce the risk of diseases or increase effective treatment. These processes aid researchers, health practitioners, and social services to use screening methods to identify communities that may need additional testing when detecting the presence or lack of diseases. Examples of screening tools used to guide the design and implementation of public health actions for supporting community health include cholesterol measurement, fecal occult blood test, Pap test, mammography, and colonoscopy.

Descriptive epidemiology focuses on organizing and describing the distributions of diseases and health determinants. Descriptive epidemiologic approaches' measures such as sample size, central tendencies (mean, mode, median), prevalence, incidence, and disease morbidity and mortality are used to examine demographic makeup and differences (person, place, time) in specified communities. For any given study, researchers engage in descriptive epidemiology by assessing various characteristics (e.g., age, sex, income, education, etc.). Descriptive epidemiology is also used in generating/formulating a hypothesis which can be tested using analytic epidemiology.

Analytic epidemiologic approaches examine the hypothesis of associations of suspected determinants/risk factors of health outcomes using various epidemiological study designs and analytic models. Further, analytic epidemiology is an approach that studies etiological (casual) associations and relationships between causative factors and health conditions amongst communities. It answers the *how and why* of disease occurrence by testing the hypotheses generated by descriptive methods. Analytic epidemiology involves analytic models, such as logistic or linear regression, to deduce the effect of a particular risk factor (i.e., exposure) on a health outcome.

The following terminologies are commonly used in epidemiologic research:

- *Prevalence*: Refers to the number of existing cases of a health outcome in a population divided by the total population.
- *Incidence*: Is a risk measure used to capture the rate at which new cases of a disease or an outcome are occurring.
- *Morbidity*: Refers to the condition of being diseased or ill.
- *Mortality*: Refers to the condition of being subject to death.
- *Etiology*: The cause of a disease or an outcome.
- *Confounders*: Other variables that are associated with the risk factor of interest as well as the health outcome, but do not exist along the causal pathway of the two.
- *Effect modifiers*: These are factors that either decrease or increase the effect of the risk factor of interest upon the health outcome.

All of these terms relate in some manner to the intertwining of demography and public health. Epidemiology both succeeded and complemented social demography, which deals with population composition and change and how they interact with sociological variables (Fang, 1981; Omran, 2005). The theoretical impetus for epidemiology can be traced back to the work of Kurt Mayer (1962), who noted that the causes and effects of population changes should go beyond statistical measurement of the basic components (e.g., mortality or fertility), calling for the incorporation of theoretical frameworks from other fields. Subsequently, epidemiology theory evolved, which takes into account the biological, economic, sociologic, psychologic, and demographic ramifications of transitional processes (Omran, 2005).

Hippocrates in fourth century B.C.E., proposed that the human body was made up of four different substances (i.e., humors), namely, black bile, yellow bile, blood, and phlegm (Krieger, 2011). Ideal health equated to a balance of all four, and vice versa (see Fig. 8.1). The humors theory was eventually replaced with the miasma and contagion (i.e., germ) theory by the late 1800s (Last, 2001; Sterner, 2007). However, with the advances in biological sciences especially related to genetic predisposition and infectious diseases, these nineteenth-century theories were exchanged to biomedical, lifestyle, and social epidemiological theories of disease distribution (Krieger, 2014). See Fig. 8.1.

Time period	Theories of disease causation
4th century B.C.E.	*Humors* theory by Hippocrates (Hippocrates, 400 B.C.E.)
19th century A.D.	Miasma versus contagion theories (Sterner, 2007)
20th century A.D. – current	Biomedical, lifestyle (Krieger, 2014)
	Social epidemiology (Krieger, 2014) • Social production • Health and human rights • Social determinants • Psychosocial • Ecosocial

Fig. 8.1 Key milestones and major theories

More recent theorizing revolves around biomedical mechanism and individual lifestyles (Krieger, 2014). Alternative theories are somewhat less concrete, and oftentimes used in epidemiologic research; these include political, life course, and historical theories (Krieger, 2014). When discussing sustainable health in the context of epidemiology and creating policies, a more encompassing theoretical structure must be employed. One such example is Nancy Krieger's eco-social theory of disease distribution, which recognizes the social, political, and economic processes in shaping epidemiological profiles, and defines community epidemiology (Krieger, 2011, 2014). Thus, community epidemiology is an interdisciplinary science for understanding the broader dynamics of changes in different health systems, and for formulating policies to improve overall community health (Alvarez-Dardet et al., 2015; Berkman, Kawachi, & Glymour, 2014; Marmot, 2005).

In summary, community epidemiology theory relates to social epidemiology and relies on engaging community members in order to define social risk-driven disease mortality and morbidity (Smith, 1998). Understandably, engaging communities in epidemiologic studies requires a massive amount of time, relationship building, and finances. However, successful community assets-oriented epidemiologic interventions would reduce the reductionist tendencies of traditional epidemiologic studies that rely heavily on biomedical and lifestyle theories (Alvarez-Dardet et al., 2015; Smith, 1998).

History of Research and Practice

Previously, we observed that epidemiologic approaches have been applied in addressing disease occurrences as early as before 500 CE (Krieger, 2011). The history of epidemiologic approaches has involved historical figures whose efforts sought to understand and explain diseases, death occurrences, and provide information for the prevention and control of diseases (Gordis, 2014). Though not an exhaustive list, these figures include:

- John Graunt (1620–1647): Summarized the pattern of mortality in the seventeenth century.
- James Lind (1716–1794): Discovered the cause and prevention of scurvy.
- Edward Jenner (1749–1823): Developed vaccine that provided immunity to smallpox.
- William Farr (1807–1883): Developed a system for codifying medical conditions.
- Alexander Fleming (1881–1955): Discovered the antibiotic penicillin.
- John Snow (1800–1899): Showed that cholera was transmitted by fecal contamination of drinking water and made recommendations for public health action.

A major impetus to the development of epidemiology in modern times stemmed from eighteenth-century epidemics challenging "the belief that epidemic disease posed only occasional threats to an otherwise healthy social order" and occurred from "high levels of lethal infections" to people in the emerging industrial centers of the western world (Porter, 2005, p. 376). Epidemiologic-oriented public health practices had little regard for social justice, and in fact, upheld discriminatory practices (Paul & Spencer, 1995; Pearson, 1901). However, pious beliefs about poverty and moral failing no longer explained or offered a means of dealing with the consequences of industrialization and urbanization. Further to this, economic growth was causing widespread social instability. Major strikes between 1877 and 1892, and the assassination of President Garfield in 1881, indicated the potential risks for civil disorder. In such an atmosphere, social reform took on a new urgency. Increasing numbers of social reform movements developed during the last quarter of the century aimed at ameliorating social crises and preventing revolution and anarchy. Health reform played a significant role in this context which exemplified the new social consciousness (Porter, 2005, p. 376).

For instance, many early public health researchers and statisticians were fervent supporters of eugenics, including Sir Francis Galton (1822–1911), Karl Pearson (1857–1936), and R.A. Fisher (1890–1962) (Goering, 2014; Paul & Spencer, 1995; Pearson, 1901). Eugenics was the practice or advocacy of controlled selective breeding of human populations to improve the population's genetic composition (Cavaliere, 2018). These mid-eighteenth-century to mid-nineteenth-century beliefs enforced practices based on patriarchal values to protect the interests of those that were of "good birth," culminating in the eugenics-oriented human breeding programs instituted by Germany's Nazi regime preceding World War II (Goering, 2014). This led to the establishment of ethical and legal frameworks that govern the use of human subjects in experimental research. One such example framework is the Nuremberg Code of 1947 which provided a set of research ethics and guidelines for human experimentation in response to the atrocities committed by Nazi researchers (Shuster, 1997). The authors of the Nuremberg Code stated that human beings must voluntarily consent to participate in research before they are enrolled (Shuster, 1997).

Later in 1964, the World Medical Association (WMA) developed and ratified the first version of the Declaration of Helsinki to further clarify the ethical principles that should be followed when conducting human subjects research (Rickham, 1964). The Association's foremost goal was to address ethical complexities associated with conducting human subjects research internationally. The Declaration states that "no national or international ethical, legal or regulatory requirement should reduce or eliminate any of the protections for research subjects" (Rickham, 1964).

In 1974, the United States government ratified the National Research Act and also created the National Commission for the Protection of Human Subjects of Biomedical and Behavioral Research. After almost four years of monthly discussions and deliberations, the Nation Commission published the Belmont Report, a document that identifies basic ethical principles and provides guidelines to address ethical issues that arise from human subjects research (National Commission for the Protection of Human Subjects of Biomedical and Behavioral Research, 1979). This came as a result of the Tuskegee Study of Untreated Syphilis in the Negro Male, conducted from 1932 to 1972, a deliberate medical neglect of several hundred African American males who suffered from syphilis (Jones & Tuskegee Institute, 1993) despite the availability of penicillin treatment for the disease. Another minority group to fall victim to medical mistreatment in the United States are Native Americans; in the 1960s and 1970s, several Indian Health Service regions sterilized over 3000 American Indian women without obtaining their permission (Lawrence, 2000). The physicians that performed the operations did so with the mindset that Native women would not be able to use birth control intelligently (Lawrence, 2000).

Current community epidemiology practices must be used to engage communities in beneficial health promotion programs premised on their specific needs (George, Duran, & Norris, 2014). The next section considers some of the pertinent approaches.

Pertinent Community Epidemiologic Approaches

Community assets-oriented epidemiological approaches prioritize public health practices, not just for disease prevention and treatment, but for sustainable community health (Botello et al., 2013; Morgan et al., 2010; Rotegård et al., 2010). These approaches can be characterized by the following categories: community network based, continuum approaches, and community diagnostic approaches.

Community Network-Based Approaches Community networks consist of a multi- and interdisciplinary group of people committed to helping communities mobilize and manage resources to improve the community's health (Parry et al., 2002; Valente, Chou, & Pentz, 2007). Surveillance community network is a community assets-oriented approach based on health network systems data to improve the health of communities, as well as address pertinent health concerns in the community (Parry et al., 2002). An example is the South African Community Epidemiology Network on Drug Use (SACENDU). Partners included clinicians, researchers, and policymakers. Since 1996, the Network has been involved in assessing drug-related problems and designing appropriate interventions by employing multi-source approaches to data collection. Using descriptive epidemiological approaches, the SACENDU gathered data on alcohol and other drug use (AOD) indicators from 50 specialist AOD treatment centers across South Africa, which included both private and non-governmental institutions. Findings from the SACENDU surveillance study have produced evidence on the different types of substance trends, uses, and abuses—including AOD patterns of the demand and supply in South Africa. These findings have aided the understanding of substance use epidemiology and the formation and implementation of AOD reduction programs in South Africa (see also Research Box 8.1 on the Community Epidemiology Workgroup Model (CEWGM) created by the National Institute of Drug Abuse).

Continuum Approaches A community-epidemiology continuum framework aims to understand the continuum distribution of health assets and risks in communities. An example is the effort to understand between-and-within community agent-host protections and transmission of diseases. The work of Fenton and Pedersen (2005) is an example of a four-stage agent-host interaction framework that has facilitated the understanding, classifying, and quantifying of between-and-within transmission rates by identifying the location of a host-pathogen system within the continuum, with a view to mitigate disease threats.

Multi-host pathogens can infect several host species (Pedersen, Altizer, Poss, Cunningham, & Nunn, 2005; Roche et al., 2013), and those that

Research Box 8.1: Community Epidemiology Work Group Approach for Substance Use (Kozel, Robertson, & Falkowski, 2002)

Background: The issue of drug abuse and misuse presents many challenges that are uniquely associated with the epidemiological issues of accurately assessing patterns, measuring problems, and so on. As such, this study employed a community network-based approach by which a Community Epidemiology Workgroup (CEWG) made up of researchers, academics, community members, and local officials sought to develop a working model for drug abuse-free communities in the United States, and that would work in multiple setting and nations.

Method: A CEWGM convened by the National Institute of Drug Abuse (NIDA) sought to understand the dynamics of community drag abuse and considered future innovations to mitigate drug abuse through continuous surveillance, information exchange, and empowering communities to mitigate drug abuse by members.

Findings: From the 21 cities that utilized a CEWG, researchers concluded that the integration of drug use surveillance provided significant benefits in understanding the changes in patterns and trends of drug abuse. For example, the CEWG was one of the first sources to identify the advent of Rohypnol as an increasing drug of abuse, which eventually became banned in the United States. Unfortunately, most traditional public health programs were found to not have drug abuse monitoring integrated into their ongoing systems and network concepts.

Conclusion and Implications: The CEWG model has been employed in numerous cities, states, and even across different countries to understand the ongoing dynamics of drug abuse and improve use of tools for monitoring drug abuse patterns at local levels.

What Do You Think?

1. What community health assets would be required for the success of a community surveillance program on drug use and abuse in US cities?
2. What is the value of CEWG community involvement to facilitate sustainability of drug abuse surveillance? What would be its relative advantage compared to alternative approaches?

cross species barriers to infect humans can be particularly deadly (i.e., COVID-19; Ebola virus, HIV, West Nile virus). A community asset approach would seek to alter the ecological characteristics of the community based on its strengths to mitigate multi-host-pathogen transmissions and prevent community spread (see also Dobson, 2004; Dobson & Meagher, 1996; Fenton & Pedersen, 2005).

Community Diagnosis Approaches Community diagnosis approaches aim to assess community risk and protective factors to community well-being (Feinberg, 2012). An example of a community diagnosis approach is the Healthy Public Housing Initiative and The Asthma Center on Community Environment and Social Stress (ACCESS) (Boston, Massachusetts). The Healthy Public Housing Initiative (HPHI) started in 2001 and aimed to improve home environments in order to improve health and quality of life for those who live in public housing, especially pediatric asthmatics (Freeman, Brugge, Bennett-Bradley, Levy, & Carrasco, 2006). This initiative was a partnership of the Boston Housing Authority, the Boston Public Health Commission, Boston University School of Public Health, Urban Habitat Initiatives, as well as residents in housing communities (Freeman et al., 2006). ACCESS began shortly after the HPHI, with the main partners being the Channing Laboratory of Brigham and Women's Hospital, the Harvard School of Public Health, and the Center for Community Health Education Research and Service, Inc. (CCHERS). ACCESS sought not only to conduct comprehensive community assessments, but to determine the role of physical and social environmental exposures, and evaluate the efficacy of existing community health center asthma interventions (Freeman et al., 2006).

In a study by Feinberg, Jones, Greenberg, Osgood, and Bontempo (2010, Feinberg, 2012), a community diagnosis model was utilized to map consistency of associations between communities health assets for preventing substance abuse by Pennsylvania teenagers. In summary, using the community diagnosis approach helped target community assets use to mitigate risk factors to health and wellbeing in a sustainable way.

Cultural, Professional, and Legal Influences on Community Epidemiology Programs

Culture is fundamental to the health behavior of populations (Ayeni, 2008) and helps to shape people's understanding of their assets and risks for wellbeing (Napier et al., 2014). As such, culture in community epidemiology can reveal the health assets and liabilities of communities based on their attributed meanings (Massé, 2006). Applied to community epidemiology, health interpretations would influence the behavioral choices of community members with regard to their community living choices (see Chap. 3, this volume), nutrition choices (see Chap. 5, this volume), and substance use (see Chap. 6, this volume). Most significant is the fact that culture would influence what communities consider to be health assets and liabilities, in addition to influencing the options a community would consider to achieve better health outcomes.

In community epidemiology, policies and laws influence community health priorities and interventions. Some of the notable public health achievements such as immunization, fluoridation of drinking water, and tobacco control have been achieved through the formulation and implementation of policies that translate into laws (Frieden, 2014; Moulton, Goodman, Cahill, & Baker Jr, 2002). Recognition of the ability of laws to shape health outcomes led to the emergence of legal epidemiology which focuses on the impacts of laws on the distribution and frequency of diseases (Ramanathan, Hulkower, Holbrook, & Penn, 2017).

Further, community epidemiologic research is subject to ethical review. Ethical concerns have implications for policy in regard to the treatment of human subjects and the design and implementation of research protocols (Piasecki, Waligora, & Dranseika, 2017). In conducting community work, researchers must acknowledge not only the scientific component of research but also the ethical principles, especially for research involving human subjects. According to the American College of Epidemiology, guidelines must include measures such as minimizing risks and maximizing benefits, obtaining informed consent, maintaining public trust, meeting obligations to communities, and submitting proposed studies for ethical review (McKeown, Weed, Kahn, & Stoto, 2003). As such, various universities maintain a Human Subjects Review Board to review all research protocols and ensure that researchers adhere and meet the ethical standards of research.

Related Disciplines Influencing Epidemiologic Approaches

The advancement of knowledge regarding population health risks and protections, as well as disease etiologies affecting communities, requires an interdisciplinary approach to the design of sustainable health systems. The related disciplines would include public health, community medicine, biostatistics, environmental sciences, econometrics, and information sciences. Historically, epidemiologists have collaborated with public health practitioners in promoting community health. Public health is the science of preventing disease, prolonging life, and promoting health through organized efforts of the public at large (Merrick, 2013). Public health is concerned with the capacities of people to meet the complex demands of health in society such as literacy, education, occupational safety, and behavioral health. Through education and safety standards of public health, the control of disease is increased as health disparities decrease. A key function of public health is the assessment of community health for the purpose of evaluating policy, interventions, and resource allocation (World Health Organization, 2018). Public health practitioners create awareness about health issues affecting community members using data produced via epidemiologic methods.

Community medicine is another discipline that complements community epidemiology. Community medicine focuses on determinants of health, local health issues, community-oriented primary health care, and organization of health care services to attain optimal quality of health in a given geometric area (Abramson & Kark, 1983; Acheson, 1980). It addresses certain aspects of health promotion, disease prevention, health restoration, and rehabilitation of people in the community (Lilienfeld, 1978). Similar to community epidemiology, community medicine applies a focus on how to help people live healthier and happier within the community, in addition to addressing inequities within the community. Some issues in community medicine can be ameliorated using community epidemiologic methods. An example issue in community medicine is the discharge of patients who often have medicine-related problems at care transitions, a problem that epidemiologists have long worked to find a solution to by reducing readmissions (Spinewine, Claeys,

Foulon, & Chevalier, 2013). A 2005 study found that the provision of community medicine, continued education, and treatment of discharged patients dramatically reduced adverse events (Cohen et al., 2005). Fluoridation of drinking water is another classic example of clinical observation leading to epidemiologic investigation and community-based intervention.

Biostatistics is a way of visualizing data and involves collecting, describing, analyzing, and interpreting data, as well as causal relationships and disease trends (Dakhale, Hiware, Shinde, & Mahatme, 2012). In community epidemiology, statistical methods are used to formulate and test hypotheses, determine appropriate sampling techniques, and coordinate data collection procedures using appropriate statistical tools. For example, analytic models such as logistic or linear regression are used to deduce the effect of a particular risk factor on a health outcome. Notably, many problems affecting community members can be addressed by data collection, which can generate the development of effective policies and interventions intended to improve community health. As such, it is important that epidemiologists and biostatisticians continue to collaborate to collect quality data in order to explain the relationship between exposures and health-related states or events affecting communities. While there may be many ways to include humans into research studies, biostatisticians utilize epidemiologic methods in controlling for bias—especially selection bias resulting from methods that cause a differential opportunity for inclusion of particular subjects based on a characteristic related to their disease or their clinical course (Szczech, Coladonato, & Owen, 2002). Biases in studies present the opportunity for misdiagnosis, incorrect influences of a disease, and the methods to treat any given disease.

Healthy environments promote community health, and unhealthy environments can have an impact on morbidity and mortality. As such, environmental science is a discipline focused on environmental issues such as pollution that impact populations, and how we can address these issues (Chiras, 2014). Given that exposure assessment is crucial in estimating and reducing air pollution impacts, advanced technologies like air pollution sensors, smartphones, and air models are being utilized in estimating air pollution exposures in communities (Larkin & Hystad, 2017). Also pertinent to the field of epidemiology and environmental health is the application of Geographic Information Systems (GIS) in modeling disease occurrences. GIS are systems that provide a framework for collecting, analyzing, and displaying spatial data (Clarke, McLafferty, & Tempalski, 1996). Dating back to John Snow's use of spot maps, epidemiologists have always used maps to analyze the interrelationship between a health outcome, the environment, and location (Clarke et al., 1996). During an investigation, GIS can provide information on communities most affected by an exposure, after which tables, maps, graphs can be created to communicate results. Information provided via GIS can be used to design health interventions for communities experiencing the worst health outcomes and social injustice as a result of inequalities in those results.

Epidemiologic approaches are also influenced by economic literature related to financial stress and burden which impact on health outcomes. For example, when different groups live in either resource-rich or resource-poor areas of a community, their life trajectory can be greatly altered by of the quality of the neighborhood, and its schools can affect an individual's health from childhood to adulthood. Research has shown that living in a clean and safe neighborhood that also allows for education attainment is a major determinant of health (Woolf, Johnson, Phillips, & Philipsen, 2007). Structural inequities impact those living in resource-poor areas for the entirety of their lives. For example, African American women are at higher risk of having low-birthweight infants, which leads to higher infant death rates, even after considering socioeconomic factors (Braveman et al., 2017). Also, one of the strongest predictors of life expectancy is educational attainment, which differs across class and racial/ethnic groups, and ultimately affects employment status, income, and individual and intergenerational wealth (Olshansky et al., 2012). Another example of the effect of structural inequities is the lending policies of banking institutions, which continue to create differences in asset development, such as home or business ownership (Pager & Shepherd, 2008). Understanding the economic burden is vital in evaluating public health policies and programs, allocating resources to communities who need it the most, and health threat preparedness. Health economists also undertake cost-effective analysis to develop econometric techniques to inform policies in the design of interventions that will benefit communities (Moscone, Siciliani, & Vittadini, 2017).

Information science is a field essentially involved with the analysis, collection, classification, storage, dissemination, and use of information. Information sciences have been employed to improve traditional public health practice within the community for many years (Koch & Hägglund, 2009). Information science skills are being applied in community epidemiology through monitoring and distribution of data online, wearable devices, and personal visits, which has created a plethora of information for health professionals to assimilate for future reduction of health disparities within every community. An additional benefit of information science on community epidemiology is the expediency and privacy of an individual's medical records. Electronic health records are defined as computerized medical information systems (Abdekhoda, Ahmadi, Dehnad, Noruzi, & Gohari, 2016). The increased adoption of electronic health record systems by community health care providers puts forth an opportunity for access to more timely information, providing the opportunity to create small area views of community health (Comer, Gibson, Zou, Rosenman, & Dixon, 2018).

Research Critical to Community Epidemiology Practice

Epidemiologic research is historically associated with extensive data mining (Krieger, 2014), and more recently, with machine learning using computers to extract community health information, which may be useful for public consumption (Krieger, 2014). Unfortunately, these tools fail to employ theory, which is necessary to clarify disease causality, especially when trying to bridge individual and community/population-level disease occurrence (Krieger, 2014). For instance, in order to understand the underlying factors of health assets disparities in communities, one must first understand the different types of inequities that cause them. Health inequities are the systematic differences in opportunities for achieving optimal health within a specific population (Braveman, 2006).

The factors that structure health inequities include race and ethnicity, gender, socioeconomic status, immigration status, geography, and so on (World Health Organization, 2007). However, it is structural inequities in community health assets that are arguably the most important for community epidemiology to study with a view to developing models for sustainable community health. Drivers of structural inequities in community health assets such as access to clean water, adequate housing, and quality of schools (National Academies of Sciences, Engineering, and Medicine; Health and Medicine Division, 2017) are clearly important to sustainable community health. Regardless, few community epidemiology studies employ a health assets-oriented approach in preference to deficit approaches framed on disease conditions.

One issue community members face when participating in community epidemiology work is language barriers. Target communities may not be with fluency in the language of the researchers and vice versa (Lai et al., 2006; Larkey, Gonzalez, Mar, & Glantz, 2009). Moreover, the host community might be suspicious of the community epidemiology team (Jones, 2000; Friis, 2010), and lengthy consent documents may preclude an individual from participating (Mills et al., 2006). Community members might have differing priorities to those of the community epidemiology program and may encounter issues such as scheduling conflicts and lack of transportation, and so on (Yancey, Ortega, & Kumanyika, 2006). Research needs to advance the science of creating community partnerships for health and wellbeing to optimize the benefit from community epidemiology approaches.

Conclusion

Historically, epidemiologic approaches have been critical in identifying communities with a disproportionate burden of diseases via disease quantification, determining disease etiology, and implementing intervention measures. These approaches are crucial tools for understanding and providing evidence that can be used to facilitate the development and implementation of effective intervention programs, especially for addressing health disparities and inequities that exist among marginalized groups. In recent times, community epidemiology has emerged with a focus on health solutions that are driven by local constituencies to identify the health assets of communities for sustainable health outcomes. Community epidemiology approaches with their focus on community engagement and health assets identification are of added value to the design and implementation of sustainable health systems.

Self-Check Questions

1. Define the term community epidemiology.
2. What are the landmark historical events in the history of theory and practice in community-oriented epidemiology?
3. Discuss current community epidemiological approaches and emerging high prospect approaches
4. How do culture, legislation, and professional issues on scope of practice and relevance influence community epidemiology practice?
5. What are the major research and practice evidence issues in community epidemiology?

Discussion Questions

1. How do you distinguish between descriptive and analytic epidemiology?
2. How is a health assets approach to community epidemiology of higher potential yield to sustainable community health compared to a health risks approach?
3. How is your field of study related to community epidemiology?

Experiential Learning Exercise

1. Visit with a local community organization to learn about their typical activities and the outcomes they seek from them. Based on your study, consider how the activities of that organization are based on the health assets of that community and how those activities contribute to the health assets of the community.
2. Think about a community you are familiar with. Make a list of their health assets important to their long-term health. Say how and why you selected each of those assets? Would any of those assets also be health liabilities? How and why?

List of Relevant Online Resources

https://www.drugsandalcohol.ie/5934/1/NIDA_Assessing_drug_abuse_2nd_ed.pdf
http://www.mclph.umn.edu/watersedge/
https://www.cdc.gov/csels/dsepd/ss1978/lesson1/index.html
https://www.cdc.gov/eis/index.html
https://dash.harvard.edu/bitstream/handle/1/3372911/Brandt_Racism.pdf

References

Abdekhoda, M., Ahmadi, M., Dehnad, A., Noruzi, A., & Gohari, M. (2016). Applying electronic medical records in health care. *Applied Clinical Informatics, 7*(2), 341–354. https://doi.org/10.4338/ACI-2015-11-RA-0165

Abramson, J., & Kark, S. (1983). Community oriented primary care: Meaning and scope. In *Community oriented primary care: New directions for health services delivery*. National Academies Press (US). https://www.ncbi.nlm.nih.gov/books/NBK234632/

Acheson, R. M. (1980). Community medicine: Discipline or topic? Profession or endeavour? *Journal of Public Health, 2*(1), 2–6.

Alvarez-Dardet, C., Morgan, A., Cantero, M. T. R., & Hernán, M. (2015). Improving the evidence base on public health assets—The way ahead: A proposed research agenda. *Journal of Epidemiology and Community Health, 69*(8), 721–723.

Ayeni, O. (2008). Epidemiology and culture. *McGill Journal of Medicine: MJM, 11*(1), 87–89.

Berkman, L. F., Kawachi, I., & Glymour, M. (2014). *Social epidemiology* (Second Edition, New to this Edition). Burlington, MA Oxford University Press.

Botello, B., Palacio, S., Garcia, M., Margolles, M., Fernandez, F., Hernan, M., … Cofino, R. (2013). Methodology for health assets mapping in a community. *Gaceta Sanitaria, 27*(2), 180.

Braveman, P. (2006). Health disparities and health equity: Concepts and measurement. *Annual Review of Public Health, 27*, 167–194. https://doi.org/10.1146/annurev.publhealth.27.021405.102103

Braveman, P., Heck, K., Egerter, S., Dominguez, T. P., Rinki, C., Marchi, K. S., & Curtis, M. (2017). Worry about racial discrimination: A missing piece of the puzzle of Black-White disparities in preterm birth? *PLoS One, 12*(10), e0186151. https://doi.org/10.1371/journal.pone.0186151

Cavaliere, G. (2018). Looking into the shadow: The eugenics argument in debates on reproductive technologies and practices. *Monash Bioethics Review, 36*(1), 1–22. https://doi.org/10.1007/s40592-018-0086-x

Chiras, D. D. (2014). *Environmental science.* Jones & Bartlett Learning.

Clarke, K. C., McLafferty, S. L., & Tempalski, B. J. (1996). On epidemiology and geographic information systems: A review and discussion of future directions. *Emerging Infectious Diseases Journal – CDC, 2*(2). https://doi.org/10.3201/eid0202.960202

Coffman, M. J., de Hernandez, B. U., Smith, H. A., McWilliams, A., Taylor, Y. J., Tapp, H., … Dulin, M. (2017). Using CBPR to decrease health disparities in a Suburban Latino Neighborhood. *Hispanic Health Care International: The Official Journal of the National Association of Hispanic Nurses, 15*(3), 121–129. https://doi.org/10.1177/1540415317727569

Cohen, M., Kimmel, N., Benage, M., Cox, M., Sanders, N., Spence, D., & Chen, J. (2005). Medication safety program reduces adverse drug events in a community hospital. *Quality & Safety in Health Care, 14*(3), 169–174. https://doi.org/10.1136/qshc.2004.010942

Collins, S. E., Clifasefi, S. L., Stanton, J., The Leap Advisory Board, null, Straits, K. J. E., Gil-Kashiwabara, E., … Wallerstein, N. (2018). Community-based participatory research (CBPR): Towards equitable involvement of community in psychology research. *The American Psychologist, 73*(7), 884–898. https://doi.org/10.1037/amp0000167

Comer, K. F., Gibson, P. J., Zou, J., Rosenman, M., & Dixon, B. E. (2018). Electronic Health Record (EHR)-based community health measures: An exploratory assessment of perceived usefulness by local health departments. *BMC Public Health, 18.* https://doi.org/10.1186/s12889-018-5550-2

Dakhale, G. N., Hiware, S. K., Shinde, A. T., & Mahatme, M. S. (2012). Basic biostatistics for post-graduate students. *Indian Journal of Pharmacology, 44*(4), 435–442. https://doi.org/10.4103/0253-7613.99297

Dobson, A. (2004). Population dynamics of pathogens with multiple host species. *The American Naturalist, 164*(Suppl 5), S64–S78. https://doi.org/10.1086/424681

Dobson, A., & Meagher, M. (1996). The population dynamics of brucellosis in the Yellowstone National Park. *Ecology, 77*(4), 1026–1036. https://doi.org/10.2307/2265573

Fang, S. L. (1981). [An elementary introduction to social demography]. *Ren Kou Yan Jiu = Renkou Yanjiu, 4*, 34–36.

Feinberg, M. E. (2012). Community epidemiology of risk and adolescent substance use: Practical questions for enhancing prevention. *American Journal of Public Health, 102*(3), 457–468. https://doi.org/10.2105/AJPH.2011.300496

Feinberg, M. E., Jones, D., Greenberg, M. T., Osgood, D. W., & Bontempo, D. (2010). Effects of the communities that care model in Pennsylvania on change in adolescent risk and problem behaviors. *Prevention Science : The Official Journal of the Society for Prevention Research, 11*(2), 163–171. https://doi.org/10.1007/s11121-009-0161-x

Fenton, A., & Pedersen, A. B. (2005). Community epidemiology framework for classifying disease threats. *Emerging Infectious Diseases, 11*(12), 1815–1821. https://doi.org/10.3201/eid1112.050306

Freeman, E. R., Brugge, D., Bennett-Bradley, W. M., Levy, J. I., & Carrasco, E. R. (2006). Challenges of conducting community-based participatory research in Boston's neighborhoods to reduce disparities in asthma. *Journal of Urban Health: Bulletin of the New York Academy of Medicine, 83*(6), 1013–1021. https://doi.org/10.1007/s11524-006-9111-0

Frieden, T. R. (2014). Six components necessary for effective public health program implementation. *American Journal of Public Health, 104*(1), 17–22.

Friis, R. H. (2010). *Epidemiology* 101. Burlington, MA: Jones & Bartlett Publishers.

George, S., Duran, N., & Norris, K. (2014). A systematic review of barriers and facilitators to minority research participation among African Americans, Latinos, Asian Americans, and Pacific islanders. *American Journal of Public Health, 104*(2), e16–e31. https://doi.org/10.2105/AJPH.2013.301706

Goering, S. (2014). Eugenics. In *The Stanford Encyclopedia of Philosophy* (Fall 2014). Metaphysics Research Lab, Stanford University. https://plato.stanford.edu/archives/fall2014/entries/eugenics/

Gordis, L. (2014). *Epidemiology (Fifth)*. Philadelphia, PA: Elsevier/Saunders.

Hornby-Turner, Y. C., Peel, N. M., & Hubbard, R. E. (2017). Health assets in older age: A systematic review. *BMJ Open, 7*(5), e013226.

Israel, B. A., Schulz, A. J., Parker, E. A., & Becker, A. B. (1998). Review of community-based research: Assessing partnership approaches to improve public health. *Annual Review of Public Health, 19*(1), 173–202. https://doi.org/10.1146/annurev.publhealth.19.1.173

John, M. L. (2001). *A dictionary of epidemiology*. Oxford, UK: Oxford University Press.

Jones, C. P. (2000). Levels of racism: A theoretic framework and a gardener's tale. *American Journal of Public Health, 90*(8), 1212–1215.

Jones, J. H., & Tuskegee Institute. (1993). *Bad blood: The Tuskegee syphilis experiment*. New York, NY: The Free Press.

Kjærgård, B., Land, B., & Bransholm Pedersen, K. (2014). Health and sustainability. *Health Promotion International, 29*(3), 558–568. https://doi.org/10.1093/heapro/das071

Klemera, E., Brooks, F. M., Chester, K. L., Magnusson, J., & Spencer, N. (2017). Self-harm in adolescence: Protective health assets in the family, school and community. *International Journal of Public Health, 62*(6), 631–638.

Koch, S., & Hägglund, M. (2009). Health informatics and the delivery of care to older people. *Maturitas, 63*(3), 195–199. https://doi.org/10.1016/j.maturitas.2009.03.023

Kozel, N. J., Robertson, E. B., & Falkowski, C. L. (2002). The community epidemiology work group approach. *Substance Use & Misuse, 37*(5–7), 783–803.

Krieger, N. (2011). *Epidemiology and the people's health: Theory and context*. Oxford, UK: Oxford University Press.

Krieger, N. (2014). Got theory? On the 21st c. CE rise of explicit use of epidemiologic theories of disease distribution: A review and ecosocial analysis. *Current Epidemiology Reports, 1*(1), 45–56. https://doi.org/10.1007/s40471-013-0001-1

Lai, G. Y., Gary, T. L., Tilburt, J., Bolen, S., Baffi, C., Wilson, R. F., … Ford, J. G. (2006). Effectiveness of strategies to recruit underrepresented populations into cancer clinical trials. *Clinical Trials (London, England), 3*(2), 133–141. https://doi.org/10.1191/1740774506cn143oa

Larkey, L. K., Gonzalez, J. A., Mar, L. E., & Glantz, N. (2009). Latina recruitment for cancer prevention education via Community Based Participatory Research strategies. *Contemporary Clinical Trials, 30*(1), 47–54. https://doi.org/10.1016/j.cct.2008.08.003

Larkin, A., & Hystad, P. (2017). Towards personal exposures: How technology is changing air pollution and health research. *Current Environmental Health Reports, 4*(4), 463–471. https://doi.org/10.1007/s40572-017-0163-y

Last, J. (2001). Miasma theory. In *Encyclopedia of public health* (p. 765). New York, NY: Macmillan.

Lawrence, J. (2000). The Indian Health Service and the sterilization of Native American women. *American Indian Quarterly, 24*(3), 400–419. https://doi.org/10.1353/aiq.2000.0008

Lilienfeld, D. E. (1978). Definitions of epidemiology. *American Journal of Epidemiology, 107*(2), 87–90. https://doi.org/10.1093/oxfordjournals.aje.a112521

Marmot, M. (2005). Social determinants of health inequalities. *Lancet (London, England), 365*(9464), 1099–1104. https://doi.org/10.1016/S0140-6736(05)71146-6

Massé, R. (2006). Epidemiology and Culture By James A. TrostleISBN 0-521-79050-6, Cambridge University Press, New York, New York (Telephone: 845-353-7500, Fax: 845-353-4141, Website: http://www.cambridge.org/us/), 2005, 208 pp., $70.00 (Hardback). *American Journal of Epidemiology, 163*(1), 97–98. https://doi.org/10.1093/aje/kwj014

Mayer, K. (1962). Developments In The Study of Population. *Social Research, 29*(3), 293–320.

Mckenzie, J. (2018). Epidemiology: The study of disease, injury, and death in the community. In *An introduction to community & public health* (pp. 62–89). Burlington, MA: Jones & Bartlett Learning.

McKeown, R. E., Weed, D. L., Kahn, J. P., & Stoto, M. A. (2003). American college of epidemiology ethics guidelines: Foundations and dissemination. *Science and Engineering Ethics, 9*(2), 207–214. https://doi.org/10.1007/s11948-003-0008-y

Merrick, J. (2013). Public health in a global context. *Frontiers in Public Health, 1*. https://doi.org/10.3389/fpubh.2013.00009

Mills, E. J., Seely, D., Rachlis, B., Griffith, L., Wu, P., Wilson, K., … Wright, J. R. (2006). Barriers to participation in clinical trials of cancer: A meta-analysis and systematic review of patient-reported factors. *The Lancet. Oncology, 7*(2), 141–148. https://doi.org/10.1016/S1470-2045(06)70576-9

Morgan, A., Ziglio, E., & Davies, M. (Eds.). (2010). *Health assets in a global context: Theory, methods, action*. New York, NY: Springer Science & Business Media.

Moscone, F., Siciliani, L., & Vittadini, G. (2017). Special issue on health economics and policy: Guest editors' introduction. *Health Economics, 26*(S2), 3–4. https://doi.org/10.1002/hec.3576

Moulton, A. D., Goodman, R. A., Cahill, K., & Baker, E. L., Jr. (2002). Public health legal preparedness for the 21st century. *The Journal of Law, Medicine & Ethics, 30*(2), 141–143.

Napier, A. D., Ancarno, C., Butler, B., Calabrese, J., Chater, A., Chatterjee, H., … Macdonald, A. (2014). Culture and health. *The Lancet, 384*(9954), 1607–1639.

National Academies of Sciences, Engineering, and Medicine; Health and Medicine Division. (2017). The root causes of health inequity. In A. Baciu

(Ed.), *Communities in action: Pathways to health equity*. National Academies Press (US). https://www.ncbi.nlm.nih.gov/books/NBK425845/

National Commission for the Protection of Human Subjects of Biomedical and Behavioral Research. (1979). Protection of human subjects; Belmont Report: Notice of report for public comment. *Federal Register, 44*(76), 23191–23197.

Olshansky, S. J., Antonucci, T., Berkman, L., Binstock, R. H., Boersch-Supan, A., Cacioppo, J. T., … Rowe, J. (2012). Differences in life expectancy due to race and educational differences are widening, and many may not catch up. *Health Affairs (Project Hope), 31*(8), 1803–1813. https://doi.org/10.1377/hlthaff.2011.0746

Omran, A. R. (2005). The epidemiologic transition: A theory of the epidemiology of population change. 1971. *The Milbank Quarterly, 83*(4), 731–757. https://doi.org/10.1111/j.1468-0009.2005.00398.x

Pager, D., & Shepherd, H. (2008). The sociology of discrimination: Racial discrimination in employment, housing, credit, and consumer markets. *Annual Review of Sociology, 34*, 181–209.

Parry, C. D. H., Bhana, A., Plüddemann, A., Myers, B., Siegfried, N., Morojele, N. K., … Kozel, N. J. (2002). The South African Community Epidemiology Network on Drug Use (SACENDU): Description, findings (1997–99) and policy implications. *Addiction, 97*(8), 969–976. https://doi.org/10.1046/j.1360-0443.2002.00145.x

Paul, D. B., & Spencer, H. G. (1995). The hidden science of eugenics. *Nature, 374*(6520), 302–304. https://doi.org/10.1038/374302a0

Pearson, K. (1901). *National life from the standpoint of science; an address delivered at Newcastle, November 19, 1900*. London, UK: A. and C. Black.

Pedersen, A. B., Altizer, S., Poss, M., Cunningham, A. A., & Nunn, C. L. (2005). Patterns of host specificity and transmission among parasites of wild primates. *International Journal for Parasitology, 35*(6), 647–657. https://doi.org/10.1016/j.ijpara.2005.01.005

Piasecki, J., Waligora, M., & Dranseika, V. (2017). What do ethical guidelines for epidemiology say about an ethics review? A qualitative systematic review. *Science and Engineering Ethics, 23*(3), 743–768. https://doi.org/10.1007/s11948-016-9829-3

Porter, D. (2005). *Health, civilization and the state: A history of public health from ancient to modern times* (p. 376). Routledge. https://doi.org/10.4324/9780203980576

Ramanathan, T., Hulkower, R., Holbrook, J., & Penn, M. (2017). Legal epidemiology: The science of law. *The Journal of Law, Medicine & Ethics: A Journal of the American Society of Law, Medicine & Ethics, 45*(1 Suppl), 69–72. https://doi.org/10.1177/1073110517703329

Rickham, P. P. (1964). Human experimentation. Code of ethics of the world medical association. Declaration of Helsinki. *British Medical Journal, 2*(5402), 177. https://doi.org/10.1136/bmj.2.5402.177

Roche, B., Benbow, M. E., Merritt, R., Kimbirauskas, R., McIntosh, M., Small, P. L. C., … Guégan, J.-F. (2013). Identifying the Achilles' heel of multi-host pathogens: The concept of keystone "host" species illustrated by Mycobacterium ulcerans transmission. *Environmental Research Letters : ERL [Web Site], 8*(4), 045009. https://doi.org/10.1088/1748-9326/8/4/045009

Rotegård, A. K., Moore, S. M., Fagermoen, M. S., & Ruland, C. M. (2010). Health assets: A concept analysis. *International Journal of Nursing Studies, 47*(4), 513–525.

Salimi, Y., Shahandeh, K., Malekafzali, H., Loori, N., Kheiltash, A., Jamshidi, E., … Majdzadeh, R. (2012). Is Community-Based Participatory Research (CBPR) useful? A systematic review on papers in a decade. *International Journal of Preventive Medicine, 3*(6), 386–393.

Shuster, E. (1997). Fifty years later: The significance of the Nuremberg Code. *The New England Journal of Medicine, 337*(20), 1436–1440. https://doi.org/10.1056/NEJM199711133372006

Smith, A., Vidal, G. A., Pritchard, E., Blue, R., Martin, M. Y., Rice, L. J., … Starlard-Davenport, A. (2018). Sistas Taking a Stand for Breast Cancer Research (STAR) Study: A Community-Based Participatory Genetic Research Study to Enhance Participation and Breast Cancer Equity among African American Women in Memphis, TN. *International Journal of Environmental Research and Public Health, 15*(12). https://doi.org/10.3390/ijerph15122899

Smith, M. H. (1998). Community-based epidemiology: Community involvement in defining social risk. *Journal of Health & Social Policy, 9*(4), 51–65. https://doi.org/10.1300/J045v09n04_05

Spinewine, A., Claeys, C., Foulon, V., & Chevalier, P. (2013). Approaches for improving continuity of care in medication management: A systematic review. *International Journal for Quality in Health Care: Journal of the International Society for Quality in Health Care, 25*(4), 403–417. https://doi.org/10.1093/intqhc/mzt032

Sterner, C. S. (2007). *A brief history of miasmic theory.* http://www.carlsterner.com/research/files/History_of_Miasmic_Theory_2007.pdf

Susser, M. (1973). *Causal thinking in the health sciences; concepts and strategies of epidemiology*. Oxford, UK: Oxford University Press.

Szczech, L. A., Coladonato, J. A., & Owen, W. F. (2002). An introduction to epidemiology and biostatistics and issues in interpretation of studies. *Seminars in Dialysis, 15*(1), 60–65. https://doi.org/10.1046/j.1525-139x.2002.00017.x

Valente, T. W., Chou, C. P., & Pentz, M. A. (2007). Community coalitions as a system: Effects of network change on adoption of evidence-based substance abuse prevention. *American Journal of Public Health, 97*(5), 880–886. https://doi.org/10.2105/AJPH.2005.063644

Vangeepuram, N., Mayer, V., Fei, K., Hanlen-Rosado, E., Andrade, C., Wright, S., & Horowitz, C. (2018). Smartphone ownership and perspectives on health apps among a vulnerable population in East Harlem, New York. *MHealth, 4*. https://doi.org/10.21037/mhealth.2018.07.02

Woolf, S. H., Johnson, R. E., Phillips, R. L., & Philipsen, M. (2007). Giving everyone the health of the educated: An examination of whether social change would save more lives than medical advances. *American Journal of Public Health, 97*(4), 679–683. https://doi.org/10.2105/AJPH.2005.084848

World Health Organization. (2007). *Everybody's business—Strengthening health systems to improve health outcomes: WHO's framework for action*. World Health Organization.

World Health Organization. (2018). *Essential public health functions, health systems, and health security*. https://extranet.who.int/sph/docs/file/2091

Yancey, A. K., Ortega, A. N., & Kumanyika, S. K. (2006). Effective recruitment and retention of minority research participants. *Annual Review of Public Health, 27*(1), 1–28. https://doi.org/10.1146/annurev.publhealth.27.021405.102113

Part III

Indicators and Outcomes of Sustainable Community Health

9

Quality Care Improvement

David Palm, Valerie Pacino, and Li-Wu Chen

Introduction

The building blocks of sustainable community health include eliminating health disparities, improving health equity, and achieving social justice. While there are many factors that influence sustainable community health, it cannot be achieved without improvements in the *quality of care* at the individual treatment level and at the community- or population-health levels. Hospital systems are required to report on several key quality measures, including preventable readmissions, patient satisfaction, and the number of heart attack patients who receive aspirin upon arrival in an emergency department (Glance, Osler, Mukamel, & Dick, 2008). In recent years, payment reforms that emphasize value of care over volume of care have created new incentives for health care providers to control costs and improve quality. These changes have also broadened the concept of QCI to include, not only improvements in the treatment of

D. Palm (✉) • V. Pacino • L.-W. Chen
University of Nebraska Medical Center, Omaha, NE, USA
e-mail: david.palm@unmc.edu; valerie.pacino@unmc.edu; liwuchen@unmc.edu

E. Mpofu (ed.), *Sustainable Community Health*,
https://doi.org/10.1007/978-3-030-59687-3_9

individual patients, but also improvements in the overall population health in the community.

Learning Objectives

After reading this chapter, the reader should be able to:

1. Define quality care improvement (QCI) applied to community health settings.
2. Describe the past and current perspectives of QCI and its role in advancing the sustainable community health model.
3. Discuss the current and emerging QCI metrics for promoting the sustainable community health.
4. Evaluate cultural, professional, and legal policies that affect QCI processes and practices in community health systems, including identifying relevant stakeholders and their prospective roles.
5. Identify research opportunities to further explore the importance of QCI for sustainable community health.

QCI encourages the medical community to develop closer working relationships and alliances with public health agencies, other agencies in non-health sectors such as transportation and housing support, and local community-based organizations to reduce health disparities, increase health equity, and improve care coordination. Greater care coordination has created new types of health workers, data sharing agreements, and flexible funding strategies. Although achieving the broader version of QCI is more challenging, it is a critical dimension of the sustainable community health model at the population level.

The definition of **population health** varies widely across the health care spectrum. For example, health care leaders in accountable care organizations (ACOs) tend to use "population health" to narrowly describe efforts to improve care for their patient populations (Noble, Greenhalgh, & Casalino, 2014). Meanwhile, public health leaders often think of "population health" in terms of all people living within a geographic area. For our purposes, we use the definition proposed by Kindig and Stoddart: population health is "the health outcomes of a group of individuals, including the distribution of such outcomes within the group" (2003, p. 3). This definition encourages thinking of population health not only in terms of geographic regions (e.g., nations, states, and communities), but also in terms of the distribution of health outcomes across different population groups within those geographic regions (e.g., immigrant groups, LGBTQ groups, and justice-involved groups). As Kindig and Stoddart note, this definition forces us to consider the multiple determinants of health, including "medical care, public health interventions, aspects of the social environment (income, education, employment, social support, culture) and the physical environment (urban design, clean air and water), genetics, and individual behavior,…as well as the resource allocation issues involved in linking determinants to [health] outcomes." To that end, we think of the social determinants for health as "the conditions in which people are born, grow, live, work and age. These circumstances are shaped by the distribution of money, power, and resources at global, national and local levels" (Kindig & Stoddart, 2003, p. 4). The social determinants of health as depicted in Fig. 9.1 are mostly responsible for *health inequities*—"the unfair and avoidable differences in health status seen within and between countries" (WHO, 2008). We cannot hope to improve population health or reduce health disparities across population groups without addressing the social determinants of health (see also Chap. 2, this volume).

Many states are experimenting with new health care delivery models and payment models that incorporate QCI as a key driver in reducing health disparities and promoting health equity. Most of these models focus on improving nonmedical factors that influence health and promote sustainable community health. While these models use different approaches to address these challenges, they share several underlying principles, including: (1) coordinating care around the needs of patients across the continuum of care (e.g., primary care, hospital, and the community); (2) broadening the scope of services (i.e., moving from a narrow focus on individual patient's conditions to a broad focus on the health of populations, including the social determinants of health); (3) using data and information systems to screen and track high-risk patients and direct them to the most appropriate care setting to avoid unnecessary hospitalizations and emergency department visits; (4) creating an effective referral system and data sharing arrangements between health care providers and community-based organizations; (5) forming multi-sector

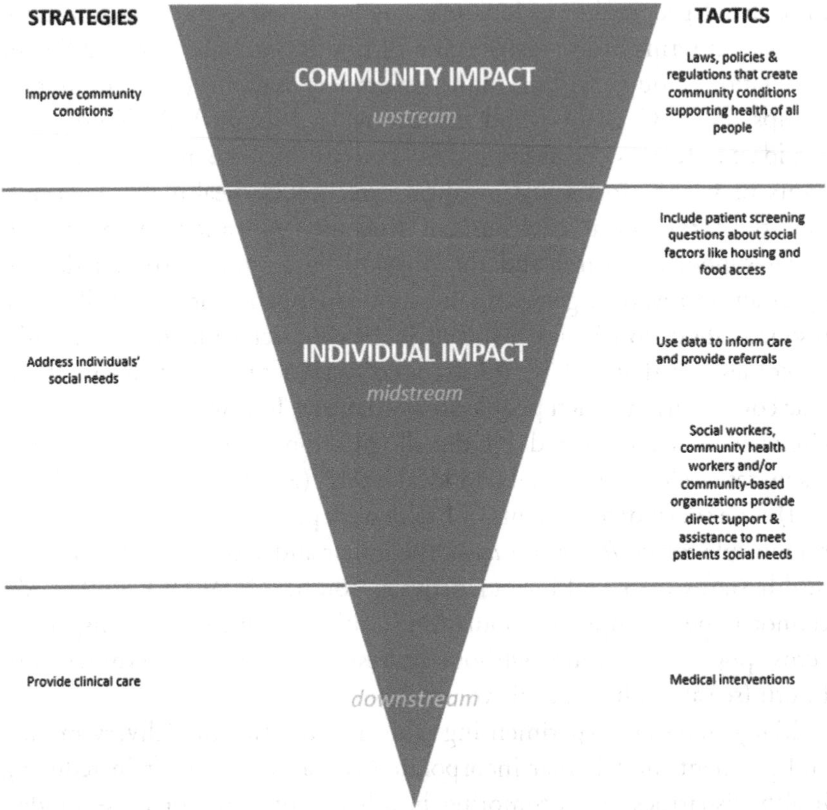

Fig. 9.1 Social determinants of health. (*Source*: Adapted from Kaiser Family Foundation, 2019)

partnerships that include public and private stakeholders to address medical and nonmedical patient needs; and (6) developing appropriate measures to track the progress of patients and populations, and to ensure that all major risk factors are addressed with evidence-based or best-practice interventions. Although states and jurisdictions have used these approaches to enhance sustainable community health, the models described in the following sections are successful when they embrace the broader concept of QCI by promoting health equity and by addressing health disparities.

Professional and Legal Definitions of QCI

Before we dive into examples, developing a common nomenclature is helpful. We propose terms and meanings related to QCI, such as quality improvement, continuous care improvement, and change management.

Quality improvement (QI) describes a systematic, formal, and iterative process of collecting and using data to test, change, and improve the performance of a system. Data are gathered to identify gaps between current quality and expected quality, changes are introduced to a system to narrow those gaps, and the effect of those changes on outcomes and performance is routinely measured. QI relies on a robust performance management strategy, iterative tests of change, and coordinated access to valid data. *Quality care improvement* (QCI) brings principles of QI to health care organizations and systems, and QCI projects assess whether care is safe, timely, effective, efficient, and equitable. Not to be confused with QCI, *continuous quality improvement* (CQI) is a component of QCI that embeds the improvement process in the delivery of services at the site level. CQI engages frontline staff in an ongoing process of addressing what and how care is delivered. While there are subtle distinctions between these definitions, QI, QCI, and CQI are used interchangeably throughout this chapter.

In recent years, the concept of QCI has moved from a relatively narrow focus to a broader emphasis (see Fig. 9.2). QCI initially focused downstream on individual treatment through clinical interventions. While this is an important aspect of QCI, its focus has shifted to move midstream to address the social needs of individuals (i.e., screening for the social needs of individuals and providing services to meet these needs). However, QCI efforts must also continue to move upstream to change community conditions and improve the health of communities and populations. These improvements require changes in laws, policies, and regulations to create more favorable health conditions (Castrucci & Auerbach, 2019). By habitually refocusing our attention upstream, there is a greater possibility of eliminating health disparities, improving health equity, and achieving social justice.

From an organizational perspective, the crux of QCI is the willingness and ability to encourage and manage both anticipated and unanticipated

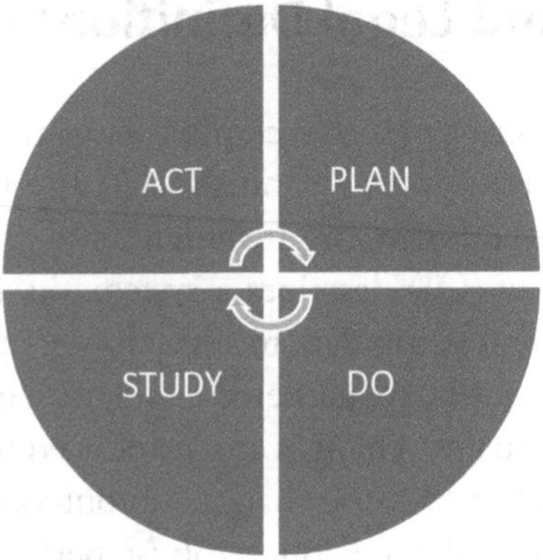

Fig. 9.2 Social determinants and social needs: moving beyond midstream. (*Source*: Adapted from Castrucci & Auerbach, 2019)

change, which is the key to a successful quality improvement effort. *Change management* is a systematic approach to preparing and supporting internal and external stakeholders to adapt to and sustain a lasting change within an organization. According to the National Association of County and City Health Officials (NACCHO), QI involves "designing system and process changes that lead to operational improvements, and an organizational culture where quality is ingrained in organizational values, goals, practices, and processes" (2017). For example, an operational change could be something as discrete as revising the approval process for contracts, or it could be as transformational as a complete shift in organizational strategy and culture that embraces quality. In both cases, structural and process changes are introduced, and change management is key to obtain buy-in from employees during and after the transition phase.

Historical Evolution of Practice Related to QCI

QCI is a relatively new concept in the delivery of health care services. In a 2001 report, the Institute of Medicine (IOM) first defined its vision for how to narrow the gap—which it astutely described as a chasm—between what care is provided and what care should be received. The IOM attributed these adverse quality issues to outmoded systems of work and recommended that all health care organizations pursue the delivery of health care according to the following six aims: safe, effective, patient-centered, timely, efficient, and equitable (IOM, 2001). This vision was later operationalized to include fostering rapid advances in health care, redesigning care delivery, furthering measurement and informed purchasing, and preventing iatrogenic injuries (foremost among them medication errors, hospital-acquired infections, and other preventable events). In response to the mandates contained in the Medicare Prescription Drug, Improvement, and Modernization Improvement Act of 2003, this novel vision of QI moved health systems forward incrementally, but it did not incentivize them to transition to sustainable community health. However, when the Affordable Care Act (ACA) was passed in 2010, it paved the way for the federal government, particularly through the Centers for Medicare and Medicaid Services (CMS), to become an incubator of innovation by funding new health care delivery such as accountable care organizations (ACOs) and payment models that reward robust QCI and proactively improve the health of populations.

Over the last decade, CMS has prioritized value (specifically, improving individual care, improving population health, and lowering costs) in many of their programs by paying providers higher rates for improving the quality of care. Until recently, these value-based payments were largely focused on clinical care but stopped short of addressing nonmedical determinants of health or amplifying truly sustainable community health. For example, CMS's Hospital Value-Based Purchasing Program pays acute care hospitals based on metrics such as rates of readmissions, rates of adverse events, adoption of evidence-based care standards, patient engagement, care transparency for consumers, and population health. CMS's Hospital-Acquired Condition Reduction Program and Hospital Readmissions Reduction Program also reward providers for improving their quality of clinical care. Unfortunately, CMS saw only modest improvements in readmission rates and 30-day mortality rates from 2013 to 2017 (Hinton, Musumeci, Rudowitz, Antonisse, & Hall, 2019). Further, these improvements were largely the result of improved clinical care and were not related to any successes in addressing the social determinants of health.

Approaches to QCI for Sustainable Community Health

In the United States, new innovative reimbursement and health care delivery models have provided the foundation for innovation at the state and community levels. These models make it easier for providers and community partners to link clinical and community approaches to health because they are based on the concept of value. Value is based on improvements in individual and population health outcomes, as well as the cost of delivering those outcomes. Value-based payment systems reimburse hospitals and providers based on patient health outcomes and shift the incentive from quantity of care to quality of care. This is a striking departure from fee-for-service or capitated approaches that have dominated health care systems until the last decade. With the passage of the Patient Protection and Affordable Care Act (ACA), a larger number of providers are adopting fee-for-value approaches that reward them when they collaborate with patients to improve individual and population health metrics (Abrams et al., 2015).

Two new health care delivery models have emerged that focus on value-based care, including ACOs and patient-centered medical homes (PCMHs). ACOs are value-based *payment* mechanisms made up of voluntary groups of providers that are contractually responsible for the total cost and quality of care for a defined patient population. Many argue that the ACO model gives providers the flexibility to address patients' nonmedical needs, as well as the incentives and funding to do so. PCMHs are value-based care *delivery* mechanisms that tend to be practice-specific. Historically, they integrate primary, specialty, and acute care, but some PCMH models are pushing beyond medical care services to embrace a model of whole-person care. In this approach, a physician does not merely treat a patient, but uses a coordinated care team to collaborate with the person, screens for nonmedical needs and social determinants of health, refers people to key resources in the community, and embraces a model of health and wellness.

Current system-level QCI approaches that are implemented at the state and community levels utilize various value-based models: Section

1115 Medicaid Waivers, New MCO Requirements, and Accountable Care Organizations. Alternative, non-value-based approaches such as accountable health communities identify and address the nonmedical determinants of health and innovations developed by the Federally Qualified Health Centers. Emerging or promising approaches to quality improvement at the organizational level include the Plan-Do-Study-Act (PDSA) cycle, clinical pathways, and changes in the patient safety culture. These approaches are considered next.

Value-Based Medicaid Programs Using the Section 1115 Waivers

Several states have used Section 1115 waivers to develop an array of value-based payment models aimed at reducing costs and improving the quality of care. Most of these models encourage multi-sector partnerships and include performance measures that link financial incentives to improvements in quality. As of June 2019, 47 waivers from 39 states have been approved by CMS, and 20 waivers from 18 states are still pending (Hinton et al., 2019). A recent report identified some common themes related to these demonstrations, which include: (1) enhancing care coordination and community partnerships to address the social determinants of health (e.g., screening for social needs, linkages to community resources, and partnerships with social service agencies and community-based organizations) and (2) using payment incentives to address the social determinants of health. Evidence suggests that investing in social services results in better community health outcomes (Bradley et al., 2016; McCullough & Leider, 2016). Early QI efforts indicate improved quality, controlled costs, and reduced disparities (McConnell et al., 2017; Muoto, Luck, Yoon, Bernell, & Snowden, 2016).

The states of Oregon, California, and North Carolina have received Section 1115 waivers to address patients' nonmedical needs by funding social interventions, including case management and care coordination services, and connecting patients with basic social supports to address transportation, housing, food, and legal needs. In 2012, Oregon received a Section 1115 waiver to create coordinated care organizations (CCOs) and fund social supports and interventions (e.g., transportation to medical services and referrals to social services) not usually covered by Medicaid. Specifically, a portion of the CCO global budget, referred to as the "quality pool," is tied to performance and quality. To receive these funds, CCOs must meet performance targets on 17 quality measures (e.g., depression screening and follow-up, childhood immunization status, developmental screening at well-child visits, dental sealants for children, effective contraception use, and satisfaction with care). The quality pool is designed to offer CCOs the flexibility to invest in social interventions. For example, some centers added community health workers (CHWs) to screen, support, and collaborate with patients with nonmedical needs. While the infrastructure to support such flexible investments exists, one study found that Oregon's CCOs spent less than 0.1% of their budgets on social interventions (Kushner & McConnell, 2018). Still, Oregon's Medicaid expenditures have grown more slowly than the projected rate, resulting in $2.2 billion cost savings from 2013 to 2017 (see Research Box 9.1).

Research Box 9.1: Oregon's Coordinated Care Organizations Show Promising Results (Kushner et al., 2017)

Background

In 2012, the Oregon Medicaid program received a Section 1115 waiver to control costs and improve access to and the quality of health care services. It also provided an opportunity to increase investment in social interventions such as housing services and food insecurity programs and improve the coordination of care between physical and behavioral health. To achieve the goals of lower costs, improved access, and higher quality, Oregon established regional coordinated care organizations (CCOs) so that Medicaid patients would have a single point of accountability for health care services. The CCOs were locally governed and included Medicaid members, health care providers, and other stakeholders. They received a global budget to pay for physical, behavioral, and oral health care services and coordinate other services to better meet their social needs (e.g., housing and economic assistance). One of the CCO directives was to provide less expensive health-related services that would replace or reduce the need for medical services. They could also receive bonus payments if they met specific quality and outcome measures.

(continued)

(continued)

Method

The evaluation of the program was conducted by the Center for Health Systems Effectiveness in the Oregon Health and Science University. The evaluation team used a variety of measures to examine the changes in health care spending, quality, and access between 2011 (the year before the project began) and 2015. For most measures, they compared the changes among CCO members to a control group, which was the Medicaid program in Washington state. The Washington Medicaid program was selected because it did not make any major changes during this period.

Results

Between 2012 and 2017, considerable progress was made in achieving the goals of the project. First, the evaluators found that total health care spending per member per month decreased among CCO members relative to the control group (Medicaid members in the State of Washington). The most significant declines were for inpatient facility spending. The changes in quality measures were mixed. On the positive side, there was a decrease in the avoidable emergency department visit rate and an increase in the percentage of adolescents with at least one well-care visit among CCO members relative to Washington Medicaid members. However, glucose testing for people with diabetes fell in comparison with Washington members. Finally, most access measures for CCO members decreased slightly relative to the Medicaid members in Washington. Although CCOs experimented with spending on social interventions, overall spending was low relative to medical services. Strong efforts were made to integrate physical and behavioral health services, but these efforts mostly failed because of regulatory and contracting issues.

Conclusions and Implications

The redesign of the Oregon Medicaid program led to decreases in spending and improvements in important health care quality measures. In addition, it appears that bonus payments for health care providers are strongly associated with improvements in quality measures.

What Do You Think?

1. How was the Oregon Medicaid program redesigned and how were the payment incentives changed to control cost and improve quality?
2. What were the goals of the Oregon Medicaid program and how successful were they in meeting these goals?
3. How are the Oregon Medicaid program initiatives for sustainable community health?

Similarly, California created Whole Person Care (WPC) Pilots in 2016 to coordinate social interventions across partnerships of local health departments, managed care plans, hospitals, and social service organizations. Incentive payments funded care coordination for successful transitions for people who were formerly incarcerated into the community (Bandara et al., 2015). Bundled payments funded intensive case management for homeless patients, as well as investments in data sharing systems (Alderwick, Hood-Ronick, & Gottlieb, 2019). The savings from these programs created a flexible housing pool, which is used to cover rental subsidies and supportive housing development (Alderwick et al., 2019; see also Research Box 9.2).

Research Box 9.2: Interim Evaluation of California's Whole Person Care (WPC) Program (Pourat et al., 2019)

Background

In 2016, the California Department of Health Care Services began implementing the Whole Person Care (WPC) Program for high-risk, high-utilizing Medicaid enrollees. Most of the California counties that participated in the program focused on improving the health and wellbeing of enrollees by coordinating care across spheres of care delivery, including physical health, behavioral health, and social services. The pilot projects had the option of targeting one or more of the target populations, including individuals experiencing homelessness and individuals at risk of homelessness. The pilots were required to provide a comprehensive assessment of the patient's needs and define individual or bundles of services. Many types of services have directly addressed the social determinants of health. For example, care coordination services included benefit support such as transportation to appointments. Almost half of the enrollees received employment assistance (e.g., support in developing skills and connections that would improve their chances of obtaining employment). In addition, almost 70% of the pilots offered housing support services because nearly half of the WPC enrollees were homeless.

Method

This project was evaluated by the UCLA Center for Health Policy Research using a mixed-methods approach. The Center analyzed the data based on reportable measures, including monthly enrollment and utilization reports, bi-annual narrative reports, and claims. In addition, surveys were conducted of the 27 lead entities and 227 involved partners as well as follow-up interviews with staff from the lead entities.

(continued)

(continued)

Results

Although the project has not been completed, some interesting results have emerged. When the WPC enrollees were compared with a control group of other California Medicaid recipients, the rates of emergency department (ED) visits did not show a significant change for either group. However, there was a significant increase in hospitalizations for the WPC enrollees as compared to the control group. When the ED visit rates were assessed after the first two years, the rates for WPC enrollees declined by 19% as compared to only an 8% drop for the control group. When an assessment was made of the approaches in the delivery of services to the homeless populations after the first two years of the project, there were early successes in the delivery of housing services but also challenges in retaining permanent housing. For example, the number of WPC enrollees who received housing services increased from 58% to 67% from year 2 to year 3. Some of the common housing challenges included coordinating care, linking enrollees to housing services, and lack of affordable housing. One of the solutions to overcome these challenges was to partner with local organizations.

Conclusions and Implications

The California pilot projects were very successful in enrolling high-risk, high-utilizing Medicaid patients who were frequently homeless. Some progress was made in reducing the number of ED visits and the delivery of housing services. However, many challenges remain and reflect the historical gaps in the management of patients with complex conditions and underlying social determinants of health. Overcoming these challenges will require time, resources, and a deliberate effort.

What Do You Think?

1. How did the WPC pilot projects attempt to address the social determinants of health in a sustainable way?
2. What were the successful outcomes of the project for population health sustainability?
3. How could the WPC pilot projects be reconfigured for sustainability?

In 2018, North Carolina began using its Section 1115 waiver to create Healthy Opportunities Pilot Programs that target social needs, including housing, transportation, and food insecurity, as well as interpersonal violence and toxic stress. These pilot projects can use their funds to cover expenses related to carpet replacement to control a child's asthma, vouchers to travel to and from a food pantry, and safe housing for a pregnant woman victimized by intimate partner violence. A rapid-cycle QCI process will be used to identify which interventions are most and least effective and best practices will be disseminated to the pilots. One of the unique features of this model is that each pilot will be anchored by a community-based health or social service organization, not a health care organization (Hinton et al., 2019).

Several states are addressing child population health needs by leveraging funding through the Children's Health Insurance Program (CHIP). Once a state's CHIP administrative costs are covered, they can apply to use remaining funds for initiatives focused on direct services or public health initiatives, including maternal care, nutrition, behavioral health, school health services, lead abatement efforts, and other prevention and intervention projects (NASHP, 2018).

Value-Based Medicaid Programs with New MCO Requirements

At least 39 states provide services to Medicaid beneficiaries through contracts with risk-based managed care organizations (MCOs). Once contracts satisfy federal rules, states have the flexibility to require or create incentives for MCOs to provide care coordination activities that address the social determinants of health. These requirements or incentives may involve several activities, including screening for nonmedical needs, connecting beneficiaries to appropriate nonmedical services, and authorizing payment for members of the nonclinical workforce involved in addressing the social determinants of health (Matulis & Lloyd, 2018). Some examples of Medicaid MCO activities related to the social determinants of health are illustrated in Table 9.1. This table shows that states have a variety of strategies to address the social determinants of health, including job counseling services, connecting members with housing support services, and health coordination between health care providers and the Women, Infants, and Children's Program (WIC).

Table 9.1 History of quality of care improvement in the United States

History of QCI in the United States	
1953	Joint Commission on Accreditation of Hospitals established to provide voluntary accreditation based on a rubric of quality standards
1966	Avendus Donabedian publishes, *Evaluating the Quality of Medical Care*
1989	Agency for Health Care Policy and Research (now known as the Agency for Healthcare Research and Quality) was created
1999	Institute of Medicine publishes, *To Err is Human*
2001	Institute of Medicine publishes, *Crossing the Quality Chasm*
2003	U.S. Congress passes and President Bush signs the Medicare Prescription Drug, Improvement, and Modernization Improvement Act
2006	Physician quality reporting system (PQRS) established to provide incentive payments for successful reporting on three quality measures
2010	U.S. Congress passes and President Obama signs the Patient Protection and Affordable Care Act (ACA)
2011	CMS releases final rules for the official implementation of accountable care organizations under the ACA
2012	Quality Reporting and Hospital Value-Based Purchasing program begins to be implemented, full implementation by 2016
2014	Medicaid expansion is funded; CMS substantially expands funding for CQI innovation projects
2015	PQRS changes from incentive-based pay-for-reporting; adds penalties for those who fail to report on quality measures
2015	Medicare Access and CHIP Reauthorization Act (MACRA)—the law requires that physician payments will be based on meeting certain quality measures

Value-Based Medicaid Programs Using Accountable Care Organizations

State Medicaid programs are also contracting with accountable care organizations (ACOs) to control costs and improve QCI. Although ACOs face several challenges such as a lack of financial resources, limited staffing capacity, competing clinical priorities, and scalability of programs, several studies have found that as ACOs gain experience and become more mature, their ability to integrate medical and nonmedical services becomes more sophisticated (Fraze, Lewis, Rodriguez, & Fisher, 2016).

The North Carolina Medicaid Program began contracting with Community Care of North Carolina (CCNC), which is a web of community networks across the state composed of practicing physicians working in partnership with hospitals, health departments, and departments of social services. CCNC networks emphasize population health management, case management and clinical support, and data and feedback. Community Care Physician Network treats one of every three Medicaid patients in North Carolina and saves the state $160 million annually. It is an effective system of care for patients with chronic illnesses.

Beginning in 1988 as a demonstration project in a small rural county in eastern North Carolina, CCNC has evolved through several iterations over the last 25 years and used CQI to refine its approach. For example, rather than simply targeting high cost/high-risk patients, CCNC used CQI to develop a more refined strategic approach to complex case management from a focus on "high risk" to a focus on "high impact." Using CCNC's Complex Care Management Impactability Scores yield twice the savings of targeting emergency department and inpatient super-utilizers and three times the savings of less discriminant case management services.

Several other states, including Colorado, Massachusetts, and Vermont, have also contracted with ACOs. All states have focused on chronic disease management and many of the social determinants of health such as transportation, housing support services, nutrition classes, and exercise equipment. Some states are also required to address food access, family/caregiver support, and social isolation (see Research Box 9.3).

Research Box 9.3: The History, Evolution, and Future of Medicaid Accountable Care Organizations (Matulis & Lloyd, 2018)

Background

Many states have begun to implement Medicaid accountable care organizations (ACOs) to control costs, collect and analyze data, particularly on high-risk patients, and improve quality and patient outcomes. Various cost and quality benchmark metrics are established by the Medicaid programs, and ACOs must report on these metrics (e.g., number of unnecessary emergency department visits, number of patients that have blood pressure rates below 130/80, or Hemoglobin A1c rates below 9%). The benchmarks are usually based on the ACOs prior performance or the performance of other ACOs. To hold providers accountable and meet these cost and quality benchmarks, financial incentives are established and usually involve a shared savings arrangement (SSA). In an SSA, providers in the ACO have an

(continued)

(continued)

opportunity to share in savings if their attributed population uses a less costly set of health care resources than a predetermined baseline. In addition to meeting the cost baseline, ACOs must also meet or exceed their quality benchmarks to share in the savings.

Method

This analysis of Medicaid ACOs was based on in-depth interviews with representatives from seven states that were early adopters of Medicaid ACOs. After the interviews were conducted, common themes and lessons learned from their experiences were identified.

Results

Although not all individual Medicaid ACOs have achieved better health outcomes at a lower cost, most state initiatives have demonstrated promising results. For example, the ACOs in Colorado have saved the Medicaid program $77 million in the first three years, and they have reduced emergency department visits, high-cost imaging, and hospital readmissions. In Vermont, two ACOs reported $17 million in savings in the first two years of the program and exceeded their quality benchmarks. The 21 ACOs in the Minnesota program saved more than $212 million over four years and consistently exceeded their quality benchmarks. While there have been some promising findings, the study also identified some key challenges and lessons learned. One of the lessons learned was that there was not a single model that was used by all states. Most of the ACOs were led by providers but in some states payers or a community organization assumed a lead role. The scope of services and the types and number of quality measures also varied. For example, all programs included physical health, but some states added behavioral health, dental health, and long-term care services. One state had 38 quality measures while another state had only 12. As the programs have evolved, states have reduced the number of quality measures by focusing more on high-impact, population health quality metrics that align with other delivery system and payment reform initiatives.

Conclusions and Implications

Although Medicaid ACOs will continue to evolve in their governance structure, scope of services, and approaches to quality improvement, these early Medicaid ACO efforts demonstrate the value of connecting provider's reimbursement to patient health outcomes. The shift to a smaller number of quality measures that focus more on population health outcomes should lead to improved care coordination and a greater focus on high-risk populations.

What Do You Think?

1. In a Medicaid ACO model, what incentive do providers have to lower costs and improve the quality of care?
2. What are some of the challenges and lessons for sustainability learned from the implementation of Medicaid ACO models?
3. What would be impact on the sustainability of a health system from reducing the number of quality measures?

Value-Based Medicaid Programs Using Accountable Communities for Health

Accountable Communities for Health (ACHs) are organizations that have also expanded the concept of QCI by placing a heavy emphasis on the social determinants of health and other nonmedical factors that play an important role in improving the overall health of the population (see Research Box 9.4). One example of an ACH model is the California Accountable Communities for Health Initiative (CACHI). This Initiative is a multi-sector alliance of major health care systems, providers, and health plans, along with public health, key community and social services organizations, schools, and other partners. CACHI receives funding from CMS and private foundation funds to develop and implement prevention strategies. In July 2016, CACHI announced awards to six communities throughout the state to "advance common health goals and create a vision for a more expansive, connected, prevention-oriented system" (The California Endowment, 2016). While each community determines its governance structure based on community needs, they must engage

Research Box 9.4: Using Data for Quality Improvement: A Case Study from St. Joseph's Hospital Health System (Centers for Medicare and Medicaid Services, 2019)

Background

In 2016, the Centers for Medicare and Medicaid Services (CMS) began accepting applications to fund an Accountable Health Communities (AHC) model. The purpose of the project was to test whether health-related social needs can be systematically identified and addressed for the Medicare and Medicaid beneficiaries in communities. The foundation of the model is to develop a universal, comprehensive screening protocol to identify social needs, including poor housing quality, difficulty in paying utility bills, food insecurity, and transportation difficulties. For the AHC program to be successful, the first step is to maximize the number of beneficiaries who are screened.

In 2018, the St. Joseph Hospital Health System, a nonprofit health system in Syracuse, New York, began implementing the AHC model. St. Joseph serves as a bridge or hub for 19 clinical delivery sites, including primary care clinics, urgent care centers, and an emergency department. All sites screen for health-related social needs, and St Joseph has developed two types of reports to

(continued)

(continued)

monitor trends. The first report tracks the aggregate number of completed screenings per day for all sites. The second report is a screening dashboard that identifies the number of beneficiaries who were screened at each site and compares these numbers with program-wide benchmarks. In its role as the hub organization, staff from St. Joseph review the underlying performance issues and work with the clinical organizations to improve the screening rates.

Methods

This project was assessed using personal interviews with project staff. The interviews, which were conducted by Mathematica, involved questions about the process for developing data monitoring reports, how the reports are reviewed and how they guide quality improvement screening efforts, and future quality improvement initiatives.

Results

The reports developed by project staff have been very effective in identifying limited screening rates in low-performing clinics. When these clinics are identified, staff review the data with clinic staff and determine possible solutions that will increase screening rates (e.g., resistance by physicians and nurses to additional screening questions and inadequate staffing). Screening processes were also reviewed to identify best practices. For example, at the highest performing sites, screeners used a script to explain to beneficiaries why the screening is offered. High-performing sites also give beneficiaries the screening form on a clipboard so they can complete it in the waiting room, and they ensure that the forms are returned to the registration desk. Increasing the screening rates across all sites is a critical first step that allows this AHC to connect a larger number of high-risk beneficiaries with community-based services. Although no results are available on the impact of these referrals and linkages at this time, the improvement in the screening rates is an important step in meeting the health-related social needs of the beneficiaries.

Conclusions and Implications

The major goal of the AHC model is to screen Medicare and Medicaid beneficiaries for social needs and then refer them to community-based services (e.g., housing authority and food bank) to meet these needs. Without a high-level screening process that is continually monitored for quality and is consistent across the clinical sites, this model will not be successful.

What Do You Think?

1. What is the purpose of the AHC model, and what types of social conditions are addressed in the model that have sustainability implications?
2. What methods does the St. Joseph Hospital Health System use to monitor the screening process across their clinical sites? How sustainable are the procedures and why?
3. What were some of the factors that led to low screening rates at some sites, and what were some of the best practices of high-performing screening sites?

across multiple sectors. Further, ACHs must describe how they will share data in support of their population health improvement activities as well as community health, clinical, and cost data to support the goals of the ACH.

Similar ACH models are being tested and implemented in 11 states, including Minnesota, Vermont, Washington, and Iowa. Minnesota is leveraging $5.6 million of CMS funds to launch 15 ACHs in the state and requires each ACH to collaborate with an ACO in an innovative coordinated care model. A new Statewide Quality Reporting and Measurement System is being used to coordinate performance management across providers and settings with specific action plans for behavioral health, long-term care, and social services providers (Vickery et al., 2018). Vermont launched ACHs across its 14 health service areas (NASHP, 2018). Healthier Washington includes nine ACHs across the state, and they are focused on behavioral health challenges in school and health care settings to connect with community-based treatment services and interventions (NASHP, 2018; see also Discussion Box 9.1).

Other State Approaches for Addressing the Nonmedical Determinants of Health

Shreya Kangovi and her team at the University of Pennsylvania have created a scalable strategy for implementing patient-centered care for vulnerable populations. The model, Individualized Management for Patient-Centered Targets or IMPaCT, trains and deploys CHWs as frontline health workers. CHWs are trusted community members who share socioeconomic backgrounds with their patients. A randomized controlled trial (RCT) found that IMPaCT improves access to primary care and quality of discharge while controlling recurrent readmissions among a population of low-socioeconomic status (SES) adults with varied conditions (Kangovi, Mitra, & Grande, 2014). Other RCTs have found that CHWs can reduce hospital stays by 65% and double the rate of patient satisfaction with primary care (Kangovi et al., 2018). Where other CHW models have been unsuccessful due to poor standardization and replicability, IMPaCT is an evidence-based, exportable model of care that improves population health outcomes (see Discussion Box 9.2).

Discussion Box 9.1: ACH Models and Other Value-Based Approaches Act in New Ways

ACH models and other value-based approaches have created new ways of thinking about quality and health equity. In the fee-for-service system, the emphasis was on clinical treatment and volume of services, often to the exclusion of the social needs of patients. In contrast, value-based models have expanded the concept of QCI to include not only clinical care but to also address the broader social needs. These new value-based models provide financial incentives and an opportunity to improve the health of individuals and communities. From a health care provider perspective, however, this shift has created a dilemma because there are still many challenges and some unanswered questions. One of the challenges is that the reimbursement levels are often insufficient to cover all expenses (e.g., data collection and analysis, time for screening, and extra staffing). Second, care coordination with behavioral health providers, local health departments, and social service agencies is difficult in many areas because they operate as separate systems with different funding mechanisms. Third, it is challenging to track the outcomes of patients that are referred to community-based services and for some patients, there may not be a workable solution (e.g., permanent housing for those experiencing homelessness). Finally, many health care organizations have been forced to change their culture, and change initiatives that are not executed well often result in resistance among staff. Despite these challenges, many health care organizations are moving forward with a broader vision of QCI because they understand that it will lead to improved patient outcomes.

What Do You Think?

1. How have ACH and other value-based models expanded the concept of QCI?
2. What are some of the sustainability-related challenges of adopting one or more of these models from a provider perspective?
3. Do you think value-based models will ultimately be successful in broadening the concept of QCI in the long-term? Say why and how.

QCI in Federally Qualified Health Centers

Funded by the U.S. Department of Health & Human Services, Federally Qualified Health Centers (FQHCs) provide community-based health care in underserved areas. Many FQHCs have used a variety of QCI projects to improve patient outcomes (see Discussion Box 9.3). For

Discussion Box 9.2: Community Health Workers (CHWs) to Assist Patients in Meeting Their Needs

Many patients with complex clinical and social needs often have a greater proportion of emergency department visits and hospital readmissions. To address these challenges, health organizations are using community health workers (CHWs) to assist patients in meeting their needs. CHWs are trusted laypeople within a community, have durable relationships with other community members, and understand the landscape of community assets, services, and needs. They are trained by clinics, local health departments, universities, and other organizations to serve as a liaison between patients and health and social service organizations to reduce health disparities and improve access to and the quality of health-related services. CHWs provide a range of services, including interpretation and translation, culturally appropriate health education, informal counseling, and motivational interviewing, as well as offer some direct services such as blood pressure screening.

The Individualized Management for Patient-Centered Targets (IMPaCT) is a unique model that uses CHWs. This model has been quite successful in reducing readmissions for high-risk populations because a CHW helps patients create individualized health goals for recovery during the hospital admission. After discharge, they work with the patients for a minimum of two weeks to achieve these goals. When the CHW intervention group was compared with a control group, the intervention patients were more likely to receive timely post-hospital primary care, report higher quality discharge communication, and show greater improvements in mental health.

What Do You Think?

1. What are some of the roles and functions of CHWs, and how do they make for sustainability of health services?
2. How did the IMPaCT model use CHWs and what were the specific ways they were able to improve the quality of care for patients? How sustainable would be the related quality of care improvement initiatives?

example, Harrison Community Health Center (HCHC) is an FQHC located in a midsized city in rural Virginia, which has a major refugee resettlement population as well as many rural regions. HCHC used results from a community needs assessment, which identified mental health as the top health concern, to implement a rapid-cycle improvement process to improve depression screening and community-based follow-up. This QCI project increased depression screening from 9% to 71%. Adherence to follow-up with community mental health services increased from a baseline of 33.3 to 60.0%. Stakeholders influenced

Discussion Box 9.3: Patient Screening Protocols

Federally Qualified Health Centers (FQHCs) provide comprehensive primary care services, including mental health and dental care, to low-income and uninsured populations. In recent years, FQHCs have expanded their patient screening protocols to identify mental health conditions, particularly depression, and the social determinants of health. The two examples described above show how FQHCs have been successful in not only screening for depression and social problems, but also how they have improved the quality of care through their follow-up efforts.

What Do You Think?

1. What are the sustainable health system quality implications of early screening for mental health conditions and the social determinants of health?
2. Why is it important to follow-up on screening results if problems are detected?

process changes to make screening and follow-up care culturally appropriate for this community (Schaeffer & Joelles, 2019).

Other health centers are developing and implementing protocols to screen for social determinants of health as part of QCI projects. For example, Albuquerque's WellRx pilot systematically screened for and addressed patients' social needs during every visit. Using an 11-question instrument, the multidisciplinary team screened all patients at all visits for social determinants in three family medicine clinics over 90 days. They found that nearly half (46%) of patients screened positive for at least one area of social need. Among those, nearly two-thirds (63%) screened positive for multiple needs, most of which were previously unknown to the clinicians. Medical assistants (MAs) and CHWs connected patients with appropriate community services and resources. Using MAs to identify social needs and CHWs to intervene and refer led to a lighter workload for providers and more insight into the complex needs of patients. This QCI project demonstrated that it is feasible to implement social determinant assessments at all patient visits in a busy general practice setting. Subsequently, a university teaching hospital adopted the WellRx model, and the New Mexico Department of Health now requires MCOs to use CHWs for Medicaid patients (Page-Reeves et al., 2016).

Prospective Organizational Approaches

According to the U.S. Department of Health and Human Services, "QI activities provide an organization with opportunities to 'think outside the box' and promote creativity and innovation" (HRSA, 2011). The outside-the-box thinking, creativity, and innovation that underpin QCI are critical for us to test and sustain new approaches to community health.

The *Plan-Do-Study-Act* (PDSA) cycle is used by many health organizations to make improvements in quality (see Fig. 9.3). It offers an iterative process of developing a plan to test a small-scale change, implementing the change, observing and learning from the change, and determining what modifications should be made and whether the change is scalable. The PDSA cycle illustrates that QCI involves proactive problem solving and a culture of learning. The PDSA cycle answers the following questions:

1. What are the data telling us about how things work?
2. Can we test changes to improve quality?
3. What do the data tell us about whether we should scale up those changes?

Clinical pathways are another strategy to improve quality at the organizational level, and they have been used frequently in the United States and western Europe. Clinical pathways are based on evidence-based studies and adapted by physicians and other health care professionals to the culture of the organization. They can be used to improve processes for a

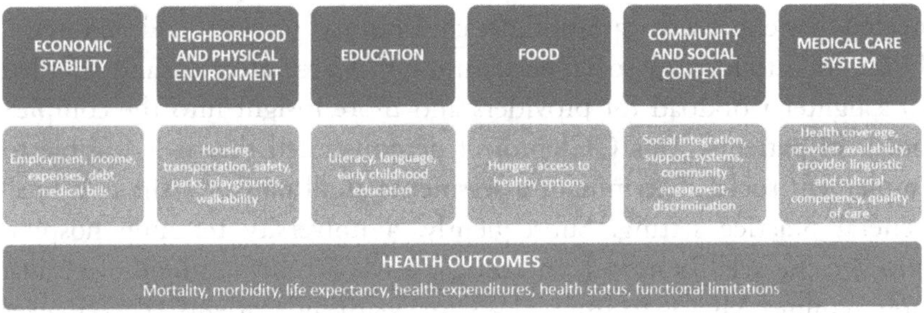

Fig. 9.3 Plan-Do-Study-Act cycle

variety of treatments, including stroke therapy, infection controls, follow-up of hospital discharges, and patient malnourishment. Clinical pathways have improved patient outcomes and reduced readmissions (European Observatory on Health Systems and Policies, 2019).

Another organizational quality strategy is *patient safety culture*. The goal is to change the culture and patterns of behavior in the organization so there is a strong commitment to and proficiency in the organization's health and safety. The patient safety culture is characterized by shared behavioral patterns involving communication, teamwork, working conditions, and outcome measures (e.g., frequency of adverse reporting). It can contribute to the quality of care by creating a new environment for safety and behaviors by developing new structures and processes. One European study found positive associations between the implementation of quality management systems and a teamwork and safety climate. The most effective interventions were team training and communication initiatives and executive or interdisciplinary walk-rounds (European Observatory on Health Systems and Policies, 2019).

The QCI framework also involves meeting patient's needs. In developing countries, the effectiveness of quality strategies may depend on meeting their primary prevention needs (see also Chap. 7, this volume). For example, in Zimbabwe, lay workers are used to screen and identify common mental disorders (Mangezi & Chibanda, 2010). In India, a mobile blood monitoring device is being tested to provide cost-effective diabetes management. This device allows any mobile phone to monitor blood glucose by lay workers and patients. It is anticipated that up to 3 million people will use this device (Grand Challenges Canada, 2019).

In most developed countries other than the United States, there is universal health insurance coverage. Although universal coverage does not assure high quality, it provides a more centralized data collection system that can track the health status of patients and better identify health disparities. In Taiwan, for example, providers receive extra bonuses for serving patients in remote or mountainous areas (Cheng, Chen, & Hou, 2010). Taiwan also has payment incentives that are tied to QCI for illnesses such as asthma and diabetes, as well as widespread information sharing and transparency that help to identify high-risk patients, improve quality, and reduce waste (Cheng, 2015). Finally, the government is attempting to change incentives to address continuity of care problems related to "physician shopping behaviors." Patients in Taiwan tend to seek medical help frequently, leading to a high number of physician visits and less time with each patient. As a result, specialists may spend less time with patients that have serious problems. These patients often end up in the hospital when it could have been avoided (Cheng et al., 2010). To overcome this problem and reduce avoidable hospitalizations, financial incentives are provided in the form of lower copayments if a patient first sees a primary care physician and is then referred for specialty care (Cheng et al., 2010).

Germany also has universal coverage and robust data to track the health of patients (Nasser & Sawicki, 2019). All hospitals are required to report findings on various indicators, allowing hospital comparisons. Volume thresholds have also been established to assure that they are performing a minimum number of complex procedures. Germany relies on its public health system to address health disparities, and care coordination projects are underway to treat patients with two or more chronic conditions.

Cultural, Professional, Legislative, and Capacity Issues Impacting QCI

Some experts would be concerned that using the health care system and clinicians to address nonmedical needs (through Section 1115 waivers or otherwise) runs the risk of medicalizing complex social issues. Similarly, some medical staff would be skeptical and question the value of nonclinical services or the use of nonclinical staff. For example, CHWs are effective because they are trusted members of the community where they serve (Grant et al., 2017). As health care organizations begin to integrate CHWs into their teams, it often creates a fundamental tension between clinical and nonclinical staff. "The marriage of community health and formal health care is powerful, but it's also tricky. If CHWs lose their identity and become medicalized, their effectiveness in the community is lessened. Health care leaders must grapple with a fundamental question:

How do we integrate a grassroots workforce into health care without totally co-opting it?" (Garfield & Kangovi, 2019).

Another major issue is how to share data between health care organizations and public health, social service agencies, and other community-based organizations (e.g., the housing authority; Walport & Brest, 2011). Major challenges include regulatory issues, privacy concerns, and interoperability of systems. In addition, some community-based organizations do not have electronic records for sharing their data important to community health (see also Chap. 10, this volume).

Capacity building is another significant and ongoing challenge. Building care coordination models that integrate medical and nonmedical services requires both health care and non-health care organizations to build capacity by investing funds upfront, but it may be several months or years before these investments are paid back (see also Chap. 6, this volume). In many cases, both health care and nonmedical organizations may have to hire new staff or at least train old staff. They may have to upgrade their data and analytic information systems and offer new types of services. Many of these services are costly and some organizations may be unwilling or not have the ability to make these necessary investments, especially when a positive return on their investment may take several years.

Additionally, the wide variation between and within states presents challenges in moving toward a robust model of sustainable community health. Although some states such as Oregon, Vermont, and California have used their Medicaid programs to pursue innovative solutions very aggressively, other states—many of which bear the brunt of significant health disparities—have moved forward at a much slower pace.

Finally, the successful implementation of a sustainable community health model that incorporates a broad concept of QCI often depends on scale or volume. In many rural areas, for example, low population densities make it very difficult to maintain a sufficient volume of patients to make this model economically sustainable. In addition, many health care and community-based organizations in rural areas also have a difficult time recruiting and retaining health professionals and other staff (Struber, 2004). Without adequate and qualified staff, it is less likely that innovative value-based Medicaid models will be implemented.

Relevance to Disciplines and Specialty Areas

Sustainable community health depends on many factors, including economic vitality, education, the environment, and community safety for all. In pursuit of these objectives, quality must be the major focus of a sustainable community health model. QCI—in the context of population and community health metrics—should be of interest to formal health care providers, such as physicians, nurses, and hospital administrators. Allied health professionals, including those working in mental and behavioral health, social work, nutritional science, physical therapy, and occupational therapy, also have a clear stake in using the tools and practices of QCI to promote sustainable community health. Those working outside of the health care system to address the nonmedical needs of their constituencies, especially in community-based organizations, are also invested in this nexus. More broadly, those who promote health by focusing on interventions at the program, policy, and system levels have a unique opportunity to advance sustainable community health through QCI. Specifically, those working in local and regional governments, including public health departments, social service agencies, housing authorities, and public transportation, among many others, should integrate these concepts into their work.

Public Health To move the needle upstream and to improve community health, local and state public health agencies need to assume a leadership role and act as the Chief Health Strategists for their communities. As described in the *Public Health 3.0 Call to Action*, Chief Health Strategists form vibrant, structured, cross-sector partnerships that leverage the strengths of each organization (DeSalvo et al., 2017). Researchers agree that this collective power of diverse organizations and individuals has the capacity to improve the health of a community. Thus, building collaborative partnerships is a widely promoted strategy to improve the community health outcomes through coordination of services and sharing of information, expertise, and resources (DeSalvo et al., 2017). The sort of multi-sector collaboration envisioned within the Public Health 3.0 framework is uniquely suited to the local level, and those working to improve the social, environmental, and economic conditions that influence health must be forward-thinking catalysts of change who anchor their interventions in a QCI approach.

QCI should be of great interest to *Health Economists* because most of the innovative models are not only designed to improve the quality of care for individual patients and health outcomes for populations but also to control the cost of health care services. There is also an opportunity to examine the cost-effectiveness and value of the nonmedical interventions.

QCI is very relevant to *Epidemiologists* because many of these models described in this chapter have the potential to improve the health of the populations across communities and reduce health disparities across populations. Epidemiologists have the knowledge and tools to investigate how effective these models have been in making improvements in population health. Those working in *Health Services Administration* have the opportunity to compare and test the effectiveness and efficiency of these models across regions and states. It also provides an opportunity to refine and suggest changes in health policy.

Issues for Research and Other Forms of Scholarship on QCI

There are many research issues related to QCI best practices and their impact on health equity, health disparities, and social justice. First, many studies have found that both medical and nonmedical factors influence the health of populations (Fraze et al. 2016). In many instances, however, there is limited evidence about what specific strategic interventions should be implemented to address the social determinants of health. For example, it is well documented that housing supports are important and even critical to improve the health of many low-income people (Jacobs, Wilson, Dixon, Smith, & Evens, 2009). Unfortunately, there is less knowledge about what specific housing supports are most cost-effective in improving health outcomes (Fraze et al., 2016). Implementing the most cost-effective programs is essential because only limited resources are available in the society to address unmet social needs.

Another research issue is to determine the most effective care coordination strategies and the most appropriate balance in the delivery of programs and services between health care, public health, and social services. Ideally, all patients should have a medical home and be screened for risk factors, including the social determinants of health (see Chap. 10, this volume). High-risk patients should then receive some combination of medical services, behavioral health services, public health services, and social services. More research is needed about the most effective combination of services for specific types of patients (e.g., chronic disease, pregnant women, and adolescent children).

A third research issue relates to building effective collaborative partnerships between health care and community-based organizations. Past research studies have found that collaborative partnerships that have strong leaders and a shared vision can lead to more positive health outcomes (Roussos & Fawcett, 2000). However, many partnerships fail because of the cultural divide and the tension that exists between health care and community-based organizations. For example, there may be differences in language or approaches (individual treatment vs whole populations) to supporting low-income populations (Alderwick et al., 2019). Since one of the keys to better population health outcomes is effective collaborative partnerships, it is critical to have a better understanding of the barriers that exist between health care and community organizations and what strategies can be used to overcome them.

Finally, more research is needed about what types of training, education, and competencies are needed for the staff who serve on the health care and community health teams (Mitchell et al., 2012). Providing high-quality care that addresses health equity and health disparities requires new skills and competencies (screening, home visitation, data analysis, and coalition building) and new types of workers, such as CHWs. Future research can help communities to understand the most appropriate balance and the most essential workers in various geographical areas (e.g., underserved rural and urban areas).

Using several different approaches, many states have used their Medicaid programs to drive change in the way health care is delivered and reimbursed. All of these models emphasize a broad QCI approach that goes beyond individual treatment outcomes and moves upstream to address many of the social determinants of health. States are using new models such as ACOs and ACHs and retooling old models (e.g., MCOs) to generate quality improvements across the individual and population health spectrum. Although these new models are driving change and producing many favorable results, especially in QCI, it is also clear that more time is needed before major shifts in population health outcomes can be seen (Matulis & Lloyd, 2018). An evaluation of the investment in social needs in Oregon and California found that successfully addressing the social needs in the community depends on the availability of services in the community. In each of the sites studied, "the scale of Medicaid patients' unmet social needs – for housing, food, income, and more – outstripped the resources available to address them" (Alderwick et al., 2019: 779). Many community-based organizations also have limited capacity, including adequate staff, training and competencies, and IT technology (Alderwick et al., 2019). On the positive side, many strong partnerships have been formed between health care providers and public health and other community-based organizations. These partnerships have created not only a greater awareness of the nonmedical determinants of health, but have also led to the implementation of intervention strategies to address high-need patients.

Summary and Conclusions

Quality must underpin any vision for sustainable community health. Increasing the effectiveness and efficiency of the infrastructure that supports community health demands robust and evidence-based quality improvement principles and practices that are safe, effective, patient-centered, timely, efficient, and equitable. New health care delivery models and payment reforms have been implemented primarily through state Medicaid and Medicare programs, and these have greatly expanded the concept of QCI. Many of these new health delivery approaches provide strong incentives to improve health care treatment for individuals and screen for and address the social determinants of health. Although the QCI concept has been expanded, many barriers still exist and must be overcome before the goals of sustainable community health can be achieved. However, this broader QCI vision and the multi-sector partnerships that have been formed to achieve the vision have changed the health system from a focus on improving the outcomes of individual patients to improving the outcomes of populations.

References

Abrams, M. K., Nuzum, R., Zezza, M. A., Ryan, J., Kiszla, J., & Guterman, S. (2015). *The Affordable Care Act's payment and delivery system reforms: A progress report at five years*. The Commonwealth Fund. Retrieved on September 16, 2019 from https://www.commonwealthfund.org/publications/issue-briefs/2015/may/affordable-care-acts-payment-and-delivery-system-reforms

Alderwick, H., Hood-Ronick, C., & Gottlieb, L. (2019). Medicaid investments to address social needs in Oregon and California. *Health Affairs, 38*(5), 774–781.

Bandara, S. N., Huskamp, H. A., Riedel, L. E., McGinty, E. E., Webster, D., Toone, R. E., & Barry, C. L. (2015). Leveraging the Affordable Care Act to enroll justice-involved populations in Medicaid: State and local efforts. *Health Affairs, 34*(12), 2044–2051.

Bradley, E. H., Canavan, M., Rogan, E., Talbert-Slagle, K., Ndumele, C., & Taylor, L. (2016). Variation in health outcomes: The role of spending on social services, public health, and health care, 2000–09. *Health Affairs, 35*(5), 760–768.

Castrucci, B., & Auerbach, J. (2019). *Health affairs: Meeting individual social needs falls short of addressing social determinants of health*. De Beaumont. Retrieved on August 5, 2019 from https://www.debeaumont.org/news/2019/meeting-individual-social-needs-falls-short-of-addressing-social-determinants-of-health/

Centers for Medicare and Medicaid Services. (2019). *Using data for quality improvement: A case study from St. Joseph's Hospital Health System*. Retrieved on April 17, 2020 from https://innovation.cms.gov/files/x/ahcm-casestudy-stjoseph.pdf

Cheng, S. H., Chen, C. C., & Hou, Y. F. (2010). A longitudinal examination of continuity of care and avoidable hospitalization. *Archives of Internal Medicine, 170*(18), 1671–1677.

Cheng, T. M. (2015). Reflections on the 20th anniversary of Taiwan's single-payer National Health Insurance System. *Health Affairs, 34*(3), 502–510.

DeSalvo, K. B., Wang, Y. C., Harris, A., Auerbach, J., Koo, D., & O'Carroll, P. (2017). Public health 3.0: A call to action for public health to meet the challenges of the 21st century. *Preventing Chronic Disease, 14*.

European Observatory on Health Systems and Policies. (2019). *Quality in health care cannot be taken for granted*. Retrieved on April 17, 2020 from http://www.euro.who.int/en/about-us/partners/observatory/news/news/2019/11/quality-in-health-care-cannot-be-taken-for-granted

Fraze, T., Lewis, V. A., Rodriguez, H. P., & Fisher, E. S. (2016). Housing, transportation, and food: How ACOs seek to improve population health by addressing nonmedical needs of patients. *Health Affairs, 35*(11), 2109–2115.

Garfield, C., & Kangovi, S. (2019). *Integrating community health workers into health care teams without coopting them*. Health Affairs Blog: Workforce in the Community. Retrieved on April 17, 2020 from https://www.healthaffairs.org/do/10.1377/hblog20190507.746358/full/but

Glance, L. G., Osler, T. M., Mukamel, D. B., & Dick, A. W. (2008). Impact of the present-on-admission indicator on hospital quality measurement: Experience with the Agency for Healthcare Research and Quality (AHRQ) Inpatient Quality Indicators. *Medical Care*, 112–119.

Grand Challenges Canada. (2019). *Bold ideas with big impact*. Retrieved on April 18, 2020 from https://www.grandchallenges.ca/

Grant, M., Wilford, A., Haskins, L., Phakathi, S., Mntambo, N., & Horwood, C. M. (2017). Trust of community health workers influences the acceptance of community-based maternal and child health services. *African Journal of Primary Health Care & Family Medicine, 9*(1), 1–8.

Hinton, E., Musumeci, M., Rudowitz, R., Antonisse, L., & Hall, C. (2019). *Section 1115 Medicaid demonstration waivers: The current landscape of approved and pending waivers*. Henry J. Kaiser Family Foundation. Retrieved on June 26 from https://www.kff.org/medicaid/issue-brief/section-1115-medicaid-demonstration-waivers-the-current-landscape-of-approved-and-pending-waivers/

Human Resources and Services Administration, U.S. Department of Health and Human Services (HRSA). (2011). *Quality improvement*. Retrieved on June 26, 2019 from https://www.hrsa.gov/sites/default/files/quality/toolbox/508pdfs/qualityimprovement.pdf

Institute of Medicine (IOM). (2001). *Crossing the quality chasm: A new health system for the 21st century*. Washington, DC: National Academy Press.

Jacobs, D. E., Wilson, J., Dixon, S. L., Smith, J., & Evens, A. (2009). The relationship of housing and population health: A 30-year retrospective analysis. *Environmental Health Perspectives, 117*(4), 597–604.

Kangovi, S., Mitra, N., & Grande, D. (2014). Patient-centered community health worker intervention to improve posthospital outcomes: A randomized clinical trial. *Journal of American Medical Association Internal Medicine, 174*(4), 535–543.

Kangovi, S., Mitra, N., Norton, L., Harte, R., Zhao, X., Carter, T., … Long, J. A. (2018). Effect of community health worker support on clinical outcomes of low-income patients across primary care facilities: A randomized clinical trial. *JAMA Internal Medicine, 178*(12), 1635–1643.

Kindig, D., & Stoddart, G. (2003). What is population health? *American Journal of Public Health, 93*(3), 380–383.

Kushner, J., & McConnell, J. (2018, August 16). *Oregon's CCOs: What do we know so far?* Paper presented at: Annual Meeting of the Oregon Primary Care Association, Lebanon, OR.

Kusher, J., Tracy, K., Lind, B., Renfro, S,, Rowland, R., & McConnell K. (2017). Evaluation of Oregon's 2012–2017 Medicaid Waiver. Portland, OR: Center for Health Systems Effectiveness, Oregon Health and Science University. https://www.oregon.gov/oha/HPA/ANALYTICS/Evaluation%20docs/Summative%20Medicaid%20Waiver%20Evaluation%20-%20Final%20Report.pdf

Mangezi, W., & Chibanda, D. (2010). Mental health in Zimbabwe. *International Psychiatry, 7*(4), 93–94.

Matulis, R., & Lloyd, J. (2018). *The history, evolution, and future of Medicaid Accountable Care Organizations.* Center for Health Care Strategies. Retrieved on April 17, 2020 from https://www.chcs.org/resource/history-evolution-future-medicaid-accountable-care-organizations/

McConnell, K. J., Renfro, S., Chan, B. K., Meath, T. H., Mendelson, A., & Cohen, D. (2017). Early performance in Medicaid accountable care organizations: A comparison of Oregon and Colorado. *Journal of American Medical Association Internal Medicine, 177*(4), 538–545.

McCullough, J. M., & Leider, J. P. (2016). Government spending in health and nonhealth sectors associated with improvement in county health rankings. *Health Affairs, 35*(11).

Mitchell, P., Wynia, M., Golden, R., McNellis, B., Okun, S., Webb, C. E., … & Von Kohorn, I. (2012). *Core principles & values of effective team-based health care.* Discussion Paper, Institute of Medicine, Washington, DC. www.iom.edu/tbc

Muoto, I., Luck, J., Yoon, J., Bernell, S., & Snowden, J. M. (2016). Oregon's coordinated care organizations increased timely prenatal care initiation and decreased disparities. *Health Affairs, 35*(9), 1625–1632.

Nasser, M., & Sawicki, P. T. (2019). *Institute for Quality & Efficiency in Health Care: Germany*. Retrieved on April 17, 2020 from https://www.commonwealthfund.org/publications/issue-briefs/2009/jul/institute-quality-and-efficiency-health-care-germany

National Academy for State Health Policies (NASHP). (2018). *States develop new approaches to improve population health through accountable health models*. Retrieved on April 17, 2020 from https://nashp.org/policy/population-health/accountable-health/

National Association of County and City Health Officials (NACCHO). (2017). *Quality improvement*. Retrieved on April 17, 20200 from https://www.naccho.org/programs/public-health-infrastructure/performance-improvement/quality-improvement

Noble, D. J., Greenhalgh, T., & Casalino, L. P. (2014). Improving population health one person at a time? Accountable care organisations: Perceptions of population health—A qualitative interview study. *BMJ Open, 4*(4).

Page-Reeves, J., Kaufman, W., Bleecker, M., Norris, J., McCalmont, K., Ianakieva, V., & Kaufman, A. (2016). Addressing social determinants of health in a clinic setting: The WellRx Pilot in Albuquerque, New Mexico. *The Journal of the American Board of Family Medicine, 29*(3), 414–418.

Pourat, N., Chuang, E., Chen, X., O'Masta, B., Haley, L. A., Lu, C., … & Huerta, D. M. (2019). *Interim evaluation of California's Whole Person Care (WPC) Program*. California Department of Health Care Services. Retrieved on April 17, 2020 from https://healthpolicy.ucla.edu/publications/Documents/PDF/2020/wholepersoncare-report-jan2020.pdf

Roussos, S. T., & Fawcett, S. B. (2000). A review of collaborative partnerships as a strategy for improving community health. *Annual Review of Public Health, 21*(1), 369–402.

Schaeffer, A. M., & Joelles, D. (2019). Not missing the opportunity: Improving depression screening and follow-up in a multicultural community. *The Joint Commission Journal on Quality and Patient Safety, 45*(1), 31–39.

Struber, J. C. (2004). Recruiting and retaining allied health professionals in rural Australia: Why is it so difficult? *Internet Journal of Allied Health Sciences and Practice, 2*(2), 2.

The California Endowment. (2016). *California Accountable Communities for Health Initiative selects six communities to receive $5.1 million*. Retrieved on April 17, 2020 from https://www.calendow.org/press-release/california-accountable-communities-health-initiative-selects-six-communities-receive-5-1-million/

Vickery, K. D., Bodurtha, P., Winkelman, T. N. A., Hougham, C., Owen, R., & Legler, M. (2018). Cross-sector service use among high health care utilizers in Minnesota after Medicaid expansion. *Health Affairs, 37*(1), 62–69.

Walport, M., & Brest, P. (2011). Sharing research data to improve public health. *The Lancet, 377*(9765), 537–539.

World Health Organization (WHO). (2008). *Closing the gap in a generation: Health equity through action on the social determinants of health*. Commission on Social Determinants of Health Final Report. Retrieved on July 10, 2019 from https://www.who.int/social_determinants/thecommission/finalreport/en/

10

Community Health Informatics

Gayle Prybutok

Introduction

Communities seek to be active participants in their health care. At the community level, community coalitions have become effective tools for improving local health over the last 20 years (Zakocs & Edwards, 2006). Community coalitions are multi-organizational alliances that include nonprofit organizations, local businesses, engaged citizens, and local government representatives, all invested in solving an issue that is a significant community concern over some time. Care environments have expanded beyond the hospital systems, necessitating the use of information systems for community health planning for the long-term health of populations (see also Chap. 1, this volume). Moreover, in medical practice, managed care works to control rising healthcare costs. While community health information tools provide both healthcare practitioners and community members platforms upon which to exchange information at little cost. This cost-effectiveness is from the fact that community health information database archives allow for resourcing both current

G. Prybutok (✉)
University of North Texas, Denton, TX, USA
e-mail: gayle.prybutok@unt.edu

E. Mpofu (ed.), *Sustainable Community Health*,
https://doi.org/10.1007/978-3-030-59687-3_10

and prospective health needs. Enabling the design and implementation of futuristic health systems. Critical to sustainable community health is the need for health information sources to evolve and provide health information in easy-to-understand terms. While maintaining accuracy, reliability, timeliness, and usefulness at the same time. As a support to both individual and community-based health decision-making processes. In addition to supporting cutting-edge research on sustainable community health. These needs are met by a variety of consumer health informatics (CHI) applications considered in this chapter.

Learning Objectives

By the end of this chapter, the reader should be able to:

1. Define community health informatics for sustainable community health
2. Outline the history of community health informatics to improving population health outcomes
3. Evaluate the relative merits of community health informatics tools and approaches for sustainable community health
4. Analyze influences of culture, professional practices and legislation on the utilization of community health information for population health
5. Examine the research and practice issues important to enhancing the use of community health informatics for sustainable community health

Consumer health informatics (CHI), is the science of online systems, their development and implementation in healthcare contexts. As a resource for delivering information on the effectiveness of health management, and interventions in public spaces of care (Cleveland & Cleveland, 2009; Rodrigues, 2009). Increasingly, many community organizations have user-oriented websites to host and provide consumers with on-demand health information (Ricciardi, Mostashari, Murphy, Daniel, & Siminerio, 2013). With CHI, users can access the most current health information available when making health decisions. CHI users can also communicate with others afflicted by the same illnesses and can assume responsibility for their health management (Suggs, 2006). CHI is person-centered, relying on person-leveled health services data to improve health literacy in health risk management and self-care.

Community health informatics collects community-based health information on needs for creative community partnerships and multi-organizational coalitions to improve community-level health service delivery (Carney & Kong, 2017; Klein et al., 2011). CHI contributes to sustainable community health by ensuring cost-effective delivery of health information in real-time that can reach a high volume of users simultaneously (Kahn, 2008). A health literacy accomplishment is crucial to maintaining community health and reducing public health risks.

One negative consequence of the big data era to both information access and consumer health informatics is low literacy on eHealth among its users. eHealth literacy includes a wider cluster of skills that users must possess to benefit from its applications and use it effectively. It is a priority to identify skills and access barriers in eHealth literacy, for the resources to assist underserved populations, in their use of it. So they might use the technology to increase their participation in managing health care sustainably (Chan & Kaufman, 2011; see also Chap. 11, this volume). Moreover, increases in the sharing of health information among users have brought persistent concern about the accuracy of exchanged health information, in clinical settings.

Community health informatics is designed to meet the health information needs of specific community stakeholders, who can be diverse in geography, ethnicity, race, religion, creed, sex, gender, health conditions, and their interactions in broader populations. The greatest challenge in community health informatics is in providing timely, accurate, and complete data that effectively targets the needs of consumer communities (Lewis, Chang, & Friedman, 2005; Unertl et al., 2016). The data sources must be complete and accurate to be useful for the design and implementation of interventions to reduce health disparities and improve efforts directed at social justice (Carney & Kong, 2017; Frost & Massagli, 2008). This chapter serves as a resource for health educators and clinicians seeking to use health informatics data, in designing and implementing community-oriented health information for sustainable health practice.

Professional and Legal Definitions of Community Health Informatics

The field of community health informatics spans public health informatics, population health informatics, and consumer health informatics (Carney & Kong, 2017; Gamache, Kharrazi, & Weiner, 2018). These informatics specializations share the common goal of promoting community health through data-informed interventions, which aim to reduce health disparities and social injustice.

Public health informatics is the study of national health data, such as data from national disease registries and data from state and local health departments (Carney & Kong, 2017). Health informatics and public health informatics rely on public health research databases and health-related performance reports, to create health national guidelines and standards. This information is then used from its research bases as evidence-based clinical practice. Population health informatics uses a variety of data sources on community health promotion. Including health information exchanges, electronic health records, regional research data, and hospital-based registries. Population health informatics is focused on improving communication between patients and providers, reducing emergency care by emphasizing primary care and prevention (DiClemente, Crosby, & Kegler, 2009), and improving the coordination of healthcare service delivery. An example of population health information systems is the Interactive Health Communication (IHC; Cline & Haynes, 2001; Eng, Gustafson, Henderson, Jimison, & Patrick, 1999). It is for facilitating interaction between patients, consumers, caregivers, and professionals; mediated by a communication technology, like the internet. EHealth communication is a population health informatics application on the internet and social media platforms. It provides health information that is necessary to make decisions on managing personal health issues. When it works as a system, for interactive health communication, population health informatics has the unique advantage of providing "Peer to Peer Health", where people want to share their experiences with wellness issues, with being sick, and about treatment options (Fox, 2011).

Consumer health informatics focuses on eHealth literacy (Chan & Kaufman, 2011; Norman and Skinner 2006). Chan and Kaufman (2011) defined eHealth literacy as "a set of skills and knowledge that are essential for productive interaction with technology-based health tools" (p e94). With eHealth literacy functions, consumer health informatics takes a patient-centered approach to delivering health information to users in need. This is because eHealth literacy emphasizes individual responsibility in managing personal health issues, collaboration with healthcare providers (Deering & Harris, 1996), and staying informed on current evidence-based practices. EHealth literacy as a function of consumer health informatics aims to make complex health information understandable at the consumer level Houston, Chang, Brown, and Kukafka (2001). The CHI approaches addressed in this chapter span the intersection of public health, population health, and consumer informatics in sustainable community health.

History of Research and Practice Community Health Informatics

In a seminal work recounting the history of health informatics, Cesnik (1996) reports that the discipline is rooted in the marriage of information technology and medicine, and includes a variety of technologies, research activities, and products. Significant events in the early history of health informatics include World War II. At which time, there was a need to manage vast amounts of information spurred in the development of mainframe computers linked to dumb terminals. After World War II, computers continued to evolve and increased in storage capacity (memory), speed, and reliability, advantaging their use in health systems management. As they continued to develop, it became clear that computers could be effective in supporting medical decision-making. This section considers signature historical developments in technology-focused health informatics. Including the evolution of mainframes and web technologies, internet-based approaches, as well as health-workforce-focused health informatics, such as knowledge management and dissemination, and use of community health workers.

Mainframes and Web Technologies As mainframes transitioned to microcomputers in the 1960s and 1970s, medical practitioners came to incorporate information technology into their office practices, and other businesses did as well. By the 1970s, health informatics as a discipline was formalized under Medical and Nursing Informatics. The personal computer was introduced in the 1980s, and medical researchers recognized the importance of sharing information electronically through local area networks (LAN) which connected personal computers, facilitating health information exchange. The internet was created, first as Web 1.0, which lasted from 1989 to 2003, and users were limited to seeking and reading information posted by the information's creator (Cotten & Gupta, 2004). Nonetheless, it was a significant health informatics capability development.

As the Web transitioned to Web 2.0 in 2004, it now became possible for users to interact with information and with other users (Cotten & Gupta, 2004; Kahn, 2008). Web 2.0 technology evolved quickly and were eagerly embraced by internet users who were becoming technology adept, and excited about the informational benefits, and opportunities for personal creativity that were now available online (Liu, 2010). Utilizing Web 2.0 technology, the U.S. National Library of Medicine (NLM), and its network of regional libraries of medicine, began to play a role in the management and distribution of health information. They forged critical alliances, demanded common scientific and health vocabularies, created Medline as a source for accurate and reliable health information for the public, and participated in the development of technologies that facilitate health information exchange (Humphreys, 2007). The maturation of the internet as an information platform beyond Web 2.0 was even more consequential.

Internet-Based Approaches As the Web transitioned to beyond Web 2.0, users could interact utilizing text, videos, blogs, and audio files, as well as participate in online gatherings and social networks (Kahn, 2008). Pew Internet Project reported that 95% of those between the ages of 18 and 29 use the internet, and 87% of those between the ages of 30 and 49 are also active users (Pew Research Center's Internet and American Life Project, Tracking Survey, 2010; see also Greenwood, 2009). The new access to a vast health information source like the internet dramatically changed individuals' and healthcare providers' environments for health-care delivery (Greenwood, 2009).

The digital divide is a major limitation in communities with less digital literacy and poorer broadband access. As they would be left behind in their health supports needs (Chang et al., 2004). To improve access, Chang et al. (2004) recommend legislating health information access for all into health information policy, standardizing health information vocabulary, distributing health information in multiple formats, and ensuring that available health information is current, accurate, and of high quality. They also stressed that incentives to improve user access to reliable health information must be a funding priority. Community libraries have helped to address the digital divide by providing both computers for internet access and instruction on how to find clinically reliable health information through an internet search. As an example, every US library now provides access to computers and the internet as a part of its public service mission to the communities in which they are located (Kinney, 2010).

Knowledge Management and Dissemination The dramatic and internet-driven evolution of health informatics brought increased demands for effective public health knowledge management and dissemination, and requisite training of the community health workforce. For instance, with information access via internet platforms, specialized interdisciplinary care teams can deliver evidence-based practices in collaboration with family physicians. Moreover, community members increasingly participate in their health management, advantaged by online health information systems (Morahan-Martin, 2004; Silver, 2015). As the state and federal levels of community health informatics, knowledge management and dissemination has enabled healthcare reform to improve community preventive health approaches and improved access to primary care (Karamitri, Talias, & Bellali, 2017). Current community health information systems are increasingly available in different languages to facilitate community access in diverse populations (Hohl et al., 2016).

Community Health Workers Increasingly, community health workers are a critical community health informatics resource. They have the advantage to come from the same community they serve, knowing first-hand the community health information. As an example, as far back as the 1960s, community health workers provided health information and care services to underserved populations in the United States (Witmer, Seifer, Finocchio, Leslie, & O'Neil, 1995). The community health workers typically work with migrant healthcare centers, community health centers, healthcare programs serving the homeless population, and even local public health departments. More recently, managed care organizations and academic medical facilities also engage community health workers as information collection and dissemination resources, complementing digital sources of data.

Sustainable Community Health Informatics Approaches

The leading approaches to community health informatics include the use of common scientific and healthcare vocabulary (Humphreys, 2007), community health literacy promotion, and enhancing community health access. These are considered below, with illustrative case examples.

Development of Common Scientific and Medical Vocabularies

The National Library of Medicine (NLM), and its network of regional libraries of medicine, have played a significant role in the management and dissemination of health information to both individuals and communities. The NLM has forged critical alliances, insisted on common scientific and health vocabularies, created Medline as a source for accurate and reliable health information for the public, and participated in the development of technologies that facilitate health information exchange (Humphreys, 2007). Also, health informatics professionals, capable of locating and facilitating scientific and health information exchange for scientists and health literacy among consumers, must be trained and accessible. They must also maintain an up-to-the-minute awareness of new health information as it emerges, and the body of knowledge expands. Health science librarians contribute significantly to these efforts as they meet standardized competencies, and serve as educators, technology resources, and information search specialists in the community and the general public (Cleveland & Cleveland, 2009). Today's mobile community populations seek to access reliable health information from any location (Vest & Gamm, 2010). With the public's interest and use of social media, the use of social media platforms to communicate health information became a viable option to broaden access to critical health information promptly.

An example of accessible medical vocabulary in use is the Centers for Disease Control's social media presence to effectively communicate with the public during a health emergency (Reynolds, 2010). The CDC's use of social media to inform and educate the public as the expanding flu outbreak utilizing Twitter, Facebook, YouTube, and podcasts had an 82% approval rating (Reynolds, 2010).

The Institute of Medicine initiative to improve patient safety required the use of commonly understood language between health providers and community members (Kohn, Corrigan, & Donaldson, 1999). Organizations like the CDC and WHO consider accessible health vocabulary critical to combatting the spread of an epidemic by disseminating accurate and reliable public health information (Peeri et al., 2020).

Health Literacy Approaches

Norman and Skinner (2006) outlined six types of eHealth literacy required of consumers and communities in the internet age. These include computer literacy (skills related to using a computer); information literacy (skills related to expressing information on health needs and finding that information, assessing, and using it to build knowledge and share it); media literacy (skills related to choosing, understanding, and using information represented in audio or visual forms); traditional literacy and numeracy (the skills to read, write, and verbally communicate information, as well as the skills to interpret scales and graphical representations of information); science literacy (the skills to communicate science concepts, the scientific method and to apply scientific reasoning); and health literacy (the skills to find, assess and apply relevant health information when using health services and making health decisions). This is a tall order and community health education and research are high priorities in facilitating this level of eHealth literacy among the general public. The need for eHealth literacy in order to distinguish between clinically reliable and unreliable health information found online is highlighted in the case study by Tang, Pang, Chan, Yeung, and Yeung (2008), which focuses on the negative impact of traditional literacy in users' ability to make use of health information found online, and traditional literacy may be the simplest of the types of literacy required by Norman and Skinner (2006) See Discussion Box 10.1.

To reduce the harm to communities from low health literacy is to require collaborations with healthcare providers and community health workers, while also preparing easy-to-understand materials, free of medical jargon, as well as healthcare provider training in communication skills (Koh et al. 2012). With basic health literacy, community members can utilize physicians' "information prescriptions" that direct patients to clinically accurate internet sources of health information for safe and sustainable health management (Timm & Jones, 2011).

Discussion Box 10.1: Population Health Literacy and Diabetes Control

Lack of health literacy can harm community health and especially for people living with chronic health conditions like Diabetes Mellitus. Poorly managed diabetes can increase the cost of health care for the entire community. For example, in a study by Tang et al. (2008) suggest lower levels of education were associated with lower levels of health literacy due to poor reading skills. Diabetes education and health education would increase health literacy to combat the disease, as would the required production of educational materials that could be used independently by community members. This highlights the need to present information in different formats, including videos, that allow individuals to learn by listening, rather than reading.

What Do You Think?

1. How is basic community education related to health literacy?
2. How would the poor management of chronic health diseases be a cost to sustainable community health?

Health Information Accuracy on the Internet

The internet offers communities new access to medical information, other users, provider scorecards, and choice. Medical and scientific research continues at a breakneck pace, and information systems that manage knowledge and facilitate information exchange and sharing are essential. Critical to all of the changes in evolving healthcare environments is the intersection of scientific and medical information and IT. Data must evolve into understandable information that is accurate, reliable, timely, and useful to respond to the health information needs of users and communities and to support informed medical decision-making at the individual and community levels (Cleveland & Cleveland, 2009).

Hughes, Joshi, and Wareham (2008) specify concerns with information accuracy, information ownership, and legal/ethical complications that can result from online health information sharing. The advent of Peer to Peer Health Information exchange (Fox, 1997), where people want to share personal health experiences, information about their illnesses, wellness information, and information about treatment options with others online, has been facilitated by new online technologies and is a growing trend. This trend introduced concerns about health information accuracy when online health information sources are not

Discussion Box 10.2: The Accuracy of Internet Information and Community Health

The accuracy of health information is critical to community health practices. As a case in point, Quinn, Corrigan, McHugh, Murphy, O'Mullane, Hill, and Redmond (2012) observed 368,000 Google searches on the topic of breast cancer in one month using the commercial program "Wordtracker". However, only about 22% of the 500 webpages contained verifiably accurate health information about breast cancer. Community members would vary in their skills to assess the quality of health information that they find during an internet search.

What Do You Think?

1. What would be the best ways to educate community members on evaluating the reliability of online health information?
2. How online information resources be designed to provide clinically accurate health information to lay community persons?

monitored by a clinical professional. The concern about the clinical accuracy and reliability of online health information has been addressed by NIH's Center for Complementary and Integrative Health, found here: https://www.nccih.nih.gov/health/finding-and-evaluating-online-resources, who provide five important questions for users of online health information to use to assess the quality of online health information. The study in Discussion Box 10.2 shows why this is so important.

Communities Access to High-Quality Online Health Information

Several approaches have been reported on ways to enhance community access to high-quality health information. The Information Rx or Information Prescription project (Leisey & Shipman, 2007) is an example. The project provides patients with health information they can understand and apply when making decisions about their health management. Similarly, the DISCERN instrument (Charnock, 1998; - the name of the tool, not an acronym) provides a standardized set of quality criteria for health information access by community members: clarity, relevance, balanced and unbiased, the credibility of the source, health action options, and shared decision-making.

Interdisciplinary Community Care

Specialized interdisciplinary care teams work with communities encouraging their active participation in their health care. The teams provide care and information systems that manage knowledge and facilitate information exchange and sharing that are essential to community health (Hohl et al., 2016). Critical to all of the changes in the evolving online healthcare environment is the effective merger of easy-to-understand, evidence-based medical information or health information and IT. The interdisciplinary teams provide healthcare information that is understandable to communities. Fisher, Durrance, and Hinton (2004) highlighted the role of the public library in bridging the digital divide and providing internet access to those unable to afford a computer in the home. Libraries can often provide resources in multiple languages and have staff trained to assist community members who are searching for reliable and specific kinds of health information (Birkenhead, 2012). Public libraries are indeed social service agencies that can address unmet needs for reliable health information in local communities.

Cultural, Legislative, and Professional Issues That Impact Sustainable Community Health

Cultural issues have always had a significant influence on community health practices and health outcomes. Culture is such an important factor that a wide variety of information resources about cultural competency are available. Alpi (2001) reported that cultural competency incorporated seven unique domains: communication patterns, attitudes and values, the physical environment and the associated resources and materials, consumer or community participation, policies and procedures, training and professional development, and population-based clinical practice.

Role of Culture Ingram (2012) highlighted the impact of culture on health information access and use, another important consideration in the communication of health information. Ingrim believes that health literacy is also affected by the cultural competence of the information source. Ingrim defined cultural competence as the drive to understand the traditions, customs, beliefs, and values of the user group for which health information is designed. Culturally related religious views and ways of relating to the world, language, and health norms must be understood before health information can be provided by members of the healthcare community to culturally diverse target audiences. This extends to views related to appropriate medical treatment as well as customs related to coping with a patient's death.

Legal Resources These vary by jurisdiction. As examples, the US federal government has taken a variety of steps to address the problem of low health literacy across the country. For instance, the Affordable Care Act (ACA) addresses issues of low health literacy in caring for underserved populations. The National Action Plan to Improve Health Literacy (U.S. Department of Health and Human Services, 2010) aims to create and disseminate health and safety information that is accurate, accessible, and actionable for the general public. Also, the Plain Writing Act of 2010 mandated that federal agencies generate documents written in clear and understandable language to make them easy to read and digest, and that clarifies the services and benefits that beneficiaries are eligible to receive, which can include disability services, health insurance, and nutritional support.

The privacy of patient health information is a major issue in the delivery of healthcare services, and in many settings, such information was protected by the US Health Insurance Portability and Accountability Act of 1996, known as HIPAA. The HIPAA Privacy Rule regulates the appropriate disclosure and use of each person's protected health information (Hader

& Brown, 2010). HIPAA applies to health insurance providers, healthcare clearinghouses, and to any healthcare provider who uses electronic media to transmit private health information (Hader & Brown, 2010).

Ethics in Health Information Sources Ethics remains a pressing issue in the use of social media platforms and the internet as a whole. In 1986, Richard Mason identified four ethical issues concerning digital information sharing, and they were: accuracy, accessibility, privacy, and property of information (Mason, 1986). Safeguarding the information individuals have available to share, and ensuring their rights to control the accuracy of that information and its release continue to remain key ethical issues with the use of all internet and social media sites. These controls are not yet fully in place, despite recent efforts to give users more control over who can see or share the information that they post.

Related Disciplines to Community Health Information

Health informatics has applications across many disciplines to improve the quality of care, producing better patient outcomes, and controlling the cost of healthcare delivery (Charnock, Shepperd, Needham, & Gann, 1999). In the health domain, informatics has a variety of additional applications in addition to community health informatics which, as discussed, includes public health informatics, population health informatics, and consumer health informatics. Providers rely on clinical informatics to manage the care of individual patients; nursing informatics seeks evidence-based solutions when making patient care decisions; pharmacy informatics examines medication-related data, and the treatment outcomes from various medications for specific populations; and imaging informatics bridges many of the health disciplines to present important images across an electronic health record to various providers. Informatics is essential to the healthcare research community. Informatics systems store, manage, and report on data collected during research conducted in a variety of disciplines. In the health economics arena, informatics is critical to the analysis of economic variables, and to mathematically model economic operations of healthcare systems. In the legislative domain, informatics helps to organize healthcare legislation and retrieve it at the anticipation of an impact from new healthcare legislation on society. Historical informatics collects and organizes important historical data like marriage certificates, birth and death records, property records, city records, and even newspapers. Informatics is a multidisciplinary science. It is often used in combination with other information systems that deliver critical information wherever and whenever it is needed. Provided internet accessibility, broadband connectivity is as freely available as is possible, and website-censorship as minimal as is possible.

Issues for Research and Practice in Consumer Informatics

The advent of health informatics in all of its disciplines, including population health informatics, public health informatics, community health informatics, and consumer health informatics, has improved our ability to evaluate health data, and identify and implement evidence-based practices for improving public health (Carney & Kong, 2017; Kindig & Stoddart, 2003). Kindig and Stoddart (2003) emphasize that population health informatics are essential in the drive to reduce health disparities and improve social justice because it considers the social determinants of health, including education, employment status, socioeconomic status, networks for social support, working conditions, physical health genetic makeup, and individual health behaviors and practices. As previously noted, public health informatics is focused on keeping the public healthy by eliminating environmental hazards, preventing injuries, preventing the expansion of disease, responding to health disasters and emergencies, promoting healthy behaviors and health habits, and making sure that health services are high quality and accessible. Two issues critical to research and practice in community health informatics include: identifying the community health information needs, effective message design.

An essential responsibility of health practitioners, health educators, and community health workers is to first assist people in defining their health information needs, and then to further assist them in meeting those needs.

Yet, community users' health information needs are diverse by demographics, emotional and environmental factors, stress, physical and psychological status, and role in society (Ormandy, 2011; Wilson, 1997). Purpose influences the user's need for information and dictates the user's response to the information they find (Case, 2002). In development are algorithms for mapping community health information that is needed by users, for them to make informed decisions, in part of sustainable health interventions. The design of health messages must be customized for the target audience, and must reflect the target community's beliefs, communication patterns, customs associated with managing health issues, familial roles, racial characteristics, and ethnic customs to be credible and deemed trustworthy.

Witte, Meyer, and Martell (2001) propose this as a multi-stage approach as follows: first, message designers must identify their audience, and the goals and objectives of the message they plan to convey. Next, it is important to identify beliefs, contextual features, and message preferences for the target audience. Once the data are collected and analyzed, effective messages may then be created from the gathered information. Questions remain on how best to customize messages for communities (Brewer et al., 2007; Kreuter & Wray, 2003). An understanding of the target community's culture would be mutually beneficial for service users and providers. More still is that more studies are needed on community message customization of eHealth message (Choi, Kim, Sung, & Sohn, 2011).

Several important considerations are associated with using the internet for healthcare information that has not yet been resolved of its issues. These include maintaining the privacy of personal health information (Awsumb, 2010), ethical use of information posted on the internet and social media, inaccuracies of health information posted on sites gone unmonitored by health care, or health information professionals. Nonetheless, the application of CHI in all aspects of health care is at the foundation for sustainable community health interventions, improvements to public health, and increases to social justice.

Summary and Conclusion

Self-Check Questions

1. Define and describe community health informatics in the context of sustainable community health.
2. Outline how community health informatics contributes to improvements in community health.
3. Evaluate the evidence for alternative community informatics approaches to sustainable community health.
4. What are the cultural, professional, and legal practice influences on community health informatics implementation? Discuss concerning examples.
5. How is community health informatics an interdisciplinary science?

Field-Based Experiential Exercises

1. Choose an illness community spread (e.g. COVID-19) and visit a public library to find information on the illness that would be accessible to members of the community. Ask to speak to the reference librarian and summarize your findings.
2. Look up the Social Media Toolkit published by the Centers for Disease Control (found here: https://www.cdc.gov/socialmedia/tools/guidelines/socialmediatoolkit.html) to find information on messaging effectively to identified communities, utilizing a CHI approach. Summarize your findings and suggest how the information accessibility features could be enhanced for your identified community.

Discussion Questions

1. What is your view about the parameters of successful community health informatics coalitions for reducing the health disparities that impact communities?
2. How would you improve the accessibility and utilization of CHI in a community familiar to you?
3. How would you improve CHI approaches to disseminating health messages in a community familiar to you?

Online Learning Resources

Finding and Evaluating Online Resources webpage (https://nccih.nih.gov/health/webresources) to provide contemporary guidelines on evaluating online health information sources for individuals and communities.

MedlinePlus https://medlineplus.gov/ Explore this repository of health information created by the National Library of Medicine.

AHRQ Agency for Health Care Research and Quality https://www.ahrq.gov/ Explore this site dedicated to making healthcare delivery safer for the public.

Joint Commission Speak Up Pamphlet Series https://www.jointcommission.org/speakup.aspx Take a look at the pamphlets available from the Joint Commission that encourage patients to "speak up" when they have a question about managing their illness or if they notice something that may lead to an error in the care they receive.

Evidence-Based Practice Read this article that explains Evidence-Based Practice and how it is improving healthcare delivery https://www.ncbi.nlm.nih.gov/pmc/articles/PMC226388/

Health informatics is a broad-ranging topic with the capacity to improve population health, reduce health disparities, and improve social justice at multiple levels. Consumer health informatics (CHI) helps users to access the most current health information available when making health decisions, to communicate with others afflicted by the same illnesses, and to assume responsibility for their health management. It is person-centered, community-centered, relying on individualized and community-leveled outcome data for responsive health services. Health informatics is interdisciplinary, involving population health informatics, public health informatics, community health informatics, and consumer health informatics. Each of these disciplines within health informatics continues to develop new and better ways to communicate essential health information, with the capacity to improve public health, reduce health disparities, and improve social justice. Community health informatics is a tool that enables collaboration between community members and public health entities on health promotion, and improvements over the coordination of healthcare service delivery.

References

Alpi, K. M. (2001). Multicultural health information seeking: Achieving cultural competency in the library. *Journal of Hospital Librarianship, 1*(2), 51–59.

Awsumb, S. (2010). Social networking sites: The next E-discovery frontier. *Computer & Internet Lawyer, 27*(4), 16–20.

Birkenhead, G. (2012). Informing the public health. *Health Information and Libraries Journal, 29*(3), 177–179.

Brewer, N. T., Chapman, G. B., Gibbons, F. X., Gerrard, M., McCaul, K. D., & Weinstein, N. D. (2007). Meta-analysis of the relationship between risk perception and health behavior: The example of vaccination. *Health Psychology, 26*(2), 136.

Carney, T. J., & Kong, A. Y. (2017). Leveraging health informatics to foster a smart systems response to health disparities and health equity challenges. *Journal of Biomedical Informatics, 68*, 184–189.

Case, D. (2002). *Looking for information: A survey of research on information seeking, needs and behavior*. San Diego, CA: Academic Press/Elsevier Science.

Cesnik, B. R. A. N. K. O. (1996). History of health informatics. *Health informatics: an overview*, 7–12.

Chan, C., & Kaufman, D. (2011). A framework for characterizing eHealth literacy demands and barriers. *Journal of Medical Internet Research, 13*(4), e94. https://doi.org/10.2196/jmir.1750

Chang, B. L., Bakken, S., Brown, S., Houston, T., Kreps, G., Kukafka, R., … Stavri, P. Z. (2004). Bridging the digital divide: Reaching vulnerable populations. *Journal of the American Medical Informatics Association, 11*, 448–457.

Charnock, D. (1998). *The DISCERN handbook. Quality criteria for consumer health information on treatment choices*. Radcliffe, UK: University of Oxford and The British Library.

Charnock, D., Shepperd, S., Needham, G., & Gann, R. (1999). DISCERN: An instrument for judging the quality of written consumer health information on treatment choices. *Journal of Epidemiology & Community Health, 53*(2), 105–111.

Choi, S. M., Kim, Y., Sung, Y., & Sohn, D. (2011). Bridging or bonding? A cross-cultural study of social relationships in social networking sites. *Information, Communication & Society, 14*(1), 107–129. https://doi.org/10.1080/13691181003792624

Cleveland, A., & Cleveland, D. (2009). *Health informatics for medical librarians*. New York, NY: Neal-Schuman Publishers, Inc.

Cline, R. J., & Haynes, K. M. (2001). Consumer health information seeking on the internet: The state of the art. *Health Education Research, 16*(6), 671–692.

Cotten, S. R., & Gupta, S. S. (2004). Characteristics of online and offline health information seekers and factors that discriminate between them. *Social Science & Medicine, 59*(9), 1795–1806.

Deering, M., & Harris, J. (1996). Consumer health information demand and delivery: Implications for libraries. *Bulletin of the Medical Library Association, 84*(2), 209.

Department of Health and Human Services. (2010). *National action plan to improve health literacy*. Washington, DC: HHS. [cited 2020 Mar 15]. Retrieved September 29, 2019, from http://www.health.gov/communication/hlactionplan/pdf/Health_Literacy_Action_Plan.pdf

DiClemente, R. J., Crosby, R. A., & Kegler, M. C. (Eds.). (2009). *Emerging theories in health promotion practice and research*. Carrollton, TX: John Wiley & Sons.

Eng, T. R., Gustafson, D. H., Henderson, J., Jimison, H., & Patrick, K. (1999). Introduction to evaluation of interactive health communication applications. Science panel on interactive communication and health. *American Journal of Preventative Medicine, 16*(1), 10–15.

Fisher, K., Durrance, J., & Hinton, M. (2004). Information grounds and the use of need based services by immigrants in Queens, New York: A context based, outcome evaluation approach. *Journal of the American Society for Information Science and Technology, 55*(8), 754–766.

Fox, S. (1997). *The internet circa 1998*. Retrieved May 12, 2020, from http://www.pewinternet.org/Commentary/2007/June/The-Internet-Circa-1998.aspx

Fox, S. (2011, August). *Mind the gap: Peer to peer healthcare*. Presented at the National Institutes of Health seminar series, medicine: Mind the gap, Washington, DC.

Frost, J., & Massagli, M. (2008). Social uses of personal health information within patients like me, an online patient community: What can happen when patients have access to one another's data. *Journal of Medical Internet Research, 10*(3), e15.

Gamache, R., Kharrazi, H., & Weiner, J. P. (2018). Public and population health informatics: The bridging of big data to benefit communities. *Yearbook of Medical Informatics, 27*(1), 199.

Greenwood, B. (2009). Pew reports: Internet age barrier vanishing. *Information Today, 26*(5), 2.

Hader, A., & Brown, E. (2010). Legal briefs: Patient privacy and social media. *AANA Journal, 78*(4), 270–274.

Hohl, S., Thompson, B., Krok-Schoen, J., Weier, R., Martin, M., Bone, L., … Paskett, E. (2016). Characterizing community health workers on research teams: Results from the centers for population health and health disparities. *American Journal of Public Health, 106*(4), 664–670.

Houston, T. K., Chang, B. L., Brown, S., & Kukafka, R. (2001). Consumer health informatics: A consensus description and commentary from American medical informatics association members. In *Proceedings of the AMIA symposium* (p. 269). Bethesda, MD: American Medical Informatics Association.

Hughes, B., Joshi, I., & Wareham, J. (2008). Health 2.0 and medicine 2.0: Tensions and controversies in the field. *Journal of Medical Internet Research, 10*(3), e23.

Humphreys, B. L. (2007). Building better connections: The national library of medicine and public health. *Journal of the Medical Library Association, 95*(3), 293–300.

Ingram, R. (2012). Using Campinha-Bacote's process of cultural competence model to examine the relationship between health literacy and cultural competence. *Journal of Advanced Nursing, 68*(3), 695–704.

Kahn, J. (2008). The wisdom of patients: Health care meets online social media. *Medical Benefits, 25*(13), 12–12.

Karamitri, I., Talias, M. A., & Bellali, T. (2017). Knowledge management practices in healthcare settings: A systematic review. *The International Journal of Health Planning and Management, 32*(1), 4–18.

Kindig, D., & Stoddart, G. (2003). What is population health? *American Journal of Public Health, 93*(3), 380–383.

Kinney, B. (2010). The internet, public libraries, and the digital divide. *Public Library Quarterly, 29*(2), 104–161.

Klein, P., Fatima, M., McEwen, L., Moser, S. C., Schmidt, D., & Zupan, S. (2011). Dismantling the ivory tower: Engaging geographers in university–community partnerships. *Journal of Geography in Higher Education, 35*(3), 425–444.

Koh, H., Berwick, D., Clancy, C., Baur, C., Brach, C., Harris, L., & Zerhusen, E. (2012). New federal policy initiatives to boost health literacy can help the nation move beyond the cycle of costly 'crisis care'. *Health Affairs, 31*(2), 434–443.

Kohn, L., Corrigan, J., & Donaldson, M. (1999). *To err is human: Building a safer health system*. Washington, DC: Committee on Quality of Health Care in America, Institute of Medicine.

Kreuter, M. W., & Wray, R. J. (2003). Tailored and targeted health communication: Strategies for enhancing information relevance. *American Journal of Health Behavior, 27*(Supplement 3), S227–S232.

Leisey, M. R., & Shipman, J. P. (2007). Information prescriptions: A barrier to fulfillment. *Journal of the Medical Library Association: JMLA, 95*(4), 435.

Lewis, D., Chang, B. L., & Friedman, C. P. (2005). Consumer health informatics. In *Consumer health informatics* (pp. 1–7). New York, NY: Springer.

Liu, Y. (2010). Social media tools as a learning resource. *Journal of Educational Technology Development and Exchange, 3*(1), 101–114.

Mason, R. O. (1986). Four ethical issues of the information age. *MIS Quarterly, 10*(1), 5–12.

Morahan-Martin, J. M. (2004). How internet users find, evaluate, and use online health information: A cross-cultural review. *Cyberpsychology & Behavior, 7*(5), 497–510.

National Center for Complimentary and Integrative Health. (2018, January). Publication Number D337.

Norman, C., & Skinner, H. (2006). eHEALS: The eHealth literacy scale. *Journal of Medical Internet Research, 8*(4), e27. https://doi.org/10.2196/jmir.8.4.e27

Ormandy, P. (2011). Defining information need in health–assimilating complex theories derived from information science. *Health Expectations, 14*(1), 92–104.

Peeri, N. C., Shrestha, N., Rahman, M. S., Zaki, R., Tan, Z., Bibi, S., … Haque, U. (2020). The SARS, MERS and novel coronavirus (COVID-19) epidemics, the newest and biggest global health threats: What lessons have we learned? *International Journal of Epidemiology*. https://doi.org/10.1093/ije/dyaa033

Pew Research Center's Internet and American Life Project. (2010, May). *Tracking survey*. Retrieved on March 28, 2011, from http://www.pewinternet.org/Static-Pages/Trend-Data/Whos-Online.aspx

Quinn, E. M., Corrigan, M. A., McHugh, S. M., Murphy, D., O'Mullane, J., Hill, A. D. K., & Redmond, H. P. (2012). Breast cancer information on the internet: analysis of accessibility and accuracy. *The Breast, 21*(4), 514–517.

Reynolds, B. J. (2010). Building trust through social media. CDC's experience during the H1N1 influenza response. *Marketing Health Services, 30*(2), 18–21.

Ricciardi, L., Mostashari, F., Murphy, J., Daniel, J. G., & Siminerio, E. P. (2013). A national action plan to support consumer engagement via e-health. *Health Affairs, 32*(2), 376–384.

Rodrigues, J. J. (Ed.). (2009). *Health information systems: Concepts, methodologies, tools, and applications: Concepts, methodologies, tools, and applications* (Vol. 1). London, UK: Igi Global.

Silver, M. P. (2015). Patient perspectives on online health information and communication with doctors: A qualitative study of patients 50 years old and over. *Journal of Medical Internet Research, 17*(1), e19.

Suggs, L. S. (2006). A 10-year retrospective of research in new technologies for health communication. *Journal of Health Communication, 11*(1), 61–74.

Tang, Y. H., Pang, S. M., Chan, M. F., Yeung, G. S., & Yeung, V. T. (2008). Health literacy, complication awareness, and diabetic control in patients with type 2 diabetes mellitus. *Journal of Advanced Nursing, 62*(1), 74–83.

Timm, D., & Jones, D. (2011). The information prescription: Just what the doctor ordered! *Journal of Hospital Librarianship, 11*(4), 358–365.

Unertl, K. M., Schaefbauer, C. L., Campbell, T. R., Senteio, C., Siek, K. A., Bakken, S., & Veinot, T. C. (2016). Integrating community-based participatory research and informatics approaches to improve the engagement and health of underserved populations. *Journal of the American Medical Informatics Association, 23*(1), 60–73.

US Department of Health and Human Services. (2010). *Understanding the impact of health IT in underserved communities and those with health disparities. Briefing paper*. Washington, DC: Author.

Vest, J. R., & Gamm, L. D. (2010). Health information exchange: Persistent challenges and new strategies. *Journal of the American Medical Informatics Association, 17*(3), 288–294.

Wilson, T. D. (1997). Information behavior: An interdisciplinary perspective. *Information Processing and Management, 33*(4), 551–572.

Witmer, A., Seifer, S., Finocchio, L., Leslie, J., & O'Neil, E. (1995). Community health workers: Integral members of the health care work force. *American Journal of Public Health, 85*(8 Pt 1), 1055–1058.

Witte, K., Meyer, G., & Martell, D. (2001). *Effective health risk messages: A step-by-step guide*. Thousand Oaks, CA: Sage.

Zakocs, R., & Edwards, E. (2006). What explains community coalition effectiveness: A review of the literature. *American Journal of Preventive Medicine, 30*(4), 351–361.

11

Telehealth Utilization in Low Resource Settings

Charles P. Bernacchio, Josephine F. Wilson, and Jeewani Anupama Ginige

Introduction

Telemedicine is the provision of health services (e.g., assessment, treatment, and education) across distances through the use of telecommunication technology (Benavides-Vaello, Strode, & Sheeran, 2013; Brown et al., 2015). Our use of the term "telehealth" is inclusive of all teletechnology uses for meeting behavioral as well as physical healthcare needs with the use of any technology-assisted delivery of care including teletherapy and tele-rehabilitation. Telemedicine as a type of telehealth service seeks to improve access to health services, including specialty care

C. P. Bernacchio (✉)
University of Southern Maine, Portland, ME, USA
e-mail: charles.bernacchio@maine.edu

J. F. Wilson
Wright State University, Dayton, OH, USA
e-mail: josephine.wilson@wright.edu

J. A. Ginige
Western Sydney University, Sydney, NSW, Australia
e-mail: j.Ginige@westernsydney.edu.au

E. Mpofu (ed.), *Sustainable Community Health*,
https://doi.org/10.1007/978-3-030-59687-3_11

access, while reducing travel costs, decreasing wait time for specialty visits, yielding higher visit rates, reducing missed visits, and improving continuity of care (Brown-Connolly, 2002; Kruse, Bouffard, Dougherty, & Parro, 2016; Saeed, Diamond, & Bloch, 2011). Increasingly, telehealth and telemedicine are utilized to provide healthcare services in low-resource settings (Kim, 2010). Low-resource settings are mostly low-income countries and also regions in middle- or high-income countries with limited infrastructure for the delivery of health and social services.

Learning Objectives

By the end of this chapter, the reader will be able to:

1. Define telehealth and telemedicine in the context of sustainable community health systems.
2. Outline the history of research and practice in telehealth and telemedicine.
3. Discuss current and prospective practices in telehealth and telemedicine in rural, low resource settings.
4. Evaluate the influence of culture and legislation on telehealth and telemedicine practices in some jurisdictions.
5. Examine the issues for research and other forms of scholarship to improve the use of telehealth and telemedicine in low resource settings.

Mobile health or *mHealth* is the use of mobile and wireless devices to improve health outcomes and healthcare services. It is a fast-emerging form of telehealth solutions for all types of healthcare needs (White, Thomas, Ezeanochie, and Bull 2016). *mHealth* or *tele-video conference* telehealth service delivery models have been implemented with efficacy around the globe (Campbell et al., 2017; Dorstyn, Mathias, & Denson, 2013; Ettinger, Pharaoh, Buckman, Conradie, & Karlen, 2016; Hall, Fottrell, Wilkinson, & Byass, 2014; Ickenstein et al., 2010; Zakus et al., 2019), although the benefits were not consistent [see Discussion Box 11.1]. One explanation for mixed research support outcomes could be from differences in infrastructure and health policy capabilities for *mHealth* in some countries limiting access to services and resources. *mHealth* has occurred concomitantly with increased cell phone usage (HRSA, 2015, as cited in White et al., 2016), making it more accessible to people in low resource settings.

Discussion Box 11.1: Community Case Management & mHealth (Zakus et al., 2019)

Under the WHO's oversight, World Vision Niger and Canada, the Niger Ministry of Public Health implemented an integrated community case management (iCCM) service in a few health districts in Niger in 2013 utilizing *mHealth* to diagnose and treat children under five years of age presenting with diarrhea, malaria, and pneumonia and to refer children with severe illness to the higher level facilities. The relative advantage of diagnosis and treatment of the three most prevalent childhood diseases using the mHealth technology, although encouraging, was relatively small. *mHealth* costs and logistics were major limitations as was the lack of trained community health worker personnel in the use of the technology.

What Do You Think?

1. What factors may have impacted the use of mHealth technology in the Niger low resource country setting?
2. How may community health workers (HWs) be a resource in the use of mHealth technologies in low resource settings?

The Millennium Development Goals (MDGs) of the World Health Organization (WHO) to increase economic, social, and health wellbeing across the world are unlikely to be met in most low- and middle-income countries without investment in infrastructure to permit telehealth care (UN, 2009; Rotheram-Borus, Tomlinson, Swendeman, Lee, & Jones, 2012). Many underdeveloped, primarily rural, and economically challenged countries realize that "due to distance from centrally located infrastructure in urban areas, rugged natural terrain that makes it difficult for service providers to economically set up the physical infrastructure and of course the economic viability of operations in terms of investment returns, these rural populations cannot economically support the investments" (Mupela, Mustarde, & Jones, 2011, p. 1). Mobile health interventions with patients from communities with low-digital literacy as in most limited-resource settings raise questions about patient safety and confidentiality with the use of mHealth interventions, which would reduce acceptability (Campbell et al., 2017). Nonetheless, digital health platforms are here to stay and their propagation in low resource settings guaranteed (Wootton, Patil, Scott, & Ho, 2009) [see Discussion Box 11.1]. This chapter provides some guidance on promising practices utilizing telehealth for sustainable community health in low resource settings.

Professional and/or Legal Definitions and Theories on Sustainable Telehealth and Telemedicine

The World Health Organization (WHO) is playing a growing role in helping developing nations use telehealth in rural and underserved areas (Schneider, 2011). WHO is currently assessing projects globally and overseeing successful programs in Asia and Africa as well as several Latin American countries. The projects are managed at the national level, with WHO largely playing an advisory and facilitation role (WHO, 2016). The WHO defines telemedicine as:

[t]he delivery of healthcare services, where distance is a critical factor, by health care professionals using information and communications technologies for the exchange of valid information for the diagnosis, treatment and prevention of diseases and injuries, research and evaluation, and for the continuing education of healthcare providers, all in the interest of advancing the health of individuals and their communities (WHO, 2010).

The term *telemedicine* generally describes using telecommunications to enhance the delivery of medical care by allowing a physician or a practitioner at one location to observe a patient or data concerning that patient at another location (Coleman, 2002). It is not a single technology or discrete set of related technologies but rather, it is a large and very heterogeneous collection of clinical practices, technologies, and organizational arrangements (Flowers, 1999 as cited in Coleman, 2002). The term *Telemedicine* involves a subset of telehealth including many medical specialties (e.g., teleradiology, telepsychiatry, telecardiology, telepathology, or teleoncology) (American Nursing Association [ANA], 1997).

According to the American Nursing Association (ANA), several definitions relate to a new area of practice called *Telehealth*, which is the removal of time and distance barriers to deliver healthcare services or related activities (ANA, 1997). Many technologies used in telehealth include telephones, computers, direct links to healthcare instruments, interactive video transmissions, and the transmission of images and teleconferencing by telephone or video. At least three distinct technologies fall under the umbrella of telehealth (Stephens, 2014). These are (1) real-time (synchronous or instantaneous) that may include video conferencing and the use of peripheral devices to enable live communication; (2) store-and-forward that is data being captured locally, then stored or cached for forwarding and later use [requires the use of a secure Web server, encrypted e-mail, appropriate store-and-forward software, or an electronic health record system]; and (3) remote patient monitoring that includes services to remotely collect, store, and communicate patient or client biometric health information to practitioners.

A useful differentiation between various types of telehealth involves classifying modalities as either in "real time" (synchronous) or in "store and forward" (asynchronous). "Synchronous telehealth allows for live interaction between users and includes media such as videoconferencing, while asynchronous forms of telehealth allow for storage and release of information over time and include examples such as medical image sharing or texting" (Toh, Pawlovich, & Grzybowski, 2016, p. 961).

Within the United States, the federal government created a type of "store and forward" data system as part of its e-government called *Telehealth* that aids the Centers for Medicare and Medicaid Services in bringing critical public health information to potential beneficiaries and service recipients (Schmeida & McNeal, 2007). This web site[1] is a government outreach effort to enroll underserved populations (rural, poor, children, and persons with CID) into a program, and to inform beneficiaries about benefit changes as they occur. We utilize the terms telehealth and telemedicine interchangeably in the rest of this chapter.

History of Research and Development of Telehealth

In 1997, WHO's former director-general, Dr Hiroshi Nakajima, announced the organization's implementation of a telehealth global strategy for the twenty-first century by improving collaboration with

[1] http://www.pewinternet.org/report_display.asp?r=26

international organizations including the World Bank and the International Telecommunications Union, local ministries of health and universities, while creating an advisory committee on health telematics (Schneider, 2011). WHO's mission is to promote telehealth for use in disease surveillance, prevention, health education, and training, "giving priority to the poorest countries," according to a WHO document on the telehealth policy. However, the history of telehealth and telemedicine with rural and remote communities predates the WHO initiatives by about half a century (Baumann & Scales, 2016; Nickelson, 1998; Stephens, 2014). Notable developments include telemedicine services to the Australian remote outback (Stephens, 2014). As early as 1926, an electrical engineer was asked to solve the problem of communicating with isolated farms and communities in Australia. The first telemedicine approach was called the "Flying Doctor" that was essentially an emergency air ambulance, with one aircraft, one pilot, and one doctor. Patients in the outback made emergency calls in Morse code using two-way bicycle, pedal-powered radios (Stephens, 2014).

In the United States, resurgent telehealth and telemedicine initiatives to provide mental and physical health to rural communities can be traced back to the 1950s. One early initiative by the University of Nebraska's School of Medicine for telemental health was delivered over closed-circuit television in 1959 (Nickelson, 1998). As another example, in the late 1960s an urban medical center in Boston partnered with its international airport medical station connecting by a remote-controlled camera that used interactive video communications for transmitting a series of diagnostic data (Coleman, 2002). By the 1970s, telehealth and telemedicine programs that utilized satellites or dedicated video connections were in operation supported by federal financial and technical assistance. For example, in 1976 the Canada-US Hermes satellite was used to carry out brief trials of telemedicine for rural and remote areas in northern Canada. Nonetheless, few health care providers had utilized teletechnologies into their practice as these were still new and many health centers were not equipped to deliver telehealth or telemedicine.

By the 1980s, the US military and certain correctional institutions began experimenting with telehealth to address the costs for accessing specialty health services in remote regions. These efforts were facilitated by the increasing availability of less costly technologies and delivery systems (Nickelson, 1998) and through the expansion of digital communication (Sikka, Paradise, & Shu, 2014). These initiatives included exploratory use of the internet and small telehealth systems involved only three programs in 1989, which provided medical consultations by telemedicine, with a phenomenal increase to roughly 139 telemedicine programs that were devoted to patient care by 1997 [excluding teleradiology] (Grigsby, 1998 as cited in Coleman, 2002). At this time, Arthur Little (1992) stressed the potential for telecommunications under the national healthcare reform debate that leads to a net savings of billions of dollars in overall healthcare costs and caught the attention of policymakers (as cited in Nickelson, 1998). The mushrooming of telehealth and telemedicine in the United States was primarily on the back of National Information Infrastructure Initiatives for computing and telecommunications networks, and incentives to technology industries to rapidly develop low-cost, telecommunications products for individuals and businesses. At the start of the new millennium, telemedicine appeared in several countries, ranging from low-level forms (e.g., e-mail or phone) to high-level forms (e.g., specialized equipment for telediagnosis). Consultations using telemedicine across the state, national, and continental borders were becoming an everyday occurrence (Coleman, 2002).

The US Department of Veterans Affairs (VA), in 2001, was one of the first healthcare systems in the United States to implement home telehealth services. The VA developed several early telehealth programs targeted at adults with complex medical conditions such as diabetes, stroke, and congestive heart failure, as well as an in-home monitoring program using a text messaging device, so physical therapists and occupational therapists were able to monitor the activities of daily living and safety of frail older adults living in their own homes (Lee & Harada, 2012). Unfortunately, the incorporation of telehealth into practice was slow to be adopted by physicians and hospitals due to concerns regarding cost, privacy, reimbursement, as well as logistics of setting up a telehealth network (Sikka et al., 2014). Nonetheless, significant US Federal investment in telehealth has played an active role in the implementation of telemedicine by (1) authorizing federal grant programs to support local and statewide telemedicine networks; (2) reimbursing telemedicine services under

Medicare and Medicaid; and (3) using telemedicine to provide direct medical care for military service, veteran, Native American, and correctional care populations. Multiple federal agencies are playing a significant role in the development and implementation of telemedicine, including:

- The Office for the Advancement of Telehealth, Department of Health & Human Services (HHS, 2016)
- The Telemedicine and Advanced Technology Research Center, at the US Army Medical Research and Materiel Command (USAMRMC), Ft. Detrick, MD
- The Veterans Health Administration, at the Department of Veterans Affairs
- The National Library of Medicine, at the Department of HHS
- The Federal Communications Commission (Universal Services Order)

Over the past several years, the Department of Homeland Security, Health Resources and Services Administration (HRSA), government-supported research organizations, and nongovernmental organizations (NGOs) have started providing funds to explore the utilization of advanced communications to improve situational awareness, surveillance, and medical response (Balch, 2008, p. 608).

European Union initiatives include implementation of telehealth such as the teleneuromedicine network for quicker access to neuromedical expertise in Germany, which networked comprehensive stroke unit centers and mostly was supported by the state system (Ickenstein et al., 2010). The federal Medicare Benefits Scheme (MBS) of the Department of Health of Australia includes a range of incentives to providers for delivering specialists telehealth services to Australians in remote, regional, and outer metropolitan areas, with higher reimbursements for video-based telehealth services, compared to face-to-face consultations (Bursell, Zang, Keech, & Jenkins, 2016). Telehealth has been reliable for patient diagnosis in the low-income country of Nepal, for cervical cancer screening in Botswana, and for health education in South Africa (Wootton & Bonnardot, 2015; see also Kim, 2010).

Current and Prospective Practices in Telehealth and Telemedicine

Current approaches to the delivery of telehealth and telemedicine for sustainable community health include the **store and forward** mode of telemedicine, **electronic health records, agency-wide intranet and video conferencing systems, and mHealth devices**. Prospective or futuristic approaches are those that aim to develop a seamless, hierarchical network that links health information and medical expertise to points of need from points of care at the national, regional, state, and local levels. We consider the use of telehealth and telemedicine systems for supporting the medical healthcare needs of rural and remote communities rather than for community health mapping and diagnostics (see Chap. 10, this volume for community health informatics approaches).

Store and Forward Telemedicine

As previously observed, the "store and forward" method is an older type of telehealth that occurs through online web access that provides search access (e.g., WebMD®), where the user enters symptoms through a web search platform and this stored information database will yield possible associated medical diagnoses and the prescribed treatment for identified conditions. For example, Australia's Queensland Hospital and Children's Health[2] permits sending the images of suspected physical conditions to clinicians via web/email, then the clinician examines these images later to provide a possible diagnosis and respond with appropriate treatment and care.

The project application of telemedicine, the "Virtual Doctor Project" is a store and forward example intended to alleviate primary healthcare problems in the Eastern province of Zambia. The project uses off-road vehicles fitted with satellite communication devices and modern medical equipment for a mobile clinic to deliver primary healthcare services to

[2] https://www.childrens.health.qld.gov.au/chq/health-professionals/referring-patients/telehealth-store-and-forward/

some of the neediest areas of the country (Mupela et al., 2011). This conceptual approach is modeled around the "store and forward" mode of telemedicine, where images and information are collected at remote health delivery sites and transmitted via email to doctors based in distant locations. The project uses a collaborative and constructive process applying the Zambia Ministry of Health's (MoH) preferred health indicators to monitor patient outcomes. The MoH's system allows these "virtual doctors" to offer ongoing diagnostic assistance, when required, to the healthcare staff on the ground.

Médecins Sans Frontières, AKA "Doctors Without Borders" is an international, independent, and medical humanitarian organization that delivers emergency aid to people affected by armed conflict, epidemics, natural disasters, and exclusion from healthcare (MSF, 2020). Approximately ten years ago, the MSF developed a multilingual telemedicine network to assist its field medical staff by providing direct access to specialist advice that expanded access to service worldwide (Bonnardot et al., 2014). Store and forward telehealth approaches have improved on patient management by field physicians in low-resource settings (Delaigue et al., 2018).

Electronic Health Records, Agency-Wide Intranet, and Video Conferencing Systems

Electronic health records (EHR) are medical record documentation that requires a digital unidirectional transfer, principally text information that is compact and needs little bandwidth. Most medical records can be transferred before the telemedicine session. For instance, the Connecticut's Community Health Center, Inc. (CHCI) uses a centralized practice management EHR system, an agency-wide intranet and a video conferencing system. The use of EHR by the CHCI increased patient access to specialty care, thereby reducing the impact of advanced disease (Khatri, Haddad, & Anderson, 2013). Project Extension for Community Healthcare Outcomes (Project ECHO™) is an innovative telemedicine program that improves patient care by developing and supporting the competence of primary care providers (Khatri et al., 2013). Moreover, by incorporating ECHO's model into its EHR and video conferencing systems, the CHCI developed a fully integrated healthcare delivery platform with remote access capabilities, allowing healthcare access to a vulnerable population similar to that of a community health center.

Telehealth services work best when the services are provided within an organization or network (i.e., when a healthcare system or hospital system provides telehealth to outlying units in its network; Lambert, Gale, Hartley, Croll, & Hansen, 2016). This enables "remote patient monitoring," reducing visits to hospitals or doctors' offices, while increasing the efficiency of services. Remote patient monitoring can improve patient health via lifestyle coaching (Shandle, 2008). Highly sophisticated sensors that are attached to or implanted in the patients can also communicate physical data to prevent and reduce risks by relaying life-saving information to physicians. Remote patient monitoring provides crucial information to providers before gaps in care create crises and helps to sustain home-care settings. Research data from a study on remote monitoring in a rural Midwestern state in the United States suggest that when remote monitoring telehealth technology was utilized in the home-care setting, both patients and providers were very satisfied with services (Hicks, Fleming, & Desaulnier, 2009). Both the patients and healthcare staff felt it was easy to communicate and that the technology was convenient and user friendly. The patients also felt that home telehealth technology had a very positive impact on the provider–patient relationship and improved quality of care. The study's findings also suggest that home care monitoring reduces hospitalizations and decreases personnel expenses.

The use of a web real-time communication (WebRTC) video conferencing system was a part of a large telehome-monitoring project that was implemented over six remote locations in Australia (Jaccard, Nepal, Cellar, and Yan, 2016). It yielded evidence to show the system works well for a relatively small number of users in rural settings. However, video consultation (VC) services were an acceptable model of care for indigenous patients, with high levels of satisfaction reported from patients, families, and HWs (Mooi, Whop, Valery, & Sabesan, 2012).

Hierarchical Networking of Health Information and Medical Expertise

The telehealth service provides for health information and a medical expertise network enabling care coordination as well as responder coordination in large-scale disaster situations. The American Telemedicine Association (ATA) Emergency Preparedness Special Interest Group (Balch, 2008) is an example. The ATA identifies existing telemedicine networks regionally and identifies a minimal data set for the networks to develop a telemedicine response that can increase medium and long-term surge capacity in affected Mass Casualty Index areas (Balch, 2008). Such a network of existing medical and health facilities would be available to complement and strengthen other existing efforts related to public health, emergency response, and threat detection in the United States, such as the Center for Disease Control's state-based Health Alert Network, as well as local and regional Emergency Medical Operation Centers.

Cultural, Legislative, Professional, and Ethical Issues Impacting Telehealth

Cultural influences on the use of telehealth with low-resource communities include the socioeconomic digital divide. Legal and professional influences on the adoption and use of telehealth services in low resource settings include regulatory requirements, health insurance buy-in for the use of telehealth services, and the lack of training by health providers in the use of telehealth technologies. We briefly discuss each of these next.

Socioeconomic Factors Influencing Access to Telehealth

The socioeconomic gradient by geographic location influences use adoption of telehealth services (Lee, Black, & Held, 2019; Schmeida & McNeal, 2007). This has the effect to increase health risks among rural community residents while failing to receive timely health services.

People without internet access and experience (perhaps the oldest and poorest) remain disadvantaged with respect to accessing critical information that can link them to needed healthcare services. As with traditional means of citizen-initiated contact, individuals who are elderly or less affluent are more likely than their counterparts to take part in online searches for Medicare and Medicaid information, a fact that may arise from their greater need for services (Schmeida & McNeal).

Legal and Professional Influences Regulations to guide the use of telehealth with rural and remote communities are evolving, leaving health providers unsure as to the limits to their responsibilities. Telehealth is a remote service resource, and which can be delivered across state lines. Yet, in the US setting, for example, regulatory instruments would vary across states making, complicating consultation where (1) physician and patient being located in different states, (2) physician and patient are in the same state while a consulting physician is out-of-state, or (3) patient, physician, and consulting physician are each in different states (Gupta & Sao, 2011). In order to meet the varying requirements across different states, compliance increases the costs for out-of-state providers and consequently fosters a monopoly for in-state providers. The deference to state authority primarily represents states' interests and not necessarily what is the most ideal approach in facilitating the practice of telemedicine.

Medical insurance companies would be reluctant to reimburse telehealth and telemedicine consultation to the same rate as face-to-face consultation, which would discourage the sustainable use of these services. For this reason, the sustainable use of telehealth and telemedicine systems would depend on applying the same reimbursement rate for services as with face-to-face consultation. Moreover, telemedicine carries a significant risk for medical malpractice insurance (Gupta & Sao, 2011), which would further discourage health providers from using this high prospect health support resource (Lee et al., 2019).

Few providers are competent in the use of telehealth and telemedicine systems (Brennan, Holtz, Chumbler, Kobb, & Rabinowitz, 2008; Makaroun et al. 2017). The professional culture inertia to prefer the use of face-to-face consultation over telehealth and telemedicine platforms is a major barrier to the use of telehealth and telemedicine for sustainable community health. The fact that providers easily acquire requisite skills for brief training is a hopeful sign for the future of telehealth and telemedicine as resources for sustainable community health. Community members may be reticent to use telehealth and telemedicine for their health consulting as a result of digital illiteracy (Campbell et al., 2017; Fitzner et al. 2014) and also concerns about the privacy of their data transmitted digitally, which would be hacked (Ettinger et al., 2016; Hall & McGraw, 2014) or acquired by third parties for commercial use without the consent of the community members (Mathieson et al. 2017; Nakamura et al. 2019) [see Discussion Box 11.2].

Discussion Box 11.2: Telehealth Privacy and Security Protections

Hall and McGraw explain that existing regulations are insufficient to provide strong privacy and risk protections for telehealth users. Currently, the Health Insurance Portability and Accounting Act (HIPAA) contains the primary set of regulations that guide the privacy and security of health information. HIPAA requires that identifiable health information be encrypted so that only those authorized to read it can do so. HIPAA, however, applies only to "covered entities"—healthcare providers and insurers—not to patients. The Food and Drug Administration (FDA) regulates medical devices but not consumer-facing devices and apps, focusing on technical issues related to the security and integrity of information. In this way, the FDA will ensure patient safety but not patient privacy.

What Do You Think?

1. How would the regulatory complexities between HIPAA and FDA influence the use of telehealth with rural and remote communities?
2. What level of security and privacy protection is needed for the underlying telehealth data and system to instill trust in the use of telehealth solutions?
3. Why is action by policymakers needed to ensure an adequate security and privacy policy exists for patients and the playing field is level for companies implementing telehealth?

Related Behavioral Health Disciplines Influencing Community-Oriented Telehealth

Several disciplines are involved in the research and implementation of telehealth and telemedicine for sustainable community health. Examples include community psychiatry, health services administration, public health, and information science and engineering. We briefly outline their roles below.

Community Psychiatry Community psychiatry is focused on preventing and treating mental illness in populations that are exposed to harmful biopsychosocial factors (Caplan & Caplan, 2000). Community psychiatrists provide crisis care to people facing traumatic events to help them address their current difficulties. They organize support groups that include other professional and non-professional caregivers to meets the needs of the population in crisis. In addition, community psychiatrists provide educational support to individual healthcare professionals as well as caregiving agencies. Community psychiatrists using telehealth would for medication management, working closely with primary care providers and psychotherapists to ensure that patients receive medication monitoring and well-coordinated care (Bashshur, Shannon, Bashshur, & Yellowlees, 2016; Caplan & Caplan, 2000). Additionally, forensic psychiatry uses telehealth consultations providing care to rural child psychiatric outpatient clinics, rural juvenile detention centers, and rural school districts (McLennan, 2018; Miller, Clark, Veltkamp, Burton, & Swope, 2008).

Health Services Administration The field of health services administration combines policy, leadership, business management, and science in directing the human and fiscal resources needed to deliver effective health services in *hospitals, hospital networks,* and/or *healthcare systems* (Bohmer & Edmondson, 2001). Practitioners in health services administration combine telehealth, and telemedicine leadership and collaboration, with opportunistic use of technical and financial resources, to address local health needs. They typically must (1) pursue initiating as well as follow-up grants to support telehealth; (2) create independent entities with appropriate local telehealth expertise; (3) tailor telehealth innovations to growing needs and accessible technology options; and (4) facilitate participation within the rural health institution, and collaboration with the local community and external partners, to make innovations sustainable (Myers, 2019; Singh, Mathiassen, Stachura & Astapova, 2010). As an example, health services administrators in Canada have created sustainable telehealth services for individuals in remote, rural locations in Canada, many of whom are of Aboriginal descent with poorer health status than that of the non-Aboriginal population, by addressing issues related to geography, technical infrastructure, human resources, cross-jurisdictional services, and community readiness (Muttitt, Vigneault & Loewen, 2004).

Public Health The public health field provides disease prevention and health promotion services at the population level, in addition to individual-level services, such as well-child visits, prenatal care, and primary care (Slifkin, Silberman, & Reif, 2001). Public health institutions are primarily focused on providing accessible clinical services to low-income populations, such as in rural areas, where residents are disproportionately poorer and have less education (Ricketts, 2000). Most recently, public health officials have been working to promote the adoption of telehealth services in order to provide sustainable healthcare services to rural populations (Singh et al., 2010). Public health policymakers who work with government regulatory offices continue to study the impact of broadband access on the nation's health (Bauerly, McCord, Hulkover, & Pepin, 2019; Connect2Health, 2018).

Information Science and Engineering Information Science is a social science discipline concerned with how we use and manage information including the technology behind it. Information Technology (IT) is the engineering side of information science and is primarily concerned with computer hardware and software and telecommunications, which are engineered into information systems (Kun, 2001). The technologies produced by information scientists and engineers have enabled public health officials to improve access to health care for all populations, especially those in rural areas. Software engineers have developed online platforms and software applications for computers, smartphones, and tablets. Network security engineers have developed software to protect patients' private health information, as well as medical and billing data, to provide privacy, confidentiality, and security of personal information (Kun,

2001). Technological developments introduced by information scientists and engineers are changing the delivery of health care in rural areas. These technological developments include mobile telehealth services, remote monitoring, social networking, and wearable devices (Institute of Medicine, 2012). In Malaysia, health services administrators, public health officials, and IT engineers are working together to develop an integrated telehealth system for telehealth services across Malaysia. The information scientists developed an information flow model prior to telehealth integration to strengthen the integration of telehealth services in Malaysia (Sugijarto, Safie, Mukhtar, & Sulaiman, 2013) Other information scientists use data-driven prognostic techniques to analyze the condition of telemedicine networks, which must operate reliably in order to support vital medical services, and to prevent failure of these networks in harsh weather conditions or due to wear and aging (Fong, Ansari, & Fong, 2012).

Rural Health Rural health is an interdisciplinary study of health and healthcare delivery in rural environments (Chan, 2010). Individuals living in rural areas have different healthcare needs from those in urban areas and are more likely to suffer from a lack of access to health care. People living in rural areas tend to be under the age of 18 or elderly, are poorer, and have less education, higher rates of tobacco and alcohol use, and higher *mortality rates* when compared to their urban counterparts (Chan, 2010). Most significantly, transportation barriers prevent individuals living in rural and geographically isolated areas from tapping into essential healthcare services enjoyed by their urban counterparts. Telehealth has been promoted as a way to overcome transportation barriers for patients and healthcare providers in rural and geographically isolated areas, as well as providing clinical, educational, and administrative benefits for rural areas (Chan, 2010). In rural British Columbia, Canada, the health benefits of telehealth services have been evaluated and elucidated (Moehr et al., 2006). Strong program management and addressing the needs identified in rigorous evaluation processes improves the sustainability of telehealth projects, including emergency and trauma services in remote regions (Moehr et al., 2006). In rural health, telemedicine can be an effective approach for communication and counseling, allowing physicians to monitor their patients' chronic conditions, which can improve their patients' quality of life and reduce hospital admissions and deaths from chronic diseases.

Community Medicine Similar to community psychiatry, the discipline of community medicine seeks to provide medical care to populations in specific geographic areas (Joseph et al., 2018). Community medicine works in close partnership with public health and other community healthcare agencies to provide preventive and treatment services to a given community, often a low-income population. Among its chief functions, community medicine focuses on the identification and prioritization of health needs of a defined community, the provision of interventions to address the health needs and health determinants of that community, and the delivery of health care to address identified health needs (Joseph et al., 2018). In response to calls for telehealth support by community medicine personnel in rural areas, the Office for the Advancement of Telehealth in the US has funded 12 regional and 2 national Telehealth Resource Centers, which assist healthcare organizations, networks, and providers with implementing cost-effective telehealth programs to serve rural and medically underserved areas and populations (Rural Health Information Hub, 2020). Community medicine specialists at two medical schools in Ohio have collaborated to accelerate telehealth expansion to provide remote virtual primary care during the COVID-19 pandemic across Ohio, including in rural areas where access to care is limited (Olayiwola et al., 2020). In this Ohio-based project, telehealth care has been used for a wide range of primary care needs, including chronic disease management, physical exams, well-child visits, wellness checks, mental health follow-up, medication management, new patient encounters, acute non-emergent complaints, and lifestyle counseling.

Issues for Research and Other Forms of Scholarship in Sustainable Telehealth and Telemedicine in Low-Resource Areas

There is a very limited number of published international studies on the use of telehealth and telemedicine system uses in the developing countries (Augusterfer, Mollica, & Lavelle, 2015). However, the consensus continues to be that telehealth and telemedicine bring the promise of evidence-based best practices in community medicine to underserved and difficult to reach regions of the world. Three research and practice issues require ongoing studies in the context of the use of telehealth and telemedicine as resources for sustainable community health.

Financial and Logistical Challenges Among those major challenges to the telehealth delivery of care, a majority are financial in nature and may require considerable investment, particularly in low resource settings (e.g., resource-limited), which more often are located within rural regions. The National Consortium of Telehealth Resource Centers[3] identified the following financial hurdles to developing effective telehealth services: start-up costs (includes technology equipment) and related fees; broadband internet or availability of an alternative; increase in staffing (e.g., monitoring or facilitating care); and need for training and workforce development (Louder & Solomon, 2017). The "inadequacy of fee-for-service reimbursement to sustain the expanded use of telehealth and telemedicine programs is complicated by system-level issues including workforce supply challenges, recruitment and retention challenges, and high rates of un-insurance and under-insurance" (Lambert et al., 2016, p. 377). Studies will need to focus and examine the economic impact of the telemental health in rural communities that will provide decision makers with a clearer understanding of the impacts associated with a behavioral telehealth approach as well as implementation investments that contribute toward building sustainable, intelligent rural communities (Holland et al., 2018). The integration of behavioral health into telehealth and telemedicine is essential for supporting the wellness and health needs of rural communities. Research should also be conducted to study the complicated system-level issues including workforce supply challenges, recruitment and retention challenges, and high rates of uninsurance and underinsurance that will need to be resolved in order to sustain the expanded use of telemental health programs (Lambert et al., 2016; Lee et al., 2019).

Privacy Concern One aspect of telemedicine that is a by-product of advancing technologies involves concerns about the privacy and security of telehealth systems. These concerns may adversely affect people's trust

[3] https://www.telehealthresourcecenter.org/

in telehealth and threaten the ability of these systems to improve the accessibility, quality, and effectiveness of health care (Hale & Kvedar, 2014).

The quality of care may demand attention to issues such as privacy and security of stored data and sessions, for example, tele-video conference sessions require private rooms where technology support personnel or any other staff who should not have access to personal health data are not present during a session. Medical and healthcare practitioners are being advised that comprehensive standards and regulations may be needed that ensure strong privacy and security protections for telehealth and all electronic consumer information (Hale & Kvedar, 2014) [see Discussion Box 11.2].

Further studies should explore more factors that influence rural populations' telehealth service use. Such research could examine factors that correlate with different forms of telehealth services, in addition to gaining data on patient perceptions of possible risks and barriers to the use of telehealth services, to gain increased knowledge about which approaches offer the most benefit (Lee et al., 2019). Research should also ascertain where more comprehensive standards and regulations are needed to ensure strong privacy and security protections not only for telehealth but also for all electronic consumer information (Hale & Kvedar, 2014). A greater understanding of what concerns and risks patients may have that would influence public access to telehealth and telemedicine will be key in supporting and facilitating sustainable holistic health for rural communities.

Buy-in by Practitioners As previously noted, there may also be provider-level "push-back" to the idea of a telehealth option due to expressed patient fears that the quality of care might be compromised when *mHealth* delivery is being substituted for office visits. Research has noted that strengthening provider acceptance and the use of telehealth services could benefit patients in rural communities (Lee et al., 2019). Yet, a telehealth approach to service delivery is a vast shift from in-person care for both providers and community members. Hence, engaging high-risk populations in telehealth services with specialty care providers early in the disease process could be a vital step toward improving long-term health status and comfort with telehealth services. Research on the use of telehealth and telemedicine in low resource settings should address the staff and patient comfort level using telehealth tools and to assess community satisfaction with telehealth services since addressing challenges and concerns for risks that arise will increase acceptance.

Need for Complementary Services According to Myers (2019), the use of telehealth technologies is a viable option for addressing community health needs as it provides expanded access to services, permits more effective care management, and fosters integrating primary and mental health care services. "Telehealth diffusion should be complemented with other needed changes in the education, training, and licensing of healthcare professionals; care delivery models; and reimbursement and funding methodologies" (p. 237). The implementation of telehealth and telemedicine will increasingly involve reconciling issues in coverage and reimbursement, licensure, broadband access, and adequacy, in addition to resolving privacy and security barriers and advancing key policy changes.

Universality Versus Locality of the Evidence The challenges in telehealth research to clearly ascertain the direct relationship among categories or generalizability of findings are driving a need to create an evaluation framework to explicitly separate the structural and outcome variables as individual and organizational (Hebert, 2001). Using this design helps to extract commonalities and differences that will allow drawing conclusions about where telehealth is effective, as well as what variables are indicators of success (e.g., quality of care) and starts an inquiry into other issues. This evaluation approach focuses on several questions: (1) As telehealth technology is introduced, what are the expected relationships between structure, process, and outcomes? (2) What can be deemed explicit and what is implied? (3) How does using the technology compare to the process of traditional care? (4) Are the outcomes of using the technology understood for clients, providers, and organizations? Answers to these and similar questions are critical to the long-term use of telehealth and telemedicine systems for sustainable community health.

Summary and Conclusion

In this chapter, we have identified multiple factors that challenge the provision of health services in rural and other low-resource areas, including large geographic distances for patients to travel for care, inadequate numbers of health professionals, and social support services that are available across rural areas, recruitment and retention of qualified professional staff, and the difficulty in enrolling a sufficient number of patients in rural areas to sustain medical services financially. Major concerns in rural areas that contribute to a risk of poorer health include the lack of specialty care providers, including psychiatry, healthcare facilities (e.g., limited beds/long waitlists), and case management services. These disparities that disadvantage the rural poor are of concern because this at-risk population exhibits a higher rate of many diseases than their urban counterparts while having low health resources. This chapter has introduced the use of "telehealth," which broadly includes technology-assisted delivery of care including teletherapy, tele-rehabilitation, and other "tele" technology use. The delivery of home health care facilitated through newer technologies will be a common occurrence as telemedicine evolves. Increased use of mobile technology will connect patients in low resource settings to specialty care, and the use of digital recording devices for continuous and intermittent transfer of biophysical signs for tracking healthcare quality at the community rather than clinic levels.

Self-Check Questions

1. Define telehealth and telemedicine in the context of sustainable community health systems.
2. Outline the history of research and practice in telehealth and telemedicine.
3. Discuss current and prospective practices in telehealth and telemedicine in low resource settings, explaining their use for both *remote monitoring* and *mHealth* platforms.
4. How could state regulations influence the delivery of telehealth services in low resource settings?
5. Examine the issues for research and other forms of scholarship to improve the use of telehealth and telemedicine in low resource settings.

Discussion Questions

1. What options do low resource setting communities have for use of telehealth services?
2. How would you address the digital divide in providing telehealth services to low resource setting communities? Discuss with reference to examples.
3. Of ethical issues in the use of telehealth and telemedicine in low resource settings, which of the issues would be critical to sustainable community health and how?

Field-Based Experiential Exercises

1. It is important to understand the nature and norms of the locations you will be working with remotely. Service expectations can be quite different in different regions, as can medical services purchasing power, reimbursement options, and access to other non-telehealth caregivers. The first step requires your leadership to actually travel to the region or location where the telehealth will be delivered. There is simply no substitute for taking the time to visit your remote sites, meet your colleagues, and learn firsthand about their lives, patients, local opportunities, challenges, and concerns. What are three more considerations that will be important when developing the telehealth initiative that is needed for the targeted population in that geographic area?
2. The implementation of electronic medical records and other health information technology (HIT) is taking place at a rapid rate. Telehealth systems should be designed and structured to support health information exchange. Substantial seed funding opportunities are increasingly available to support HIT deployment and integration, often focused on the establishment of high speed (T1 and above) network infrastructure. This same network can form the backbone of your telehealth program. What might be two more expectations your organization should have as you address the privacy and security risks with the telehealth communications that are congruent with HIT recordkeeping?
3. Telehealth activities should be designed to complement your standard practices and working methods, not complicate or interrupt them. Telehealth should be integrated alongside your face-to-face clinical activities. Telehealth examination rooms (both patient and provider sites) should be located in close proximity to the clinical staff. Foremost, you should keep it simple. What are two other bits of advice that experts would recommend to assure a more seamless system of telehealth operations?

Internet Resources

World Health Organization Telehealth—Health and sustainable development with a range of health activities and strategies for telehealth.
https://www.who.int/sustainable-development/health-sector/strategies/telehealth/en/

National Telehealth Resource Centers—A collaborative of 12 regional and 2 national Telehealth Resource Centers committed to implementing telehealth programs for rural and underserved communities.
telehealthresourcecenters.org

Center for Connected Health Policy—Non-profit, non-partisan organization working to maximize telehealth's ability to improve health outcomes, care delivery, and cost-effectiveness.
www.cchpca.org

Telehealth Technology Assessment Center— National center offering a variety of services in the area of technology assessment to provide answers to questions about selecting appropriate technologies for your telehealth program.
www.telehealthtechnology.org

American Telemedicine Association—An organization completely focused on accelerating the adoption of telehealth, which is working to change the way the world thinks about health care.
www.americantelemed.org

Center for Telehealth and e-Health Law—An organization that supports healthcare providers, law firms, associations, universities insurance companies, and venture capital firms that work to overcome legal and regulatory issues related to telehealth.
www.ctel.org

References

American Nurses Association. (1997). Telehealth: A tool for nursing practice. *Nursing Trends Issues, 2*(4), 1–3.

Little, A. D. (1992). *Telecommunications: can it help solve America's health care problems?*. Boston, MA: Arthur D. Little, Incorporated.

Augusterfer, E. A., Mollica, R. F., & Lavelle, J. (2015). A review of telemental health in international and post-disaster settings. *International Review of Psychiatry, 27*(6), 540–546. https://doi.org/10.3109/09540261.2015.1082985

Balch, D. (2008). Developing a national inventory of telehealth resources for rapid and effective emergency medical care: A white paper developed by the

American telemedicine association emergency preparedness and response special interest group. *Telemedicine Journal and e-Health: The Official Journal of the American Telemedicine Association, 14*(6), 606–610. https://doi.org/10.1089/tmj.2007.0127

Bashshur, R. L., Shannon, G. W., Bashshur, N., & Yellowlees, P. M. (2016). The empirical evidence for telemedicine interventions in mental disorders. *Telemedicine Journal and e-Health: The Official Journal of the American Telemedicine Association, 22*(2), 87–113. https://doi.org/10.1089/tmj.2015.0206

Bauerly, B. C., McCord, R. F., Hulkover, R., & Pepin, D. (2019). Broadband access as a public health issue: The role of law in expanding broadband access and connecting underserved communities for better health outcomes. *The Journal of Law, Medicine & Ethics, 47*(2), 39–42.

Baumann, P. K., & Scales, T. (2016). History of information communication technology and telehealth. *Academy of Business Research Journal, 3*, 48.

Benavides-Vaello, S., Strode, A., & Sheeran, B. C. (2013). Using technology in the delivery of mental health and substance abuse treatment in rural communities: A review. *The Journal of Behavioral Health Services & Research, 40*(1), 111–120.

Bohmer, R., & Edmondson, A. C. (2001). Organizational learning in health care. *Health Forum Journal, 44*(2), 32–35.

Bonnardot, L., Liu, J., Wootton, E., Amoros, I., Olson, D., Wong, S., & Wootton, R. (2014). The development of a multilingual tool for facilitating the primary-specialty care interface in low resource settings: The MSF tele-expertise system. *Frontiers in Public Health, 2*, 126.

Brennan, D. M., Holtz, B. E., Chumbler, N. R., Kobb, R., & Rabinowitz, T. (2008). Visioning technology for the future of telehealth. *Telemedicine Journal and e-Health, 14*(9), 982–985.

Brown-Connolly, N. E. (2002). Patient satisfaction with telemedical access to specialty services in rural California. *Journal of Telemedicine & Telecare*. 8 Suppl(2), 7–10. https://doi.org/10.1177/1357633X020080S204

Brown, R. A., Marshall, G. N., Breslau, J., Farris, C., Osilla, K. C., Pincus, H. A., … Miyashiro, L. (2015). Access to behavioral health care for geographically remote service members and dependents in the US. *Rand Health Quarterly, 5*(1), 249–254.

Bursell, S. E., Zang, S., Keech, A. C., & Jenkins, A. J. (2016). Evolving telehealth reimbursement in Australia. *Internal Medicine Journal, 46*(8), 977–981.

Campbell, J. I., Aturinda, I., Mwesigwa, E., Burns, B., Santorino, D., Haberer, J. E., … Siedner, M. J. (2017). The technology acceptance model for resource-limited settings (TAM-RLS): A novel framework for mobile health interventions targeted to low-literacy end-users in resource-limited settings. *AIDS and Behavior, 21*(11), 3129–3140. https://doi.org/10.1007/s10461-017-1765-y

Caplan, G., & Caplan, R. (2000). Principles of community psychiatry. *Community Mental Health Journal, 36*(1), 7–24. https://doi.org/10.1023/a:1001894709715

Chan, M. (2010). Global policy recommendations. *France: Graphic Design: Rasmussen/CH*, 14–18. ISBN 9789241564014

Coleman, J. (2002). HMOs and the future of telemedicine and telehealth part 1, MCO trends.Connect2Health Task Force. (2018). Federal Communications Commission. *Mapping Broadband Health in America 2017*. Available at https://www.fcc.gov/reports-research/maps/connect2health/index.html#ll=48.004625,-92.460937&z=4&t=insights&inb=in_bb_access&inh=in_diabetes_rate&dmf=none&inc=none&slb=90,100&slh=10,22. Last visited 4 Oct 2020.

Delaigue, S., Bonnardot, L., Steichen, O., Garcia, D. M., Venugopal, R., Saint-Sauveur, J., & Wootton, R. (2018). Seven years of telemedicine in médecins sans frontières demonstrate that offering direct specialist expertise in the frontline brings clinical and educational value. *Journal of Global Health, 8*(2), 020414. https://doi.org/10.7189/jogh.08.020414

Department of Health and Human Services [HHS]. (2016). *E-Health and telemedicine: Report to congress.* https://aspe.hhs.gov/system/files/pdf/206751/TelemedicineE-HealthReport.pdf. Accessed 10 Mar 2020

Dorstyn, D., Mathias, J., & Denson, L. (2013). Applications of telecounselling in spinal cord injury rehabilitation: A systematic review with effect sizes. *Clinical Rehabilitation, 27*(12), 1072–1083. https://doi.org/10.1177/0269215513488001

Ettinger, K. M., Pharaoh, H., Buckman, R. Y., Conradie, H., & Karlen, W. (2016). Building quality mHealth for low resource settings. *Journal of Medical Engineering & Technology, 40*(7–8), 431–443. https://doi.org/10.1080/03091902.2016.1213906

Fitzner, K. K., Heckinger, E., Tulas, K. M., Specker, J., & McKoy, J. (2014). Telehealth Technologies: Changing the Way We Deliver Efficacious and Cost-Effective Diabetes Self-Management Education. *Journal of Health Care for the Poor and Underserved, 25*(4), 1853–1897. https://doi.org/10.1353/hpu.2014.0157

Fong, B., Ansari, N., & Fong, A. C. M. (2012). Prognostics and health management for wireless telemedicine networks. *IEEE Wireless Communications, 2012*, 83–89. Accessed at: https://web.njit.edu/~ansari/papers/12WC.pdf

Gupta, A., & Sao, D. (2011). The constitutionality of current legal barriers to telemedicine in the United States: Analysis and future directions of its relationship to national and international health care reform. *Health Matrix, 21*(2), 385. (Cleveland, Ohio: 1991).

Hale, T. M., & Kvedar, J. C. (2014). Privacy and security concerns in telehealth. *Virtual Mentor, 16*(12), 981–985.

Hall, C. S., Fottrell, E., Wilkinson, S., & Byass, P. (2014). Assessing the impact of mHealth interventions in low- and middle-income countries – what has been shown to work? *Global Health Action, 7*, 25606. Medline: 25361730. https://doi.org/10.3402/gha.v7.25606

Hall, J. L., & McGraw, D. (2014). For telehealth to succeed, privacy and security risks must be identified and addressed. *Health Affiliation (Millwood), 33*(2), 216–221.

Hebert, M. (2001). Telehealth success: Evaluation framework development. *Studies in Health Technology and Informatics*, 1145–1151. https://doi.org/10.3233/978-1-60750-928-8-1145

Hicks, L. L., Fleming, D. A., & Desaulnier, A. (2009). The application of remote monitoring to improve health outcomes to a rural area. *Telemedicine Journal and E-Health, 15*(7), 664–671. https://doi.org/10.1089/tmj.2009.0009

Holland, J., Hatcher, W., & Meares, W. L. (2018). Understanding the Implementation of Telemental Health in Rural Mississippi: An Exploratory Study of Using Technology to Improve Health Outcomes in Impoverished Communities. *Journal of Health & Human Services Administration, 41*(1), 52–86.

Ickenstein, G. W., Groß, S., Tenckhoff, D., Hausn, P., Becker, U., Klisch, J., & Isenmann, S. (2010). An empirical analysis of the current need for teleneuromedical care in german hospitals without neurology departments. *International Journal of Telemedicine and Applications, 2010*, 916868–916810. https://doi.org/10.1155/2010/916868

Institute of Medicine. (2012). *The role of telehealth in an evolving health care environment: Workshop summary.* Washington, DC: The National Academies Press. https://doi.org/10.17226/13466

Jaccard, J. J., Nepal, S., Cellar, B., & Yan, B. (2016). WebRTC-based video conferencing service for telehealth. *Computational Informatics, 98*, 169–193. https://doi.org/10.1007/s00607-014-0429-2

Joseph, A., Kadri, A. M., Krishnan, A., Garg, B. S., Ahmed, F. U., Kumar, P., … Srivastava, V. K. (2018). IAPSM declaration 2018: Definition, role, scope of community medicine and functions of community medicine specialists. *Indian Journal of Community Medicine: Official Publication of Indian Association of Preventive & Social Medicine, 43*(2), 120–121. https://doi.org/10.4103/ijcm.IJCM_115_18

Khatri, K., Haddad, M., & Anderson, D. (2013). Project ECHO: Replicating a novel model to enhance access to hepatitis c care in a community health center. *Journal of Health Care for the Poor and Underserved, 24*(2), 850–858. https://doi.org/10.1353/hpu.2013.0093

Kim, J. A. (2010). Telehealth in the developing world. *Healthcare Informatics Research, 16*(2), 140–141.

Kruse, C., Bouffard, S., Dougherty, M., & Parro, J. (2016). Telemedicine Use in Rural Native American Communities in the Era of the ACA: a Systematic Literature Review. *Journal of Medical Systems, 40*(6), 1–9. https://doi.org/10.1007/s10916-016-0503-8

Kun, L. G. (2001). Telehealth and the global health network in the 21st century. From homecare to public health informatics. *Computer Methods and Programs in Biomedicine, 64*(3), 155–167. https://doi.org/10.1016/s0169-2607(00)00135-8

Lambert, D., Gale, J., Hartley, D., Croll, Z., & Hansen, A. (2016). Understanding the business case for telemental health in rural communities. *The Journal of Behavioral Health Services & Research, 43*, 366. https://doi.org/10.1007/s11414-015-9490-7

Lee, A. C., & Harada, N. (2012). Telehealth as a means of health care delivery for physical therapy practice. *Health Policy in Perspective, 92*(3), 463–469.

Lee, S., Black, D., & Held, M. L. (2019). Factors associated with telehealth service utilization among rural populations. *Journal of Health Care for the Poor and Underserved, 30*(4), 1259–1272. https://doi.org/10.1353/hpu.2019.0104

Louder, D. & Solomon, A. (2017). Review of the telehealth landscape and fundamentals of telehealth program development, telehealth overview. *Making connections with technology: The issues, benefits and barriers to using telehealth in behavioral.* Healthcare, Maine Behavioral Healthcare Workforce Development Collaborative.

Makaroun, L. K., Bowman, C., Duan, K., Handley, N., Wheeler, D. J., Pierluissi, E., ... Chen, A. H. (2017). Specialty Care Access in the Safety Net—the Role of Public Hospitals and Health Systems. *Journal of Health Care for the Poor and Underserved, 28*(1), 566–581. https://doi.org/10.1353/hpu.2017.0040

Mathieson, K., Leafman, J. S., & Horton, M. B. (2017). Access to Digital Communication Technology and Perceptions of Telemedicine for Patient Education among American Indian Patients with Diabetes. *Journal of Health Care for the Poor and Underserved, 28*(4), 1522–1536. https://doi.org/10.1353/hpu.2017.0131

McLennan, J. D. (2018). Video-conferencing telehealth linkage attempts to schools to facilitate mental health consultation. *Journal of the Canadian Academy of Child and Adolescent Psychiatry = Journal De l'Academie Canadienne De Psychiatrie De l'Enfant Et De l'Adolescent, 27*(2), 137–141.

Médecins Sans Frontières (MSF) International. Available: http://www.msf.org/en/about-msf. Accessed 4 Jan 2020.

Miller, T. W., Clark, J., Veltkamp, L. J., Burton, D. C., & Swope, M. (2008). Teleconferencing model for forensic consultation, court testimony, and continuing education. *Behavioral Sciences & the Law, 26*(3), 301–313. https://doi.org/10.1002/bsl.809. Accessed 4 Jan 2020.

Moehr, J. R., Schaafsma, J., Anglin, C., Pantazi, S. V., Grimm, N. A., & Anglin, S. (2006). Success factors for telehealth—A case study. *International Journal of Medical Informatics, 75*(10–11), 755–763. https://doi.org/10.1016/j.ijmedinf.2005.11.001

Mooi, J. K., Whop, L. J., Valery, P. C., & Sabesan, S. S. (2012). Teleoncology for indigenous patients: The responses of patients and health workers. *Australian Journal of Rural Health, 20*(5), 265–269. https://doi.org/10.1111/j.1440-1584.2012.01302.x

Mupela, E. N., Mustarde, P., & Jones, H. L. C. (2011). Telemedicine in primary health: The virtual doctor project Zambia. *Philosophy, Ethics, and Humanities in Medicine, 6*(1), 9–9. https://doi.org/10.1186/1747-5341-6-9

Muttitt, S., Vigneault, R., & Loewen, L. (2004). Integrating telehealth into aboriginal healthcare: The Canadian experience. *International Journal of Circumpolar Health, 63*(4), 401–414. https://doi.org/10.3402/ijch.v63i4.17757

Myers, C. R. (2019). Using telehealth to remediate rural mental health and healthcare disparities. *Issues in Mental Health Nursing, 40*(3), 233–239.

Nakamura, Y., Laberge, M., Davis, A., & Formoso, A. (2019). Barriers and Strategies for Specialty Care Access through Federally Qualified Health Centers: A Scoping Review. *Journal of Health Care for the Poor and Underserved, 30*(3), 910–933. https://doi.org/10.1353/hpu.2019.0064

Nickelson, D. W. (1998). Telehealth and the evolving health care system: Strategic opportunities for professional psychology. *Professional Psychology: Research and Practice, 29*(6), 527.

Olayiwola, J. N., Magaña, C., Harmon, A., Nair, S., Esposito, E., Harsh, C., … Wexler, R. (2020). Telehealth as a bright spot of the COVID-19 pandemic: Recommendations from the virtual frontlines ("Frontweb"). *JMIR Public Health and Surveillance, 6*(2). https://doi.org/10.2196/19045

Ricketts, T. C. (2000). The changing nature of rural health care. *Annual Review of Public Health, 21*(1), 639–657.

Rotheram-Borus, M. J., Tomlinson, M., Swendeman, D., Lee, A., & Jones, E. (2012). Standardized functions for smartphone applications: Examples from maternal and child health. *International Journal of Telemedicine and Applications*, 1–16. https://doi.org/10.1155/2012/973237

Rural Health Information Hub. (2020). *Telehealth use in rural healthcare*. Accessed at: https://www.ruralhealthinfo.org/topics/telehealth#resource-centers

Saeed, S. A., Diamond, J., & Bloch, R. M. (2011). Use of telepsychiatry to improve care for people with mental illness in rural North Carolina. *NC Med J, 72*(3), 219–222.

Schmeida, M., & McNeal, R. S. (2007). The telehealth divide: Disparities in searching public health information online. *Journal of Health Care for the Poor and Underserved, 18*(3), 637–647. https://doi.org/10.1353/hpu.2007.0068

Schneider, P. (2011). Telehealth core to WHO's missions. *Healthcare Informatics: The Business Magazine for Information and Communication Systems, 15*(7), 59.

Shandle, J. (2008). A new telehealth dawns: Remote medical monitoring. *Electronic Engineering Times*, October, 24–26.

Sikka, N., Paradise, S., & Shu, M. (June, 2014). *Telehealth in emergency medicine: A primer* (pp. 1–11). American College of Emergency Physicians: Telemedicine Primer.

Singh, R., Mathiassen, L., Stachura, M. E., & Astapova, E. V. (2010). Sustainable rural telehealth innovation: a public health case study. *Health Services Research, 45*(4), 985–1004. https://doi.org/10.1111/j.1475-6773.2010.01116

Slifkin, R. T., Silberman, P., & Reif, S. (2001). The effect of Medicaid managed care on rural public health departments. *The Journal of Rural Health, 17*(3), 187–196.

Stephens, S. (2014). Telehealth is calling. *PT in Motion, 6*(4), 30–38.

Sugijarto, D. P., Safie, N., Mukhtar, M., & Sulaiman, R. (2013). Telehealth model information flow: A case study on laboratory information system. *Procedia Technology, 11*, 740–747.

Toh, N., Pawlovich, J., & Grzybowski, S. (2016). Telehealth and patient-doctor relationships in rural and remote communities. *Canadian Family Physician, 62*, 961–963.

United Nations [UN]. (2009). *The millennium development goals report*. http://www.un.org/millenniumgoals/pdf/MDG.Report.2009.ENG.pdf. Accessed 4 Jan 2020.

White, A., Thomas, D. S. K., Ezeanochie, N., & Bull, S. (2016). Health worker mHealth utilization: A systematic review. *CIN: Computers, Informatics, Nursing, 34*(5), 206–213. https://doi.org/10.1097/CIN.0000000000000231

Wootton, R., & Bonnardot, L. (2015). Telemedicine in low-resource settings. *Frontiers in Public Health, 3*, 3. https://doi.org/10.3389/fpubh.2015.00003

Wootton, R., Patil, N. G., Scott, R. E., & Ho, K. (Eds.). (2009). *Telehealth in the developing world*. New Brunswick, NJ: IDRC.

World Health Organization. (2010). *World health report*. Geneva, Switzerland: World Health Organization.

World Health Organization. (2016). *Be healthy, be mobile: Annual report 2018*. Geneva, Switzerland: World Health Organization.

Zakus, D., Moussa, M., Ezechiel, M., Yimbesalu, J. P., Orkar, P., Damecour, C., & Nganga, G. (2019). Clinical evaluation of the use of an mhealth intervention on quality of care provided by community health workers in Southwest Niger. *Journal of Global Health, 9*(1), 010812. https://doi.org/10.7189/jogh.09.010812

12

Metrics and Evaluation Tools for Communicable and Non-communicable Diseases

Lu Liang

Introduction

In the context of community health, metrics are predictable, designable, and testable measures of health determinants and outcomes (Goldman & Coussens, 2004). Metrics are useful in defining health problems, prioritizing community needs, driving policy development, evaluating health inequalities, and monitoring progress in reaching short-term and long-term community health goals (Jakubowski & Frumkin, 2010). Using quantifiable metrics to inform community health status on communicable and non-communicable diseases (NCDs) has a long and productive history (de Martel, Georges, Bray, Ferlay, & Clifford, 2020). Moreover, health metrics and evaluation tools have the advantage to produce robust and archivable data that is available for prospective community health planning efforts, while also allowing for comparisons across different

L. Liang (✉)
University of North Texas, Denton, TX, USA
e-mail: lu.liang@unt.edu

E. Mpofu (ed.), *Sustainable Community Health*,
https://doi.org/10.1007/978-3-030-59687-3_12

communities (cross-sectional) and over time (longitudinal) for best community health practices (e.g., Didzun et al., 2019; Jiwani et al., 2019).

Learning Objectives

After completing this chapter, the reader should be able to:

1. Define health metrics and evaluation tools at the community level.
2. Outline the history of research on metrics for communicable and non-communicable diseases.
3. Describe current key approaches to the use of metrics and evaluation tools for preventing and controlling communicable and non-communicable diseases.
4. Discuss the matters for research and practice innovations in the use of metrics and tools for communicable and non-communicable diseases applied to promote community health.
5. Highlight the needs of interdisciplinary efforts in advancing the research on health metrics and evaluation tools for sustainable community health.

Non-communicable diseases are not transmissible directly from one person to another. Examples of NCD include cardiovascular diseases, cancer, diabetes, and chronic respiratory diseases, which are typically linked by common preventable risk factors related to lifestyle, mainly tobacco and alcohol use, unhealthy diet, and physical inactivity (World Health Organization, 2009). Non-communicable diseases mostly develop from family history, genetic disorders, and harmful environmental exposures (e.g., drugs, industrial chemicals, tobacco smoke). Communicable diseases are caused by microorganisms that can spread from one person to another (Edemekong & Huang, 2019) and have a high risk for community spread. They tend to occur when hosts, infectious agents, and an environment predispose to the agent's transmission. Examples include COVID-19, tuberculosis, avian flu, valley fever, malaria, and salmonella. However, chronic exposure to NCD increases the risk of communicable diseases to hosts because of their weaker immune systems (Ackland, Choi, & Puska, 2003). For example, older adults and people with certain medical conditions (e.g., cancer, chronic kidney disease, obesity, serious heart conditions) are at increased risk of severe illness from COVID-19 (Dietz & Santos-Burgoa, 2020; Tolksdorf, Buda, Schuler, Wieler, &

Haas, 2020). Social determinants of health grounded in the socio-economy and culture create pathways for the risk for NCD, so that while not technically transmittable in the sense of infection, would enable NCD to pass on in a community population based on these vulnerabilities (Ackland et al., 2003; Harries et al., 2015).

Communicable diseases are largely prevalent in developing countries with their underdeveloped health systems (World Health Organization, 2016b) and overburdened health systems from managing a high prevalence of communicable and non-communicable diseases (Boutayeb, 2006). As an example, an estimated 40·5 million (71%) of the 56·9 million worldwide deaths were from NCDs (Countdown, 2018). Developing countries also carry a higher burden of NCD, and the leading causes of death are respiratory tract infections, diarrheal diseases, tuberculosis, malaria, and AIDS (Gavazzi, Herrmann, & Krause, 2004). For instance, about 23·4 million (or 64% of the total) deaths in the 23 low- to middle-income countries were from NCDs, mostly associated with tobacco use and being overweight (Alwan et al., 2010), the management of which would impoverish 6–11% of the total population of developing nations in the absence of sustainable community health implementation (Jaspers et al., 2015). Within developed and developing countries, risks for communicable diseases and NCDs vary by social, economic, and environmental factors so that the relatively socio-economically advantaged communities have a lower risk for both communicable diseases and NCDs (Adler & Newman, 2002). The relatively socio-economically advantaged communities also live in environmentally healthier neighborhoods with less pollution, better sanitation, nutrition security, and walkability for wellness (Tessum et al., 2019).

Strategies to prevent NCDs are therefore mainly focused on controlling risk factors in an integrated manner, particularly at the family and community levels, because of the similarity in the environmental, economic, social, and behavioral determinants of the risk factors (World Health Organization, 2010). Communicable diseases typically mandate prevention and control strategies at the community, national, and international levels (Taylor, 1996). Prevention refers to measures that are applied to prevent disease occurrence, whereas control refers to measures that are applied to prevent transmission after the disease outbreak (Communicable Diseases Module 2, 2020). Public health laws have the authority to prevent diseases through vaccinations and individual screening. When outbreaks happen, laws may authorize the isolation of individuals and communities who have been exposed to diseases, and the closure of businesses to mitigate community spread (World Health Organization, 2016a). A most recent global example is the travel ban and stay-at-home policies caused by COVID-19. For individuals, implementing good hygiene procedures and disease prevention education are important for preventing community spread of communicable diseases. Increasingly, health systems seek to utilize integrative approaches to prevent and control for communicable diseases and NCDs insofar as share predispositions in vulnerable populations (Harries et al., 2015).

Metrics and evaluation tools are critically important for mitigating both communicable diseases and NCDs. The selection of appropriate metrics and tools discussed in this chapter considers several factors. Priority is given to metrics that would be applicable and relevant to a variety of communities. While communities are diverse in terms of their population health needs and priorities, they would find the use of core metrics and tools for evaluating community health status useful to sustainable health initiatives. Moreover, community health initiatives increasingly rely on quantitative modeling, which requires applying metrics and evaluation tools to optimize community health planning, implementation, and evaluation efforts.

The natural and human environments are altering phenomenally from the global warming climate and urbanization (National Research Council, 2010). Substantial evidence has proved that whether disease transmission occurs during contact between susceptible and infectious hosts depends heavily on environmental conditions (Fang et al., 2015; Liang et al., 2010; 2014). Increasingly, research on sustainable community health is focused on investigating the impact of environmental changes on disease transmission and the evaluation metrics and tools for disease conditions.

Professional and Legal Definitions for Metrics and Evaluation Tools

Metrics are predefined measures or indicators that are used to monitor, analyze, and optimize certain dimensions of the community health. For example, the Lancet Countdown Committee established 41 trackable and quantifiable indicators in five health-related domains in response to climate change: climate change impacts, exposures, and vulnerability; adaptation, planning, and resilience for health; mitigation actions and health co-benefits; economics and finance; and public and political engagement (Watts et al., 2018). At which time duration and geographical coverage can metrics on these five core domains can be calculated largely depends on the data availability and data resolution. For instance, in low-resource settings, high temperature and small water bodies are ideal breeding sites for mosquitos, which is a major vector for many communicable diseases (e.g., malaria). While malaria vector data from small ponds and high temperatures would typically be hard to collect at the very fine spatial resolution, these data could be available at the local scale through ground/aerial habitat surveys (e.g., Fillinger et al., 2009), citizen science efforts (e.g., NASA Citizen Science App), and satellite observations (e.g., the Global Observer Mosquito Health Mapper).

Evaluation toolkits are a collection of information, resources, tools, and advice for a specific subject area or activity. According to the Centers for Disease Control and Prevention (CDC, 2012) evaluation tools tend to be for specific surveillance programs, although they can be adapted for use by other programs (see also CDC Surveillance Resource Center, n.d.). Evaluation tools are diverse in their formats and contents, ranging from program codes and software, survey questionnaires, to interactive data visualization and query platform. For instance, a software that runs Ecological Niche modeling for malaria has been used to model the habitat suitability for West Nile, with some modifications on the input variables and parameter setting. The Community Assessment for Public Health Emergency Response (CASPER) Toolkit (CDC, 2012) is an example of an evaluation toolkit for household-level community health. Many cities, states, federal agencies, international laws and conventions, research entities, as well as non-profit organizations, have developed evaluation tools for health risk assessment (Hosseinpoor, Schlotheuber, Nambiar, & Ross, 2018; Reis et al., 2019).

History of Research and Practice of Metrics and Evaluation Tools for Communicable and Non-communicable Diseases

Metrics and tools are not new concepts in community health. However, growing attention is being paid to metrics and evaluation tools in community health for mitigating risk for communicable diseases and NCDs. The evolution of the research and applications of metrics and evaluation tools for communicable diseases and NCDs has been quite evident over the past decades, especially since the emergence of big data and computer technology (Dolley, 2018; Hay, George, Moyes, & Brownstein, 2013). Big data is the term applied to data sets whose size or type is beyond the ability of traditional relational databases to capture, manage, and process the data with low latency. They are characterized by volume (consisting of enormous quantities of data), velocity (created in real time), variety (being structured, semi-structured, and unstructured), exhaustivity (an entire system is captured), fine grained (in resolution) and uniquely indexical (in identification), relationality (containing common fields that enable the conjoining of different data sets), extensionality (can add/change new fields easily), and scalability (can expand in size rapidly; Kitchin & McArdle, 2016).

Here, the history and trends in the research on metrics and evaluation tools for communicable diseases and NCDs can be summarized into two parts: (1) the shift from single disciplinary effort to synergized, interdisciplinary collaboration, and (2) qualitative measures to big data-driven approaches.

From the Single Disciplinary Effort to Synergized, Interdisciplinary Collaboration No one can argue the importance of interdisciplinary research in the context of the present community health context. Single disciplines emphasize the efforts to promote a coherent and ordered focus of investigation and study (Eisenberg & Pellmar, 2000). In the past, with the less focus on interdisciplinary collaborative research, there were limitations to "not knowing how," even though with consensus on "how important" a health issue was. In contrast, the interdisciplinary approach focuses primarily on the different disciplines and the diverse perspectives they bring to illustrate a theme (International Bureau of Education).

One of the earliest interdisciplinary community health science breakthroughs is exemplified by the work of Dr. John Snow's cholera map (Hempel, 2007). In the nineteenth century, Cholera was one of the most threatening diseases in Britain, which was believed to be transmitted by bad air. It was not until 1854 that a physician John Snow visually identified an overlapped pattern between cholera outbreak points and the drinking well. He superimposed all the cholera deaths on a map of public water pumps and noted that those addresses were mostly close to a water well on Broad Street. Dr. Snow's integrated usage of epidemiology and geography made a major contribution to recognizing Cholera was a water-borne communicable disease. This is one of the earliest documented examples highlighting the application of Geographic Information System (GIS) in public health. Nowadays, the integration of these two fields fosters a new field, Medical Geography, which focuses on using geographic techniques to study the impact of a person's surroundings on their health (e.g., Hui et al., 2009). Similar cases can also be found in geophylogeny (e.g., Liang et al., 2010, 2014) and environmental health (e.g., Fang et al., 2015).

From Qualitative Measures to Big Data-Driven Approaches Increasingly large volumes of information are available for communicable and non-communicable disease research and decision-making, with the boosting data collection, storage, and analytical capacity (Mooney & Pejaver, 2018). Big data has been successfully applied in surveillance and signal detection, predicting future risk, targeted interventions, and understanding disease (Dolley, 2018). The big data theory and methods greatly extend health research in the dimension of place, person, and time. Geographic information has been greatly enabled by the Global Positioning System and smartphone technology. As an example, with the COVID-19 outbreak, many high-tech companies released a batch of mobility data that is aggregated from requests for location service in their Apps, such as Google Map, Apple Map. Many studies have used those mobility maps as a measure of social distancing (Soucy et al., 2020). The temporal frequency of data availability also improved significantly.

Environmental and social data is typically easier to collect than disease data and thus has been used to estimate the disease activity. The booming of social media, which is used by one-third of global citizens, is rapidly changing the way how researchers are monitoring diseases. Social media data has been well documented in addressing critical issues in communicable diseases, such as outbreak trend (Ginsberg et al., 2009), and in non-communicable diseases, such as food desert, dietary choices, and physical activities (De Choudhury, Emre, Dredze, Coppersmith, & Kumar, 2016; Widener & Li, 2014).

Current and Emerging Practices in Metrics and Tools for Communicable and Non-communicable Diseases

Metrics for communicable diseases are selected based on relevance to the impacts of environmental changes on disease transmission, including climate, biological environment, population density, and human contact network. Metrics and tools for evaluating the key determinants of NCD include psychosocial and genetic factors, environmental factors, and health risk behaviors. This chapter discusses current metrics and evaluation tools for communicable diseases and NCDs separately for clarity and comparison.

Metrics for Communicable Diseases

Our community health systems are continually challenged by the emergence or re-emergence of communicable disease outbreaks, whether a flu pandemic or hand-foot-and-mouth disease (Association of State and Territorial Health Officials, 2020). What makes communicable diseases threatening is the risk for a community spread through epidemiologically significant contacts. As an example, the COVID-19 virus is mainly spread from close contact (i.e., within about six feet) with a person who is currently asymptomatic or sick with COVID-19 (CDC Frequently Asked Questions, n.d.). COVID-19 transmission occurs mainly via respiratory droplets that are astoundingly virulent, especially in enclosed spaces and on commonly shared surfaces. For COVID-19, although its epidemiology is not fully understood as of now, the elderly, with chronic lung diseases, are considered to be vulnerable hosts (Bi et al., 2020). Regarding the optimal environment, some early studies found that COVID-19 did not spread as efficiently in warmer and more humid regions as it did in colder areas (Bukhari & Jameel, 2020). Regardless, pandemic spikes in COVID-19 transmission have been observed in the warmer and humid Southern states of the United States in the summer months of the year 2020, suggesting evolving rather than definitive knowledge about the virus.

There are four key determinants in the transmission of communicable diseases: the presence of infectious agents and susceptible hosts, contacts between them, and optimal environmental conditions to result in pathology. This section will highlight metrics and tools in measuring factors that can impact virus transmission and spread. The major factors to be discussed include climate, biological environment, population density, and human contact network (Fig. 12.1).

Climatic Metrics

Climatic conditions present a direct environmental risk to human communicable diseases (Watts et al., 2015). Globally, 23% of all deaths in 2012 were attributable to the environment, and an additional 250,000

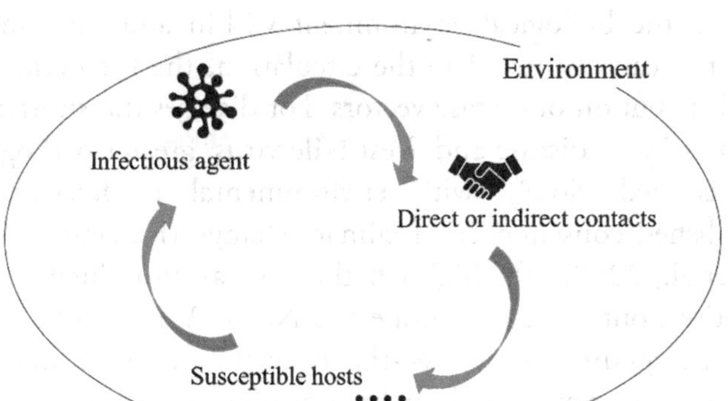

Fig. 12.1 Host-agent-environment interplay

potential deaths annually from 2030 to 2050 would be associated with global warming (Hales, Kovats, Lloyd, & Campbell-Lendrum, 2014). Temperature conditions affect the survival rates of a large number of known pathogens, although there is little knowledge of emerging pathogens. Also, excessive bursts of rainfall could cause sanitary sewer overflow and promote the emergence and spread of communicable diseases (Environmental Protection Agency, 1996), indicating the importance of the environment to preventing and controlling communicable diseases.

Contemporary and historical climate measurements, including temperature, precipitation, and relative humidity, are typically collected by weather stations. Their site locations are sparsely distributed on every continent and the density is typically determined by population density and government subsidies. Satellite-derived climatic data is estimated from radiances measured in various wavelength bands. The wavelength band approach provides wall-to-wall estimation over the Earth's surface. For instance, Moderate Resolution Imaging Spectroradiometer (MODIS) is one of the polar-orbit satellites that can provide the land surface temperature at a 1 km resolution. However, since remote sensing is not based on direct measurement, the data accuracy depends on the sensor quality and retrieval algorithms.

Biological Environment

Changes in the biological environment within and surrounding the communities can also mediate the circulation, the life cycles, and the shifting distribution of disease vectors. For diseases transmitted by vectors, such as Lyme disease and West Nile virus, greater pathogen transmission is tied closely with environmental biodiversity loss—a well-established consequence of climate change (Keesing et al., 2010; Pongsiri et al., 2009). The highly pathogenic avian influenza virus that caused serious outbreaks in Europe and North American poultry farms came from migrant waterfowls—the avian flu virus's natural reservoir, whose migration timing is heavily controlled by temperature (Liang et al., 2010; Lycett et al., 2016, see Research Box 12.1 for an example of avian flu). Social practices can potentiate or harm environmental safety

Research Box 12.1: Combining the Spatial–Temporal and Phylogenetic Analysis Approach for Improved Understanding of Global H5N1 Transmission (Liang et al., 2010)

Background

Since late 2003, the highly pathogenic influenza A H5N1 had initiated several outbreak waves that swept across the Eurasia and Africa continents. Getting prepared for reassortment or mutation of H5N1 viruses has become a global priority. Although the spreading mechanism of H5N1 has been studied from different perspectives, its main transmission agents and spread route problems remain unsolved.

Methods and Results

Based on a compilation of the time and location of global H5N1 outbreaks from November 2003 to December 2006, this study reports an interdisciplinary effort that combines the geospatial informatics approach with a bioinformatics approach to form an improved understanding on the transmission mechanisms of the H5N1 virus. Aspherical coordinate-based analysis revealed spatial and temporal clusters of global H5N1 cases on different scales, associated with two different transmission modes of H5N1 viruses: poultry transportation mode and wild bird migration model. In this study, major residential areas—the important poultry production, trading, and transportation centers—were used as a substitute for characterizing poultry transportation patterns. The wild bird migration pattern was quantified by the global wetland distribution, as wetlands are the most important breeding, wintering, and stopping sites for waterfowl. Utilizing both geographic and phylogenetic analysis, an H5N1 spreading route map was obtained.

Of the two transmission modes, the poultry transportation pattern was hard to quantify at a very fine spatial scale, as the global poultry trading data is not available at the local community level, or lower.

What Do You Think?

1. How can knowledge of wild bird migratory patterns assist in mitigating avian flu community spread at the host, vectors, transmission, or environment levels?
2. What procedures could incentivize the collection of poultry transportation data for enhancing community avian flu mitigation?

mechanisms from communicable diseases. For example, many rural Asian communities keep the practice of small-scale backyard poultry farming, and wet food markets that stockpile various species in cages next to or on top of each other are associated with the increased risks of bird flu virus-type infections (Paul et al., 2013; Woo, Lau, & Yuen,

2006). Schistosomiasis is a disease that is transmitted through contact with freshwater bodies that contain the parasite, such as in the act of collecting water, along with bathing, washing, or swimming. As a result, water-based interventions, such as constraining access to infectious water, have proven to be effective in preventing transmission of schistosome infection (Evan Secor, 2014).

Measures of Population Density

Population density provides spatial distribution information, an important metric in epidemiological models (e.g., SIR model). The most typical way to acquire population density data is from population surveys and census. For example, in the United States, such information is collected and maintained by the United States Census Bureau, which conducts decennial censuses (every ten years). The American Community Survey provides population density data on an annual basis. For developing countries, the population information usually lacks fine spatial and temporal resolution, rendering it less usable for mapping community health assets and liabilities. There are global population density data at the country level available from intergovernmental organizations, such as the World Bank (https://data.worldbank.org/indicator/EN.POP.DNST) and United Nations (https://unstats.un.org/unsd/demographic/sconcerns/densurb/default.htm).

A new approach to reflect human agglomeration is using nighttime light data from the Defense Meteorological Satellite Program's Operational Linescan System (Elvidge et al., 2001). One of its advantages exists in its ability to provide a stable and persistent data source on population estimation, urbanization, and socio-economic parameters (Anderson, Tuttle, Powell, & Sutton, 2010; Bagan & Yamagata, 2015; Tan et al., 2018).

Metrics to Quantify Human Contact Network

The pattern of human interactions is a social sustainability influence with important implications for the spread and management of communicable diseases. As previously noted, the pandemic spread of some communicable diseases, especially airborne, fecal-oral diseases, are usually transmitted via droplets during close proximity interactions (Fiore et al., 2008). A social sustainability approach for virus transmission involves micro-liquid droplet cleaning, social distancing, and wearing protective masks. Similarly, the identification of core groups of individuals with large numbers of social interaction intermixing has been the basis for controlling transmittable diseases, including sexual infection strategies (Macke & Maher, 1999), and airborne virus infections such as AIDS, COVID-19, and flu. Thus, the human contact network becomes a widely used metric in structured-population models for an enhanced understanding of the spatial transmission of epidemics in communities (Wang, Wang, Zhang, & Li, 2013).

Traditional methods of evaluating human contact networks are typically based on detailed diary-based surveys, but are often limited by small sample sizes (Read, Eames, & Edmunds, 2008) and are subject to human error (Eagle, Pentland, & Lazer, 2009). Wireless sensor network technology is employed as a promising solution to obtain high-resolution contact network data relevant to communicable disease transmission. For instance, in a study conducted at an American high school, this approach captured the vast majority (94%) of the community of interest and is proven as free of human error (Salathé et al., 2010).

Social media (e.g., Facebook, Twitter) and participatory epidemiology have emerged as new tools in gathering information quickly and improving coverage and accessibility (Freifeld et al., 2010; Schmidt, 2012). A social media application is an internet-based application where people can communicate and share resources and information. Increasingly, social network data are utilized to estimate human activities and population demographics that are relevant to epidemic spread modeling. Generally, these social network-based models capture human-to-human interactions among the community population, and social media data can quickly respond to the changes in human activity dynamics.

Google Flu Trends is perhaps the most well-known example of collective intelligence in communicable disease epidemiological studies. This internet-based surveillance tool operated by Google attempts to predict influenza activity by aggregating and analyzing Google Search queries with data-intensive algorithms (Cook, Conrad, Fowlkes, & Mohebbi,

2011; Ginsberg et al., 2009). In the 2009 flu pandemic, Google Flu Trends captured the influenza case spike two weeks prior to the CDC report being released. CDC's national surveillance program is based on weekly reports, which takes up to two weeks for these numbers to be compiled into publicly available information. Whereas web search queries are real time and thus, ideally, can bridge the CDC's two-week lag. Google Trends has also been applied to predict influenza and Middle East Respiratory Syndrome in Korea (Seo & Shin, 2017).

Although predictions based on search or social media data on diseases have become commonplace, it is still doubtful whether they can supplant more traditional methods, such as surveys. The two biggest concerns are big data hubris and algorithm dynamics (Lazer, Kennedy, King, & Vespignani, 2014). Besides, using mobile phones to detect spatial proximity of subjects is limited in population representation, with the young generation as the dominant population. For these reasons, social media tools should be used as a reasonable proxy for trending estimates of disease prevalence in populations.

Metrics and Tools for Evaluating Non-communicable Diseases

As previously noted, non-communicable diseases are mostly chronic diseases such as cardiovascular diseases, cancers, and diabetes. Globally, NCDs account for about six out of ten deaths (World Health Organization, 2008). Different from epidemics of communicable diseases that follow predictable patterns, NCDs are not spread by disease agents or vectors (hence the term non-communicable). Nonetheless, gradients in socioeconomic factors including health risk behaviors (e.g., indoor cooking, physical inactivity, smoking, dietary, and substance use) are directly implicated as causal factors for many NCDs (Fig. 12.2). For instance, environmental risk factors such as air pollution and waste disposal can contribute to a range of NCDs including asthma and other chronic respiratory diseases. Psychosocial and genetic factors also play a role (Fig. 12.2), highlighting the significance of social factors in the community prevalence of NCDs.

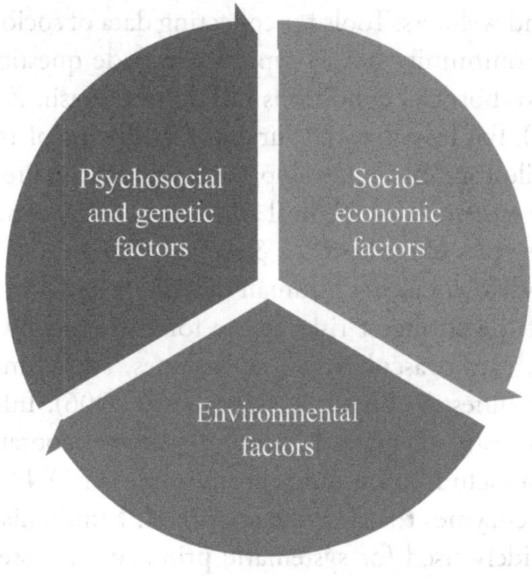

Fig. 12.2 Key determinants of non-communicable diseases in community health

Within communities, individuals tend to choose a similar lifestyle to those who share the same environment, and mostly from sharing a similar socio-economic gradient. Families also have historical links to the places they presently inhabit, which is from the influence of kinship relationships and also economics, such as when families invest in long-term, shared housing in a particular geographical local or neighborhood. The natural socio-economic and genetic linkages would accentuate the social–environmental determinants of NCD. Community health interventions targeting NCD should measure and track the socio-economic and environmental gradients behind community health status profiles for intervention design and implementation. This section discusses the metrics and tools to quantify some key socio-psychological and genetic factors that would influence community NCD prevalence and mitigation.

Socio-psychosocial and Genetic Factors

Socio-psychological factors include individual-level processes and meanings that influence mental states impacting health outcomes (Upton, 2013). At the community level, these operate as cultural ways of knowing about health and wellness. Tools for collecting data of socio-psychological influences on community health typically include questions to identify medical and psychosocial conditions in families (Nasir, Zimmer, Taylor, & Santo, 2019). For instance, the Survey of Wellbeing of Young Children is a 54-item milestone-based developmental questionnaire that conducts comprehensive screening on medical and psychosocial conditions in children under five years old (Sheldrick & Perrin, 2013).

Among the various causes of human disease, genetics or family history is often one of the strongest risk factors for common NCD complexes such as cancer, cardiovascular disease, diabetes, autoimmune disorders, and psychiatric illnesses (Blazer & Hernandez, 2006). Inherited genetic disorders are mostly diagnosed through genetic laboratory screening tests, which can include examining chromosomes or DNA, or blood testing for certain enzymes that may be abnormal. Family history questionnaire is also widely used for systematic primary care assessment of the family history of many chronic diseases, such as diabetes, ischemic heart disease, breast cancer, and colorectal cancer (Walter et al., 2013).

Environmental Factors

There is growing evidence that environmental factors have a significant impact on long-term health conditions (Black, O'Loughlin, Kendig, & Wilson, 2012). For instance, environmental contingencies that cause stress including noise, poor air quality, high-density housing, poor or inadequate road and urban design, and lack of green space.

Sound Noise and Water Pollution Decibels are a way to measure the volume of a sound. Instantaneous noise levels can be easily measured with many smartphone applications or a computer. The most direct, accurate way to find the decibel level of a sound is to use a decibel meter, but the cost can be expensive. Without any tools, noise levels can be estimated by the decibel-level table of common noise sources.

Air Pollution Air pollution represents a prominent threat to global society by causing cascading effects on community health in both developing and developed countries. About 90% of global citizens live in areas that exceed the safe exposure level in the World Health Organization (WHO) air quality guidelines (Health Effects Institute, 2019). The concentration of particle pollutants at a specific location and time is determined by many factors, including emissions, meteorology, microenvironment, and physicochemical transformations. Those factors are constantly changing over the location (e.g., building block apart) and time (e.g., rush hour versus midnight), and thus result in highly variable spatiotemporal patterns of $PM_{2.5}$ within cities (Dionisio et al., 2010; Van Vliet & Kinney, 2007). The fine particulate matter (PM) with a diameter of less than 10 μm has raised particular concerns in public and officials. In addition, human mobility tends to be higher in this globalizing society (Montanari, 2005), which makes it even more difficult to account for a person's lifetime exposure to environmental stressors.

The most typical and conventional way of measuring the critical air pollutants continuously is through monitoring stations. For instance, the Environmental Protection Agency (EPA) examines air pollution trends of the six principal pollutants in the United States. However, the number of these stations is usually too low to capture neighborhood-scale spatial variability in the pollutants' distribution (Dimakopoulou, Gryparis, & Katsouyanni, 2017). The Dallas-Fort Worth Metroplex, home to ~7 million people, has only six stations monitoring $PM_{2.5}$, and most are located in old city areas. To fill the data gaps due to insufficient ground monitor density, several approaches have been proposed and implemented, including the use of Geographic Information System (GIS).

Geographic Information System (GIS) plays a prominent role in enhancing efforts in spatial modeling of air pollution distribution. Some studies have used spatial interpolation techniques, such as kriging (Singh, Carnevale, Finzi, Pisoni, & Volta, 2011). The most recent advancement in this field is Land Use Regression (LUR) modeling. LURs are advantageous because they link environmental characteristics—especially those that influence pollutant emission intensity and dispersion efficiency—to concentrations at the measurement site (Adams & Kanaroglou, 2016). However, some overlooked facets in the development of LURs are the quantity and quality of input air pollution data, three-dimensional (3D) representation of predictor variables, and the wind effects (Hart et al., 2020).

Some recent studies have used near real-time satellite observations of columnar aerosol optical depth (AOD)—a measure of light extinction by aerosols in the entire atmospheric column, to provide spatially and temporally resolved predictions of ground-level $PM_{2.5}$ concentrations at regional and global scales (e.g., Xiao et al., 2017; Xu et al., 2016). Satellite-based $PM_{2.5}$ prediction typically rests on the hypothesis that satellite AOD is related to ground-level $PM_{2.5}$ in a linear form, assuming a single aerosol layer in a well-mixed boundary layer of height H, and with uniform aerosol optical properties. However, studies have shown that the relationship between $PM_{2.5}$ and AOD is not always linear, but rather determined by a multivariate function of a large number of parameters, including humidity, temperature, boundary layer height, surface pressure, population density, topography, wind speed, land cover type, surface reflectivity, season, land use, a normalized variance of rainfall events, size spectrum and phase of cloud particles, cloud cover, cloud optical depth, cloud top pressure, and the proximity to particulate sources releasing $PM_{2.5}$ (Lary et al., 2014; Lary, Lary, & Sattler, 2015). Moreover, most models are often built based on data collected from fixed monitoring stations that are limited in quantity and spatial coverage, which may further restrict the models' prediction capability.

With recent advances in geospatial and sensor technologies, the expanded use of low-cost air quality sensors is recommended by EPA for air quality monitoring (Williams et al., 2018). Unlike traditional stationary measurements that require high operational cost and expertise, low-cost sensors are easy to use, portable, and affordable. They present an enormous opportunity for personal sampling studies by tracking air pollution exposure and time-activity patterns at the individual level in real time. The low-cost sensors has the ability to capture the degree of variability over space and time, and enable citizens to engage directly in health monitoring (In Research Box 12.2, a case study is presented to illustrate the usage of the participatory monitoring). There is an unequal distribution of PM exposures among different populations, especially in bad air conditions (Liang et al., 2019). As low-cost sensor technologies are increasingly adopted for air pollution monitoring (Williams et al., 2018), they add to the resources for studies on community health, environmental justice, protocol development, and geospatial air quality mapping and modeling.

Research Box 12.2: City Health Outlook: Participatory Air Quality Monitoring Campaigns (Liang et al., 2019)

The fine particulate matter (PM) with a diameter of less than 10 μm has raised public health concerns. A spatiotemporal personal exposure assessment allows for a realistic appraisal of the risks the populations are facing. A key outcome of a spatiotemporal personal exposure assessment with high compliance in sensor validation, personal sampler wearing, data retrieval, and validation is desirable for air pollution reduction interventions.

These data can be collected utilizing citizen science approaches by which volunteers who could be ordinary community members can wear portable environmental sensors to record their real-time personal air pollution exposure and routes by their travel behavior, living conditions, geolocations, commute route and time, and self-reported chronic diseases they might have. Citizen science that involves public participation in scientific research is an emerging community health research practice. The citizen scientists use a portable environmental monitoring device for data collection, which has an aerosol nephelometer, a GPS receiver, a humidity sensor, and a temperature sensor. The device is housed in a 90 mm × 90 mm × 22 mm box with a weight of 150 g, which makes it highly portable. All logged data are wirelessly transmitted to a platform every 30 minutes using the integrated 4G model.

What Do You Think?

1. What air pollution monitoring devices are you familiar with? How are they used and by whom?
2. How is citizen science a sustainable environmental health practice?
3. Identity and describe a citizen science environmental health program in your community or another community of choice. Evaluate its sustainability qualities.

Urban Planning and Public Health Urban planning and public health share common missions and perspectives, as both aim to improve human wellbeing, emphasize needs assessment and service delivery, manage complex social systems, and focus at the population level (Kochtitzky et al., 2006; see also Chap. 3, this volume). The synergies between urban planning and public health are diverse, from the creation of green space to promote physical activity, transportation planning to mitigate air pollution, to clustering of fast-food establishments around schools (Kochtitzky et al., 2006). A key component in understanding the influence of urban planning on community health is to have a detailed assessment of the economics of the major land use and the built environment, in both landscape and vertical dimensions.

Many cities use zoning maps to show the typical land use or zoning categories in the districts. While this offers an accurate and fine-scale characterization of landscape patterns, it costs a large amount of time and labor efforts to maintain and update. Remote sensing has been recognized widely as a cost-effective technology for monitoring and mapping the urban environment (Li, Gong, & Liang, 2015; Reynolds, Liang, Li, & Dennis, 2017). There is overwhelming evidence to suggest that vegetation and built-up are significant landscape elements in determining air quality and population health (Wu, Xie, Li, & Li, 2015). Urban green space can provide cleaner air, especially in its vicinity areas (Łowicki, 2019; Zupancic, Westmacott, & Bulthuis, 2015), and thus enhance community health. The density of urban built-up is found to be positively correlated with air pollution levels and population health (Weng & Yang, 2006). Consequently, the footprints of the buildings and green space are increasingly used to quantify the urban morphology effects on community health at the horizontal plane. Scale issue needs to be accounted carefully when investigating the urban planning effects on community health (see Discussion Box 12.1). With the rising urban landscape, not only the coverage of urban surface, but also height plays a critical role in individual and community health, such as physical activity, and pollution dispersion (see Discussion Box 12.2).

Nutrition and Diet Nutrition and diet factors contribute significantly to the burden of preventable illnesses and premature deaths (see also Chap. 5, this volume). Worldwide, 11 million deaths and 255 million disability-adjusted life-years were attributable to dietary risk factors (Afshin et al., 2019). In the United States, four of the ten leading causes of death have been linked to diet (Harnack, Block, & Lane, 1997). Some studies used focus groups to identify factors that influence dietary behavior, such as barriers to fruit and vegetable consumption (Harnack et al., 1997). Mobile applications brought dietary health research new opportunities. In 2020, approximately 45.12% of the world's population used a

Discussion Box 12.1: Metrics Scales

Scale is a widely used concept across historical environmental science disciplines, such as ecology and hydrology (Blöschl & Sivapalan, 1995; Levin, 1992), and translatable to community health to study spatiotemporal data on diseases and their environmental correlates. With the use of spatial–temporal data, it is possible to represent a heterogeneous landscape with a variety of land cover types captured or derived from remotely sensed data at different resolutions ranging from a few centimeters to a few kilometers. A very fine spatial scale can provide more accurate information in community health research, although spatial–temporal data mismatches may occur. As an example, in Malaria research, a very important biological control of its transmission in an urban setting is the small water ponds around residential areas. In contrast, with course resolution data, for example, 1 km or lower, it will be very difficult to fully capture those important mosquito breeding environments. But the challenge with fine spatial resolution data is that they are less available in low-resource settings, which limits their use as a resource for community health research.

Self-Check Questions

1. What are practicable ways to collect health risk data on community settings with low or no access to remote sensing data collection tools? Justify your choices with examples.
2. How could the spatial–temporal scale mismatch influence the metrics and evaluation tools in community health research?

Discussion Box 12.2: Three-Dimensional Metrics

Most metrics are 2D, that is, their spatial coverage is in the *x* and *y* directions, and rarely do they account for the *z*-direction. However, there is a pressing need for community health to extend from 2D to 3D. One example can be found in population estimation. For both communicable diseases and NCDs, human population data is an essential variable in the formula for calculating their global disease burden or estimating the susceptible population. When gridded population data is needed, remotely sensed, human settlement data sets rank the most important geographic factors to estimate population densities and distributions at regional and global scales (Palacios-Lopez et al., 2019). However, the dwelling floor area, which accounts for the 3D living environment, should be a more accurate measure than settlement footprint in population size calculation.

Self-Check Questions

1. What other metrics in community health studies should account for the 3D effect?
2. What is the trade-off between 3D effects and scale?

smartphone. The prevalence of smartphones makes an analysis of the food intake more popular. Dietary monitoring applications are designed to help users manage portion control and stay with calorie and carbohydrate limits through built-in cameras and weight sensors.

Insufficient physical activity is also one of the leading risk factors for diseases such as cardiovascular diseases, cancer, and diabetes (see also Chap. 16, this volume). Three types of measures are practical for measuring physical activity in community health research: self-reporting, monitors, and direct observation. Self-reporting is the most widely used method that is cost-effective but the least accurate. Monitors such as pedometers, accelerometers, and heart rate monitors can provide accurate information, but the devices can be expensive and expertise is needed to collect and manage the data. Third, direct observation can be used to evaluate school physical education programs and assess how people are using parks and other physical activity facilities (Sallis, 2010).

Cultural, Professional, and Legislation Influences on the Use of Metrics and Tools for Communicable and Non-communicable Diseases

The influence of culture on the prevention and control of both communicable and non-communicable diseases is also evident. In the discussion box below, we provide COVID-19 examples to reveal the cultural differences amid the combat of this virus (see Discussion Box 12.3).

International agreements, national and subnational legislation, regulations, and other executive instruments, and decisions of courts and tribunals determine the success of prevention and control of communicable and non-communicable diseases (Magnusson 2009). The World Health Organization (WHO)—a specialized agency of the United Nations responsible for international public health—offers technical support to countries on appropriate legal strategies. For example, the WHO initiated a global action plan for the prevention and control of NCDs with

Discussion Box 12.3: Cultural Differences Affect the Usage of Different Strategies in Combating COVID-19

The COVID-19 has raised many interesting cultural differences in how they combat the epidemic among countries, especially between Eastern and Western countries. One evident phenomenon is how people and the country authority react to the idea of wearing face masks. Authorities in many Western countries were initially reluctant to recommend the use of face masks to inhibit the spread of the coronavirus, partly because of a cultural view that face masks are for sick people. However, it is common for people in many East Asian countries to wear a face mask for a variety of health reasons, including their much denser air pollution levels and other everyday health reasons. In normal days, like cold days or polluted days, Asian people would put on a mask to keep warm or reduce the inhalation of dirty air. They also operate under obligatory political systems that require compliance with public health edicts, as was the case with the COVID-19. Thus, it is relatively easy for Asian countries to adopt and adhere to wearing masks as a preventive strategy.

What Do You Think? Self-Check Questions

1. What are some cultural beliefs that you perceive would influence preferences for the pandemic community spread in a community you are familiar with?
2. How might the cultural beliefs you identified to influence community spread mitigation be altered for wider adoption of evidence-based community safety practices?

work plans and evaluations. Many of the recommended policies are legal interventions, requiring legislation or executive actions for effective implementation. However, the WHO is not mandatorily applying international law due to its organizational culture, which is dominated by scientists, doctors, and medical experts (Fidler, 2004). Each country has its legal authorities to govern disease control. In the United States, the secretary of the Department of Health and Human Services has statutory responsibility for communicable disease prevention and control, and the Division of Global Migration and Quarantine works to fulfill this responsibility through a variety of activities (CDC Quarantine and Isolation page, n.d.).

Related Disciplines on Metrics and Evaluation Tools for Communicable and Non-communicable Diseases

The science on metrics and evaluation tools for communicable and non-communicable diseases is interdisciplinary in nature. Disciplines such as epidemiology, geospatial science, data science, social and political sciences, and economics contribute to the observation, analysis, understanding, and interpretation of different facets of communicable and non-communicable diseases. Epidemiology is the study that concentrates on the occurrence of disease among individuals in relation to possible risk factors (Lilienfeld & Stolley, 1994). A classical epidemiological model is the Susceptible, Infected, and Recovered (SIR) model that computes the theoretical number of people infected with a contagious illness in a closed population over time (Bjørnstad, Finkenstädt, & Grenfell, 2002). Geospatial science is a discipline that focuses on using information technology to understand people, places, and processes of the earth. It has been widely used in the visualization, exploration, and analysis of community health status indicators and metrics (e.g., Heitgerd et al., 2008; Sopan et al., 2012). Data science is a field that uses scientific methods, processes, algorithms, and systems to extract knowledge and insights from structural and unstructured data (Van Der Aalst, 2016). Data has always been an indispensable part of the quantification of metrics and evaluation tools.

Social and political sciences have been exploring different facets of public health policy (Gagnon et al., 2017). They share the common interest with community health in promoting the public good. Social and political sciences influence the community health by identifying the questions and key challenges faced by public health, and thus affect the selection of metrics and evaluation tools. Economics has been widely and most commonly applied to assess the health care cost and productivity losses of disease control, prevention, and intervention for both communicable diseases and NCDs (American Diabetes Association, 2018; Jacob et al., 2019).

The need for interdisciplinary research can be supported by two main arguments. First, metrics and evaluation tools are quantitative and data driven. The current big data era provides an unprecedented opportunity for community health-related research. Very often, we do not lack data, but lack data in need and the technique that can distill the essential information from the data. This highlights the importance of any state-of-the-art data acquisition (e.g., geospatial, sensor technology) and big analytics fields (e.g., computer and data science) in community health research.

Second, today's health challenges are in demand for global and synergistic approaches (Kivits, Ricci, & Minary, 2019). As an example, tackling the worldwide COVID-19 epidemic needs collaborative efforts around the world. The reasearch on the travel restriction effects on the spread of COVID-19 outbreak calls for the knowledge and data from transportation, biostatistics, and public health (Chinazzi et al., 2020). Besides medical concerns, another important social issue raised amid COVID-19 is the exacerbating inequities, as more evidence shows the virus hits harder in low-income and underserved communities (van Dorn, Cooney, & Sabin, 2020). The research and solution to this issue would thus need the interdisciplinary efforts from health care professionals, physicians, communication, law enforcement, sociology, and public health agencies.

Issues for Research and Other Forms of Scholarship

Disease does not have a country boundary, although prevalence patterns in communicable diseases and NCDs are clear by regions The unique threat of communicable diseases exists in the fact that it can emerge anywhere on the planet and spread quickly to other regions through human activities (e.g., trade, travel), natural environment (e.g., shared watershed or underground water), and wildlife (e.g., migratory birds; Liang et al., 2010). As globalization gets intensified, the need for global cooperation in use of metrics and evaluation tools for disease mitigation increases (Fidler, 2004).

However, the challenges of implementing sustainable health system to address communicable diseases and NCDs are real (Beaglehole et al.,

2011). Despite the compelling evidence that international collaboration would reduce mortality from communicable diseases and NCDs at a fraction of the present cost by 2030, prevention attention focused on the environmental and health financing policies (Gluckman, Hanson, & Mitchell, 2010). First, the effective enforcement of health international conventions and law is lacking, without which the commitment to fund metrics and evaluation tools for disease mapping for intervention would be low. It is quite often that countries agree to an international legal obligation but remain without fulfilling their obligations. Intersectoral and multilateral agency collaborations, with or without national government involvement, would achieve some progress in reducing the global burden from communicable diseases and NCDs to a degree more than with no action taken (Magnusson, 2009). For multilateral agencies to enhance their collaborations toward the reduction of communicable diseases and NCDs, they may need to develop and share area level and global big data sets to which they have access. This would be a futuristic plan toward sustainable global health led by multilateral organizations invested in global health citizenship science.

A second challenge exists in the lack of medical and scientific resources and infrastructure to investigate the relevant measures and evaluation tools and undertake such measures. This is particularly difficult for low-income countries. The term “10/90 gap” is used to describe this phenomenon: only 10% of global health research resources are devoted to developing countries, where over 90% of the global disease burden occurred (Luchetti, 2014). There has been little progress by the international community to raise the per capita, for sustainable health, of the developing world population from the estimated $20 to about $44–60, which would be sufficient to sustain the health of the entire developing country populations (Luchetti, 2014). Achieving this per capita for sustainable health of developing country population would require a commitment of only 1% of the global GDP to health resourcing (World Health Organization, 2003). There seems to be no political will among the developed countries to bring this about, although very cost-effective in the long run.

The third challenge echoes to the aforementioned cultural differences and the power of constitutional authority. National and international responses to the COVID-19 exhibit a vivid example. Since its emergence, China took far-reaching, aggressive, yet effective measures of a national lockdown (Cyranoski, 2020). Following China, other countries, including Western democracies such as Italy, Spain, and France, implemented progressive restrictions after the first case confirmed (Gatto et al., 2020). Meanwhile, while the United States is now top on the confirmed cases and deaths, this mitigation approach has been piecemeal and uncoordinated. Some states have directed "shelter in place" order but other states are much loose. Different from countries such as China, Japan, and Italy, whose national authorities have the constitutional power of ordering major public-health interventions, such as national lockdown and mass quarantines, this authority in the United States lies primarily with states and localities. These differences at the world stage have implications for the prevention and control of communicable diseases and NCDs, beyond the constraints of fiscal resource capabilities. The differences in health management regimes would also impact what data are collected about communicable diseases and NCDs, how and with whom the data are shared, and to what purpose.

Summary and Conclusion

Designing, selecting, and calculating metrics is a crucial component in community health research, management, and control. Carefully designed and quantified metrics and tools can assist policymakers in making the best. They also offer standardized methods for the users to make cross-comparisons among populations, locations, and periods. Failure to invest in the metrics and tools for evaluating community health may leave communities and government entities poorly prepared for disease control and mitigation, thus increasing the probability of severe adverse consequences. Yet, how to determine a set of metrics for a particular community setting and implement them in the communities is a challenging task as it requires interdisciplinary collaboration among decision-makers, researchers, health policymakers, and communities.

Self-Check Questions

1. Define metrics and evaluation tools as they apply to community health? How are they alike and different from each other?
2. Identity two leading developments the development of the science of metrics and evaluation tools applied to community health?
3. What are the current and emerging community-oriented approaches to the use of metrics and evaluation tools for preventing and controlling communicable and non-communicable diseases?
4. How may cultural and professional practices influence the adoption and use of metrics and evaluation tools as resources for community health?
5. What are key research and practice issues you perceive to influence the use of metrics and evaluation tools to support community health initiatives?

Discussion Questions

1. Which health metrics and evaluation tools are you familiar with? What are they used for in a community you are familiar with?
2. What metrics and evaluation tools are used in your field of study and how? If you are unaware, what metrics and evaluation tools could your field utilize, and how?
3. What are the pressing community health issues in your community, and what types of interdisciplinary tools or approaches are engaged in addressing those needs?

Internet/Website Resources

https://www.lancetcountdown.org/

https://www.cdc.gov/eval/tools/developmenttools/index.html

https://www.who.int/healthinfo/topics_standards_tools_data_collection/en/

https://www.ruralhealthinfo.org/toolkits/community-health-workers/6/metrics

http://www.healthdata.org/data-tools

References

Ackland, M., Choi, B.C.K., & Puska, P. (2003). Rethinking the terms non-communicable disease and chronic disease. *Journal of Epidemiology & Community Health, 57*(11), 838–839.

Adams, M. D., & Kanaroglou, P. S. (2016). Mapping real-time air pollution health risk for environmental management: Combining mobile and stationary air pollution monitoring with neural network models. *Journal of Environmental Management, 168*, 133–141.

Adler, N. E., & Newman, K. (2002). Socioeconomic disparities in health: Pathways and policies. *Health Affairs, 21*(2), 60–76.

Afshin, A., Sur, P. J., Fay, K. A., Cornaby, L., Ferrara, G., Salama, J. S., … Afarideh, M. (2019). Health effects of dietary risks in 195 countries, 1990–2017: a systematic analysis for the Global Burden of Disease Study 2017. *The Lancet, 393*(10184), 1958–1972.

Alwan, A., MacLean, D. R., Riley, L. M., d'Espaignet, E. T., Mathers, C. D., Stevens, G. A., & Bettcher, D. (2010). Monitoring and surveillance of chronic non-communicable diseases: Progress and capacity in high-burden countries. *The Lancet, 376*(9755), 1861–1868.

American Diabetes Association. (2018). Economic costs of diabetes in the US in 2017. *Diabetes Care, 41*(5), 917–928.

Anderson, S. J., Tuttle, B. T., Powell, R. L., & Sutton, P. C. (2010). Characterizing relationships between population density and nighttime imagery for Denver, Colorado: Issues of scale and representation. *International Journal of Remote Sensing, 31*(21), 5733–5746.

Association of State and Territorial Health Officials. (2020). *State health department framework: Preventing infectious diseases through healthcare.* Retrieved on April 28, 2020 from https://www.astho.org/Programs/Infectious-Disease/Integration/Preventing-Infectious-Diseases-through-Healthcare/

Bagan, H., & Yamagata, Y. (2015). Analysis of urban growth and estimating population density using satellite images of nighttime lights and land-use and population data. *GIScience & Remote Sensing, 52*(6), 765–780.

Beaglehole, R., Bonita, R., Horton, R., Adams, C., Alleyne, G., Asaria, P., Baugh, V., Bekedam, H., Billo, N., Casswell, S., & Cecchini, M. (2011). Priority actions for the noncommunicable disease crisis. *The Lancet, 377*(9775), 1438–1447.

Bi, Q., Wu, Y., Mei, S., Ye, C., Zou, X., Zhang, Z., Liu, X., Wei, L., Truelove, S.A., Zhang, T., and Gao, W. (2020). Epidemiology and transmission of

COVID-19 in 391 cases and 1286 of their close contacts in Shenzhen, China: a retrospective cohort study. *The Lancet Infectious Diseases*.
Bjørnstad, O. N., Finkenstädt, B. F., & Grenfell, B. T. (2002). Dynamics of measles epidemics: Estimating scaling of transmission rates using a time series SIR model. *Ecological Monographs, 72*(2), 169–184.
Black, D., O'Loughlin, K., Kendig, H., & Wilson, L. (2012). Cities, environmental stressors, aging, and chronic disease. *Australasian Journal on Aging, 31*(3), 147–151.
Blazer, D. G., & Hernandez, L. M. (Eds.). (2006). *Genes, behavior, and the social environment: Moving beyond the nature/nurture debate*. Washington, DC: National Academies Press.
Blöschl, G., & Sivapalan, M. (1995). Scale issues in hydrological modeling: A review. *Hydrological Processes, 9*(3–4), 251–290.
Boutayeb, A. (2006). The double burden of communicable and non-communicable diseases in developing countries. *Transactions of the Royal Society of Tropical Medicine and Hygiene, 100*(3), 191–199.
Bukhari, Q., & Jameel, Y. (2020). *Will coronavirus pandemic diminish by summer?*. Retrieved on May 10, 2020, from https://papers.ssrn.com/sol3/papers.cfm?abstract_id=3556998
CDC Surveillance Resource Center (n.d.). Retrieved on Oct 23, 2020 from https://www.cdc.gov/surveillancepractice/index.html
Centers for Disease Control and Prevention (CDC). (2012). *Community Assessment for Public Health Emergency Response (CASPER) toolkit: Second edition*. Atlanta, GA: Author.
Chinazzi, M., Davis, J. T., Ajelli, M., Gioannini, C., Litvinova, M., Merler, S., … Viboud, C. (2020). The effect of travel restrictions on the spread of the 2019 novel coronavirus (COVID-19) outbreak. *Science, 368*(6489), 395–400.
Communicable Diseases Module: 2. (2020). *Prevention and control of communicable diseases and community diagnosis*. Retrieved on May 8 from URL: https://www.open.edu/openlearncreate/mod/oucontent/view.php?id=85&printable=1
Cook, S., Conrad, C., Fowlkes, A. L., & Mohebbi, M. H. (2011). Assessing Google flu trends performance in the United States during the 2009 influenza virus A (H1N1) pandemic. *PLoS One, 6*(8).
Countdown, N. C. D. (2018). NCD Countdown 2030: Worldwide trends in non-communicable disease mortality and progress towards Sustainable Development Goal target 3.4. *The Lancet, 392*(10152), 1072–1088.
Cyranoski, D. (2020). What China's coronavirus response can teach the rest of the world. *Nature, 579*(7800), 479–480.

De Choudhury, M., Emre, K., Dredze, M., Coppersmith, G., & Kumar, M. (2016). Discovering shifts to suicidal ideation from mental health content in social media. In *Proceedings of the 2016 CHI conference on human factors in computing systems* (pp. 2098–2110).

de Martel, C., Georges, D., Bray, F., Ferlay, J., & Clifford, G. M. (2020). Global burden of cancer attributable to infections in 2018: A worldwide incidence analysis. *The Lancet Global Health, 8*(2), e180–e190.

Didzun, O., De Neve, J. W., Awasthi, A., Dubey, M., Theilmann, M., Bärnighausen, T., ... Geldsetzer, P. (2019). Anaemia among men in India: A nationally representative cross-sectional study. *The Lancet Global Health, 7*(12), e1685–e1694.

Dietz, W., & Santos-Burgoa, C. (2020). Obesity and its implications for COVID-19 mortality. *Obesity, 28*(6), 1005–1005.

Dimakopoulou, K., Gryparis, A., & Katsouyanni, K. (2017). Using spatio-temporal land-use regression models to address spatial variation in air pollution concentrations in time-series studies. *Air Quality, Atmosphere & Health, 10*(9), 1139–1149.

Dionisio, K. L., Rooney, M. S., Arku, R. E., Friedman, A. B., Hughes, A. F., Vallarino, J., ... Ezzati, M. (2010). Within-neighborhood patterns and sources of particle pollution: Mobile monitoring and geographic information system analysis in four communities in Accra, Ghana. *Environmental Health Perspectives, 118*(5), 607.

Dolley, S. (2018). Big data's role in precision public health. *Frontiers in Public Health, 6*, 68.

Eagle, N., Pentland, A., & Lazer, D. (2009). Inferring friendship network structure by using mobile phone data. *Proceedings of the National l Academy of Science, 106*, 15274–15278.

Edemekong, P. F., & Huang, B. (2019). *Epidemiology of prevention of communicable diseases*. StatPearls [Internet]. Treasure Island, FL: StatPearls Publishing.

Eisenberg, L., & Pellmar, T. C. (Eds.). (2000). *Bridging disciplines in the brain, behavioral, and clinical sciences*. Washington, DC: National Academies Press.

Elvidge, C. D., Imhoff, M. L., Baugh, K. E., Hobson, V. R., Nelson, I., Safran, J., ... Tuttle, B. T. (2001). Night-time lights of the world: 1994–1995. *ISPRS Journal of Photogrammetry and Remote Sensing, 56*(2), 81–99.

Environmental Protection Agency: EPA (1996). *Sanitary sewer overflows what are they and how can we reduce them*? Retrieved on May 10 from https://www3.epa.gov/npdes/pubs/ssodesc.pdf

Evan Secor, W. (2014). Water-based interventions for schistosomiasis control. *Pathogens and Global Health, 108*(5), 246–254.

Fang, L. Q., Goeijenbier, M., Zuo, S. Q., Wang, L. P., Liang, S., Klein, S. L., … Glass, G. E. (2015). The association between hantavirus infection and selenium deficiency in mainland China. *Viruses, 7*(1), 333–351.

Fidler, D. P. (2004). Revision of the World Health Organization's International Health Regulations. *American Society of International Law Insights, 8*(8).

Fillinger, U., Sombroek, H., Majambere, S., van Loon, E., Takken, W., & Lindsay, S. W. (2009). Identifying the most productive breeding sites for malaria mosquitoes in The Gambia. *Malaria Journal, 8*(1), 62.

Fiore, A. E., Shay, D. K., Broder, K., Iskander, J. K., Uyeki, T. M., Mootrey, G., … Cox, N. S. (2008). Prevention and control of influenza: Recommendations of the Advisory Committee on Immunization Practices (ACIP), 2008. MMWR. Recommendations and reports: Morbidity and mortality weekly report. *Recommendations and Reports, 57*(RR-7), 1–60.

Freifeld, C. C., Chunara, R., Mekaru, S. R., Chan, E. H., Kass-Hout, T., Iacucci, A. A., & Brownstein, J. S. (2010). Participatory epidemiology: Use of mobile phones for community-based health reporting. *PLoS Medicine, 7*(12), e1000376.

Gagnon, F., Bergeron, P., Clavier, C., Fafard, P., Martin, E., & Blouin, C. (2017). Why and how political science can contribute to public health? Proposals for collaborative research avenues. *International Journal of Health Policy and Management, 6*(9), 495.

Gatto, M., Bertuzzo, E., Mari, L., Miccoli, S., Carraro, L., Casagrandi, R., & Rinaldo, A. (2020). *Spread and dynamics of the COVID-19 epidemic in Italy: Effects of emergency containment measures.* Proceedings of the National Academy of Sciences. Preprint.

Gavazzi, G., Herrmann, F., & Krause, K. H. (2004). Aging and infectious diseases in the developing world. *Clinical Infectious Diseases, 39*(1), 83–91.

Ginsberg, J., Mohebbi, M. H., Patel, R. S., Brammer, L., Smolinski, M. S., & Brilliant, L. (2009). Detecting influenza epidemics using search engine query data. *Nature, 457*(7232), 1012–1014.

Gluckman, P. D., Hanson, M. A., & Mitchell, M. D. (2010). Developmental origins of health and disease: Reducing the burden of chronic disease in the next generation. *Genome Medicine, 2*(2), 14. https://doi.org/10.1186/gm135

Goldman, L., & Coussens, C. M. (2004). *Environmental health indicators: Bridging the chasm of public health and the environment.* Washington, DC: National Academies Press.

Hales, S., Kovats, S., Lloyd, S., & Campbell-Lendrum, D. (2014). *Quantitative risk assessment of the effects of climate change on selected causes of death, 2030s, and 2050s*. Geneva, Switzerland: World Health Organization.

Harnack, L., Block, G., & Lane, S. (1997). Influence of selected environmental and personal factors on dietary behavior for chronic disease prevention: A review of the literature. *Journal of Nutrition Education, 29*(6), 306–312.

Harries, A. D., Kumar, A. M. V., Satyanarayana, S., Lin, Y., Takarinda, K. C., Tweya, H., … Zachariah, R. (2015). Communicable and non-communicable diseases: Connections, synergies, and benefits of integrating care. *Public Health Action, 5*(3), 156–157.

Hart, R., Liang, L., & Dong, P. (2020). Monitoring, Mapping, and Modeling Spatial–Temporal Patterns of PM2.5 for Improved Understanding of Air Pollution Dynamics Using Portable Sensing Technologies. *International Journal of Environmental Research and Public Health, 17*(14), 4914.

Hay, S. I., George, D. B., Moyes, C. L., & Brownstein, J. S. (2013). Big data opportunities for global infectious disease surveillance. *PLoS Medicine, 10*(4), e1001413.

Health Effects Institute. (2019). *State of Global Air 2018. Special Report*. Boston, MA: Health Effects Institute.

Heitgerd, J. L., Dent, A. L., Elmore, K. A., Kaplan, B., Holt, J. B., Metzler, M. M., … Comer, K. F. (2008). Peer-reviewed: Community health status indicators: Adding a geospatial component. *Preventing Chronic Disease, 5*, 3. https://www.ncbi.nlm.nih.gov/pmc/articles/PMC2483562/

Hempel, S. (2007). *The strange case of the Broad Street pump: John Snow and the mystery of cholera*. Los Angeles, CA: University of California Press.

Hosseinpoor, A. R., Schlotheuber, A., Nambiar, D., & Ross, Z. (2018). Health Equity Assessment Toolkit Plus (HEAT Plus): Software for exploring and comparing health inequalities using uploaded datasets. *Global Health Action, 11*(sup1), 20–30.

Hui, F. M., Xu, B., Chen, Z. W., Cheng, X., Liang, L., Huang, H. B., … Zhou, X. N. (2009). Spatio-temporal distribution of malaria in Yunnan Province, China. *The American Journal of Tropical Medicine and Hygiene, 81*(3), 503–509.

Jacob, V., Chattopadhyay, S. K., Hopkins, D. P., Reynolds, J. A., Xiong, K. Z., Jones, C. D., … Goetzel, R. Z. (2019). Economics of community health workers for chronic disease: Findings from community guide systematic reviews. *American Journal of Preventive Medicine, 56*(3), e95–e106.

Jakubowski, B., & Frumkin, H. (2010). Peer-reviewed: Environmental metrics for community health improvement. *Preventing Chronic Disease, 7*, 4.

Jaspers, L., Colpani, V., Chaker, L., van der Lee, S. J., Muka, T., Imo, D., … Pazoki, R. (2015). The global impact of non-communicable diseases on households and impoverishment: A systematic review. *European Journal of Epidemiology, 30*(3), 163–188.

Jiwani, S. S., Carrillo-Larco, R. M., Hernández-Vásquez, A., Barrientos-Gutiérrez, T., Basto-Abreu, A., Gutierrez, L., … Miranda, J. J. (2019). The shift of obesity burden by socioeconomic status between 1998 and 2017 in Latin America and the Caribbean: A cross-sectional series study. *The Lancet Global Health, 7*(12), e1644–e1654.

Keesing, F., Belden, L. K., Daszak, P., Dobson, A., Harvell, C. D., Holt, R. D., … Myers, S. S. (2010). Impacts of biodiversity on the emergence and transmission of infectious diseases. *Nature, 468*(7324), 647–652.

Kitchin, R., & McArdle, G. (2016). What makes Big Data, Big Data? Exploring the ontological characteristics of 26 datasets. *Big Data & Society, 3*(1), 2053951716631130.

Kivits, J., Ricci, L., & Minary, L. (2019). Interdisciplinary research in public health: The 'why' and the 'how'. *Journal of Epidemiology and Community Health, 73*(12). https://doi.org/10.1136/jech-2019-212511

Kochtitzky, C. S., Frumkin, H., Rodriguez, R., Dannenberg, A. L., Rayman, J., Rose, K., … Kanter, T. (2006). Urban planning and public health at CDC. *MMWR Supplements, 55*(2), 34–38.

Lary, D. J., Faruque, F. S., Malakar, N., Moore, A., Roscoe, B., Adams, Z. L., & Eggelston, Y. (2014). Estimating the global abundance of ground-level presence of particulate matter ($PM_{2.5}$). *Geospatial Health*, S611–S630.

Lary, D. J., Lary, T., & Sattler, B. (2015). Using machine learning to estimate global $PM_{2.5}$ for environmental health studies. *Environmental Health Insights, 9*, EHI-S15664.

Lazer, D., Kennedy, R., King, G., & Vespignani, A. (2014). The parable of Google Flu: Traps in big data analysis. *Science, 343*(6176), 1203–1205.

Levin, S. A. (1992). The problem of pattern and scale in ecology: The Robert H. MacArthur award lecture. *Ecology, 73*(6), 1943–1967.

Li, X., Gong, P., & Liang, L. (2015). A 30-year (1984–2013) record of annual urban dynamics of Beijing City derived from Landsat data. *Remote Sensing of Environment, 166*, 78–90.

Liang, L., Gong, P., Cong, N., Li, Z., Zhao, Y., & Chen, Y. (2019). Assessment of personal exposure to particulate air pollution: The first result of City Health Outlook (CHO) project. *BMC Public Health, 19*(1), 711.

Liang, L., Liu, Y., Liao, J., & Gong, P. (2014). Wetlands explain most of the genetic divergence patterns of Oncomelania hupensis. *Infection, Genetics and Evolution, 27*, 436–444.

Liang, L., Xu, B., Chen, Y., Liu, Y., Cao, W., Fang, L., … Gong, P. (2010). Combining spatial-temporal and phylogenetic analysis approaches for improved understanding of global H5N1 transmission. *PLoS One, 5*(10).

Lilienfeld, D. E., & Stolley, P. D. (1994). *Foundations of epidemiology*. New York, NY: Oxford University Press.

Łowicki, D. (2019). Landscape pattern as an indicator of urban air pollution of particulate matter in Poland. *Ecological Indicators, 97*, 17–24.

Luchetti, M. (2014). Global Health and the 10/90 gap. *British Journal of Medical Practitioners, 7*(4), a731.

Lycett, S. J., Bodewes, R., Pohlmann, A., Banks, J., Bányai, K., Boni, M. F., … Dán, Á. (2016). Role for migratory wild birds in the global spread of avian influenza H5N8. *Science, 354*, 6309.

Macke, B. A., & Maher, J. E. (1999). Partner notification in the United States: An evidence-based review. *American Journal of Preventive Medicine, 17*(3), 230–242.

Magnusson, R. S. (2009). Rethinking global health challenges: Towards a 'global compact' for reducing the burden of chronic disease. *Public Health, 123*(3), 265–274.

Montanari, A. (2005). Human mobility, global change, and local development. Contribution to the Italian PRIN 2002, Research Programme on Tourism and development: Local peculiarity and territorial competitivity. *Belgeo. Revue belge de géographie*, (1–2), 7–18.

Mooney, S. J., & Pejaver, V. (2018). Big data in public health: Terminology, machine learning, and privacy. *Annual Review of Public Health, 39*, 95–112.

Nasir, A., Zimmer, A., Taylor, D., & Santo, J. (2019). Psychosocial assessment of the family in the clinical setting. *BMC Psychology, 7*(1), 3.

National Research Council. (2010). *Understanding the changing planet: Strategic directions for the geographical sciences*. Washington, DC: National Academies Press.

Palacios-Lopez, D., Bachofer, F., Esch, T., Heldens, W., Hirner, A., Marconcini, M., … Tatem, A. J. (2019). New perspectives for mapping global population distribution using world settlement footprint products. *Sustainability, 11*(21), 6056.

Paul, M., Baritaux, V., Wongnarkpet, S., Poolkhet, C., Thanapongtharm, W., Roger, F., … Ducrot, C. (2013). Practices associated with highly pathogenic avian influenza spread in traditional poultry marketing chains: Social and economic perspectives. *Acta Tropica, 126*(1), 43–53.

Pongsiri, M. J., Roman, J., Ezenwa, V. O., Goldberg, T. L., Koren, H. S., Newbold, S. C., … Salkeld, D. J. (2009). Biodiversity loss affects global disease ecology. *Bioscience, 59*(11), 945–954.

Read, J. M., Eames, K. T., & Edmunds, W. J. (2008). Dynamic social networks and the implications for the spread of infectious disease. *Journal of the Royal Society Interface, 5*(26), 1001–1007.

Reis, F., Moura, B., Guardado, D., Couceiro, P., Catarino, L., Mota-Pinto, A., … Palavra, F. (2019). Development of a healthy lifestyle assessment toolkit for the general public. *Frontiers in Medicine, 6*, 134.

Reynolds, R., Liang, L., Li, X., & Dennis, J. (2017). Monitoring annual urban changes in a rapidly growing portion of Northwest Arkansas with a 20-year landsat record. *Remote Sensing, 9*(1), 71.

Salathé, M., Kazandjieva, M., Lee, J. W., Levis, P., Feldman, M. W., & Jones, J. H. (2010). A high-resolution human contact network for infectious disease transmission. *Proceedings of the National Academy of Sciences, 107*(51), 22020–22025.

Sallis, J. F. (2010). Measuring physical activity: Practical approaches for program evaluation in Native American communities. *Journal of Public Health Management and Practice, 16*(5), 404–410.

Schmidt, C. W. (2012). Trending now: Using social media to predict and track disease outbreaks. *Environmental Health Perspectives, 120*(1). https://doi.org/10.1289/ehp.120-a30

Seo, D. W., & Shin, S. Y. (2017). Methods using social media and search queries to predict infectious disease outbreaks. *Healthcare Informatics Research, 23*(4), 343–348.

Sheldrick, R. C., & Perrin, E. C. (2013). Evidence-based milestones for surveillance of cognitive, language, and motor development. *Academic Pediatrics, 13*(6), 577–586.

Singh, V., Carnevale, C., Finzi, G., Pisoni, E., & Volta, M. (2011). A cokriging based approach to reconstruct air pollution maps, processing measurement station concentrations, and deterministic model simulations. *Environmental Modelling & Software, 26*(6), 778–786.

Sopan, A., Noh, A. S. I., Karol, S., Rosenfeld, P., Lee, G., & Shneiderman, B. (2012). Community health map: A geospatial and multivariate data visualization tool for public health datasets. *Government Information Quarterly, 29*(2), 223–234.

Soucy, J. P. R., Sturrock, S. L., Berry, I., Daneman, N., MacFadden, D. R., & Brown, K. A. (2020). Estimating the effect of physical distancing on the COVID-19 pandemic using an urban mobility index. *medRxiv*.

Tan, M., Li, X., Li, S., Xin, L., Wang, X., Li, Q., … Xiang, W. (2018). Modeling population density based on nighttime light images and land use data in China. *Applied Geography, 90*, 239–247.

Taylor, A. L. (1996). Controlling the global spread of infectious diseases: Toward a reinforced role for the International Health Regulations. *Houston Law Review, 33*, 1327.

Tessum, C. W., Apte, J. S., Goodkind, A. L., Muller, N. Z., Mullins, K. A., Paolella, D. A., … Hill, J. D. (2019). Inequity in consumption of goods and services adds to racial-ethnic disparities in air pollution exposure. *Proceedings of the National Academy of Sciences, 116*(13), 6001–6006.

Tolksdorf, K., Buda, S., Schuler, E., Wieler, L. H., & Haas, W. (2020). Influenza-associated pneumonia as a reference to assess the seriousness of coronavirus disease (COVID-19). *Eurosurveillance, 25*(11), 2000258.

Upton, J. (2013). Psychosocial factors. In M. D. Gellman & J. R. Turner (Eds.), *Encyclopedia of behavioral medicine*. New York, NY: Springer.

Van Der Aalst, W. (2016). Data science in action. In *Process mining* (pp. 3–23). Berlin, Heidelberg: Springer.

van Dorn, A., Cooney, R. E., & Sabin, M. L. (2020). COVID-19 exacerbating inequalities in the US. *The Lancet, 395*(10232), 1243–1244.

Van Vliet, E. D. S., & Kinney, P. L. (2007). Impacts of roadway emissions on urban particulate matter concentrations in sub-Saharan Africa: New evidence from Nairobi, Kenya. *Environmental Research Letters, 2*(4), 045028.

Walter, F. M., Prevost, A. T., Birt, L., Grehan, N., Restarick, K., Morris, H. C., … Emery, J. D. (2013). Development and evaluation of a brief self-completed family history screening tool for common chronic disease prevention in primary care. *The British Journal of General Practice, 63*(611), e393–e400.

Wang, L., Wang, Z., Zhang, Y., & Li, X. (2013). How human location-specific contact patterns impact spatial transmission between populations? *Scientific Reports, 3*(1), 1–10.

Watts, N., Adger, W. N., Agnolucci, P., Blackstock, J., Byass, P., Cai, W., … Cox, P. M. (2015). Health and climate change: Policy responses to protect public health. *The Lancet, 386*(10006), 1861–1914.

Watts, N., Amann, M., Arnell, N., Ayeb-Karlsson, S., Belesova, K., Berry, H., … Campbell-Lendrum, D. (2018). The 2018 report of the Lancet Countdown on health and climate change: Shaping the health of nations for centuries to come. *The Lancet, 392*(10163), 2479–2514.

Weng, Q., & Yang, S. (2006). Urban air pollution patterns, land use, and thermal landscape: An examination of the linkage using GIS. *Environmental Monitoring and Assessment, 117*(1-3), 463–489.

Widener, M. J., & Li, W. (2014). Using geolocated Twitter data to monitor the prevalence of healthy and unhealthy food references across the US. *Applied Geography, 54*, 189–197.

Williams, R., Kilaru, V., Conner, T., Clements, A., Colon, M., Breen, M., … & Feinberg, S. (2018). *New paradigm for air pollution monitoring: Emerging sensor technologies 2014-2018 progress report.* Presented at ACE Webinar, Research Triangle Park, NC, April 30, 2018.

Woo, P. C., Lau, S. K., & Yuen, K. Y. (2006). Infectious diseases emerging from Chinese wet-markets: Zoonotic origins of severe respiratory viral infections. *Current Opinion in Infectious Diseases, 19*(5), 401–407.

World Health Organization. (2003). *Macroeconomics and health: An update: Increasing investments in health outcomes for the poor: Second consultation on macroeconomics and health* (No. WHO/SDE/CMH/03.1). Geneva, Switzerland: Author.

World Health Organization. (2008). *The global burden of disease: 2004 update.* Geneva, Switzerland: Author.

World Health Organization. (2009). *2008-2013 action plan for the global strategy for the prevention and control of noncommunicable diseases: Prevent and control cardiovascular diseases, cancers, chronic respiratory diseases, and diabetes.* Geneva, Switzerland: Author.

World Health Organization. (2010). *Global strategy for the prevention and control of noncommunicable diseases.* Geneva, Switzerland: Author.

World Health Organization. (2016a). *Advancing the right to health: the vital role of law.* World Health Organization. Retrieved on July 23 from https://apps.who.int/iris/handle/10665/252815. License: CC BY-NC-SA 3.0 IGO

World Health Organization. (2016b). *Disease burden and mortality estimates.* Retrieved on July 23, 20202 from https://www.who.int/healthinfo/global_burden_disease/estimates/en/index1.html

Wu, J., Xie, W., Li, W., & Li, J. (2015). Effects of urban landscape pattern on PM2. 5 pollution—A Beijing case study. *PLoS One, 10*(11), e0142449.

Xiao, Q., Wang, Y., Chang, H. H., Meng, X., Geng, G., Lyapustin, A., & Liu, Y. (2017). Full-coverage high-resolution daily $PM_{2.5}$ estimation using MAIAC AOD in the Yangtze River Delta of China. *Remote Sensing of Environment, 199*, 437–446.

Xu, X., Wang, G., Chen, N., Lu, T., Nie, S., Xu, G., … Schwartz, J. (2016). Long-term exposure to air pollution and increased risk of membranous nephropathy in China. *Journal of the American Society of Nephrology*, ASN-2016010093.

Zupancic, T., Westmacott, C., & Bulthuis, M. (2015). *The impact of green space on heat and air pollution in urban communities: A meta-narrative systematic review* (p. 67). Vancouver, BC: David Suzuki Foundation.

Part IV

Sustainable Community Health in Populations

13

Older Adults' Wellbeing

Maidei Machina, Elias Mpofu, Solymar Rivera-Torres, Rebekah Knight, and Theresa Abah

Introduction

The world is experiencing dramatically increased numbers of people living to an advanced old age. That is known as the "gray tsunami" (Longman, 2010). According to the United Nations Department of Economic and Social Affairs (2019), World Health Organization [WHO] (2019), and the National Institute on Aging [NIA] (2019), the population of people aged 65 or older is anticipated to grow from 524 million in 2010 to approximately 1.5 billion in 2050 worldwide (see Table13.1).

M. Machina (✉)
Westmead Hospital, Sydney, NSW, Australia
e-mail: Maidei.Machina@health.nsw.gov.au

E. Mpofu
University of North Texas, Denton, TX, USA

University of Sydney, Sydney, NSW, Australia

University of Johannesburg, Johannesburg, South Africa
e-mail: Elias.Mpofu@unt.edu

E. Mpofu (ed.), *Sustainable Community Health*,
https://doi.org/10.1007/978-3-030-59687-3_13

Learning Objectives

By the end of the chapter, the reader should be able to:

1. Define the population of older adults in their diversity of community health needs.
2. Outline the history of research and practice on older adults' community wellbeing.
3. Discuss the current and evolving practices in sustainable community health systems for older adults.
4. Explore the role of culture, professional practices, and legislation on sustainable health approaches for older adults.
5. Describe the role of interdisciplinary approaches in designing and implementing sustainable health systems for older adults.
6. Propose areas for research and practice in sustainable health systems for older adults in contemporary society.

The population of older adults in developing countries is expected to increase by more than 250% by 2050 (UNWPA, 2019), compared to a 71% increase in developed countries (NIA, 2011).

Older adults require housing, healthcare, social security, caregiving, and long-term care services in order to experience sustainable aging and wellbeing (Rowe & Kahn, 2015; World Health Organization [WHO], 2015). Poverty, food insecurity, malnutrition, deteriorating health, poor mental health, elder abuse, functional illiteracy, reduced productivity, gender inequality, lack of shelter, lack of accessible communities, isolation, and ageism are amongst the most common inequalities and injustices that the aging population face today (WHO, 2015). Older adult age also comes with chronic illness and disability, requiring long-term health

S. Rivera-Torres • R. Knight
University of North Texas, Denton, TX, USA
e-mail: solymarrivera@my.unt.edu

T. Abah
University of Sacramento, Sacramento, CA, USA
e-mail: t.abah@csus.edu

Table 13.1 World population prospects for years 2019 and 2050 by geographic region, for the age-group 65 years and older

Region	Number of persons aged 65 or over in 2019 (millions)	Number of persons aged 65 or over in 2050 (millions)	Percentage change between 2019 and 2050
Sub-Saharan Africa	31.9	101.4	218%
Northern Africa and Western Asia	29.4	95.8	226%
Central and Southern Asia	119.0	328.1	176%
Latin America and the Caribbean	56.4	114.6	156%
Australia and New Zealand	4.8	8.8	84%
Oceania, excluding Australia and New Zealand	0.5	1.5	190%
Europe and Northern America	200.4	296.2	48%

Source: United Nations, Department of Economic and Social Affairs, Population Division (2019). *World Population Prospects 2019*

system supports (WHO, 2020). Moreover, the deterioration of physical, cognitive, and mental faculties that comes with the aging process, increase the risk of age discrimination, morbidity, the loss of independence, withdrawal, and marginalization from productive and meaningful social activities (Centers for Disease Control [CDC], 2019; Quadagno, 2018). Older adults may correctly perceive that the successful resolution of their health and wellness needs is dependent on the broader political, economic, physical, and social environments in which they live; which, if not well structured, may expose them to avoidable health disparities, inequities, and social injustices (Ayalon et al., 2020; Quadagno, 2018; WHO, 2015). Establishing sustained older adult health would require an increased focus on enhancing the wellness of older adults, in addition to providing comprehensive physical-medical care. This would result in the development of vibrant communities in which older adults are key partners to the overall wellbeing of the community, with a futuristic and engaged older adult population (Brothers, Gabrian, Wahl, & Diehl, 2016; Lubitz, Cai, Kramarow, & Lentzner, 2003; Shelton et al., 2019).

When this vibrancy in older adult wellbeing is sustained and multiplied throughout a community, then wellbeing benefits to the whole community would increase exponentially.

With the world population living longer today than ever before, the world is contending with developing and implementing best national health policies and programs that ensure the long-term health and wellness of the older adult population (Kuruvilla et al., 2018; Quadagno, 2015). Further to this, how healthy the length in years will be depends on how well the society plans for the social, economic, and environmental needs of the aging population, while aiming to improve inclusivity in their diverse social ecologies (Cheadle, Egger, LoGerfo, Schwartz, & Harris, 2010; Jackson, Roberts, & McKay, 2019; Wurm, Diehl, Kornadt, Westerhof, & Wahl, 2017). In this regard, the sustainable health of older adults would be secured if policies and interventions are designed to focus on community system environment level supports, as opposed to disease or population specific policies and interventions.

Meeting the health and wellbeing needs of older adults aging with, or into, disability presents a unique mix of needs which are best addressed by a long-term focus rather than short-term stop-gap measures (Ayalon et al., 2020; National Research Council, 2001). The solutions to these challenges are complex and depend on the collaboration of families and communities, as well as social, political, economic, and health service delivery systems, to provide optimal support to older adults across their life-space capabilities. While research scholarship in global aging is a recent phenomenon, the development and adoption of programs that cater for the wellbeing of the burgeoning population of older adults will transform the social structure of many societies (Rowland, 2009). This chapter presents some approaches to sustainable community wellbeing of older adults that would have long-term benefits at lower fiscal cost when compared to approaches premised on taking care of their medical care needs primarily.

Professional and Legal Definitions and Relevant Theories

When does someone become "old" or "elderly" is a common question. Standard terms refer to the older generation as seniors, retirees, older people, older adults, and elders. These multiple referent terms would be partly true for some older adults but not others, given the diversity in the population profile of older people (Czaja, Boot, Charness, & Rogers, 2019). Nonetheless, there is consensus in referring to those whose ages span from 50 or more years as older adults. The World Health Organization (WHO, 2010) defined older adults to encompass *"a range of characteristics including chronological age, change in social role, and changes in functional capabilities."* Older age commonly refers to people over 50 years. On the spectrum of the aging older adult population, those 50–59 years old would be the very young old, those 60–74 years are the young old, those 75–84 years are the old-old, and those 85 years and older as the very old (Diehl & Wahl, 2020). However, these age-related gradations would not reflect the subjective sense of aging across the older adult population; as some very young old may perceive themselves to be old-old, while some very old adults have a sense of wellbeing associated with the younger-old age groups (Westerhof et al., 2014).

This process of aging is known by the term senescence, which refers to the biological, emotional, intellectual, social, and spiritual changes that come with old age (Xu & Larbi, 2017). Older adults experience senescence differently which influences their health and wellbeing, with some having a higher sense of wellness than others (Crews & Ice, 2012). The tendency from ageism to perceive all older adults as a homogenous and burdensome population would be harmful to their long-term wellbeing by marginalizing and denying them the ability to age healthily in their communities.

The WHO defined healthy aging as "*the process of developing and maintaining the functional ability... [or] capabilities to carry out actions a person has reason to value.... to meet their basic needs; to learn, grow and make decisions; to be mobile; to build and maintain relationships; and to contribute to society... including the home, community the built environment, and services infrastructure"* (WHO, 2015).

Healthy aging is associated with (1) maintenance of the key activities of independent living; (2) cognitive capacity; (3) social connections and communication; (4) personal mobility; (5) transportation; and (6) improved access to healthcare (U.S. Health and Human Services Department, 2017; WHO, 2015). Among older adults, healthy life expectancy (HLE) is the number of remaining years of life that would be spent in good health or with a sense of wellness (Stiefel, Perla, & Zell, 2010). While life expectancy is undoubtedly increasing globally, older adults are at higher risk of having a lower HLE due to the processes of aging, as well as from gaps in the inclusivity of community health social policies, which result in older adults living longer but with shortened periods of wellness. Older adults may live longer in ill health or experience a prolongation of the process of dying (Bernd, Doyle, Grundy, & McKee, 2009). Structuring community health systems to allow for the "compression of morbidity" in which illness and disability occur much later in life by reducing the burden of illness and minimizing the "expansion of morbidity" where older adults live longer but in poor health, should be the goal of older-age-friendly communities. A growing number of studies suggest that the promotion of "healthy aging" earlier in life reduces lifetime healthcare expenditure, while a failure to do so increases them (Australian Institute of Health & Welfare, 2014).

History of Research and Practice

Anthropological and economic research shows that since the industrial revolution, there has been a steady demographic transition of people surviving in increased numbers into older age where their capabilities deteriorate to the point of becoming dependent on the younger generation (Laslett, 1995). This demographic transition fostered the foundations of aged care services as an increased expectation was placed on family members to provide care for relatives that were no longer capable of sustaining themselves. Neighbors stepped in when family was unavailable, and town officials intervened when all primary familial-based systems demonstrably failed. This pattern of dependency on family members and receiving assistance from local authorities persisted well into the twentieth century. The need for formalized aged care policies, supports, and services was subsequently recognized and contributed to the creation of statutory pension systems and other elements of what make up the modern welfare states and systems that we have today.

In the past century, many nations around the globe have passed legislation or enacted policies for the sustenance of older adults, the maintenance of personal control of their activities, including for their protection from abuse (Daly, 2011; Faulkner, 1982; Martin & Smith, 1993; Montgomery et al., 2016). As examples, in the United States, three federal acts were signed into law that altered the system of aged care services and created federal safety nets for the elderly. The first was the Social Security Act of 1935, which established an old-age insurance program that would provide financial benefits to retirees based on tax contributions by employers and employees. This measure protects older adults from falling into poverty and provides them with a means to sustain life and their independence. The Social Security Act was amended in 1965 and resulted in the creation of the Medicare and Medicaid programs, which provide federal health insurance for the elderly and low-income Americans. The second was the Older Americans Act (OAA) of 1965 which aimed to meet the objective of assisting older adults to secure basic livelihoods including accessible and affordable housing and supported living arrangements; community-based long-term care services options; opportunities for employment with no discriminatory personnel practices because of age; ongoing civic participation and full community participation; and low-cost transport (United State, 1978). The third—the Recognize, Assist, Include, Support, and Engage Family Caregivers Act of 2017—provides for support of family caregivers for older adults aging with, and into, disability.

Globally, the trend is toward the creation and implementation of age-friendly communities encompassing services, political and economic systems, built environments (i.e., physical structures such as streets and

Case Illustration 13.1: Age-Friendly Environments

Rynell is a 90-year-old woman living independently in a medium-sized U.S. city that has a very active community life in the downtown area. Because she lives alone, Rynell enjoys the social atmosphere that community events promote. She is still ambulatory but uses a rolling walker when she goes out of her home to support her balance. Unfortunately, the sidewalks and curbs in Rynell's community were built in the 1950s and are very hard to navigate due to the disintegration of the cement. Additionally, there are no universal cuts or ramps to assist those using rolling devices or those who cannot step up and down curbs easily. While Rynell can easily procure rides to the events that she wants to attend downtown, she has an increased fear of falling due to the condition of the sidewalks and curbs downtown. When a meeting on proposed future bond money budgets was scheduled, Rynell went to advocate for bond money to be used to repair and update the downtown sidewalk area to enable easier access for those who use rolling devices. Impolite comments were made by younger audience members stating, "we do not need to spend all of that money on sidewalks just to help the old people."

What Do You Think?

1. What does the case of Rynell say to you about the social pillars of age friendliness of a community?
2. How would self-advocacy enhance the wellbeing of older adults in a community?
3. What resources would be necessary to optimize successful civic engagement wellbeing of older adults?

buildings), and social environments that promote inclusion and provide opportunities to older adults to contribute to community life (see Case Illustration 13.1). The implementation of these older-adult-friendly policies vary widely across the globe by the socio-economic and cultural resources of each setting.

Sustainable health practices for older adults are trending toward the implementation of policies, programs, and strategies for healthy aging and full community inclusion. Community-oriented approaches have the strength to engage the older adults as partners in their wellbeing choices, including opportunities to advocate on their own behalf. Best practice older adult wellbeing programs would be those that seek to foster age-friendly households, workplaces, transportation systems, and access to the wider community environment.

Pertinent Sustainable Community Health Approaches

This chapter considers some multifactorial and interdisciplinary older adults' health and wellbeing programs with evidence or promise of sustainability. In doing so, we would like to underscore the caveat that the older adults themselves should be equal partners with their families, communities, and health professionals in the development and implementation of older-age-friendly health and wellbeing policies. Above all, the preferences and priorities of the older adults in their health and wellbeing needs, and the solutions for them, would be critical to helping older people to achieve optimal HLE—overcoming the impact health inequities, injustices, and discrimination have on their health, behaviors, and lives. We consider different types of sustainable health systems for older adults by living arrangements (aging-in-place, assistive living), physical activity, mobility, leisure and recreation enablers, transportation supports, and nutrition and diet choices.

Living Arrangements Living arrangements matter to the sense of wellbeing of older adults. Yet, many older adults may be restricted in the choices they have for their living arrangements by the contingencies of their family members and others involved with their care. Moreover, older adults may have higher personal living costs which may erode their employer-provided benefits much quicker than their HLE, thus limiting their residential options.

Aging-in-Place Aging in place refers to older adults choosing to live in their homes as they continue to age. It is aligned with self-management, self-determination, and prioritization of personal values for preferred lifestyle in older age with independent community living and participation. Independent living refers to self-managing in the community, or with support from a partner or family member, as opposed to living in a high level of care facility (Maddox & Gaus, 2019; Perkins & Berkman, 2012). While living independently in their own homes, older adults may, over time, require increased support with everyday activities of daily living

(ADLs) and instrumental activities of daily living (IADLs) such as self-care tasks, housekeeping, and financial management. This support is often provided by family or paid caregivers. Caring for older family members living within their home environment can, and frequently does, place significant demands and stress on unpaid family caregivers. See Case Illustration 13.2.

When independent living needs are dependent on a family member, that support may be lost if that family member died or became incapacitated. The resulting adverse health effects, social isolation, and lack of meaningful inclusion could prove detrimental to the community of adults aging with chronic illness and disability in the virus pandemic era. As such, with uncertainty in personal relationships increasing with age, the subjective demands and prominence older adults place on initiating and maintaining close relationships are amplified.

Case Illustration 13.2: Decreasing Demand for Care Through Increased Self-Management

John is a 64-year-old man who has chronic obstructive pulmonary disease (COPD). COPD is a progressive lung condition that makes it difficult to breathe. John was diagnosed with COPD five years ago, and his condition is deteriorating. Secondary to the chronic disease, John has progressively decreased his level of physical activity and is primarily housebound, which in turn has significantly reduced John's exercise tolerance. John is highly dependent on his two daughters and paid carers to provide assistance with some of his Activities of Daily Living (showering) and to complete all his Instrumental Activities of Daily Living (laundry, cleaning, shopping, and cooking). Although John has a basic understanding of COPD, he is highly anxious and fearful of exacerbating his condition. The above case illustration exemplifies the interplay between declines in physical functioning and family carer availability to the sense of wellbeing on an older adult with a progressive health condition.

What Do You Think?

1. What living arrangements would be sustainable for John's sense of wellness?
2. What community activity choices could John engage in for higher HLE, and how?

eHealth Technologies such as sensors, robots, and apps may also help sustain and even improve the health and quality of life of older adults preferring to age in place. Other assistive devices such as chair lifts, video doorbells, medication dispensers, and voice command tools may help older adults to age in place long-term, avoiding residential care, even when incapacitated.

Assisted Living As people age, they may require an increased amount of care to preserve their physical safety, health, wellbeing, and quality of life due to changes in their functional capabilities and cognitive capacity. They may have health conditions such as hypertension, arthritis, vision impairment, and heart disease (Boyle, Naganathan, & Cumming, 2010; Ming & Zecevic, 2018), which would complicate their self-management without the use of appropriate supports. Older adults may consequently be placed in aged care facilities secondary to acopia—usually due to financial hardship, having limited social supports, and being unable to afford or access adequate services.

Assisted living facilities are a type of aged care facility that aim to help older adults with basic activities of daily living, such as meal preparation, laundry, cleaning, shopping, or transport in a setting that resembles a personal home. These types of aged care facilities are best suited for older adults who are mostly independent but require occasional assistance with daily tasks, cannot live safely at home, and would benefit from an active social community. After a median stay of approximately two years, over half of older adults residing in Assisted Living facilities transition to a nursing home (National Center for Health Statistics, 2019).

Older adults transitioning from living in their homes to living in Assisted Living facilities may experience uncertainty and anxiety. Further to this, a study by Roberts and Adams (2017) showed that the quality of life of residents in Assisted Living facilities deteriorates over time in the face of declining health, mobility, and social losses. However, it is increasingly recognized that successful aging in Assisted Living facilities is associated with the development of sufficient social supports and being actively engaged in meaningful life activities (leisure, social, and instrumental activities of daily living) to promote the quality of life of older adults (Howie, Troutman-Jordan, & Newman, 2013). These findings support the literature which advocates for the development of client-centered activity programs to promote the participation of older adults in Assisted Living facilities in diverse activities to maintain their functional abilities and manage functional decline (Cummings & Cockerham, 2004; Horowitz & Vanner, 2010).

Although aged care facilities are the fastest growing residential care option for older adults, consideration should also be taken to incentivize and encourage families to provide care to older family members within the community. Innovative solutions may include modifying existing tax systems to allow for reduced taxation for carers and further subsidizing the cost of services where adult children provide material support to their elderly parents (Hope et al., 2012).

Physical Activity, Leisure and Recreation, and Transportation Despite evidence highlighting the importance of active living, healthy eating, and mindfulness, the majority of older adults aged 65 and over adopt sedentary lifestyles and spend the most time sitting. This places older people at an increased risk of physical deconditioning, falls, accelerated bone loss, Type 2 Diabetes Mellitus, and other adverse health conditions. Leisure and recreation programs and services are sustainable options to promote, maintain, and enhance the health status and wellbeing of older adults in a community.

Leisure and recreation activity reduce the social isolation of older adult's cohort (Chang, Wray, & Lin, 2014). They achieve a sense of wellness in older adults by enhancing their feelings of happiness, personal development, and community involvement (Lamanes & Deacon, 2019).

Travel is a vital contributing factor to older people's quality of life, their sense of freedom, and independence. Access to public and government-subsidized transport can help older adults avail themselves for employment, shopping, services, social and recreational activities, and community engagement. In order to prevent the loss of mobility in addition to reducing the isolating and physical degenerating impacts of driving cessation due to changes in older adults' health, physical, and cognitive status, transportation services need to be convenient, frequent, accessible, affordable, and comfortable with access to an extensive range of destinations (Browning & Sims, 2007; Harrison & Ragland, 2003). Meeting the transportation needs of an aging population is essential to achieving the goal of sustainable mobility.

Nutrition and Diet Older adults are at increased risk of malnutrition and experience a higher rate of food insecurity, which compromises their chances of aging healthily. Yet, nutrition security is a key factor to healthy living (Chernoff, 2001; see also Chap. 5, this volume). Programs that promote healthy eating in older adults should ideally have the ability to address the unique needs, in addition to the social and environmental circumstances of each older person (i.e., an individual's dietary requirements and restrictions, access to shopping, cooking resources, and feeding skills). An exemplar program that promotes healthy eating is the Australian-based program Meals On Wheels. It provides home delivery services of hot and cold meals that cater to a wide range of dietary requirements and restrictions. Meals on Wheels is also staffed by volunteers who can take the time to enjoy a meal with socially isolated individuals to promote social engagement (Australian Meals on Wheels Association Inc., n.d.).

Assistive Technology Supports

The use of smart devices with sensors by older adults living independently or with assistance in order to monitor their health status, and support a variety of activities of daily living (ADLs) and instrumental activities of daily living (IADLs) is increasing. This increase will enable their successful aging in place in the comfort of their homes and with a sense of community safety. During the aging process, there are declines in physical and cognitive function that are associated with declines in the performance of ADLs. Thus, the use of sensor surveillance technologies for tracking older adults' functional status is crucial for successful aging in place, either independently or with assistance.

Sensors can be wearable, allowing for ongoing monitoring of the older adult's physiological and accelerometric activity statuses (to measure balance, gait, falls). While non-wearable, environmentally embedded devices utilize interconnected devices with or without internet connectivity to support participation in life situations (although some environmental devices can monitor falls, movement, gait speed, as well as activity). Smart technologies can be worn or embedded in the physical environment to monitor health status, activity, and participation (i.e., an Apple watch), monitor falls, monitor the environment (i.e., a Nest thermostat), monitor movement (i.e., a Ring doorbell), and monitor interactions with others.

Cultural, Legislative, and Professional Issues

The association between healthy aging, equity, and social justice is relatively novel, given the age-old cultural prejudice toward older adult populations (Diehl & Wahl, 2020; Quadagno, 2018). Yet, age-friendly healthcare policies and practices would make a difference (Dodds, 2005). We briefly consider some influences on culture and social policies on the community wellbeing of older adult populations for addressing the health inequities older adults would experience.

Across cultures, older adults tended to be misperceived as less competent compared to younger adults, adding to their sense of marginalization and exclusion from their communities, and restricted access to social services (Diehl & Wahl, 2020). This is particularly true for those residing in low- to middle-income countries around the world. Within the vast majority of countries, inadequate action has been taken to develop targeted and well-funded policies, interventions, and programs that are designed to contend with the perceived and recognized inequities and injustices of older persons (Quadagno, 2018).

Policies for financing the costs of health and social care of older adults have evidence for the community wellbeing of older adults. Such policies include implementation of retirement income from an earlier age, consideration of raising the retirement age, promoting societal cultures that encourage early financial planning and investment, and allowing asset-rich but income-poor individuals to turn their housing assets into income to help pay for their own care needs (Hope et al., 2012). In Europe, for example, the pension reform in the early 1980s encouraged the inclusion of the private insurance players in the management of pension funds to increase earnings to pay for disability benefits and widows' benefits (Kohli & Arza, 2011). In the USA, under the Older Americans Act (OAA), adults become eligible for some services and protections at age 60 and generally become qualified at 65 for additional benefits from the Medicare program (United States, 1978).

Most important for sustainable health and wellbeing is the involvement of family caregivers and the older adults themselves in taking responsibility for the older adult's wellbeing. With the increasing cost of aged care services, families and their older adult members cover the gap in wellness support provided by public welfare services (Quadagno, 2018).

Related Disciplines Influencing Community Aspects

Older adult's wellbeing is an interdisciplinary science (Geriatrics Interdisciplinary Advisory Group, 2006; Partnership for Health in Aging, 2014; Young et al., 2011). An example of an effective interdisciplinary care service is the Programs of All-Inclusive Care for the Elderly (PACE) based in the United States. PACE consists of interdisciplinary geriatric care teams that develop individualized care plans, coordinate, and provide comprehensive healthcare and social services that meet the needs of participating older adults (Programs of All-Inclusive Care for the Elderly, n.d.). Older adult interdisciplinary teams are comprised, minimally, of medical clinicians, nurses, physiotherapists, occupational therapists, speech therapists, dieticians, clinical pharmacists, social workers, and recreational therapists. The PACE provides services in seniors day centers and is a capitated program that has reduced the use of institutional care and medical services, in addition to demonstrating enhanced customer satisfaction. Ongoing research and education are required to train healthcare providers on how to work collaboratively and implement best practices within interdisciplinary care teams in diverse settings, including the community, hospitals, clinics, and aged care facilities (Elliott, Stolee, Boscart, Giangregorio, & Heckman, 2018).

Healthcare and social professions have traditionally placed the most importance and value on providing interdisciplinary geriatric assessment, care, and management services. However, to achieve sustainable and healthy aging, multi-sectoral partnerships beyond health and social services must be formed. The identification, networking, and engagement of stakeholders in multiple [civil, private, government, non-government, international and humanitarian] sectors can result in leveraging resources, the sharing of key learnings and experiences, supporting policy development and dissemination, and inspire decisive action to transform and re-orient current healthcare systems and processes (WHO, 2020).

Research Critical to Sustainable Health of Older Adults

Despite available evidence on the benefits of community health services in promoting the health of the aging community (Jones & Wells, 2007; Mpofu, 2014; Pérez, Ro, & Treadwell, 2009), little is known about the mediating role of community-oriented services in reducing the influence of social determinants of health on community health outcomes (Campbell & Jovchelovitch, 2000). Moreover, healthcare practitioners have little knowledge about what services are available and beneficial to their older adults (Siegler, Lama, Knight, Laureano, & Reid, 2015).

To facilitate sustainable aging, healthcare systems and practitioners must progressively adapt to evolving population demographics, altered health risk profiles, shifting disease burdens, and increased rates of aging with disability. The summary effect of these factors has influenced and necessitated a reorientation in the way national healthcare systems provide health services. To illustrate, although chronic diseases have been the leading causes of ill health and death worldwide for many decades, the risk factors for these conditions have become increasingly more common. As such, the future older population may have a substantially significant burden of lifestyle-related diseases than in the past (Australian Institute of Health & Welfare, 2014). Within aging populations, the increase in prevalence of non-communicable diseases, multi-morbidities, and rates of chronic disease highlights the need and increased demand for services that are proactive, holistic, highly coordinated, and have a preventative approach to care; as opposed to reactive, episodic and disease-specific interventions. Aging adults may also experience improved quality of care when engaged with services that are founded on lasting patient–provider relationships rather than incidental, provider-led care.

Most modern healthcare systems must contend with shrinking funding streams, ever-changing political environments, and inefficient policies. This, in turn, results in the slowed reorientation of current healthcare systems to providing adequate preventative health and social care to older adults. Consequently, older adults remain high consumers of health services, and an increase in demand for health services is expected. With increased research and investment in sustainable community approaches for the older adults, people in aging populations would experience increased longevity with reduced disease severity secondary to medical advances, new technologies, and modifications in behavior that promote healthier lifestyles; people would also increase their expectations regarding the quality and quantity of care they should receive (Bernd et al., 2009).

Taking action to foster healthy and sustainable aging is everyone's business—with healthcare professionals having to take the primary onus to recognize, research, develop, and implement strategies to meet the needs of the aging population. Regardless of care setting, the health profiles of older adults are often complex, with multiple interacting comorbidities that require the expertise of healthcare professionals from multiple disciplines and extensive care coordination services. When healthcare providers, care coordination services, and social stakeholders (i.e., family members, the local community) fail to work cohesively, care can be unduly fragmented and fall short of addressing a person's needs (Mpofu, Machina, Wang, & Knight, 2019). Research is needed to provide the evidence on successful older adult, family, and community partnerships for the sustainable health of older adults.

To achieve sustainable health for older adult populations, adequate numbers of highly trained healthcare practitioners and effective health promotion programs are required to support the needs of the aging population worldwide. To illustrate, there were approximately 7000 geriatricians in practice during the year 2014 in the US. Hafner (2016) estimates that medical schools must train a minimum of 1500 additional geriatricians per year over the next 15 years to meet the targeted need for 30,000 geriatricians by the year 2030, thus projecting a workforce shortage that needs to be addressed (Hafner, 2016).

Summary and Conclusion

As global life expectancy continues to rise, facilitating sustainable and healthy aging with special emphasis on addressing the inequities, injustices, and discrimination suffered by older people is essential. Modern society is functioning in a bid to create equitable and sustainable healthcare systems in order to support healthy aging. However, these systems are often confronted with numerous complex and multidimensional challenges that require novel innovation, in addition to the restructuring and reorientation of the way national healthcare systems provide health services. Further to this, a shift in service culture that focuses on caring and improving the quality of life of elderly people rather than basic care provision is required to ensure that people age in [physical, social, economic, and political] environments that are right for them and enable them to continue to contribute to their communities by preserving their independence and retaining autonomy over their own health management. The establishment and viability of sustainable community health for the aging population is dependent on the successful coalition between individuals, communities, and the health agencies responsible for planning and coordinating services.

Self-Check Questions

1. Define the population of older adults and their diversities in community health assets and needs.
2. Briefly outline the historical evolution of older adult wellbeing as a practice.
3. Identify and discuss the significance of types of community wellbeing approaches with older adults. Outline possible influences of culture and social policy on older adults' wellbeing practices.
4. Identify and describe three priority areas for research on older adults' wellbeing practices. Give reasons for your selection.

Discussion Questions

1. Older women are more likely to live in poverty, have poor health, be socially isolated, experience financial insecurity, and have weak social connections. Suggest what can be done to support older adult women from experiencing an unfair degree level of inequality, injustice, and discrimination in health.
2. The number of older adults is increasing and with not enough of the workforce to work with the older adult population for community wellness. Suggest some innovative ideas on how the workforce shortage of geriatricians can be addressed?

Field-Based Experiential Exercises

1. Visit your local older adults center to learn about their community experiences important for their wellbeing and how.
2. Take time to engage an older family member on policies, programs, and initiatives they believe have improved the health and wellbeing of older adults in their community.

Online Learning Resources

1. Administration for Community Living (ACL) – The Administration for Community Living is a branch of the United States Federal Government dedicated to All Americans – including people with disabilities and older adults. The official website can be found at: https://acl.gov/about-acl
2. The National Institute on Aging (NIA) – The NIA is one of the 27 institutes and centers of the National Institutes of Health (NIH). The NIA leads a broad scientific effort to understand the nature of aging and to extend the healthy, active years of life. NIA is the primary federal agency supporting and conducting Alzheimer's disease research. The official website can be found at: https://www.nia.nih.gov/about
3. The National Center for Chronic Disease Prevention and Health Promotion was created to help people and communities prevent chronic diseases and promote health and wellness for all. Their programs include national, regional, and local programs for diabetes, heart disease and stroke, oral health, tobacco-free living, and others. You can find more information about these programs at: https://nccd.cdc.gov/nccdsuccessstories/searchstories.aspx

References

Australian Institute of Health and Welfare. (2014). *Australia's health 2014* [Ebook] (14th ed., pp. 256–271). Australian Institute of Health and Welfare. Retrieved on May 3, 2020, from https://www.aihw.gov.au.

Ayalon, L., Chasteen, A., Diehl, M., Levy, B., Neupert, S. D., Rothermund, K., … Wahl, H. W. (2020). Aging in times of the COVID-19 pandemic: Avoiding ageism and fostering intergenerational solidarity. *The Journals of Gerontology: Series B*. https://doi.org/10.1093/geronb/gbaa051

Bernd, R., Doyle, Y., Grundy, E., & McKee, M. (2009). How can health systems respond to population ageing? (Policy brief 10). *The regional office for Europe of the World Health Organization*, 4–24. Retrieved from http://www.euro.who.int/en/health-topics/Life-stages/healthy-ageing/publications/2009/how-can-health-systems-respond-to-population-ageing

Boyle, N., Naganathan, V., & Cumming, R. G. (2010). Medication and falls: Risk and optimization. *Clinics in Geriatric Medicine, 26*(4), 583–605. https://doi.org/10.1016/j.cger.2010.06.007

Brothers, A., Gabrian, M., Wahl, H. W., & Diehl, M. (2016). Future time perspective and awareness of age-related change: Examining their role in predicting psychological well-being. *Psychology and Aging, 31*(6), 605–617.

Browning, C., & Sims, J. (2007). Ageing without driving: Keeping older people connected. In *No way to go: Transport and social disadvantage in Australian communities* (pp. 1–10). Melbourne, VIC: Monash University.

Campbell, C., & Jovchelovitch, S. (2000). Health, community and development: Towards a social psychology of participation. *Journal of Community and Applied Social Psychology, 10*(4), 255–270. https://doi.org/10.1002/1099-1298(200007/08)10:4<255::AID-CASP582>3.0.CO;2-M.

Castro, C. M. (2020). *Active choices: Telephone-assisted physical activity self-management program for older adults*. Arlington, VA: National Council on Aging. Retrieved on May 11, from https://d2mkcg26uvg1cz.cloudfront.net/wp-content/uploads/F_Active-Choices-FINAL.pdf

Centers for Disease Control and Prevention. (2019). *Identifying vulnerable older adults and legal options for increasing their protection during all-hazards emergencies: A cross-sector guide for states and communities*. Atlanta, GA: U.S. Department of Health and Human Services.

Chang, P. J., Wray, L., & Lin, Y. (2014). Social relationships, leisure activity, and health in older adults. *Health Psychology, 33*(6), 516.

Cheadle, A., Egger, R., LoGerfo, J. P., Schwartz, S., & Harris, J. R. (2010). Promoting sustainable community change in support of older adult physical activity: Evaluation findings from the Southeast Seattle senior physical activity network (SESPAN). *Journal of Urban Health, 87*(1), 67–75.

Chernoff, R. (2001). Nutrition. In E. A. Swanson, T. Tripp-Reimer, & K. Buckwalter (Eds.), *Health promotion and disease prevention in the older adult* (pp. 43–55). New York, NY: Springer Publishing Company.

Crews, D. E., & Ice, G. H. (2012). *Aging, senescence, and human variation. Human evolution: An evolutionary and biocultural perspective.* New York, NY: John Wiley & Sons.

Cummings, S., & Cockerham, C. (2004). Depression and life satisfaction in assisted living residents. *Clinical Gerontologist, 27*(1–2), 25–42. https://doi.org/10.1300/j018v27n01_04

Czaja, S. J., Boot, W. R., Charness, N., & Rogers, W. A. (2019). *Designing for older adults: Principles and creative human factors approaches*. Boca Raton, FL: CRC Press.

Daly, J. M. (2011). Domestic and institutional elder abuse legislation. *Nursing Clinics of North America, 46*(4), 477.

Diehl, M., & Wahl, H. W. (2020). *The psychology of later life: A contextual perspective*. Washington, DC: American Psychological Association.

Dodds, S. (2005). Gender, ageing, and injustice: Social and political contexts of bioethics. *Journal of Medical Ethics, 31*(5), 295–298. https://doi.org/10.1136/jme.2003.006726

Elliott, J., Stolee, P., Boscart, V., Giangregorio, L., & Heckman, G. (2018). Coordinating care for older adults in primary care settings: Understanding the current context. *BMC Family Practice, 19*(1). https://doi.org/10.1186/s12875-018-0821-7

Faulkner, L. R. (1982). Mandating the reporting of suspected cases of elder abuse: An inappropriate, ineffective and ageist response to the abuse of older adults. *Family Law Quarterly, 16*, 69–91.

Geriatrics Interdisciplinary Advisory Group. (2006). Interdisciplinary care for older adults with complex needs: American Geriatrics Society position statement. *Journal of the American Geriatrics Society, 54*(5), 849–852. https://doi.org/10.1111/j.1532-5415.2006.00707.x

Harrison, A., & Ragland, D. (2003). Consequences of driving reduction or cessation for older adults. *Transportation Research Record: Journal of the Transportation Research Board, 1843*(1), 96–104. https://doi.org/10.3141/1843-12

Hafner, K. (2016). *As population ages, where are the geriatricians?* Retrieved on July 10, 2020 from https://www.nytimes.com/2016/01/26/health/where-are-the-geriatricians.html

Hope, P., Bamford, S., Beales, S., Brett, K., Kneale, D., Macdonnell, M., & McKeon, A. (2012). *Creating sustainable health and care systems in ageing societies.* Retrieved from http://www.cpahq.org/cpahq/cpadocs/Creating%20Sustainable%20Health%20and%20Care%20Systems%20in%20Ageing%20Societies.pdf

Horowitz, B., & Vanner, E. (2010). Relationships among active engagement in life activities and quality of life for assisted-living residents. *Journal of Housing for the Elderly, 24*(2), 130–150. https://doi.org/10.1080/02763891003757056

Howie, L., Troutman-Jordan, M., & Newman, A. (2013). Social support and successful aging in assisted living residents. *Educational Gerontology, 40*(1), 61–70. https://doi.org/10.1080/03601277.2013.768085

Jackson, K., Roberts, R., & McKay, R. (2019). Older people's mental health in rural areas: Converting policy into service development, service access and a sustainable workforce. *Australian Journal of Rural Health, 27*(4), 358–365.

Jones, L., & Wells, K. (2007). Strategies for academic and clinician engagement in community-participatory partnered research. *JAMA, 297*(4), 407. https://doi.org/10.1001/jama.297.4.407

Kohli, M., & Arza, C. (2011). The political economy of pension reform in Europe. *Handbook of Aging and the Social Sciences*, 251–264. https://doi.org/10.1016/b978-0-12-380880-6.00018-6

Kuruvilla, S., Sadana, R., Montesinos, E. V., Beard, J., Vasdeki, J. F., de Carvalho, I. A., … Koller, T. (2018). A life-course approach to health: Synergy with sustainable development goals. *Bulletin of the World Health Organization, 96*(1), 42.

Lamanes, T., & Deacon, L. (2019). Supporting social sustainability in resource-based communities through leisure and recreation. *The Canadian Geographer/Le Géographe canadien, 63*(1), 145–158.

Laslett, P. (1995). Necessary knowledge: Age and aging in the societies of the past. In P. Laslett & D. Kertzer (Eds.), *Ageing in the past: Demography, society, and old age* (1st ed., pp. 1–79). Berkeley, CA: University of California Press. Available at http://ark.cdlib.org/ark:/13030/ft096n99tf/

Longman, P. (2010, November). Think again global aging: A gray tsunami is sweeping the planet—And not just in the places you expect. How did the world get so old, so fast? *Foreign Policy*, (182), 52+. Retrieved from https://libproxy.library.unt.edu:7175/apps/doc/A241179403/OVIC?u=txshracd2679&sid=OVIC&xid=29cdbb6b

Lubitz, J., Cai, L., Kramarow, E., & Lentzner, H. (2003). Health, life expectancy, and healthcare spending among the elderly. *New England Journal of Medicine, 349*(11), 1048–1055.

Maddox, B. B., & Gaus, V. L. (2019). Community mental health services for autistic adults: Good news and bad news. *Autism in Adulthood, 1*(1), 15–19. https://doi.org/10.1089/aut.2018.0006

Martin, S., & Smith, R. W. (1993). OBRA legislation and recreational activities: Enhancing personal control in nursing homes. *Activities, Adaptation, & Aging, 17*(3), 1–14.

Ming, Y., & Zecevic, A. (2018). Medications & polypharmacy influence on recurrent fallers in community: A systematic review. In *Canadian Geriatrics Journal, 21*(1), 14–25. Canadian Geriatrics Society. https://doi.org/10.5770/cgj.21.268.

Montgomery, L., Anand, J., Mackay, K., Taylor, B., Pearson, K. C., & Harper, C. M. (2016). Implications of divergences in adult protection legislation. *The Journal of Adult Protection, 18*(3), 149–160.

Mpofu, E. (2014) (Ed). *Community-oriented health services* (1st ed.). New York, NY: Springer Publishing Company.

Mpofu, E., Machina, M., Wang, B., & Knight, R. (2019). Allied and clinical mental health systems-of-care and strength-based approaches. In L. Levers & D. Hyatt-Burkhart (Eds.), *Clinical mental health counseling: Practicing in integrated systems of care* (1st ed., pp. 43–68). New York, NY: Springer Publishing Company.

National Centre for Health Statistics. (2019). *Long-term care providers and services users in the United States, 2015–2016* (pp. 1–38). Washington, DC: U.S. Government Publishing Office.

National Institute on Aging. (2011). *Global health and aging*. Retrieved from https://www.nia.nih.gov/sites/default/files/2017-06/global_health_aging.pdf

National Institute on Aging. (2019). *Global aging*. Retrieved on July 20, 2020 from https://www.nia.nih.gov/research/dbsr/global-aging

National Research Council. (2001). *Preparing for an aging world: The case for cross-national research*. Washington, DC: The National Academies Press. https://doi.org/10.17226/10120

Partnership for Health in Aging. (2014). Position statement on interdisciplinary team training in geriatrics: An essential component of quality health care for older adults. *Journal of the American Geriatrics Society, 62*(5), 961–965. https://doi.org/10.1111/jgs.12822

Pérez, L., Ro, M., & Treadwell, H. (2009). Vulnerable populations, prison, and federal and state medicaid policies: Avoiding the loss of a right to care. *Journal of Correctional Health Care, 15*(2), 142–149. https://doi.org/10.1177/1078345808330040

Perkins, E. A., & Berkman, K. A. (2012). Into the unknown: Aging with autism spectrum disorders. *American Journal on Intellectual and Developmental Disabilities, 117*(6), 478–496. https://doi.org/10.1352/1944-7558-117.6.478

Quadagno J . (2015). The transformation of Medicaid from poor law legacy to middle class entitlement? In A. B. Cohen D. C. Colby K. A. Wailoo , & J. E. Zelizer (Eds.), *Medicare and medicaid at 50: America's entitlement programs in the age of affordable care* (pp. 77–94). New York: Oxford University Press.

Quadagno, J. (2018). *Aging and the life course: An introduction to social gerontology*. McGraw-Hill Higher Education.

Roberts, A., & Adams, K. (2017). Quality of life trajectories of older adults living in senior housing. *Research on Aging, 40*(6), 511–534. https://doi.org/10.1177/0164027517713313

Rowe, J. W., & Kahn, R. L. (2015). Successful aging 2.0: Conceptual expansions for the 21st century. *Journals of Gerontology. Series B, Psychological Sciences and Social Sciences*. https://doi.org/10.1093/geronb/gbv025

Rowland, D. T. (2009). Global population aging: History and prospects. In P. Uhlenberg (Ed.), *International handbook of population aging*. International handbooks of population (Vol. 1). Dordrecht, Netherlands: Springer.

Shelton, R. C., Lee, M., Brotzman, L. E., Crookes, D. M., Jandorf, L., Erwin, D., & Gage-Bouchard, E. A. (2019). Use of social network analysis in the development, dissemination, implementation, and sustainability of health behavior interventions for adults: A systematic review. *Social Science & Medicine, 220*, 81–101.

Siegler, E. L., Lama, S. D., Knight, M. G., Laureano, E., & Reid, M. C. (2015). Community-based supports and services for older adults: A primer for clinicians. *Journal of Geriatrics, 2015*, 1–6. https://doi.org/10.1155/2015/678625

Stiefel, M. C., Perla, R. J., & Zell, B. L. (2010). A healthy bottom line: Healthy life expectancy as an outcome measure for health improvement efforts. *The Milbank Quarterly, 88*(1), 30–53.

United Nations, Department of Economic and Social Affairs, Population Division (2019). *World Population Prospects 2019: Highlights*. New York, NY: Author.

U.S. Department of Health and Human Services. (2017). *HHS action plan to reduce racial and ethnic health disparities: A nation free of disparities in health and health care* (pp. 197–230). National Strategy for Achieving Health Equity and a Plan to Reduce Racial and Ethnic Health Disparities. https://doi.org/10.1037/e553842012-001

United States. (1978). *Older Americans act of 1965, as amended.* Washington, DC: Administration on Aging, Office of Human Development Services, U.S. Department of Health, Education, and Welfare.

Westerhof, G. J., Miche, M., Brothers, A. F., Barrett, A. E., Diehl, M., Montepare, J. M., … Wurm, S. (2014). The influence of subjective aging on health and longevity: A meta-analysis of longitudinal data. *Psychology and Aging, 29*(4), 793.

World Health Organisation. (2020). *Decade of healthy aging 2020–2030.* Geneva, Switzerland: The World Health Organisation. Retrieved on May 11, 2020, from www.who.int/ageing/decade-of-healthy-ageing

World Health Organization. (2010). *Definition of an older or elderly person.* Retrieved on May 11, 2020, from http://libproxy.library.unt.edu:2506/healthinfo/survey/ageingdefnolder/en/index.html

World Health Organization. (2015). *World report on ageing and health.* Geneva: The World Health Organization. Retrieved on May 11, 2020 from www.who.int/ageing/events/world-report-2015-launch/en/

World Health Organization (2019). *Global strategy and action plan on ageing and health.* Retrieved on July 20, 2020 from from: https://www.who.int/ageing/WHO-GSAP-2019.pdf?ua=1

Wurm, S., Diehl, M., Kornadt, A. E., Westerhof, G. J., & Wahl, H. W. (2017). How do views on aging affect health outcomes in adulthood and late life? Explanations for an established connection. *Developmental Review, 46*, 27–43.

Xu, W., & Larbi, A. (2017). Markers ofT cell senescence in humans. *International Journal of Molecular Sciences, 18*(8), 1742.

Young, H., Siegel, E., McCormick, W., Fulmer, T., Harootyan, L., & Dorr, D. (2011). Interdisciplinary collaboration in geriatrics: Advancing health for older adults. *Nursing Outlook, 59*(4), 243–250. https://doi.org/10.1016/j.outlook.2011.05.006

14

Intellectual and Developmental Disabilities Wellbeing

Elias Mpofu, Elizabeth Houck, April Linden, and Crystal Fernandez

Introduction

The World Health Organization (WHO, 2020) reports that 15% of the world's population lives with some type of disability. This is a total of over one billion people worldwide. The global prevalence of intellectual and developmental disabilities (IDD) ranges from 6.2 to 10.4 per 1000 people and with a higher prevalence in low- and middle-income countries (LMIC) (Tomlinson et al., 2014; WHO, 2007; WHO and World Bank,

E. Mpofu
University of North Texas, Denton, TX, USA

University of Sydney, Sydney, NSW, Australia

University of Johannesburg, Johannesburg, South Africa
e-mail: Elias.Mpofu@unt.edu

E. Houck (✉) • A. Linden • C. Fernandez
University of North Texas, Denton, TX, USA
e-mail: bethjoyhouck@my.unt.edu; CrystalFernandez@my.unt.edu

E. Mpofu (ed.), *Sustainable Community Health*,
https://doi.org/10.1007/978-3-030-59687-3_14

2011), with significant variations among countries (Boyle et al., 2011; Maulik, Mascarenhas, Mathers, Dua, & Saxena, 2011; McConkey, Mulvany, & Barron, 2006; McKenzie, Milton, Smith, & Ouellette-Kuntz, 2016). The prevalence of IDD in developing countries is less certain due to lack of accurate reporting and documentation (Friedman, Gibson Parrish, & Fox, 2018; Mpofu, 2016; Tomlinson et al. 2014; WHO, 2007). It is likely considerably higher than in developed countries from severe population-level health disparities such as poverty, malnutrition, and insufficient maternal and child health services (Graham, 2005; Maulik et al., 2011; WHO, 2008).

Learning Objectives

On studying this chapter, the reader should be able to:

1. Define intellectual and developmental disabilities (IDD) and the community wellness needs of people with IDD.
2. Outline historical, ethical, and professional issues in the development of IDD inclusive community health systems.
3. Discuss contemporary and emerging health and wellbeing practices for the sustainable health of people with IDD.
4. Outline cultural, professional, and legislative/policy issues that influence sustainable community health practices for people with IDD.
5. Identify gaps in research and practices in community health for individuals with IDD to be addressed for sustainability.

Intellectual disability (ID) comprises the 10.4 per 1000 global prevalence and falls under the umbrella term of intellectual and developmental disability (IDD) (Robinson, Dauenhauer, Bishop, & Baxter, 2012; Schalock & Luckasson, 2014), and pervasive developmental disabilities, inclusive of autism, attention deficit disorder, and Asperger's syndrome, comprise the 6.2 per 1000 (Tomlinson et al. 2014). While some countries use the broad term developmental disability (DD), the term ID is more globally used to refer to persons with lifelong cognitive impairments (Robinson et al., 2012). Developmental disabilities are also lifelong impairments, occur before the age of 22, and can include cognitive impairments, physical impairments, or both (Friedman et al., 2018). The within country variations in IDD prevalence are in part from countries using different data sources inclusive of population data and administrative data, with or without other case authentication approaches. Population data would include those from birth registries and national census, although most countries do not include consensus questions with the level of specificity needed to identify IDD. Administrative data sets, which are among the more commonly used for prevalence estimates, are from case enrolment health care, education, social services, income support services, and for determining eligibility for publicly funded services and/or tracking use of such services. People with IDD who do not access publicly funded programs for any reason will be missing from administrative data sets. Reliable IDD prevalence data are critical to programs for meeting the health needs of this vulnerable population in a sustainable way, and yet, even in developed countries, these data sources often are incomplete (Friedman et al., 2018; Tomlinson et al., 2014).

Overall, people with disabilities have lower health-related quality of life than the general population (Allerton, Welch, & Emerson, 2011; Emerson & Hatton, 2014; Krahn, Walker, & Correa-De-Araujo, 2015; WHO, 2020; WHO & World Bank, 2011). For instance, people with IDD are less likely to have access to appropriate preventative care and primary care systems (Parish, Swaine, Son, & Luken, 2013; Parish et al., 2015; Robinson et al., 2012; WHO, 2020). With few exceptions (see Strouse, Sherman, & Sheldon, 2013), community health systems tend to lack inclusivity of people with IDD and lack sustainability qualities (Roll-Petterson, Olsson, & Rosales, 2017). Sustainable community health practices aim to maintain and improve health and wellbeing of community members (Bloch et al., 2014; Šiška, Beadle-Brown, Káňová, & Šumníková, 2018), with an emphasis on social justice issues and health equity (Bloch et al., 2014; Centers for Disease Control and Prevention [CDC], 2011). Sustainable community health systems are critical to the health and wellbeing of people with IDD, their families, and the larger community.

Historically, people with IDD have been an unrecognized population in terms of their wellness needs, creating health disparities that put them at a disadvantage (Krahn & Fox, 2014; Krahn et al., 2015; Robertson, Hatton, Baines, & Emerson, 2015; WHO, 2011). There is a wide range of health support needs for people with IDD (Ward, Nichols, & Freedman, 2010). Moreover, given the diverse needs of people with IDD, sustainable health systems serving this population should be cost-effective and have a long-term impact (Neely-Barnes, Marcenko, & Weber,

2008). Lack of sustainable health systems for people with IDD result from failure to build on present success and identify lessons for long-term success, which can perpetuate health inequities and social injustices.

Sustainable health systems are designed to optimize the control of individuals with IDD for choices such as where and with whom the person lives; what the person does during the day; their quality of relationships with others; what and with whom the person does things of personal interest; and if, where and with whom they meet their spiritual needs (Healthy People, 2020). With health systems sustainability, the community of people with IDD is empowered to pursue their health promoting interests and opportunities for personal growth, including work-wellness and community citizenship. Few of the currently available public health systems would provide these qualities for people with IDD (McCausland, McCallion, Cleary, & McCarron, 2016). This makes it critical to explore the structure and qualities of community health systems that would enhance the overall health and wellness of people with IDD, beyond the services provided by community medicine.

As people with IDD get older, and specifically as they age out of public-school settings, maintaining social networks becomes increasingly difficult. Social health for people with IDD becomes more difficult to maintain with increasing age (McCausland et al., 2016; Simplican, Leader, Kosciulek, & Leahy, 2015; Young-Southward et al., 2017). For instance, one study found that adults with IDD living in communities had a social network on the average of two people, excluding staff (McCausland et al., 2016). While people with IDD living independently in the community are reported to have more frequent contact with family members than those in institutional settings, this is often only monthly visits (Kilroy, Egan, Walsh, McManus, & Sarma, 2015). For individuals with severe-to-profound IDD, staff reported 40.3% had not had contact with a friend in over a year and 49.8% visited with a family member less than once per month (McCausland et al., 2016). Sustainable health systems would improve the social connectedness—the social health and wellbeing—of people IDD.

Despite the many barriers they face, people with IDD are now more likely than ever before to live in inclusive communities, access educational opportunities, access work opportunities in mainstream settings

(Mirenda, 2014), and access healthcare services in the communities in which they live (Spassiani, Parker Harris, & Hammel, 2016). These affordances are significant to the social and economic wellbeing of the community of people with IDD. While significant progress has been made in inclusive living with IDD, many communities lack in health inclusivity for people with IDD and health disparities remain for this population (Baumbusch, Moody, Hole, Jokinen, & Stainton, 2019; Spassiani et al., 2016). However, increasingly, state and federal agencies are implementing initiatives to reduce these disparities, promoting equity for all their citizens (Spassiani et al., 2016; WHO, 2020).

We discuss some community-level solutions for the sustainable health of people with IDD in a later section.

Professional and Legal Definitions of IDD

The American Association on Intellectual and Developmental Disabilities (AAIDD, 2019) defines intellectual disabilities as "a disability characterized by significant limitations in both intellectual functioning and in adaptive behavior, which covers many everyday social and practical skills. This disability originates before the age of 18" (para 1). Similarly, the World Health Organization (WHO, 2020) defines ID as "a significantly reduced ability to understand new or complex information and to learn and apply new skills (impaired intelligence). This results in a reduced ability to cope independently (impaired social functioning), and begins before adulthood, with a lasting effect on development" (para 1). The American Psychiatric Association (APA) defines IDD as "a disorder with onset during the developmental period that includes both intellectual and adaptive functioning deficits in conceptual, social, and practical domains" (American Psychiatric Association, 2013, p. 33). Although each organization's definition of ID may slightly differ, there is a consensus that ID affects:

- Intellectual functioning
- Adaptive functioning
- And symptoms must arise before adulthood

Intellectual Functioning ID can affect intellectual functioning in a variety of ways, and by convention, the levels of mild, moderate, severe, and profound are commonly used to describe levels of ID (DSM-5, APA, 2013), and based on the results from ability measures such as IQ tests (these types of tests have social justice risks, depending on how they are used: Braddock, 1998; Tassé, 2016; see also Discussion Box 14.1). Mild ID is defined by scoring 55–70 on an IQ test (Boat & Wu, 2015). The person with mild ID demonstrates adequate functioning in most life domains, requiring only occasional support from others to live independently (Friedman et al., 2018). With moderate ID, the person is described as having an IQ score between 40 and 50 (Boat & Wu, 2015) and would require significant assistance managing personal health, communication, and social participation (McCreary, 2005). Those with severe ID are classified as having an IQ score between 25 and 35 (Boat & Wu, 2015) and needing ongoing support by caregivers in most areas of personal functioning. Those with profound ID, classified by an IQ score under 25 (Boat & Wu, 2015), require constant support and supervision to main-

Discussion Box 14.1: The Social Construction of Intellectual Ability Measures

Our current technologies of "IQ" testing, which are the basis for identifying an intellectual disability, were developed originally for the purposes of identifying people with who were believed to be "deviant social menaces...[with] an incurable disease" (Braddock, 1998, pg. 5). The identification of ID was the justification for "rampant abuse" especially among poor minority Americans and unjust deportation of immigrants deemed "mentally deficient" (Braddock, 1998, pg. 5). These ascriptions risk human rights violations. People with ID/DD are often denied many or all of their rights, frequently based only on deference to the convenience of others or in the name of "rehabilitating" people with ID (Bannerman, Sheldon, Sherman, & Harchik, 1990).

What Do You Think?

1. What human rights would be impaired by applying disvaluing categories on the abilities of people with ID?
2. How would the implementation of health policies based on ability limitations influence inclusive health and wellbeing practices with people with ID?

tain safety (McCreary, 2005). Alternatives to the use of these terms include using categories to delineate subgroups based on the level of support needed. For example, categories including "Supports needed", "Substantial supports needed", and "Very substantial supports needed" (Schalock & Luckasson, 2014).

Adaptive Functioning Impairment in adaptive functioning is a common symptom of all IDD-related disorders (DSM, APA, 2013). Adaptive functioning includes communication, social skills, personal independence, and school/work functioning. Adaptive functioning assessments are not used as often as intellectual functioning assessments in school placement, developmental research, and other disability statistics (Obi et al., 2011). Although this is the case, adaptive functioning assessments such as the assessment of Functional Living Skills developed by Partington and Mueller (2012) or the Essential for Living assessment developed by McGreevy, Fry, and Cornwall (2012) may provide more information about individual and community-based interventions and their effectiveness with specific individuals than intellectual functioning measures.

Comorbid Conditions People with IDD may present with comorbidities, which "refers to the presence of at least two distinct and separate disabilities or pathologies in the same individual" (White, Chant, Edwards, Townsend, & Waghorn, 2005, p. 396) and affects between 14.3 and 67.3% of individuals diagnosed with IDD. Common comorbidities linked with IDD include mood disorder, obsessive-compulsive disorder, depression, obesity (Melville, Hamilton, Hankey, Miller, & Boyle, 2007), heart conditions (Patja, Mlsk, & Iivanainen, 2001), and seizures (Oeseburg, Dijkstra, Groothoff, Reijneveld, & Jansen, 2011). These comorbidities would need to be considered when implementing sustainable health systems for people with IDD.

IDD and Aging People with IDD have historically had much shorter life expectancies than other populations (Coppus, 2013; McCarron, McCallion, Fahey-McCarthy, & Connaire, 2010; Ryan, Guerin, Dodd, & McEvoy, 2011). In the 1920s, the average life expectancy for people with IDD was 19 years old (Coppus, 2013). Currently, the average age of death for people with IDD is 66 years old (Coppus, 2013). As people with IDD age, they encounter the same health issues as other populations but are also at an increased risk for early-onset dementia, cardiovascular disease, hearing and vision loss, mobility decline, and gastrointestinal conditions (Robinson et al., 2012).

IDD as a Social Construct IDD is a social construct, which has implications for the design of inclusive community-centric health systems. It is a socially constructed classification of human ability attributes based on "the expressions of limitations in individual functioning that represents a substantial disadvantage to the individual with a social context" (Schalock & Luckasson, 2013, p. 87). Moreover, the narratives about IDD will continue to change and evolve as communities become more inclusive of the diversity in human ability attributes beyond those premised on neurological indicators only (Angrosino et al., 1998; Emerson, 2012).

History of Research and Practice in Community Living for People with IDD

Historically, research and practice with people with IDD has been deficit rather than asset oriented. For instance, beginning in the late nineteenth century and up to until the 1960s, an intellectual disability was conceptualized as a condition that was permanent, for which there was no cure, no treatment, and no hope. Such conceptualizations supported acts of institutionalization (Ervin, Hennen, Merrick, & Morad, 2014; Keith & Keith, 2013) and, in many cases, involuntary sterilization (Smith & Polloway, 1993). Although some people with IDD continued living with family, they had no access to educational opportunities outside of the home (Shorter, 2000). Thus, families of children with IDD were left with the choice to live in their homes with their child in isolation, or to drop their child off at an institution for life, on the premise that people with IDD could not learn any functional skills (Keith & Keith, 2013). Additionally, medical personnel lacked training in providing services to people with IDD making it difficult or impossible for many families to find healthcare providers who could meet the needs of their loved ones (Friedman, 2019). This was set to change with the deinstitutionalization movement beginning in the 1960s following recognition of the dehumanizing, dangerous, and often rampantly abusive conditions in institutions for people with IDD.[1]

Deinstitutionalization Movement Deinstitutionalization has been one of the largest scale initiatives worldwide seeking to decrease health disparities for people with IDD (Kilroy et al., 2015; Larson, Lakin, & Hill, 2012; McCausland et al., 2016; Spassiani et al., 2016). The goals of this movement were to increase freedom and choice for people with IDD in where to live, activities in which to engage, and service providers by re-integrating people with IDD back into existing communities (Larson et al., 2012; Saloviita, 2000; Spassiani et al., 2016). Since 1970, this movement has decreased the number of individuals living in institutional settings in the United States by 70%. Following deinstitutionalization, many studies have documented improvements in quality of life indicators for people with IDD including activity access, mobility, choice, and relationships, (Kilroy et al., 2015; Larson et al., 2012; McCausland et al., 2016; Saloviita, 2000).

Despite progress in many areas, there are still disparities for people with IDD across life domains, and many interventions that have been implemented have lacked sustainability (Kilroy et al., 2015; Larson et al., 2012; McCausland et al., 2016). In some cases, the lack of sustainability has led to re-institutionalization (Baumbusch et al., 2019). Re-institutionalization, which is also known as long-term care, suggests that people with IDD are not valued members of society and that they do not have a real choice in most areas of their lives (Spassiani et al., 2016).

The deinstitutionalization movement led to several new policies and legislative action, including Public Law 88-156 that provided funding for

[1] A photographic essay, *Christmas in Purgatory* (Blatt & Kaplan, 1966), and a documentary of *Willowbrook state institution* by Geraldo Rivera in 1972 are the most well-known works that captured the horrific conditions of state institutions and brought them to public attention (Rothman, 2017). In September 1962, Eunice Kennedy Shriver addressed the nation in the *Saturday Evening Post* to tell the story of her sister, Rosemary, who had IDD. Eunice ends the letter emphasizing that people with IDD have a chance for a useful life—that people with IDD could benefit from educational and treatment programs and efforts in science could advance such work.

children with IDD, and Public Law 90-538 that provided funding for the first early intervention programs in 1968. The Education for All Handicapped Children Act of 1975, provided for the "free and appropriate public education for all children" (Zettel, 1977). Public Law 94-1432: The Education for All Handicapped Children Act of 1975 (EAHCA, 1975; Zettel, 1977), established federal funding for the support of special education and related services, and Rehabilitation Act of 1973 increased provisions for the healthcare, educational, and vocational needs of people with disabilities. As an example of a historic vocational rehabilitation initiative, the development of supported employment starting in the 1980s was one of the factors that led to increased health-related quality of life for people with IDD (Certo & Luecking, 2011; Fesko, Hall, Quinlan, & Jockell, 2012; Nord & Hepperlen, 2016. Supported employment services offer people with IDD assistance through training programs, job search assistance, and training in the workplace. This is in contrast to sheltered employment services that are in segregated settings specifically designed for people with IDD in which only people with IDD work (Nord & Hepperlen, 2016). Recently, in the United States, many states have reduced funding for sheltered workshops following lawsuits citing that these settings unnecessarily facilitate continued segregation of people with IDD (Sulewski et al., 2017). Nevertheless, these workshops have remained a serviceable model for people with more severe levels of IDD, providing for safe and supervised daytime activity (Sulewski et al., 2017; Young-Southward, Cooper, & Philo, 2017).

Several policies were enacted in the United States to address the longer lifespans and unmet needs of individuals with IDD in community settings. The Home and Community Based (HCBS) Medicaid services provide individuals with IDD the option of receiving necessary services in integrated community settings, as opposed to care through separate institutions (Braddock et al., 2013). The Americans with Disabilities Act (ADA: 1990) requires environmental supports to promote full community inclusion of people with disabilities. Such legal instruments and policies support the movement toward community living for people with IDD, and thus influence community-based healthcare initiatives.

Sustainable Community Health Approaches with IDD

The current leading sustainable community health approaches focused on improving the health and wellbeing of people with IDD and decreasing health disparities are informed by social-ecological models (Bas, 2019; CDC, 2011; Szporluk, 2015). Social–ecological approaches recognize the importance of the physical environment and social networks in maintaining healthy community living (Beange & Durvasula, 2001). They prioritize the environment and social networks for the community wellness of people with IDD (Cooper et al., 2011), and promote wellbeing at all levels including individual health, family/organizational health, community health, and population health. Specifically, they focus on the health and wellbeing benefits of housing, employment and work, neighborhood citizenship, physical activity, recreation and leisure, healthcare consultation, and long-term care planning.

Housing Support Services Housing is essential to the community wellbeing of populations (Oswald & Wahl, 2004; see Chap. 3, this volume). Historically, individuals with disabilities were not expected to outlive their parents, so many remained in the family home until death. With increasing population longevity, people with IDD and their family members contend with long-term housing needs of adult family members with IDD (Levine, Halper, Peist, & Gould, 2010; Taggart, Truesdale-Kennedy, Ryan, & McConkey, 2012). Housing arrangements often include "home" style apartment setups (Saloviita, 2000), group home settings or family homes (Perry & Felce, 2003), and independent home living with or without carer assistance. Historically, housing needs of people with IDD would be managed by a service agency informing caregivers of services that the individual with IDD needs and caregivers making guardianship decisions for the individual (Bowey, McGlaughlin, with Claire Saul, 2005). Increasingly, people with IDD are involved in their own housing decisions and participate in local, state, and federal housing support provisions (Harrison, 2004). As an example, the Community for Permanent Supported Housing (CPSH) of North Texas, a 501(c) (3) non-profit organization, offers a "Road to Hope" program to help individuals with IDD and their family caregivers navigate housing assistance. They provide caregivers step-by-step resources to assist their adult children in planning for personalized home living. The CPSH offers workshops and other information about locating suitable housing, assessing the need for in-home, and community-based services, identifying service providers, and equipping the home with safety devices. The CPHS also provides training to people with IDD and their family caregivers on budgeting and funding sources such as the Department of Housing and Urban Development (HUD) Housing. CPHS is inclusive in its home living supports for people with IDD, providing training to families of people with IDD in housing options and promoting input by individuals with IDD.

Employment and Work Work participation has health promotion effects for those with disability similar to those in the general population (Beyer, Brown, Akandi, & Rapley, 2010; Van Campen & Iedema, 2007; Vickerstaff & Phillipson, 2011). Community employment is associated with high social integration, self-esteem, and overall quality of life (Fesko et al., 2012; Young-Southward et al., 2017). Barriers to employment for people with IDD include fear of losing disability-related social assistance, complexity of employment support services (Certo & Luecking, 2011), and continued prejudice from community members (Certo & Luecking, 2011; Nord & Hepperlen, 2016).

There is great variability in the employment status of persons with IDD. Overall, 80% of people with IDD utilize segregated daytime programs that were designated specifically to serve people with IDD (Mirenda, 2014). Additionally, data from the United States, Canada, and Ireland suggest that over 70% of people with IDD are unemployed (Mirenda, 2014). People with mild IDD were most likely to be employed, with about 65% employed in full or part-time positions (Verdonschot, De Witte, Reichrath, Buntinx, & Curfs, 2009). People with IDD with competitive employment had higher health-related quality of life compared to peers sheltered workshop or other segregated work support settings (Mirenda, 2014; Nord & Hepperlen, 2016).

Health effects of work participation for people with IDD include better access to social support, higher engagement in physical activity, and better mental health outcomes (Young-Southward et al., 2017). The more successful communities with the employment of people with IDD have employment service providers who have strong relationships with the local business community, enabling targeted job propositioning to prospective employers (Migliore, Butterworth, Nord, Cox, & Gelb, 2012; Nord & Hepperlen, 2016), while family networking remains a major bridge to the employment opportunities of individuals with IDD (Hall, Bose, Winsor, & Migliore, 2014).

Starfire is an organization based out of Cincinnati, Ohio, providing supported employment services for people with disabilities with a community citizenship focus (Sulewski et al., 2017; About 92% of their participants achieved increases in their social network from the employment connections developed through their program (Sulewski et al., 2017).

Neighborhood Citizenship Development of sustainable relationships and true community is extremely important aspects of sustained health for people with IDD. Neighborhood network approaches (Neely-Barnes et al., 2008; Strouse et al., 2013) have evidence for developing sustainable community living models to support the full range of functioning levels of people with IDD. Neighborhood network approaches seek to promote health equity and social justice for people with IDD by enhancing their access to (1) pleasant and safe surroundings, (2) resources to safeguard legal and personal rights, (3) building and maintaining positive relationships with others, (4) engaging in living healthy lifestyles, (5) opportunities for choice and control, (6) effective learning opportunities, and (7) high level of participation in community life (Strouse et al., 2013).

Physical Activity, Leisure, and Recreation

Access to and engagement in physical activity, leisure, and recreation opportunities influences the sustainable health of populations (Bartlo & Klein, 2011; Válková, 2015; see also Chap. 16, this volume). Several studies suggest that individuals with IDD are less likely to be physically active as compared to other community members (Cartwright, Reid, Hammersley, & Walley, 2017; Einarsson et al., 2015; Hilgenkamp, Reis, van Wijck, & Evenhuis, 2012; Lin et al., 2010), from a lack of inclusive community programs (Amado, Stancliffe, McCarron, & McCallion, 2013; Patterson & Pegg, 2009; Rimmer & Yamaki, 2006; Stebbins, 1992)

For instance, community programs may not include physical activity options for people with IDD (Cartwright et al., 2017), and when they do, the activities show poor maintenance across time (Patterson & Pegg, 2009). Exceptions are the physical activity programs that incorporated caregivers in the development of new routines for people with IDD (Hassan, Landorf, Shields, & Munteanu, 2019), suggesting that these social aspects of physical activity engagement are extremely important for this population. Proven physical activity programs for the sustainable health of people with IDD provide family health forums, free health screenings for athletes, fitness training, training of healthcare providers, and people with disabilities having an active voice in the decisions and solutions. Such programs have evidence to improve perceived health, reductions in body weight, increases in self-confidence, and decreased barriers to exercise (Marks, Sisirak, Heller, & Wagner, 2010).

Long-Term Care Planning Many families with adults with IDD must contend with planning for the future of the member with IDD when the primary caregiver is no longer able to provide ongoing care from aging-related limitations, change in financial capability, or passing away (Taggart et al., 2012). Regrettably, few families plan for the care needs of their adult members with IDD (Burke, Arnold, & Owen, 2018; Heller & Caldwell, 2006), which leaves the adult member with IDD in a very precarious if not life-threatening situation (McCarron et al., 2010). Moreover, community-wide life events such as natural disasters and pandemics do occur and would impact the wellbeing of people with IDD disproportionally more than that of typical community members (see Case Study 14.1a and 14.1b).

Long-term care planning would enable people with IDD to access needed resources as early as possible (McKenzie, Mirfin-Veitch, Conder, & Brandford, 2017), as access to some needed resources can require networking and helping providers identify trainings to help providers

Case Study 14.1a Natural Disasters and Disability

"Decision-making regarding who, when, and why disaster survivors with disabilities end up in segregated shelters requires those making decisions to interpret complex and contradictory sets of guidance from multiple federal agencies (DOJ, DHS, HHS). It is difficult to imagine how lawmakers, local government emergency managers, and shelter operators could arrive at decisions about who goes to segregated shelters in anything but an arbitrary way" (National Council on Disability [NCD], 2019, pg. 19).

Following natural disasters, there is evidence that people with IDD who had previously been living in community settings are at increased risk of being transitioned to institutional settings. Transitioning to institutional settings significantly decreases the opportunities for an individual to expand social networks and access community resources. This suggests both that people with IDD have less access to appropriate supports following disasters which leads to either decline in functioning or degradation of supports in the community to such an extent that institutionalization of this population increases overall. The National Council on Disability identified barriers such as having less access to disaster and emergency-related programs and a lack of training for recipients of federal funds on how to interact with people with disabilities appropriately. This leads to significant degradation of safety and dignity for people with disabilities following disasters (NCD, 2019).

What Do You Think?

1. What would be the qualities of sustainable community natural disaster response programs to meet the needs of people with IDD?
2. What are the costs and benefits of institution living for managing natural disasters with IDD?

successfully extend their services to people with IDD (McCallion et al., 2017; McCarron et al., 2010). Second, aging appears to affect people with IDD earlier than typical populations (McKenzie et al., 2017; Ryan et al., 2011), and early long-term care planning by people with IDD, family carers, and guardians would be critical. These plans would include planning the end of their lives (McKenzie et al., 2017) and starting sooner allows time for conversations that may be difficult, and avoids reactive decisions being made too quickly, which can dramatically decrease health and quality of life (Voss et al., 2017)

Planning for end of life care should include (1) talking with the individual with IDD about their preferences and wishes; (2) preparing and

Case Study 14.1b Pandemics and IDD

There were significant reductions in social support during the Covid-19 pandemic in the spring of 2020 for people with disabilities in the United States who were already living in institutional settings. People in State Supported Living Centers across Texas lost contact with their loved ones without notice. Additionally, many of their friends disappeared from their environments as a result of "non-essential" personnel canceling services due to the pandemic. At a time in history when social relationships moved to digital platforms for most other people, one of the most vulnerable populations lost most outside social contact. Because of privacy restrictions, widespread systems were not set up for people to be able to even see or hear their loved one's voices even over digital platforms on a regular basis. This dramatically increased the stress of the pandemic and removed critical accountability of having families be able to check on their loved ones in a setting that has historically been associated with a significant increased risk of stress.

What Do You Think?

1. What are the vulnerabilities of people with IDD in a pandemic to be addressed for their sustainable community health?
2. How would community pandemic preparedness address the diversity in people with IDD and their living arrangements?

executing written advance directives including what medical interventions a person wants or does not want and under what conditions; (3) planning for who will take responsibility for managing day-to-day care of the individual when parents or siblings own aging prevents them from continuing as the primary caregiver for people with IDD (McCallion et al., 2017).

Cultural, Legislative, and Professional Issues That Impact Sustainable Community Health for People with IDD

As the twenty-first century saw cultural movements to address historical segregation and inequities for people with IDD, initiatives from professional organizations promoted social inclusion and breaking down barriers of access to social, economic, and health opportunities for people with IDD. However, as health disparities, inequities, and injustices persistently affect people with IDD compared to other community members, meeting the healthcare needs of individuals with IDD involves consideration of current professional, legal, and ethical issues that influence sustainable community health options for this population.

Culture, Health, and Wellbeing with IDD Although much progress has been made by professional organizations to promote more equitable healthcare access and social opportunities for people with IDD, it should be noted that these movements are largely reflective of the dominant Western culture (Blacher, Neece, & Paczkowski, 2005). Recent ideological movements have promoted social inclusion, but people with IDD from minority ethnic communities often experience stigma, isolation, racism, negative attitudes from professionals, and lack of cultural sensitivity in service delivery (Raghavan & Small, 2004). Lack of culturally sensitive services poses a challenge in service access for individuals and their families (Bogenschutz, 2014; Kirsch, 2013; Mitter, Ali, & Scior, 2019; Raghavan & Small, 2004; Scott & Havercamp, 2014; Shapiro, Monzó, Rueda, Gomez, & Blacher, 2004).

As an example, Shapiro (2004) and colleagues identified that Latina mothers of individuals with IDD often feel isolated from transition planning service systems, and that professionals exhibited poor communication and negative attitudes toward family. While some efforts have been made to examine if established programs are also feasible and effective for culturally diverse populations (see Burke, Magaña, Garcia, & Mello, 2016), more research is warranted to examine how cultural practices impact utilization of healthcare services and effectiveness of existing services.

Professional Practice and Intellectual Disability As people with IDD live and access services in community settings, professionals support individuals with IDD across a range of settings. The primary professionals who support individuals with IDD and their families are direct support providers (DSPs) or Disability Support Workers (DSWs), mental health professionals, and primary care health professionals (e.g., physicians, nurses). DSWs and DSPs are the primary service provider of long-term services and supports (LTSS) for people with IDD (Forster & Iacono, 2008; Friedman, 2018). As people with IDD typically have limited social networks, the professionals who provide support also might be one of their only sources for social relationships (Friedman, 2018; McVilly,

Stancliffe, Parmenter, & Burton-Smith, 2006; Petry, Maes, & Vlaskamp, 2007). However, professionals who work with individuals with disabilities often report insufficient training, high rates of stress and burnout (typically direct care staff) (Bogenschutz, Hewitt, Nord, & Hepperlen, 2014; Friedman, 2018; Iezzoni & Long-Bellil, 2012; Sharby, Martire, & Iversen, 2015; Weise, Fisher, & Trollor, 2018; Wilkinson, Dreyfus, Cerreto, & Bokhour, 2012), and high turnover (Gray & Muramatsu, 2013).

Ethics and Legal Issues In addition to considering competence of healthcare providers, there are a variety of other ethical issues that exist when caring for individuals with IDD, such as involvement of multiple parties, guardianship, and access to services (Adams & Boyd, 2010). First, the issue of involvement of multiple parties, such as guardians, care staff, and the individual themselves, is a critical issue (Adams & Boyd, 2010). Fisher, Orkin, Green, Chinchilli, and Bhattacharya (2009) recommend identifying the responsibility to each party (e.g., client, family, direct care staff), and how to integrate them and address them within care plans. Kerr (2003) and colleagues identified that meeting the medical needs of individuals with IDD involves frequent re-assessment and full integration of direct care staff and caregivers at home into the assessment process and care plan.

Second, guardianship and the question of *who* makes decisions for individuals with people with IDD can also arise in the context of healthcare and other economic decisions (Adams & Boyd, 2010; Suto, Clare, Holland, & Watson, 2005). While children under the age of 18 do not give informed consent due to being a minor, a formal legal proceeding is necessary to determine if an individual with IDD can give informed consent for themselves, based on their competencies.

Related Disciplines for Sustainable Community Health with for Persons with IDD

As the old saying goes "it takes a village." Establishing equitable, sustainable community health for individuals with IDD requires continual collaboration and communication between a variety of disciplines. Community initiatives to increase access to and decrease barriers to vital individual-level services can, in turn, increase general functioning and happiness at the community-level. These vital, individual-level services provided by such disciplines as occupational therapists, speech therapists, physical therapists, and behavior analysts can prepare individuals to be self-sufficient and collaborative members of their communities. We briefly consider the role of each of these disciplines as part of a sustainable health system for people with IDD.

Occupational therapy is a discipline that uses a variety of procedures and evidence-based devices to help people with injuries, illnesses, and disabilities to strengthen fine and gross motor skills to help them function in daily life. Occupational therapists often target skills such as feeding, leisure activities, dressing, and daily living routines (Friedman & VanPuymbrouck, 2018; Umeda et al., 2017). They also can help teach individuals with IDD how to communicate via assistive devices such as tablet devices in conjunction with speech therapists and behavior analysts (Wilkinson, 2011). Speech therapy (conducted by speech-language pathologists) is a discipline that assesses, diagnoses, and treats communication delays and feeding/swallowing disorders in children and adults caused by disability, injury, or illness. As described above, speech therapists often work with adults and children with IDD to establish functional communication through voice, sign language, picture communication, and/or alternative assistive devices. For vocal communication, common targets are articulation, fluency, receptive/expressive targets, and aphasia (Friedman & McNamara, 2018; Mirenda, 2014).

Physical therapy is a discipline that helps injured or physically disabled persons to improve movement and manage pain. Often, persons with IDD have limited mobility due to muscular disorders, injuries, or a lack of enriched environments. Physical therapists often work to improve mobility in walking or wheelchairs via exercise and stretching procedures. They can work individually with clients or in groups to establish community-oriented fitness and wellness programs (Friedman & Feldner, 2018). Applied behavior analysis is a discipline based on the natural science of behavior. Applied behavior analysts develop systems, strategies, and tactics based on the principles of behavior to experimentally identify and verify the environmental and behavioral changes that are most important to people (Baer, Wolf, & Risley, 1968). Assessment of functional relations allows behavior analysts to identify preferences, teach skills from a wide variety of domains, and work in conjunction with occupational and speech therapists to teach functional communication of wants, needs, and dislikes. They can work individually with clients or in groups to target communication, self-help, independence, leisure, and social skills in clinics, homes, and in the community (Baer et al., 1968; Rotholz, Moseley, & Carlson, 2013). In addition to occupational therapists, speech-language pathologists, physical therapists and behavior analysts, social workers and vocational rehabilitation, and employment services can prepare individuals to live independently, become active members of their community, and ultimately enjoy a fulfilling career or volunteer opportunity.

Issues for Research and Other Forms of Scholarship

While there is a growing body of research on improving participation in community access for people with IDD, there is often a large gap between research and practice in community health with IDD. A most significant limitation is that the present research base has not prioritized issues of gender, culture, and ethnicity in the community health and wellbeing of people with IDD (Mpofu, 2016), their family members, and other caregivers (Catalano, Holloway, & Mpofu, 2018). The current body of research evidence is thin on contributions from LMIC countries, which also have the highest prevalence of IDD (Yasamy et al., 2011). Needless to say, the research evidence from industrialized countries would not transport to developing countries without further study. We discuss the following gaps in the research and practice scholarship for the health of the community of people with IDD: policy frameworks, provider training and preparation, communication supports and dignity, respect and advocacy.

Policy Frameworks There is a practice gap guiding policymakers on best practices providing sustainable health to people with IDD. While "most people with ID do not receive the services and supports they require, ... [and] in part, attributable to the absence of relevant scientific knowledge, at present, we simply do not know what the most cost-effective services are and which services models are applicable in different contexts" (Tomlinson et al., 2014, p. 1122). Moreover, public health policy makers and practitioners lack the data needed to advocate for community programs that enhance the independence and wellbeing of individuals with IDD. One of the reasons for this is the mere incompleteness of data on the community prevalence of IDD for population health planning purposes, and especially as administrative data are increasingly incomplete as people with IDD age out of public support programs (Friedman et al., 2018). Incomplete administrative data and poor monitoring and tracking of IDD populations create an indeterminate hidden pool of people who miss out on their right to health support by omission or commission. In addition, programs for the community wellbeing of people with IDD lack in sustainability due to budget cuts, and lack of training of providers (Spassiani, Meisner, Abou Chacra, Heller, & Hammel, 2019). However, while the required infrastructure of such programs can cost more upfront for the community, the long-term savings to the community from people with IDD obtaining jobs, paying taxes, buying goods, and volunteering within their community are huge. Nonetheless, research is needed on best sustainable community health practices with a focus to identify the components of these programs that have the greatest effect in improving health outcomes for people with IDD and how to make these programs more accessible and easy to navigate for people with IDD and their families (Mirenda, 2014; Nord & Hepperlen, 2016; Young-Southward et al., 2017). Research on sustainability of programs that prioritize the voices of individuals with IDD, their families, direct support staff, and direct care staff would be of high yield (Mpofu, 2016; Tomlinson et al., 2014).

Provider Training and Preparation As previously noted, community health providers are often not prepared to meet the needs of people with IDD. Stigma against people with IDD is an ever-present issue in their access of community health services (Friedman, 2018). As an example, caregivers reported feeling that the doctors they interacted with did not understand IDD and how to interact with people with IDD (Baumbusch et al., 2019). Moreover, people with IDD reported feeling that in interactions with healthcare providers, the convenience of the providers typically took precedence over their needs as patients (Baumbusch et al., 2019). These reports are concerning as barriers to accessing healthcare and poor health outcomes for people with IDD who are living in integrated community settings is cited as a major reason for re-institutionalization in this population (Baumbusch et al., 2019). Therefore, research is needed on community health provider training models for serving people with IDD with equity and sustainability. Such training would need to address communication differences and the supports needed to fully integrate people with IDD as valued participants in healthcare systems (Bas, 2019; CDC, 2011; Faridi, Grunbaum, Gray, Franks, & Simoes, 2007; Martin, O'Connor-Fenelon, & Lyons, 2010; Mirenda, 2014; Neely-Barnes et al., 2008; Strouse et al., 2013).

Summary and Conclusion

As we have seen in this chapter, people with IDD are one of the world's most overlooked minority populations in terms of their community health and wellbeing needs, creating avoidable health disparities, inequities, and social injustices. A history across many cultures and countries of discrimination and injustice toward individuals with disabilities has resulted in a wide range of disparities in community access and participation for wellness. Community interventions and structural resources for and created by people with IDD should focus on inclusive community living aimed to enhance the health and wellbeing of people in this population. The participation of people with IDD in the development of community wellness interventions for them would lead to lasting changes in their health outcomes and those of their communities. Policy instruments and actions understandable and accessible to this population will be essential to resolving the health social injustices faced by this population. Development of sustainable communities that are fully inclusive of all members including the range of people of all intellectual abilities will result in healthier, stronger, and more resilient communities for all.

Self-Check Questions

1. Define intellectual and developmental disabilities (IDD) and their community wellness needs.
2. Outline historical developments in the field if IDD, highlighting the community health and wellbeing implications.
3. What are the most common preventable causes of poor health among people with IDD?
4. Discuss the merits and prospects of contemporary health and wellbeing practices for the sustainable health of people with IDD.
5. Outline cultural, professional, and legislative/policy issues that influence sustainable community health practices with regard to addressing the needs of people with IDD.
6. Identity and discuss the gaps in research and practices in community health for individuals with IDD focusing on sustainability.

Discussion Questions

1. How can health systems improve in early detection of developmental disabilities in low- and middle-income countries?
2. How is disability inclusivity a quality of sustainable health practices? What are its benefits and limitations as a strategy for developing sustainable community health programs?
3. The community of people with IDD is diverse as are their residence communities. What would be diversity considerations in designing and implementing sustainable community health programs for this population?
4. What are the most common preventable causes of the social exclusion and reduced quality of life among people with IDD? Consider how these may vary across regions, countries and within countries.
5. Work participation is associated with superior health outcomes for people with a disability. How may this be different for an IDD population of your nomination and why?

Field-Based Experiential Exercises

1. Interview a family caregiver to an adult with IDD on their priority health and wellness supports for their family member with IDD and their priorities. Consider how the priorities would be different for a different family caregiver from the same community with a family member with similar supports needs. What would be the most efficient ways of supporting and empowering parents/families of people with IDD in the specific social and cultural context of your interviewee family?
2. Visit a community organization that provides services to people with IDD to learn about the specific community services they provide. Do a health and wellbeing audit of those activities, considering their sustainability qualities and needs.
3. Interview a group of people with IDD to learn of their community health and wellness activities, and their reasons for engaging in those activities rather than alternate ones. Consider the extent to which to which health sustainability is a quality of those activities and how.

Online Resources

United Nations. (n.d., c). *Monitoring and Evaluation of Disability-Inclusive Development.* https://www.un.org/development/desa/disabilities/resources/monitoring-and-evaluation-of-inclusive-development.html

National Council on Disability. (2019). *Preserving Our Freedom: Ending Institutionalization of People with Disabilities During and After Disasters.*

https://www.keranews.org/post/family-faces-extremely-challenging-situation-covid-19-hits-denton-state-living-center

https://www.statesman.com/photogallery/TX/20130519/PHOTOGALLERY/305199833/PH/1

Resources for families to guide them through talking with their family members and planning for end-of-life-care can be found at:

https://www.nia.nih.gov/health/advance-care-planning-healthcare-directives

https://www.aaidd.org/news-policy/policy/position-statements/caring-at-the-end-of-life

For more tools and resources on Informed Consent for health care professionals, see: https://iddtoolkit.vkcsites.org/general-issues/informed-consent/. An overview of steps involved in the process and a checklist that can be used by medical professionals to assess an individual's capability for informed consent is included.

References

Adams, Z. W., & Boyd, S. E. (2010). Ethical challenges in the treatment of individuals with intellectual disabilities. *Ethics & Behavior, 20*(6), 407–418.

Allerton, L. A., Welch, V., & Emerson, E. (2011). Health inequalities experienced by children and young people with intellectual disabilities: A review of literature from the United Kingdom. *Journal of Intellectual Disabilities, 15*(4), 269–278.

Amado, A. N., Stancliffe, R. J., McCarron, M., & McCallion, P. (2013). Social inclusion and community participation of individuals with intellectual/developmental disabilities. *Intellectual and Developmental Disabilities, 51*(5), 360–375.

American Association on Intellectual and Developmental Disabilities (AAIDD). (2019). *Definition of intellectual disability. Intellectual disability.* Retrieved May 10, 2020, from www.aaidd.org/intellectual-disability/definition

American Psychiatric Association. (2013). *Diagnostic and statistical manual of mental disorders* (5th ed.). Washington, DC: Author.

Angrosino, M. V., Devlieger, P. J., Booth, T., Booth, W., Davies, C. A., Van Maastricht, S., … Nuttall, M. (1998). *Questions of competence: Culture, classification and intellectual disability.* Cambridge, UK: Cambridge University Press.

Baer, D. M., Wolf, M. M., & Risley, T. R. (1968). Some current dimensions of applied behavior analysis 1. *Journal of Applied Behavior Analysis, 1*(1), 91–97.

Bannerman, D. J., Sheldon, J. B., Sherman, J. A., & Harchik, A. E. (1990). Balancing the right to habilitation with the right to personal liberties: The rights of people with developmental disabilities to eat too many doughnuts and take a nap. *Journal of Applied Behavior Analysis, 23*(1), 79–89.

Bartlo, P., & Klein, P. J. (2011). Physical activity benefits and needs in adults with intellectual disabilities: Systematic review of the literature. *American Journal on Intellectual and Developmental Disabilities, 116*(3), 220–232.

Bas, D. (2019). *Disability and development report: Realizing the sustainable development goals by, for and with persons with disabilities.* New York, NY: United Nations Publications.

Baumbusch, J., Moody, E., Hole, R., Jokinen, N., & Stainton, T. (2019). Using healthcare services: Perspectives of community-dwelling aging adults with intellectual disabilities and family members. *Journal of Policy and Practice in Intellectual Disabilities, 16*(1), 4–12.

Beange, H., & Durvasula, S. (2001). Health inequalities in people with intellectual disability: Strategies for improvement. *Health Promotion Journal of*

Australia: Official Journal of Australian Association of Health Promotion Professionals, 11(1), 27.

Beyer, S., Brown, T., Akandi, R., & Rapley, M. (2010). A comparison of quality of life outcomes for people with intellectual disabilities in supported employment, day services and employment enterprises. *Journal of Applied Research in Intellectual Disabilities, 23*(3), 290–295.

Blacher, J., Neece, C. L., & Paczkowski, E. (2005). Families and intellectual disability. *Current Opinion in Psychiatry, 18*(5), 507–513.

Blatt, B., & Kaplan, F. (1966). *Christmas in purgatory*. Boston, MA: Allen & Bacon.

Bloch, P., Toft, U., Reinbach, H. C., Clausen, L. T., Mikkelsen, B. E., Poulsen, K., & Jensen, B. B. (2014). Revitalizing the setting approach–supersettings for sustainable impact in community health promotion. *International Journal of Behavioral Nutrition and Physical Activity, 11*(1), 118.

Boat, T. F., Wu, J. T., Sciences, S., & National Academies of Sciences, Engineering, and Medicine. (2015). Clinical characteristics of intellectual disabilities. In *Mental disorders and disabilities among low-income children*. Washington, DC: National Academies Press (US)..

Bogenschutz, M. (2014). "We find a way": Challenges and facilitators for health care access among immigrants and refugees with intellectual and developmental disabilities. *Medical Care, 52*, S64–S70.

Bogenschutz, M. D., Hewitt, A., Nord, D., & Hepperlen, R. (2014). Direct support workforce supporting individuals with IDD: Current wages, benefits, and stability. *Intellectual and Developmental Disabilities, 52*(5), 317–329.

Bowey, L., McGlaughlin, A., & with Claire Saul. (2005). Assessing the barriers to achieving genuine housing choice for adults with a learning disability: The views of family carers and professionals. *British Journal of Social Work, 35*(1), 139–148.

Boyle, C. A., Boulet, S., Schieve, L. A., Cohen, R. A., Blumberg, S. J., Yeargin-Allsopp, M., … Kogan, M. D. (2011). Trends in the prevalence of developmental disabilities in US children, 1997–2008. *Pediatrics, 127*(6), 1034–1042.

Braddock, D. (1998). Mental retardation and developmental disabilities: Historical and contemporary perspectives. *The state of the states in developmental disabilities* (5th edition, pp. 3–22). Washington DC: American Association on Mental Retardation.

Braddock, D., Hemp, R., Rizzolo, M. C., Tanis, E. S., Haffer, L., Lulinski, A., & Wu, J. (2013). *State of the states in developmental disabilities 2013: The great recession and its aftermath*. Washington, DC: American Association on Intellectual and Developmental Disabilities.

Burke, M., Arnold, C., & Owen, A. (2018). Identifying the correlates and barriers of future planning among parents of individuals with intellectual and developmental disabilities. *Intellectual and Developmental Disabilities, 56*(2), 90–100.

Burke, M. M., Magaña, S., Garcia, M., & Mello, M. P. (2016). Brief report: The feasibility and effectiveness of an advocacy program for Latino families of children with autism spectrum disorder. *Journal of Autism and Developmental Disorders, 46*(7), 2532–2538.

Cartwright, L., Reid, M., Hammersley, R., & Walley, R. M. (2017). Barriers to increasing the physical activity of people with intellectual disabilities. *British Journal of Learning Disabilities, 45*(1), 47–55.

Catalano, D., Holloway, L., & Mpofu, E. (2018). Mental health interventions for parent carers of children with autistic spectrum disorder: Practice guidelines from a critical interpretive synthesis (CIS) systematic review. *International Journal of Environmental Research and Public Health, 15*(2), 341.

Centers for Disease Control and Prevention. (2011). *A sustainability planning guide for healthy communities*. National Center for Chronic Disease Prevention and Health Promotion. Retrieved May 10, 2020, from http://www.cdc.gov/nccdphp/dch/programs/healthycommunitiesprogram/pdf/sustainability_guide.pdf

Certo, N. J., & Luecking, R. G. (2011). Transition and employment: Reflections from a 40 year perspective. *Journal of Vocational Rehabilitation, 35*(3), 157–161.

Community for Permanent Supported Housing (CPSH) of North Texas. (2020). *Road to Hope program*. Retrieved July 27, 2020, from https://www.txcpsh.org/#:~:text=The%20COMMUNITY%20FOR%20PERMANENT%20SUPPORTED,%2C%20Rockwall%2C%20and%20Tarrant%20counties

Cooper, S. A., McConnachie, A., Allan, L. M., Melville, C., Smiley, E., & Morrison, J. (2011). Neighbourhood deprivation, health inequalities and service access by adults with intellectual disabilities: A cross-sectional study. *Journal of Intellectual Disability Research, 55*(3), 313–323.

Coppus, A. M. W. (2013). People with intellectual disability: What do we know about adulthood and life expectancy? *Developmental Disabilities Research Reviews, 18*(1), 6–16.

Einarsson, I. O., Olafsson, A., Hinriksdóttir, G., Jóhannsson, E., Daly, D., & Arngrímsson, S. A. (2015). Differences in physical activity among youth with and without intellectual disability. *Medicine & Science in Sports & Exercise, 47*(2), 411–418.

Emerson, E. (2012). Deprivation, ethnicity and the prevalence of intellectual and developmental disabilities. *Journal of Epidemiology and Community Health, 66*(3), 218–224.

Emerson, E., & Hatton, C. (2014). *Health inequalities and people with intellectual disabilities*. Cambridge, UK: Cambridge University Press.

Ervin, D. A., Hennen, B., Merrick, J., & Morad, M. (2014). Healthcare for persons with intellectual and developmental disability in the community. *Frontiers in Public Health, 2*, 83.

Faridi, Z., Grunbaum, J. A., Gray, B. S., Franks, A., & Simoes, E. (2007). Community-based participatory research: Necessary next steps. *Preventing Chronic Disease, 4*(3), A70.

Fesko, S. L., Hall, A. C., Quinlan, J., & Jockell, C. (2012). Active aging for individuals with intellectual disability: Meaningful community participation through employment, retirement, service, and volunteerism. *American Journal on Intellectual and Developmental Disabilities, 117*(6), 497–508.

Fisher, K. M., Orkin, F. K., Green, M. J., Chinchilli, V. M., & Bhattacharya, A. (2009). Proxy healthcare decision-making for persons with intellectual disability: Perspectives of residential-agency directors. *American Journal on Intellectual and Developmental Disabilities, 114*(6), 401–410.

Forster, S., & Iacono, T. (2008). Disability support workers' experience of interaction with a person with profound intellectual disability. *Journal of Intellectual and Developmental Disability, 33*(2), 137–147.

Friedman, C. (2018). Direct support professionals and quality of life of people with intellectual and developmental disabilities. *Intellectual and Developmental Disabilities, 56*(4), 234–250.

Friedman, C. (2019). The relationship between disability prejudice and institutionalization of people with intellectual and developmental disabilities. *Intellectual and Developmental Disabilities, 57*(4), 263–273.

Friedman, C., & Feldner, H. A. (2018). Physical therapy Services for People with intellectual and developmental disabilities: The role of Medicaid home- and community-based service waivers. *Physical Therapy, 98*(10), 844–854.

Friedman, C., & McNamara, E. (2018). Home-and community-based speech, language, and hearing services for people with intellectual and developmental disabilities. *Research and Practice for Persons with Severe Disabilities, 43*(2), 111–125.

Friedman, C., & VanPuymbrouck, L. (2018). Occupational therapy in Medicaid home and community-based services waivers. *American Journal of Occupational Therapy, 72*(2), 7202205120p1–7202205120p9.

Friedman, D. J., Gibson Parrish, R., & Fox, M. H. (2018). A review of global literature on using administrative data to estimate prevalence of intellectual and developmental disabilities. *Journal of Policy and Practice in Intellectual Disabilities, 15*(1), 43–62.

Graham, H. (2005). Intellectual disabilities and socioeconomic inequalities in health: An overview of research. *Journal of Applied Research in Intellectual Disabilities, 18*(2), 101–111.

Gray, J. A., & Muramatsu, N. (2013). When the job has lost its appeal: Intentions to quit among direct care workers. *Journal of Intellectual and Developmental Disability, 38*(2), 124–133.

Hall, A., Bose, J., Winsor, J., & Migliore, A. (2014). Knowledge translation in job development: Strategies for involving families. *Journal of Applied Research in Intellectual Disabilities, 27*(5), 489–492.

Harrison, M. (2004). Defining housing quality and environment: Disability, standards and social factors. *Housing Studies, 19*(5), 691–708.

Hassan, N. M., Landorf, K. B., Shields, N., & Munteanu, S. E. (2019). Effectiveness of interventions to increase physical activity in individuals with intellectual disabilities: A systematic review of randomised controlled trials. *Journal of Intellectual Disability Research, 63*(2), 168–191.

Healthy People. (2020). "Discrimination." *Discrimination* | Office of Disease Prevention and Health Promotion. Retrieved May 10, 2020, from www.healthypeople.gov/2020/topics-objectives/topic/social-determinants-health/interventions-resources/discrimination

Heller, T., & Caldwell, J. (2006). Supporting aging caregivers and adults with developmental disabilities in future planning. *Mental Retardation, 44*(3), 189–202.

Hilgenkamp, T. I., Reis, D., van Wijck, R., & Evenhuis, H. M. (2012). Physical activity levels in older adults with intellectual disabilities are extremely low. *Research in Developmental Disabilities, 33*(2), 477–483.

Iezzoni, L. I., & Long-Bellil, L. M. (2012). Training physicians about caring for persons with disabilities: "Nothing about us without us!". *Disability and Health Journal, 5*(3), 136–139.

Keith, H. E., & Keith, K. D. (2013). *Intellectual disability : Ethics, dehumanization, and a new moral community*. Chichester, UK: Wiley-Blackwell.

Kerr, A. M., McCulloch, D., Oliver, K., McLean, B., Coleman, E., Law, T., … Prescott, R. J. (2003). Medical needs of people with intellectual disability require regular reassessment, and the provision of client-and carer-held reports. *Journal of Intellectual Disability Research, 47*(2), 134–145.

Kilroy, S., Egan, J., Walsh, M., McManus, S., & Sarma, K. M. (2015). Staff perceptions of the quality of life of individuals with an intellectual disability who transition from a residential campus to community living in Ireland: An exploratory study. *Journal of Intellectual and Developmental Disability, 40*(1), 68–77.

Kirsch, N. (2013). *The experience of African American parents raising a child with an intellectual disability: A qualitative study*. Doctoral dissertation, John F. Kennedy University.

Krahn, G. L., & Fox, M. H. (2014). Health disparities of adults with intellectual disabilities: What do we know? What do we do? *Journal of Applied Research in Intellectual Disabilities, 27*(5), 431–446.

Krahn, G. L., Walker, D. K., & Correa-De-Araujo, R. (2015). Persons with disabilities as an unrecognized health disparity population. *American Journal of Public Health, 105*(Suppl 2), S198–S206.

Larson, S., Lakin, C., & Hill, S. (2012). Behavioral outcomes of moving from institutional to community living for people with intellectual and developmental disabilities: US studies from 1977 to 2010. *Research and Practice for Persons with Severe Disabilities, 37*(4), 235–246.

Levine, C., Halper, D., Peist, A., & Gould, D. A. (2010). Bridging troubled waters: Family caregivers, transitions, and long-term care. *Health Affairs, 29*(1), 116–124.

Lin, J. D., Lin, P. Y., Lin, L. P., Chang, Y. Y., Wu, S. R., & Wu, J. L. (2010). Physical activity and its determinants among adolescents with intellectual disabilities. *Research in Developmental Disabilities, 31*(1), 263–269.

Marks, B., Sisirak, J., Heller, T., & Wagner, M. (2010). Evaluation of community-based health promotion programs for Special Olympics athletes. *Journal of Policy and Practice in Intellectual Disabilities, 7*(2), 119–129.

Martin, A. M., O'Connor-Fenelon, M., & Lyons, R. (2010). Non-verbal communication between nurses and people with an intellectual disability: A review of the literature. *Journal of Intellectual Disabilities, 14*(4), 303–314.

Maulik, P. K., Mascarenhas, M. N., Mathers, C. D., Dua, T., & Saxena, S. (2011). Prevalence of intellectual disability: A meta-analysis of population-based studies. *Research in Developmental Disabilities, 32*(2), 419–436.

McCallion, P., Hogan, M., Santos, F. H., McCarron, M., Service, K., Stemp, S., … Janicki, M. P. (2017). Consensus statement of the international summit on intellectual disability and dementia related to end-of-life care in advanced dementia. *Journal of Applied Research in Intellectual Disabilities, 30*(6), 1160–1164.

McCarron, M., McCallion, P., Fahey-McCarthy, E., & Connaire, K. (2010). Staff perceptions of essential prerequisites underpinning end-of-life care for persons with intellectual disability and advanced dementia. *Journal of Policy and Practice in Intellectual Disabilities, 7*(2), 143–152.

McCausland, D., McCallion, P., Cleary, E., & McCarron, M. (2016). Social connections for older people with intellectual disability in Ireland: Results from wave one of IDS-TILDA. *Journal of Applied Research in Intellectual Disabilities, 29*(1), 71–82.

McConkey, R., Mulvany, F., & Barron, S. (2006). Adult persons with intellectual disabilities on the island of Ireland. *Journal of Intellectual Disability Research, 50*(3), 227–236.

McCreary, B. D. (2005). *Developmental disabilities and dual diagnosis: A guide for Canadian psychiatrists. Developmental consulting program.* Kingston, Canada: Queen's University.

McGreevy, P., Fry, T., & Cornwall, C. (2012). *Essentials for living.* Winter Park, FL: Patrick McGreevy Publishing.

McKenzie, K., Milton, M., Smith, G., & Ouellette-Kuntz, H. (2016). Systematic review of the prevalence and incidence of intellectual disabilities: Current trends and issues. *Current Developmental Disorders Reports, 3,* 104–115.

McKenzie, N., Mirfin-Veitch, B., Conder, J., & Brandford, S. (2017). "I'm still here": Exploring what matters to people with intellectual disability during advance care planning. *Journal of Applied Research in Intellectual Disabilities, 30*(6), 1089–1098.

McVilly, K. R., Stancliffe, R. J., Parmenter, T. R., & Burton-Smith, R. M. (2006). I get by with a little help from my friends': Adults with intellectual disability discuss loneliness 1. *Journal of Applied Research in Intellectual Disabilities, 19*(2), 191–203.

Melville, C. A., Hamilton, S., Hankey, C. R., Miller, S., & Boyle, S. (2007). The prevalence and determinants of obesity in adults with intellectual disabilities. *Obesity Reviews, 8*(3), 223–230.

Migliore, A., Butterworth, J., Nord, D., Cox, M., & Gelb, A. (2012). Implementation of job development practices. *Intellectual and Developmental Disabilities, 50*(3), 207–218.

Mirenda, P. (2014). Revisiting the mosaic of supports required for including people with severe intellectual or developmental disabilities in their communities. *Augmentative and Alternative Communication, 30*(1), 19–27.

Mitter, N., Ali, A., & Scior, K. (2019). Stigma experienced by families of individuals with intellectual disabilities and autism: A systematic review. *Research in Developmental Disabilities, 89*, 10–21.

Mpofu, E. (2016). The evolution of quality of life perspectives in the developing world. In R. L. Schalock & K. D. Keith (Eds.), *Cross-cultural quality of life: Enhancing the lives of persons with intellectual disability* (2nd ed., pp. 175–180). Washington, DC: American Association on Mental Retardation.

Neely-Barnes, S., Marcenko, M., & Weber, L. (2008). Does choice influence quality of life for people with mild intellectual disabilities?. *Intellectual and Developmental Disabilities, 46*(1), 12–26.

Nord, D., & Hepperlen, R. (2016). More job services—Better employment outcomes: Increasing job attainment for people with IDD. *Intellectual and Developmental Disabilities, 54*(6), 402–411.

Obi, O., Van Naarden Braun, K., Baio, J., Drews-Botsch, C., Devine, O., & Yeargin-Allsopp, M. (2011). Effect of incorporating adaptive functioning scores on the prevalence of intellectual disability. *American Journal on Intellectual and Developmental Disabilities, 116*(5), 360–370.

Oeseburg, B., Dijkstra, G. J., Groothoff, J. W., Reijneveld, S. I., & Jansen, D. E. M. C. (2011). Prevalence of chronic health conditions in children with intellectual disability: A systematic literature review. *Intellectual and Developmental Disabilities, 49*(2), 59–85.

Oswald, F., & Wahl, H. W. (2004). Housing and health in later life. *Reviews on Environmental Health, 19*(3–4), 223–252.

Parish, S. L., Mitra, M., Son, E., Bonardi, A., Swoboda, P. T., & Igdalsky, L. (2015). Pregnancy outcomes among US women with intellectual and developmental disabilities. *American Journal on Intellectual and Developmental Disabilities, 120*(5), 433–443.

Parish, S. L., Swaine, J. G., Son, E., & Luken, K. (2013). Receipt of mammography among women with intellectual disabilities: Medical record data indicate substantial disparities for African American women. *Disability and Health Journal, 6*(1), 36–42.

Partington, J. W., & Mueller, M. (2012). *The assessment of functional living skills [AFLS]*. Pleasant Hill, CA: Behavior Analysts.

Patja, K., Mlsk, P., & Iivanainen, M. (2001). Cause-specific mortality of people with intellectual disability in a population-based, 35-year follow-up study. *Journal of Intellectual Disability Research, 45*(1), 30–40.

Patterson, I., & Pegg, S. (2009). Serious leisure and people with intellectual disabilities: Benefits and opportunities. *Leisure Studies, 28*(4), 387–402.

Perry, J., & Felce, D. (2003). Quality of life outcomes for people with intellectual disabilities living in staffed community housing services: A stratified random sample of statutory, voluntary and private agency provision. *Journal of Applied Research in Intellectual Disabilities, 16*(1), 11–28.

Petry, K., Maes, B., & Vlaskamp, C. (2007). Support characteristics associated with the quality of life of people with profound intellectual and multiple disabilities: The perspective of parents and direct support staff. *Journal of Policy and Practice in Intellectual Disabilities, 4*(2), 104–110.

Raghavan, R., & Small, N. (2004). Cultural diversity and intellectual disability. *Current Opinion in Psychiatry, 17*(5), 371–375.

Rimmer, J. H., & Yamaki, K. (2006). Obesity and intellectual disability. *Mental Retardation and Developmental Disabilities Research Reviews, 12*(1), 22–27.

Robertson, J., Hatton, C., Baines, S., & Emerson, E. (2015). Systematic reviews of the health or health care of people with intellectual disabilities: A systematic review to identify gaps in the evidence base. *Journal of Applied Research in Intellectual Disabilities, 28*(6), 455–523.

Robinson, L. M., Dauenhauer, J., Bishop, K. M., & Baxter, J. (2012). Growing health disparities for persons who are aging with intellectual and developmental disabilities: The social work linchpin. *Journal of Gerontological Social Work, 55*(2), 175–190.

Roll-Petterson, L., Olsson, I., & Rosales, S. A. I. (2017). Bridging the research to practice gap: A case study approach to understanding eibi supports and barriers in Swedish preschools. *International Electronic Journal of Elementary Education, 9*(2), 317–336.

Rothman, D. J. (2017). *The Willowbrook wars: Bringing the mentally disabled into the community*. London, UK: Routledge.

Rotholz, D. A., Moseley, C. R., & Carlson, K. B. (2013). State policies and practices in behavior supports for persons with intellectual and developmental disabilities in the United States: A national survey. *Mental Retardation, 51*(6), 433–445.

Ryan, K., Guerin, S., Dodd, P., & McEvoy, J. (2011). End-of-life care for people with intellectual disabilities: Paid carer perspectives. *Journal of Applied Research in Intellectual Disabilities, 24*(3), 199–207.

Saloviita, T. (2000). Improving institutions: Effects of small unit size on quality of care of people with severe intellectual disabilities. *Scandinavian Journal of Disability Research, 2*(2), 22–31.

Schalock, R. L., & Luckasson, R. (2013). What's at stake in the lives of people with intellectual disability? Part I: The power of naming, defining, diagnosing, classifying, and planning supports. *Intellectual and Developmental Disabilities, 51*(2), 86–93.

Schalock, R. L., & Luckasson, R. (2014). *Clinical judgment* (2nd ed.). Washington, DC: American Association on Intellectual and Developmental Disabilities.

Scott, H. M., & Havercamp, S. M. (2014). Race and health disparities in adults with intellectual and developmental disabilities living in the United States. *Intellectual and Developmental Disabilities, 52*(6), 409–418.

Shapiro, J., Monzó, L. D., Rueda, R., Gomez, J. A., & Blacher, J. (2004). Alienated advocacy: Perspectives of Latina mothers of young adults with developmental disabilities on service systems. *Mental Retardation, 42*(1), 37–54.

Sharby, N., Martire, K., & Iversen, M. D. (2015). Decreasing health disparities for people with disabilities through improved communication strategies and awareness. *International Journal of Environmental Research and Public Health, 12*(3), 3301–3316.

Shorter, E. (2000). *The Kennedy family and the story of mental retardation.* Philadelphia, PA: Temple University Press.

Simplican, S. C., Leader, G., Kosciulek, J., & Leahy, M. (2015). Defining social inclusion of people with intellectual and developmental disabilities: An ecological model of social networks and community participation. *Research in Developmental Disabilities, 38*, 18–29.

Šiška, J., Beadle-Brown, J., Káňová, Š., & Šumníková, P. (2018). Social inclusion through community living: Current situation, advances and gaps in policy, practice and research. *Social Inclusion, 6*(1), 94–109.

Smith, J. D., & Polloway, E. A. (1993). Institutionalization, involuntary sterilization, and mental retardation: Profiles from the history of the practice. *Mental Retardation, 31*(4), 208.

Spassiani, N. A., Meisner, B. A., Abou Chacra, M. S., Heller, T., & Hammel, J. (2019). What is and isn't working: Factors involved in sustaining community-based health and participation initiatives for people ageing with intellectual and developmental disabilities. *Journal of Applied Research in Intellectual Disabilities, 32*(6), 1465–1477.

Spassiani, N. A., Parker Harris, S., & Hammel, J. (2016). Exploring how knowledge translation can improve sustainability of community-based health initiatives for people with intellectual/developmental disabilities. *Journal of Applied Research in Intellectual Disabilities, 29*(5), 433–444.

Stebbins, R. A. (1992). *Amateurs, professionals, and serious leisure*. Montreal, Canada: McGill-Queen's Press-MQUP.

Strouse, M. C., Sherman, J. A., & Sheldon, J. B. (2013). Do Good, Take Data, Get a Life, and Make a Meaningful Difference in Providing Residential Services!. In *Handbook of Crisis Intervention and Developmental Disabilities* (pp. 424–437). New York, NY: Springer.

Sulewski, J. S., Timmons, J. C., Lyons, O., Lucas, J., Vogt, T., & Bachmeyer, K. (2017). Organizational transformation to integrated employment and community life engagement. *Journal of Vocational Rehabilitation, 46*(3), 313–320.

Suto, W. M. I., Clare, I. C. H., Holland, A. J., & Watson, P. C. (2005). Capacity to make financial decisions among people with mild intellectual disabilities. *Journal of Intellectual Disability Research, 49*(3), 199–209.

Szporluk, M. (2015). *The right to adequate housing for persons with disabilities living in cities: Towards inclusive cities*. New York, NY: United Nations Human Settlements Programme.

Taggart, L., Truesdale-Kennedy, M., Ryan, A., & McConkey, R. (2012). Examining the support needs of ageing family carers in developing future plans for a relative with an intellectual disability. *Journal of Intellectual Disabilities, 16*(3), 217–234.

Tassé, M. J. (2016). *Defining intellectual disability: Finally we all agree… almost*. Washington, DC: American Psychological Association Newsletter.

Tomlinson, M., Yasamy, M. T., Emerson, E., Officer, A., Richler, D., & Saxena, S. (2014). Setting global research priorities for developmental disabilities, including intellectual disabilities and autism. *Journal of Intellectual Disability Research, 58*(12), 1121–1130.

Umeda, C. J., Fogelberg, D. J., Jirikowic, T., Pitonyak, J. S., Mroz, T. M., & Ideishi, R. I. (2017). Expanding the implementation of the Americans with Disabilities Act for populations with intellectual and developmental disabilities: The role of organization-level occupational therapy consultation. *American Journal of Occupational Therapy, 71*(4), 7104090010p1–7104090010p6.

United States. (1975). *Education for All Handicapped Children Act of 1975*. Washington, DC: Author.

Válková, H. (2015). Physical activity, physical education and sports of persons with mental disability in relation with wellness (theoretical consideration). *Acta Salus Vitae, 2*(1), 1–22. http://odborne.casopisy.palestra.cz/index.php/actasalusvitae/article/view/25/25

Van Campen, C., & Iedema, J. (2007). Are persons with physical disabilities who participate in society healthier and happier? Structural equation modelling of objective participation and subjective well-being. *Quality of Life Research, 16*(4), 635.

Verdonschot, M. M., De Witte, L. P., Reichrath, E., Buntinx, W. H. E., & Curfs, L. M. (2009). Community participation of people with an intellectual disability: A review of empirical findings. *Journal of Intellectual Disability Research, 53*(4), 303–318.

Vickerstaff, S., & Phillipson, C. (Eds.). (2011). *Work, health and wellbeing: The challenges of managing health at work*. Bristol, UK: Policy Press.

Voss, H., Vogel, A., Wagemans, A. M., Francke, A. L., Metsemakers, J. F., Courtens, A. M., & de Veer, A. J. (2017). Advance care planning in palliative care for people with intellectual disabilities: A systematic review. *Journal of Pain and Symptom Management, 54*(6), 938–960.

Ward, R. L., Nichols, A. D., & Freedman, R. I. (2010). Uncovering health care inequalities among adults with intellectual and developmental disabilities. *Health & Social Work, 35*(4), 280–290.

Weise, J., Fisher, K. R., & Trollor, J. N. (2018). What makes generalist mental health professionals effective when working with people with an intellectual disability? A family member and support person perspective. *Journal of Applied Research in Intellectual Disabilities, 31*(3), 413–422.

White, P., Chant, D., Edwards, N., Townsend, C., & Waghorn, G. (2005). Prevalence of intellectual disability and comorbid mental illness in an Australian community sample. *Australian and New Zealand Journal of Psychiatry, 39*(5), 395–400.

Wilkinson, J., Dreyfus, D., Cerreto, M., & Bokhour, B. (2012). "Sometimes I feel overwhelmed": Educational needs of family physicians caring for people with intellectual disability. *Intellectual and Developmental Disabilities, 50*(3), 243–250.

Wilkinson, K. (2011). Answers to your biggest questions about services for people with severe disabilities. *The ASHA Leader, 16*(14), 16–19.

World Health Organization. (2007). *Atlas: Global resources for persons with intellectual disabilities: 2007*. World Health Organization.

World Health Organization. (2008). Commission on the Social Determinants of Health [CSDH]. Global health inequity—The need for action. In *Closing the gap in a generation: Health equity through action on the social determinants of health* (pp. 29–34). Geneva, Switzerland: WHO.

World Health Organization. (2020). *Disability and health*. Retrieved from https://www.who.int/en/news-room/fact-sheets/detail/disability-and-health

World Health Organization and The World Bank. (2011). *World Report on Disability*. Geneva, Switzerland: World Health Organization.

Yasamy, M. T., Maulik, P. K., Tomlinson, M., Lund, C., Van Ommeren, M., & Saxena, S. (2011). Responsible governance for mental health research in low resource countries. *PLoS Medicine, 8*(11), e1001126.

Young-Southward, G., Cooper, S. A., & Philo, C. (2017). Health and wellbeing during transition to adulthood for young people with intellectual disabilities: A qualitative study. *Research in Developmental Disabilities, 70*, 94–103.

Young-Southward, G., Philo, C., & Cooper, S. A. (2017). What effect does transition have on health and well-being in young people with intellectual disabilities? A systematic review. *Journal of Applied Research in Intellectual Disabilities, 30*(5), 805–823.

Zettel, J. J. (1977). *Public Law 94-142: The Education for All Handicapped Children Act. An overview of the Federal Law.*

15

The Wellbeing of People with Neurodiverse Conditions

Andrew M. Colombo-Dougovito, Suzanna Rocco Dillon, and Elias Mpofu

Introduction

The most consistent thing about human attributes is diversity. Regardless, other types of diversity such as race/ethnicity, sex, language are more readily recognized and accommodated for compared to less obvious ones, such as neurological differences. Admittedly, determining the expanse or prevalence of neurodiversity is difficult (Baker, 2011; Fletcher-Watson & Happé, 2019). This chapter will explain how **neurodiversity** represents the immense variety of human neurodevelopment in a similar way to how biodiversity attempts to explain the diversity of the biology of the planet. As a socially inclusive term, neurodiversity challenges the understanding of human attributes. As a term, it was never intended to be synonymous with any particular disability or even a subgroup of disabilities (Walker, 2014) or to "imply that a norm exists from which all others diverge" (Fletcher-Watson &

A. M. Colombo-Dougovito (✉)
Department of Kinesiology, Health Promotion, and Recreation,
University of North Texas, Denton, TX, USA
e-mail: andrew.colombo-dougovito@unt.edu

E. Mpofu (ed.), *Sustainable Community Health*,
https://doi.org/10.1007/978-3-030-59687-3_15

Happé, 2019, p. 25). Despite clear indications to the contrary, many, however, still use **neurodivergent**—or individuals who are **neurodiverse**—primarily to reference individuals with a diagnosis of autism. However, other neurocognitive disabilities such as attention-deficit/hyperactivity disorder (ADHD), dyslexia, dyspraxia, dyscalculia, epilepsy, Tourette syndrome (Fletcher-Watson & Happé, 2019; Kapp, Gillespie-Lynch, Sherman, & Hutman, 2013), and emotional and behavioral disorders (Armstrong, 2015) are considered under the umbrella of neurodiversity.

Learning Objectives

After reviewing this chapter, the reader should be able to:

1. Define the communities of neurodivergent people.
2. Outline the history of research and practice on community health and wellbeing with neurodiversity.
3. Discuss current and prospective approaches to sustainable health for neurodivergent individuals.
4. Identify and describe the role of relevant disciplines in providing sustainable community health services to people with neurodivergent conditions.
5. Evaluate the research and practice needs for the sustainable health of neurodivergent people.

Given the diversity that defines neurodivergence (Gray, 2002; Shtayermman, 2009), reporting on the prevalence of neurodiversity in human populations has tended to be emergent rather than definitive,

---

S. R. Dillon
School of Health Promotion and Kinesiology, Texas Woman's University, Denton, TX, USA
e-mail: sdillon@twu.edu

E. Mpofu
University of North Texas, Denton, TX, USA

University of Sydney, Sydney, NSW, Australia

University of Johannesburg, Johannesburg, South Africa
e-mail: Elias.Mpofu@unt.edu

following professional diagnostic criteria for the individual conditions. For example, the global prevalence of autistic people is estimated to be 2% (Elsabbagh et al., 2012), which translates to anywhere from 78 to 156 million people. Globally, attention-deficit/hyperactivity disorder (ADHD) prevalence among children under the age of 18 is estimated at 7.2%, or roughly 129 million (Thomas, Sanders, Doust, Beller, & Glasziou, 2015). When considering the many other conditions that fall under the neurodiversity umbrella, such as dyslexia, Tourette Syndrome, or intellectual disabilities (ID), the prevalence of "neurodiversity" likely makes it one of the larger minority groups among the disability community. Though definitive prevalence statistics are hard to find on the aggregated neurodiverse conditions, it is noted that potentially up to 17% of the global population has been diagnosed with a neurodiverse condition (Sargent 2019). Even so, estimates may be under-represented, as many conditions such as depression or anxiety remain under-diagnosed.

Within the United States (US), the Centers for Disease Control and Prevention, or CDC, estimates the prevalence of autism among 8-year-old children to be 1 in 54 (Maenner et al., 2020). Among adults, the CDC further estimates that 2.21% are on the autism spectrum (Dietz, Rose, McArthur, & Maenner, 2020) equating to roughly 5.5 million adults or about 1 in 45 individuals. Nationally and globally, males are around three times more likely to receive a diagnosis than females. For adolescents or adults seeking a diagnosis, symptoms must have been present before age 12. In the US, around 6.1 million children (9.4%) are estimated to have received an attention-deficit/hyperactive disorder (ADHD) diagnosis; and approximately 8.4% (5.4 million) have a current diagnosis (Danielson et al., 2018). Estimates among adults in the US are far more limited since around 4.4% of adults have a diagnosis of ADHD, with men comprising about 62% of those diagnosed (Kessler et al., 2006). The prevalence of intellectual disabilities (ID) is likely to be about 1.04% in the US (Maulik, Mascarenhas, Mathers, Dua, & Saxena, 2011), and between 0.05 and 1.55% globally (McKenzie, Milton, Smith, & Ouellette-Kuntz, 2016).

Although this varies widely by region, people who are neurodiverse have public health needs that are often overlooked by the health systems of their communities (World Health Organization [WHO], 2007). In the US, for example, the economic cost of supporting autistic individuals is projected to be about $461 billion by 2025 (Leigh & Du, 2015), about the same amount as those with ADHD (Gupte-Singh, Singh, & Lawson, 2017). Families of individuals with neurodiverse conditions reportedly spend four times more in healthcare costs than on a neurotypical (NT) family member (Lunsky, De Oliveira, Wilton, & Wodchis, 2019; Malcolm-Smith, Hoogenhout, Ing, Thomas, & de Vries, 2013; Matza, Paramore, & Prasad, 2005; Wang, Mandell, Lawer, Cidav, & Leslie, 2013). These statistics are troubling given the fact that people with neurodiverse conditions are at higher risk for un- or under-employment (Roux, Shattuck, Rast, Rava, & Anderson, 2015; Taylor & Seltzer, 2011), which denies them the sense of community wellness from work participation compared to other NT community members. Given the variety of presenting conditions under the neurodiversity umbrella, it is not surprising that these individuals also have diverse health needs. For instance, autistic adults have a higher prevalence of chronic conditions, including dyslipidemia, hypertension, sleep disorders, obesity, and thyroid disease (Croen et al., 2015), and autistic children have a higher likelihood of being obese compared to non-autistic peers (Broder-Fingert, Brazauskas, Lindgren, Iannuzzi, & Van Cleave, 2014; Curtin, Jojic, & Bandini, 2014). Moreover, many autistic individuals are more likely to have co-occurring psychiatric conditions such as depression or anxiety or post-traumatic stress disorder (PTSD) (Rosen, Mazefsky, Vasa, & Lerner, 2018). Potential health risk experiences by individuals with ADHD include a higher risk for obesity, binge eating, and bulimia (Fladhammer, Lyde, Meyers, Clark, & Landau, 2016; Kim, Mutyala, Agiovlasitis, & Fernhall, 2011) as well as drug and alcohol use (Whalen, Jamner, Henker, Delfino, & Lozano, 2002), which seem to begin in adolescence and traverse into adulthood (Breyer et al., 2009). Similarly, people with ID experience higher rates of co-occurring conditions such as mood disorders, obesity, diabetes, heart disease, mental illness, and early-onset dementia (Tyrer et al., 2019). The incidence of early-onset obesity among children with ID is almost double (28.9% compared to 15.5%) compared to typically developing peers, or the general population (Segal et al., 2016).

Neurodivergent people—especially those with different communication needs—are frequently excluded from many of the critical conversations or decisions that have a direct impact on their daily lives, which can erode their sense of wellness. Within these settings, they are often "erased, silenced, [and] derailed" (Hillary, 2013), which would be associated with a vast under-representation of their thoughts and preferences for wellbeing (Hughes, 2016). Many people with neurodiverse conditions require targeted support to ensure optimal community engagement and independent living. Such support includes psychoeducation for self-advocacy, school transition planning, family engagement, career development, and working with the school and civic community as stakeholders in the futures of communities (Dente & Coles, 2012; Wehmeyer & Abery, 2013; Wolgemuth et al., 2016). Neurodivergent inclusive community health and wellbeing approaches hold great promise for the sustainable health of people with neurodiverse conditions. In this chapter, we consider community health and wellness-oriented strategies for the sustainable health of people with neurodiverse conditions.

Professional Definitions and Theories of Neurodiversity

As noted earlier, neurodiversity is an inclusive term for people on the spectrum of human attributes that affect their ability to contribute to presentations and applications across a broad range of social activities. They may be predisposed to respond in non-normative ways, which add to the richness of human competencies and experiences from a diversity perspective (Austin & Pisano, 2017; Jaarsma & Welin, 2012). We present several illustrative definitions for a sample of conditions that define neurodiversity. These include autism and other neurodevelopmental conditions, such as ADHD and ID. We also maintain that these categories are social constructions that people with neurodiverse conditions may endorse differently.

Definitions of Major Neurodivergent Conditions

Autism Autism is defined by the American Psychiatric Association's (2013) Diagnostic and Statistical Manual of Mental Disorders (5th ed.; DSM-5) and the 11th edition of the World Health Organization's [WHO] (2018) International Statistical Classification of Diseases and

Related Health Problems (11th ed.; ICD-11) as one specific category, autism spectrum disorder (ASD) due to its range and variability of presentations. Diagnosis, according to both the DSM-V and ICD-11, requires evidence of both atypical social and communication behaviors, and demonstrates restrictive and repetitive behaviors. Each of these attributes should be present from birth, though diagnosis may not come until much later. Additionally, both diagnostic manuals describe hyper- and hypo-sensory sensitivities, as well as the potential for concurrent diagnoses of intellectual and/or language disability. Though recent shifts in diagnostic criteria have coalesced around the term ASD, individuals with diagnoses that pre-date this shift may have a label of *Asperger syndrome, autism, Rett's disorder, pervasive developmental disorder-not otherwise specified (PPD-NOS)* or *childhood disintegrative disorder*. However, it should be emphasized that the variability among the diagnostic criteria is "nothing compared to the variability of presentation in the autistic population" (Fletcher-Watson & Happé, 2019, p. 33).

Attention-Deficit/Hyperactivity Disorder The National Institute of Mental Health (NIMH) characterizes attention-deficit/hyperactivity disorder as an ongoing pattern of inattention and/or hyperactivity-impulsivity that interviews with development or functioning (2013). Under this definition, **inattention** describes predisposition, such as being off task, lacking in persistence, which may result in behavior that shows a lack of awareness or that seems defiant. **Hyperactivity** refers to a need to constantly move, excessive unintentional fidgeting, taps, or talks. While, **impulsivity** refers to making hasty decisions or actions that may have a high potential for harm as well as being driven by a desire for immediate rewards or failures, and the inability to delay gratification. Symptoms of ADHD may appear as early as three years of age and can continue through adolescence into adulthood.

The ICD-11 (WHO, 2018) differentiates ADHD into five categories: (1) ADHD-PI; (2) ADHD-PHI; (3) ADHD-C; (4) ADHD-Y; and ADHD-Z. **ADHD-PI** is a "predominantly inattentive presentation," in which individuals have significant difficulty sustaining attention. Hyperactive-impulsive symptoms may also be present, but not clinically significant. **ADHD-PHI** represents a "predominantly hyperactive-impulsive" presentation, in which individuals present excessive motor activity and difficulty remaining still (hyperactive) and a tendency to act to immediate stimuli without deliberation or consideration of consequences (impulsive). Some inattentive symptoms may be present but are not clinically significant. **ADHD-C** characterizes individuals that have both clinically significant inattentive and hyperactivity-impulsive symptoms. **ADHD-Y** is an "other specified" label that contains individuals that might not fit the categories mentioned above. For example, an individual may show hyperactive and inattentive but not impulsive symptoms. **ADHD-Z** identifies individuals that cannot attribute ADHD to any one or more categories, but otherwise fit the diagnosis.

Intellectual Disabilities According to the American Association on Intellectual and Developmental Disability (AAIDD, 2010), ID is characterized by "significant limitations in intellectual functioning" and "significant limitations in adaptive behavior," both of which must have an onset in childhood (before the age of 18). This definition additionally focuses on the levels of support (i.e., intermittent, limited, extensive, and pervasive) needed to maximize an individual's ability. This definition is consistent with both the *ICD-11* and *DSM-V* diagnostic criteria. Definitions of intellectual disabilities have evolved in concert with the shifting legal and social gains that encompass a shift from institutionalization to inclusive practices, self-advocacy, and self-determination (Brady et al., 2016). Accompanying this shift in terminology was a change from strictly defining ID based on intelligence quotient (IQ) to also include strengths in adaptive behavior (Schalock et al., 2007). A shift codified by US President Obama through federal legislation, known as Rosa's Law (2010), replaced "mental retardation" with "intellectual disability" in all federal health, education, and labor policy.

Theories on Neurodivergent Conditions Theories on biological, cognitive, and behavioral aspects of neurodiversity are evolving, and most consider atypical communication and attention abilities to be a commonly shared quality (Milton, 2012). This characteristic, which may (mis)construe for others as typical, may instead be unsatisfactory, reflecting a power imbalance against people with neurodiverse conditions, stemming from society's failure to create more opportunities for people with neurodiverse conditions (see Discussion Box 15.1). To create more sustainable,

Discussion Box 15.1: Strengths and Weakness of Neurodiverse Individuals

Neurodiverse individuals comprise, as the name implies, a wide variety of skill abilities. Though no one definitive set of skills exist, commonalities are present among those that fit under the "big tent" of neurodiversity. More complicating still is the fluidity in which skills, strengths, and weaknesses present themselves in different environmental contexts. See the list below to explore some strengths and weaknesses across various disabilities that are considered neurodivergent, and compare the similarities and differences.

Disability	Strengths	Weaknesses
Autism	• Sequencing • Concentration • Visual thought • Different imagination • Logical • Hyperfocus	• Social communication/integrations • Self-regulation • Obsessive • Different imagination • Hyper- and hypo-sensitivities • Speech and language difficulties
ADHD	• Intuitive • Quick-witted • Energetic • Hyperfocus • Empathetic • Good talker	• Impulsive • Self-regulation • Hyperactivity • Low frustration threshold • Distractibility as well as hyperfocus
Dyslexia	• 3D thought/spatial concepts • Visual thought • Creativity • Non-linear thought	• Difficulty reading, writing, spelling, recognizing words, and sequencing words • Distractibility

What Do You Think?

1. What do you perceive as implications for community wellness with neurodiversity?
2. How might communities leverage the strength and potential of people with neurodiverse conditions to increase opportunities for successful engagement or inclusion?

accessible community environments for the wellness of neurodivergent people, society must be inclusive. It must account for the health and wellbeing differences and needs of people with neurodiverse conditions.

Few of the existing theories are helpful in understanding autism in a community health sense; one that may is the double empathy problem (Milton, 2012). This theory makes a simplistic but revolutionary point that any social interaction requires the participation of at least two individuals. While interactions between autistic and neurotypical persons are often unsatisfactory, both parties should take responsibility for the interaction. Milton highlights the routine lack of empathy shown to autistic people by neurotypical people within the community. Thus, due to the power imbalance favoring neurotypical people, autistic people are seen as having a deficit in their communication ability. This is manifested in the inability of society to adapt behaviors (e.g., enforcing social norms like eye contact) or environments (e.g., playing music loudly in gymnasiums) to create more accessible spaces for autistic people.

As no clear pathology or biomarker exists, the presence of ADHD is measured behaviorally. Therefore, like autism, numerous theories based on decades of clinical research have been posited regarding the neurodevelopmental origins of ADHD (Bob & Konicarova, 2018). One contemporary theory, the "hot and cold" theory of ADHD, suggests that individuals with a diagnosis have strong associations with deficits in hot and cold executive functioning (Yarmolovsky, Szwarc, Schwartz, Tirosh, & Geva, 2017). One of these executive systems, the "cold executive network," is sensitive to conceptual rules and symbolic target-oriented behavior; while the other "hot executive network" is a motivational system that is dependent on social-affective information and reward (Metcalfe & Mischel, 1999). In other words, the cold system is responsible for inhibition and flexibility, while the hot system is central to enabling self-regulation, decision-making, and emotional perception. The theory posits that individuals with ADHD have atypical "disordered" responses to stimuli compared to neurotypical peers, though the argument does not assert that these responses are pathological.

ID is attributed to a multitude of causes that could occur pre-, peri-, or postnatal. These include, but are not limited to: causes related to genetics (e.g., Down syndrome and Fragile X syndrome), brain malformations (e.g., microcephaly), environmental influences (e.g., prenatal exposure alcohol or drugs, or postnatal exposure to lead or mercury), birth complications (e.g., anoxia at birth), traumatic brain injuries (occurring before age 18), infection, or severe social deprivation. Due to the variety of causes and the expansive range of impact on intellectual functioning and adaptive behavior, like autism and ADHD, there is no single pathology. Nor is there a sole set of characteristics; individuals may experience a variety of affinities and difficulties. Mckenzie (2013) proposed a "theory of (poss)ability" rooted in the social model of disability and in the understanding that impairments, like that in ID, are an interaction between the individual and their environment, postulating that competence is a function of *context,* rather than a direct, fixed property of the individual (McKenzie, 2013).

In his landmark 2015 book, *NeuroTribes*, Steve Silberman provides the following to understand neurodiversity:

> ...think in terms of *human operating systems* instead of diagnostic labels like dyslexia and ADHD. The brain is, above all, a marvelously adaptive organism, adept at maximizing its chances of success even in the face of daunting limitations. Just because a computer is not running Windows doesn't mean that it's broken (p. 471).

A consensus is building defining neurodiversity as the "limitless variability of human cognition and the uniqueness of each human mind" (Singer, 2020b, n.p.), and regarding sociability, learning, attention, and other mental functions in a non-pathological sense (Armstrong, 2015). These inclusive definitions focus on the act or behavior of being human rather than any specific means to differentiate one person from another (Baker, 2011; Runswick-Cole, 2014).

In popular culture, neurodiversity has become synonymous with "neurologically different" (i.e., centering the experiences of those without neurological impairments as the norm) (Kapp et al., 2013). Regrettably, disability-related terms often "rapidly acquire stigma, [get]. devalued, and [sic] ...lose its power as a unifying symbol for all" (Singer, 2020a, n.p.).

This difference, as supported by neurodiversity advocates, is natural and innate (see Research Box 15.1)—meaning it exists from birth, and manifests over time, though, is often associated with diagnosis by

Research Box 15.1: Deficit, Difference, or Both? Autism and Neurodiversity (Kapp et al., 2013)

Background. The neurodiversity movement challenges the medical model's interest in causation and cure, celebrating autism as an inseparable aspect of identity.

Method. Participants (N = 657) included autistic people, relatives and friends of autistic people, and people with no specified relation to autism. Using an online survey, researchers examined the perceived opposition between the medical model and the neurodiversity movement by assessing conceptions of autism and neurodiversity among people with different relations to autism.

Key findings. Self-identification as autistic and neurodiversity awareness was associated with viewing autism as a positive identity that needs no cure, suggesting core differences between the medical model and the neurodiversity movement. Nevertheless, results suggested a substantial overlap between these approaches to autism. Recognition of the negative aspects of autism and endorsement of parenting practices that celebrate and ameliorate but do not eliminate autism did not differ based on relation to autism or awareness of neurodiversity.

Conclusion and implications: These findings suggest a deficit-as-difference concept of autism wherein neurological conditions may represent equally valid pathways within human diversity. Potential areas of common ground in research and practice regarding autism are discussed.

What Do You Think?

1. How might community health and wellness systems build on the positive neurodiversity identity that has emerged among individuals with a formal or self-diagnosis?
2. What competencies would community health practitioners need for the sustainable health of people with neurodiverse conditions?

clinicians in early childhood (Kieling et al., 2010; Wilson, Hicks, Foster, McGue, & Iacono, 2015). Disparities on the socioeconomic gradient in the identification of people with neurodiverse conditions are widespread. In the US, those of lower socioeconomic background and/or racial/ethnic minorities are less likely to be identified or self-identity as neurodiverse (Baker, 2011; Sarrett, 2016; Tincani, Travers, & Boutot, 2009). **Neurodiversity**, in its proliferation, has evolved to bridge both its neurological basis and its being a social construct (Kapp, 2020; Runswick-Cole, 2014). Recognition of the neurobiological foundations of neurodiversity is not to deny the social participation implications that accompany neurodiverse conditions. For these reasons, neurodiversity advocates have always recognized the impairments inherent to medicalized diagnoses and the disabilities that are constructed by societal barriers (Kapp, 2020).

History of Research and Practice Regarding Neurodiversity and Health

Historically, neurodivergent populations were fragmented, each with their specific history from when the condition was formally recognized. For example, beginning with Leo Kanner (1943), Hans Asperger (1944) and, later, Lorna Wing, the examination and understanding of autism have been equally as varied (Wing & Gould, 1979). Early work, like that of Kanner and Asperger, focused on the deficits of autistic youth and highlighted social differences and "obsessive insistence on the preservations of sameness" (Kanner & Eisenberg, 1957) as the core characterizations that persist into adulthood. Though, radically different conceptualizations of autism have emerged since the work of Kanner and Asperger, many practitioners, clinicians, and teachers continue to refer to autism using similar descriptions. Lorna Wing and her colleague Judith Gould introduced what came to be called the "triad of impairments" (Wing & Gould, 1979): social isolations, communication, and imagination. Though, focused on impairments, Wing was careful to emphasize the variability of individuals (Wing, 1996), and that difficulties and strengths can present differently across the lifespan as well as in different environmental contexts. Though interesting, the broader history and scope of change in the diagnostic criteria of autism, or any other neurodiverse condition, is beyond the scope of this one chapter.

In 1998, to counter the heavily ingrained medical model of disability and societal misconceptions of neurodiverse conditions, Judy Singer proposed a new term—Neurodiversity (Singer, 2017). Leveraging the social model of disability and a constructivist lens, Singer proposed neurodiversity as a shift from defining disability in terms of its impairment alone, particularly with reference to "hidden" disabilities such as autism (Singer, 2017, p. 13). While being correctly credited with coining the new term, Singer was not the only person contemplating a unifying definition for individuals with neurological disabilities. Thanks in part to the democratization of the internet (Blume, 1997), many self-advocates found community and like-minded individuals with similar life experiences to their own. One such community, *InLv* (www.inlv.org), provided a space for individuals to congregate and share such experiences. Emerging from these shared experiences was a new understanding of autism as described by autistic[1] individuals—much of which ran counter to the traditional beliefs of autism guided by deficit models such as the medical model of disability.

The conversations that occurred through communities like *InLv* demonstrate that, although there were certain impairments and difficulties that individuals faced on a daily basis, there also existed many strengths—an aspect that very much is still overlooked today by neurotypical peers, as well as scholars (see also Blume, 1998). In subsequent years, neurodiversity has come to be described as "the rallying cry of the first new civil rights movement to take off in the 21st century" (Silberman, 2013, n.p.), and has created space for needed contributions from disabled scholars (Brown, Ashkenazy, & Onaiwu, 2017; Kapp, 2020).

Community-Oriented Approaches to Inclusion of Neurodiverse Populations

Like other disability groups, a concerted push by the neurodiversity movement has focused on ensuring the full community inclusion of people with neurodiverse conditions for their health and wellness. This inclusiveness would entail collaboration among the community of people with neurodiverse conditions and community partners across a broad range of community participation domains for wellness. These include physical activity (PA), social access, self-advocacy, and family support.

Physical Activity Approaches Physical inactivity is a key, modifiable risk factor for many health outcomes (see also Chap. 16, this volume), yet those with neurodiverse conditions continue to be minimally

[1] The use of identity-first language (IDL) is highly preferential to individuals on the autism spectrum (Kenny et al., 2016); therefore, the authors, out of respect for that preference, have used IDL throughout this chapter.

engaged compared to NT people. For instance, PA engagement of autistic individuals is low compared to non-autistic peers (Benson et al., 2019; Healy, Aigner, Haegele, & Patterson, 2019; Obrusnikova & Miccinello, 2012; Stanish et al., 2017) and those with other neurodevelopmental disabilities (Einarsson et al., 2015; Foley & McCubbin, 2009; Stanish & Mozzochi, 2000); see also chapter on Intellectual and Developmental Disabilities, this volume). For autistic individuals, limited engagement in PA may—as the neurodiversity movement suggests—be hindered by an array of societal or environmental barriers to engagement. Research exploring this aspect of engagement has suggested that a variety of interpersonal, intrapersonal, and community factors limit access (Blagrave & Colombo-Dougovito, 2019; Obrusnikova & Miccinello, 2012; Stanish et al., 2017). Nichols, Block, Bishop, and McIntire (2019), through the perspective of parents, identified several facilitators and barriers to PA for young autistic adults. Parents highlighted that, despite positive attitudes toward PA, lack of interest in certain physical activities, safety, physical traits (e.g., motor skill delay, aggression, and hypersensitivity), and the availability of programs and facilities act as major barriers.

Among children and adolescents with ID, evidence suggests they are less physically active and participate in lower intensity PA compared to peers without ID (Einarsson et al., 2015; Foley & McCubbin, 2009; Stanish & Mozzochi, 2000). PA levels have been shown to decline with age as sedentary behaviors increase (Phillips & Holland, 2011) despite the potential benefit to the mortality risk faced by adults with ID (Diaz, 2020). In fact, less than 50% of children and adolescents with ID meet WHO-recommendations of daily PA (60 min. of MVPA) (Wouters, Evenhuis, & Hilgenkamp, 2019). Even more troubling is the lack of effective interventions for neurodiverse populations (Colombo-Dougovito & Block, 2019; McGarty, Downs, Melville, & Harris, 2018). Limited engagement with stakeholders (Shields & Synnot, 2016) may have an impact on the development of an effective intervention, especially given the potential for decreased autonomy of this population and the likely reliance of caregivers for PA opportunities (McGarty & Melville, 2018). Unless organizations consult with key stakeholders, there is a low likelihood for opportunities to engage with the community in an impactful and sustainable way.

Social Access Approaches Societal norms often act as barriers for neurodivergent individuals, particularly those with sensory sensitivities or complex communication needs. Jessica Hughes (2016) outlines six strategies to assist with improving the accessibility of organizations: make meeting spaces and group communication accessible, presume competence, listen to people with neurodiverse conditions, acknowledge intersectionalities, reach out to under-represented neurodiverse groups, and acknowledge/question common sense perspectives.

1. ***Make meeting spaces and group communication accessible.*** One of the biggest barriers to the inclusion of individuals is accessible spaces for wellbeing. When considering accessibility, most default to considering physical barriers for those with mobility impairments. For people who are neurodivergent, accessibility is dependent on meeting sensory and communication needs. This may mean, for example, asking patrons to limit the use of strong fragrances; finding spaces without harsh lighting; or checking on the noise level of meeting spaces. Additionally, large group discussions may not work for many neurodivergent individuals; so, small groups or one-on-one conversations may need to be more common. Though it is difficult to accommodate every single need at every single event, organizations should check with potential attendees to attempt to accommodate for as much as they can. Too often, individuals with disabilities get the message that they are a burden or unwanted due to inaccessibility; organizations can take a good first step to ask what individuals need, as well as including accessibility aids as a default for any gathering.
2. ***Presume competence.*** Stemming from inclusive school movement within the field of education (Biklen & Burke, 2006), "presuming competence" rests on the principle of treating people "as if [they are] smart" (Jorgensen, 2005, p. 5). It stresses the importance of (re)constructing one's idea of "normal" by recognizing that "normal" is a social construct and abilities are fluid. For example, if an individual does not have the ability to communicate verbally in a typical manner, many would not attempt to consult that person as they presume they have nothing to "say." Yet, when given the opportunity and the appropriate accommodations, that individual may have an opinion on innumerable topics. The key is to not make presumptions based on

preconceived notions or stereotypes as this can lead to continued stigma and, ultimately, discrimination.

3. ***Listen to neurodivergent people.*** It should not be a radical suggestion to include neurodivergent people in conversations about events and opportunities that impact, directly, their daily lives. However, this rarely occurs (see also Runswick-Cole, 2014). Further, those working in with neurodiverse groups should learn to listen to people with neurodiverse conditions and to be continually learning about neurodivergent people. This information could come from academic journals and books or book chapters, like this one, but individuals must not constrain their pursuit of knowledge to the standard outlets, as these are often inaccessible to neurodivergent people. Thanks to the internet, neurodivergent individuals can compose communication in their own time away from many of the barriers they typically face.
4. ***Acknowledge and include intersectional perspectives.*** Intersectionalities (see also Kimberlé Crenshaw, 1991) define the community of people with neurodiverse conditions across gender, race, sexual orientation, religion, creed, social class, and so on. When seeking to build inclusive wellness for people with neurodiverse conditions, communities should recognize the intersection of identities and try to include a myriad of perspectives. Those that are traditionally under-represented in particular will provide unique insight about privilege, oppression, and access.
5. ***Reach out to under-represented members of the neurodivergent community.*** When designing community health and wellbeing spaces, it would be important to include as many perspectives as possible across the community of people with neurodiverse conditions. Even more difficult in this endeavor is including a diversity of voices in a hugely diverse category like neurodiversity, especially those with ID (see also Chap. 14, this volume). Therefore, individuals should seek a cross-section of representation from the breadth of neurodivergent people, including those who are autistic, with ADHD, ID, and other conditions.
6. ***Acknowledge and question "common sense" expectations.*** Communities need to continue to learn about neurodiversity and do the essential work of breaking down barriers for access. As they do so, "they will become more aware of ableist, 'common sense' expectations and the negative impacts" (Hughes, 2016, p. 14) that reduce the

accessibility and inclusiveness of community spaces for wellbeing. It is vital work for these community partners and people with neurodiverse conditions to continue to question their own assumptions and acknowledge when mistakes are made.

Self-Advocacy Approaches Self-advocacy approaches seek to support people with neurodiverse conditions to achieve full community inclusion across all health and wellbeing services (Herrera & Perry, 2014; Lam, 2016; Trainor, Lindstrom, Simon-Burroughs, Martin, & Sorrells, 2008; Wehmeyer & Abery, 2013). Examples of self-advocacy-oriented programs include the United States-based Academic Spectrum Partnership in Research and Education (AASPIRE), the Australia-based Cooperative Research Centre for Living with Autism (Autism CRC), and the United Kingdom-based Centre for Research in Autism and Education (CRAE). These are all aimed to benefit people with neurodiverse conditions and to allow their family caregivers to achieve community living options of choice (den Houting, 2018). Programs emphasizing the importance of collaborative partnerships in the implementation of best practices in community living planning are designed to optimize the fit between the person with neurodiverse conditions and the community living contexts.

Family Support Approaches Family or parental supports are critical to a successful neurodiverse community (Armstrong, 2010; Catalano, Holloway, & Mpofu, 2018; Martinez, Conroy, & Cerreto, 2012). Family settings can potentially provide a safer environment for people with neurodiverse conditions to try out a variety of social interactions and community experiences (Carter et al., 2009; Korpi, 2008; Neece, Kraemer, & Blacher, 2009). However, in order to support family members with neurodiverse conditions in healthy community living, the family members often need training in best approaches that foster opportunities for community living. The family must work with community health service providers to ensure greater prevention against maladaptive behaviors of dependency and facility-based care. Moreover, geographical location may also influence the resources families need for supporting their members who have neurodiverse conditions. For instance, those in rural communities and/or lower socioeconomic status neighborhoods are less likely to have access to basic amenities as well as work opportunities compared to higher socioeconomic status neighborhoods (Taylor & Mailick, 2014). Online family education programs are increasingly used to support family members in successfully managing community participation by members with neurodiverse conditions. Such programs offer tele-support programs to families to equip them with the competencies they need to support members with neurodiverse conditions in rural and remote areas (see Chap. 11, this volume).

The Influences of Culture, Professional Practices, and Legislation

We consider influences of culture while prioritizing the voices of people with neurodiverse conditions in their own community health and wellbeing. Through the work of dedicated advocates and individuals, neurodiversity is now more recognized as a part of the broader diversity of the schools, workplace, and community. We also consider professional practice issues involved in addressing sensory capabilities for the community health and wellbeing of people with neurodiverse conditions. Finally, we consider legal and disability rights influences that impact the community health and wellbeing of people with neurodiverse conditions, acknowledging that people with disabilities are self-determining "know what is best for themselves and their community" (Charlton, 1998, pg. 2).

Cultural and Societal Influences Empowerment, independence, integration, and self-determination are at the core of the "Nothing about Us without Us" disability rights movement, which began in the 1990s and served as the catalyst to the neurodiversity movement. The "Nothing about Us without Us" movement has advanced self-representation and control over supports and resources needed for quality of life, including health. Furthermore, the push by people with disabilities for inclusivity has challenged abled-bodied individuals to consider the implications of participation in decision-making process on the health and wellbeing of people with disabilities (Charlton, 1998). The neurodiversity movement has actively sought to push for the community equity and social justice to mitigate discriminatory practices that impair their health and wellbeing. For instance, numerous researchers (e.g., Blagrave & Colombo-

Dougovito, 2019; Buchanan, Miedema, & Frey, 2017; Nichols et al., 2019; Pan & Frey, 2006) have documented barriers that inhibit the full community participation for neurodiverse individuals within their own communities, and exclusion from the mainstream community (Buchanan et al., 2017; Nichols et al., 2019; Pan & Frey, 2006).

Professional Practice Issues Sensory and movement differences reported by and experienced by neurodivergent individuals, can have a significant impact on the ability of the individual, as well as her/his NT peers, to relate to and participate in social interactions (Colombo-Dougovito, Blagrave, & Healy, 2020; Donnellan, Hill, & Leary, 2015). Sensory issues can also "inhibit the quality of the participation" as individuals with neurodiverse conditions process sensory information differently (Healy, Msetfi, & Gallagher, 2013). For instance, professional practices focused on the development of a health-enhancing lifestyle or PA routine serving as a strength for autistic adults. Instead, enhancing their adherence and a desire for regular participation would gain more traction if they addressed hypersensitivity to sensory stimuli (e.g., lights and sound) and difficulties navigating social contexts. Professional practices for the community wellbeing of people with neurodiverse conditions should take regard of the factual preference by people with neurodiverse conditions for routine, predictability and low sensory stimulation for their sustainable health (Humphrey & Lewis, 2008). Moreover, people with neurodiverse conditions seek to achieve healthier community living through balancing their physical and mental health needs (Colombo-Dougovito et al., 2020; Demetriou, DeMayo, & Guastella, 2019). Professions must recognize that neurodiverse conditions cannot be "cured" or made "indistinguishable from peers" (Nicolaidis, 2012). Instead, disciplines should focus on understanding the accommodations and supports that individuals need to be successful, so as to increase the ease of inclusion within the community.

Legislative Influences Neurodivergent people are entitled to "comprehensive habilitation and rehabilitation services in the areas of health, employment, and education" (The United Nations, 2006). For individuals with neurodevelopmental disabilities, the Convention on the Rights of Persons with Disabilities (CPRD) provides a framework through which sustainable community-health can be developed that includes access to health care; programming to maintain or improve functioning for daily living; and equitable and accessible physical activity opportunities—all of which can enhance physical and mental wellbeing. In the US setting, civil rights laws such as the Americans with Disabilities Act (ADA) require all public entities to make their services, programs, and activities accessible to individuals with disabilities, inclusive of amenities such as golf courses, private schools, health clubs, and sports facilities. Similarly, the Rehabilitation Act of 1973 protects the work participation needs of people with disabilities, including those who identify as neurodivergent. Under the Individuals with Disabilities Education Act (IDEA) of 2004, children with neurodevelopmental disorders receive support to access multiple services including health services (34 CFR § 303.16) and "specially-designed" services, such as adapted physical education, movement education or motor development services (34 CFR § 300.39), athletics, and multidisciplinary services. These reduce the risk of acute and chronic conditions and promote health within the community. Collectively, these civil rights protections facilitate the engagement of community members with disabilities in PA and health-enhancing pursuits that optimize the health and quality of life of all community members.

Related Disciplines in Sustainable Health for Neurodiverse Individuals

A cultural shift is necessary to effectively provide accessible spaces and authentic inclusion for neurodivergent individuals. As it stands, presently, society favors NT individuals often at the detriment of those with neurodiverse conditions. It will take communities adopting adaptive and flexible routines to meet the diverse needs of individuals to ensure adequate access to community spaces. Supporting the community health and wellbeing of people with neurodiverse conditions requires interdisciplinary approaches including disciplines such as occupational therapists, speech therapists, physical therapists, and social workers. Additionally, physical educators and personal trainers can be immensely beneficial in the community health of neurodivergent individuals. Details on the role of some of these disciplines have been addressed in an earlier chapter (Chap. 14, this volume) and will not be repeated here. We note that interdisciplinary approaches lead to a better understanding of the community health needs of people with neurodiverse conditions and the necessary accommodations and supports that individuals need for wellness.

The role of the personal trainers and physical educators in providing physical activity support to people with neurodiverse conditions is particularly significant. Personal trainers work in community facilities such as health clubs, fitness or recreation centers, gyms, and yoga and Pilates studios. They help novice exercisers with the basic mechanics of exercise activities such as weight-lifting or cardiovascular exercise. They can also help motivate experienced individuals or provide new routines. They can be helpful for neurodiverse individuals by (a) building familiarity with equipment or with various exercises; (b) develop routines or exercise plans; (c) act as a source of familiarity or calm in a chaotic environment; and (d) facilitate accommodations to equipment, exercises, and even spaces to meet the varying needs of neurodiverse individuals.

Physical educators support the foundational development of motor skills and physical fitness and are a core tenet of community health (Armour, 2010). Often, physical educators serve as the first exposure that many individuals will experience to new games, sports, and physical activities. Adapted physical educators are a subspecialty within physical education who work with people with neurodiverse conditions providing foundational learning experiences. Such educators are necessary to build the skills to be physically active for a lifetime and introducing community members to new activities that one might not typically come in contact. They promote the sustainable health of communities, teaching the members about the health effects of exercise and supporting community members in the transition to community physical activity with local organizations.

Issues of Research and Other Forms of Scholarship

As a young paradigm, neurodiversity has provided a "big tent" for individuals with neurodiverse conditions to coalesce, improving on their community participation and wellbeing. We address four research and practice needs to address the gaps in evidence in the sustainable community health of people with neurodiverse conditions: research community priorities, use of community based participatory approaches, provider education and training and strength-based approaches.

Research Community Priorities Large gaps exist between the research community priorities and those of community health needs of neurodiverse individuals. This would be the case with research about neurodivergent people focused on "basic science" of the associated conditions such as neural and cognitive pathways, genetics, and other risk factors (Charman & Clare, 2004; Krahn & Fenton, 2012; Singh, Illes, Lazzeroni, & Hallmayer, 2009) rather than on their sustainable community health. This approach is at odds with the core interests of the community that individuals with neurodiverse conditions, and should be focused on successful community living including employment and acceptance (Pellicano, Dinsmore, & Charman, 2014). While important, the understanding of the causes of neurodiverse conditions does little to improve the quality of life (QOL) of such individuals. The World Health Organization (WHO) (WHOQOL group, 1995) defines QOL as an: "Individual's perception of their position in life in the context of the culture and value systems in which they live and in relation to their goals, expectations, standards, and concerns." Though gaining in popularity, QOL has received little attention historically. Robertson (2009) outlined several barriers to QOL experienced by autistic adults. They are (a) self-determination, (b) social isolation, (c) material wellbeing, (d) personal development, (e) emotional wellbeing, (f) interpersonal relationships, (g) rights, and (h) physical wellbeing. Among adults with ADHD, internalized stigma and anticipated discrimination may be limiting access to opportunities to improve their health-related QOL (Masuch, Bea, Alm, Deibler, & Sobanski, 2019). Research is desperately needed to address the barriers that are faced by neurodivergent individuals regarding their QOL. As many of the highlighted barriers are rooted in the stigmatization of these conditions, researchers will need to engage with community stakeholders and advocates to find practical solutions. Although the research community is trending toward community living and participation in recent years (Chown et al., 2017; Pellicano et al., 2018), this has been a much slower process than would be expected based on need (Warner, Cooper, & Cusack, 2019).

Community-Based Participatory Approaches Few studies on community health and wellbeing of neurodiverse people have used community-based participatory approaches (CBPA), although such approaches are vital to ensuring any action or policy is matched to the needs of the community of people with neurodiverse conditions. Community-based participatory approaches have the advantage of inclusivity, ensuring that people with neurodiverse conditions are listened to regarding their community health and wellbeing needs. As an example, the Academic Autistic Spectrum Partnership in Research and Education (AASPIRE) has utilized community-based participatory approaches for authentic inclusivity, collaborating with autistic people and individuals with ID while addressing their community living needs (Nicolaidis et al., 2011).

Provider Education and Training Professionals and other community organizations require further education in providing community health and wellbeing supports to people with neurodiverse conditions (Colombo-Dougovito, 2015; Muller-Heyndyk, 2018). As examples, professionals ill-equipped to work with people with neurodiverse conditions for healthy community participation would provide a poor experience for such individuals and a higher likelihood of avoidance in the future (Colombo-Dougovito et al., 2020). Future participation apathy can be extremely detrimental when poor experiences occur in areas related to personal health, as individuals may avoid these experiences in the future and, thus, not be able to access the benefits of those experiences. Future research should examine best practices in education and training approaches for providers of neurodiversity inclusive community health and wellbeing programs for the sustainable health of neurodivergent people.

Strength-Based Approaches Deficit-based research and practice approaches have persisted, limiting the potential implementation effectiveness of findings and have perpetuated the stigma of neurodivergent conditions. Yet, as has been noted, neurodivergent individuals possess many strengths that can be leveraged toward success (Donaldson, Krejcha, & McMillin, 2017). There is need for increased research on strength-based approaches to inform sustainable community health approaches for people with neurodiverse conditions. When providers and researchers primarily focus on deficits or impairments relating to neurodiversity, they miss out on opportunities to make a difference by identifying community wellness options for the sustainable health of people with neurodiverse conditions. By focusing on strengths over impairments, people with neurodiverse conditions, researchers, and providers can work collaboratively for the benefit of the community of people with neurodiverse conditions and their sustainable futures.

Summary and Conclusion

In this chapter, we have examined community health for neurodivergent people, recognizing their collective diversity from not only belonging to their neurodevelopmental disability groupings such as attention-deficit/hyperactivity disorder (ADHD) and autism but also recognizing how their health and wellbeing would be influenced by their intersectionalities across the socioeconomic gradient. We considered the discourse on the history and theoretical underpinnings of neurodiversity important to the sustainable community health of neurodivergent people. Of current approaches for the community health and wellbeing of people with neurodiverse conditions, those premised on full community inclusion appear to hold great promise. Particularly significant would be community-based participatory programs and strength-based sustainable health community approaches in which the community of people with neurodiverse conditions is centrally involved in the decision making. This would require interdisciplinary teamwork and collaboration with community organizations for and of people with neurodiverse conditions. Further evidence is needed on the best and most sustainable community health approaches with the community of people with neurodiverse conditions, regardless of their differentiation by biomarker functioning. By understanding these perspectives of individuals who identify as being neurodiverse, health professionals can increase community responsiveness, accessibility, and overall wellbeing.

Self-Check Questions

1. Define the community of people with neurodiverse conditions?
2. What are the major historical milestones in the recognition of neurodivergent people?
3. Outline some promising approaches in promoting the sustainable community health of people with neurodiverse conditions?
4. How is the science of neurodiversity an interdisciplinary field, and with what benefits to research and practice? What areas are in greatest need in research and practice with people with neurodiverse conditions?

Discussion Questions

1. What are some ways in which your culture has shaped your views regarding neurodiverse populations? How do you think these views might influence your development of community-health interventions and programming with people with neurodiverse conditions?
2. What are some of the barriers that individuals with neurodiverse conditions face when attempting to engage in health-enhancing behaviors? How might you employ the strategies discussed in this chapter to increase the accessibility of community health and wellbeing spaces for neurodiverse individuals, and improve the representation and co-ownership within these community spaces?

Field-Based Experiential Exercises

1. **Attend a neurodiversity event in your community to learn and participate in their priority activities.** What are the health and wellbeing implications of the neurodiversity event activities? How might these activities be enhanced for the sustainable health of people with neurodiverse conditions?
2. **Community Reflection**. Now that you have learned about neurodiversity and the various disability categories that fit under this umbrella, reflect on the health and wellbeing spaces that are in your community, such as your local parks, recreation facilities, and infrastructure. How neurodivergent inclusive are these community resources and why? What community spaces could be redeveloped for the inclusion of these individuals? Based on your observations, develop an action plan for the sustainable community health of people with neurodiverse conditions in the community space or spaces you identified.

Sample Online Resources

- Autistic Community and the Neurodiversity Movement (open source).: https://link.springer.com/book/10.1007/978-981-13-8437-0
- Autistic Self Advocacy Network: https://autisticadvocacy.org/
- Best Resources for Achievement and Intervention re Neurodiversity in Higher Education (Brain.HE): http://www.brainhe.com/index.html
- Divergents Magazine: https://www.divergents-magazine.org
- Dyslexic Advantage: https://www.dyslexicadvantage.org/
- Thinking Person's Guide to Autism: http://www.thinkingautismguide.com/
- Understood.org Neurodiversity: What You Need to Know: https://www.understood.org/en/friends-feelings/empowering-your-child/building-on-strengths/neurodiversity-what-you-need-to-know
- University/higher education partnerships
  - Stanford University's Neurodiversity Project [https://med.stanford.edu/neurodiversity.html]
  - William & Mary University Neurodiversity Initiative [https://www.wm.edu/sites/neurodiversity/index.php
  - University of North Texas's Neurodiversity Initiative: https://neurodiversity.unt.edu
- Autistic Self Advocacy Network (ASAN) *Autistic Access Needs: Notes on Accessibility* ASAN available at https://autisticadvocacy.org/wp-content/uploads/2016/06/Autistic-Access-Needs-Notes-on-Accessibility.pdf
- Blogs and personal writings from autistic and neurodivergent people such as Autistic Hoya (http://autistichoya.com), Just Stimming, (https://juststimming.wordpress.com), Ollibean (https://ollibean.com), and Thinking Person's Guide to Autism (http://www.thinkingautismguide.com) provide firsthand experiences vital to ensuring access.

References

American Association on Intellectual and Developmental Disabilities. (2010). *Intellectual disability*. West Sussex, UK: Wiley-Blackwell.

American Psychiatric Association. (2013). *Diagnostic and statistical manual of mental disorders (DSM-5®)*. Washington, DC: APA.

Americans With Disabilities Act of 1990, Pub. L. No. 101-336, 104 Stat. 328 (1990).

Armour, K. M. (2010). The physical education profession and its professional responsibility… or… why '12 weeks paid holiday'will never be enough. *Physical Education and Sport Pedagogy, 15*(1), 1–13.

Armstrong, T. (2010). *Neurodiversity: Discovering the extraordinary gifts of autism, ADHD, dyslexia, and other brain differences.* Cambridge, MA: Da Capo.

Armstrong, T. (2015). The myth of the normal brain: Embracing neurodiversity. *AMA Journal of Ethics, 17*(4), 348–352.

Austin, R. D., & Pisano, G. P. (2017). Neurodiversity as a competitive advantage. *Harvard Business Review, 95*(3), 96–103.

Baker, D. L. (2011). *The politics of neurodiversity: Why public policy matters.* Boulder, CO: Lynne Rienner Publishers.

Benson, S., Bender, A. M., Wickenheiser, H., Naylor, A., Clarke, M., Samuels, C. H., & Werthner, P. (2019). Differences in sleep patterns, sleepiness, and physical activity levels between young adults with autism spectrum disorder and typically developing controls. *Developmental Neurorehabilitation,* 22(3), 164–173.

Biklen, D., & Burke, J. (2006). Presuming competence. *Equity & Excellence in Education, 39*(2), 166–175.

Blagrave, A. J., & Colombo-Dougovito, A. M. (2019). Experiences participating in community physical activity by families with a child on the autism spectrum: A phenomenological inquiry. *Advances in Neurodevelopmental Disorders, 3*(1), 72–84.

Blume, H. (1997, June 30). Autistics, freed from face-to-face encounters, are communicating in cyberspace. *The New York Times.* https://www.nytimes.com/1997/06/30/business/autistics-freed-from-face-to-face-encounters-are-communicating-in-cyberspace.html

Blume, H. (1998, September 30). Neurodiversity. *The Atlantic.* https://www.theatlantic.com/magazine/archive/1998/09/neurodiversity/305909/

Bob, P., & Konicarova, J. (2018). Historical and recent research on ADHD. In *ADHD, stress, and development* (SpringerBriefs in Psychology series, pp. 11–19). Berlin: Springer.

Brady, N. C., Bruce, S., Goldman, A., Erickson, K., Mineo, B., Ogletree, B. T., & Wilkinson, K. (2016). Communication services and supports for individuals with severe disabilities: Guidance for assessment and intervention. *American Journal on Intellectual and Developmental Disabilities, 121,* 121–138.

Breyer, J. L., Botzet, A. M., Winters, K. C., Stinchfield, R. D., August, G., & Realmuto, G. (2009). Young adult gambling behaviors and their relationship with the persistence of ADHD. *Journal of Gambling Studies, 25*(2), 227–238.

Broder-Fingert, S., Brazauskas, K., Lindgren, K., Iannuzzi, D., & Van Cleave, J. (2014). Prevalence of overweight and obesity in a large clinical sample of children with autism. *Academic Pediatrics, 14*(4), 408–414.

Brown, L. X., Ashkenazy, E., & Onaiwu, M. G. (Eds.). (2017). *All the weight of our dreams: On living racialized autism.* Lincoln, UK: DragonBee Press.

Buchanan, A. M., Miedema, B., & Frey, G. C. (2017). 'Parents' perspectives of physical activity in their adult children with autism spectrum disorder: A social-ecological approach. *Adapted Physical Activity Quarterly, 34*(4), 401–420.

Carter, E. W., Trainor, A. A., Cakiroglu, O., Swedeen, B., & Owens, L. A. (2009). Availability of and access to career development activities for transition-age youth with disabilities. *Career Development for Exceptional Individuals, 33,* 13–24.

Catalano, D., Holloway, L., & Mpofu, E. (2018). Mental health interventions for parent carers of children with autistic spectrum disorder: Practice guidelines from a critical interpretive synthesis (CIS) systematic review. *International Journal of Environmental Research and Public Health, 15,* 341. https://doi.org/10.3390/ijerph15020341

Charlton, J. I. (1998). *Nothing about us without us: Disability oppression and empowerment.* Berkeley, CA: University of California Press. ProQuest Ebook.

Charman, T., & Clare, P. (2004). *Mapping autism research: Identifying UK priorities for the future.* London, UK: National Autistic Society.

Chown, N., Robinson, J., Beardon, L., Downing, J., Hughes, L., Leatherland, J., & MacGregor, D. (2017). Improving research about us, with us: A draft framework for inclusive autism research. *Disability & Society, 32*(5), 720–734.

Colombo-Dougovito, A. M. (2015). "Try to do the best you can": How pre-service APE specialists experience teaching students with Autism Spectrum Disorder. *International Journal of Special Education, 30*(3), 160–176.

Colombo-Dougovito, A. M., Blagrave, A. J., & Healy, S. (2020). A grounded theory of adoption and maintenance of physical activity among autistic adults. *Autism.* Advanced online publication. https://doi.org/10.1177/1362361320932444

Colombo-Dougovito, A. M., & Block, M. E. (2019). Fundamental motor skill interventions for individuals with autism spectrum disorder: A literature review. *Review Journal of Autism and Developmental Disabilities, 6*(2), 159–171.

Crenshaw, K. (1991). Mapping the margins: Identity politics, intersectionality, and violence against women. *Stanford Law Review, 43*(6), 1241–1299.

Croen, L. A., Zerbo, O., Qian, Y., Massolo, M. L., Rich, S., Sidney, S., & Kripke, C. (2015). The health status of adults on the autism spectrum. *Autism, 19*(7), 814–823.

Curtin, C., Jojic, M., & Bandini, L. G. (2014). Obesity in children with autism spectrum disorders. *Harvard Review of Psychiatry, 22*(2), 93–103.

Danielson, M. L., Bitsko, R. H., Ghandour, R. M., Holbrook, J. R., Kogan, M. D., & Blumberg, S. J. (2018). Prevalence of parent-reported ADHD diagnosis and associated treatment among US children and adolescents, 2016. *Journal of Clinical Child and Adolescent Psychology, 47*(2), 199–212.

Demetriou, E. A., DeMayo, M. M., & Guastella, A. J. (2019). Executive function in autism spectrum disorder: History, theoretical models, empirical findings and potential as an endophenotype. *Frontiers in Psychiatry, 10*, 753.

den Houting, J. (2018). Neurodiversity: An insider's perspective. *Autism, 23*, 271–273.

Dente, C. L., & Coles, K. P. (2012). Ecological approaches to transition planning for students with autism and Asperger's syndrome. *Children & Schools, 34*(1), 27–36.

Diaz, K. M. (2020). Leisure-time physical activity and all-cause mortality among adults with intellectual disability: The National Health Interview survey. *Journal of Intellectual Disability Research, 64*(2), 180–184. https://doi.org/10.1111/jir.12695

Dietz, P. M., Rose, C. E., McArthur, D., & Maenner, M. (2020). National and state estimates of adults with autism spectrum disorder. *Journal of Autism and Developmental Disorders*. Advanced online publication. https://doi.org/10.1007/s10803-020-04494-4

Donaldson, A. L., Krejcha, K., & McMillin, A. (2017). A strengths-based approach to autism: Neurodiversity and partnering with the autism community. *Perspectives of the ASHA Special Interest Groups, 2*(1), 56–68.

Donnellan, A. M., Hill, D. A., & Leary, M. R. (2015). Rethinking autism: Implications of sensory and movement differences for understanding and support. Autism: The movement perspective. *Frontiers in Integrative Neuroscience, 6*, 124. https://doi.org/10.3389/fnint.2012.00124.

Einarsson, I. O., Olafsson, A., Hinriksdóttir, G., Jóhannsson, E., Daly, D., & Arngrímsson, S. A. (2015). Differences in physical activity among youth with and without intellectual disability. *Medicine & Science in Sports & Exercise, 47*(2), 411–418.

Elsabbagh, M., Divan, G., Koh, Y. J., Kim, Y. S., Kauchali, S., Marcín, C., & Yasamy, M. T. (2012). Global prevalence of autism and other pervasive developmental disorders. *Autism Research, 5*(3), 160–179.

Fladhammer, A. B., Lyde, A. R., Meyers, A. B., Clark, J. K., & Landau, S. (2016). Health concerns regarding children and adolescents with attention-deficit/hyperactivity disorder. In *Health promotion for children and adolescents* (pp. 145–165). Berlin, Germany: Springer.

Fletcher-Watson, S., & Happé, F. (2019). *Autism: A new introduction to psychological theory and current debate*. London, UK: Routledge.

Foley, J. T., & McCubbin, J. A. (2009). An exploratory study of after-school sedentary behaviour in elementary school-age children with intellectual disability. *Journal of Intellectual and Developmental Disability, 34*(1), 3–9.

Gray, D. E. (2002). 'Everybody just freezes. Everybody is just embarrassed': Felt and enacted stigma among parents of children with high functioning autism. *Sociology of Health & Illness, 24*(6), 734–749.

Gupte-Singh, K., Singh, R. R., & Lawson, K. A. (2017). Economic burden of attention-deficit/hyperactivity disorder among pediatric patients in the United States. *Value in Health, 20*(4), 602–609.

Healy, S., Aigner, C. J., Haegele, J. A., & Patterson, F. (2019). Meeting the 24-hr movement guidelines: An update on US youth with autism spectrum disorder from the 2016 National Survey of Children's Health. *Autism Research, 12*(6), 941–951.

Healy, S., Msetfi, R., & Gallagher, S. (2013). 'Happy and a bit nervous': The experiences of children with autism in physical education. *British Journal of Learning Disabilities, 41*(3), 222–228.

Herrera, C. D., & Perry, A. (Eds.). (2014). Ethics and neurodiversity. Newcastle upon Tyne, UK: Cambridge Scholars Publishing.

Hillary, A. (2013, March 5). *Erased, silenced, derailed.* [web log post] Retrieved on May 5, 2020 from http://yesthattoo.blogspot.com/2013/03/erased-silenced-derailed.html

Hughes, J. M. (2016). *Increasing neurodiversity in disability and social justice advocacy groups*. Washington, DC: Autistic Self Advocacy Network.

Humphrey, N., & Lewis, S. (2008). 'Make me normal' the views and experiences of pupils on the autistic spectrum in mainstream secondary schools. *Autism, 12*(1), 23–46.

Jaarsma, P., & Welin, S. (2012). Autism as a natural human variation: Reflections on the claims of the neurodiversity movement. *Health Care Analysis, 20*(1), 20–30.

Jorgensen, C. (2005). The least dangerous assumption. *Disability Solutions, 6*(3), 1–9.

Kanner, L., & Eisenberg, L. (1957). Early infantile autism, 1943–1955. Psychiatric research reports. *American Psychiatric Association, 7*, 55.

Kapp, S. K., Gillespie-Lynch, K., Sherman, L. E., & Hutman, T. (2013). Deficit, difference, or both? Autism and neurodiversity. *Developmental Psychology, 49*(1), 59–71.

Kapp, S. K. (Ed.). (2020). Autistic community and the neurodiversity movement: Stories from the Frontline. London, UK: Palgrave Macmillan.

Kenny, L., Hattersley, C., Molins, B., Buckley, C., Povey, C., & Pellicano, E. (2016). Which terms should be used to describe autism? Perspectives from the UK autism community. *Autism, 20*(4), 442–462.

Kessler, R. C., Adler, L., Barkley, R., Biederman, J., Conners, C. K., Demler, O., … Spencer, T. (2006). The prevalence and correlates of adult ADHD in the United States: Results from the National Comorbidity Survey Replication. *American Journal of Psychiatry, 163*(4), 716–723.

Kieling, C., Kieling, R. R., Rohde, L. A., Frick, P. J., Moffitt, T., Nigg, J. T., … Castellanos, F. X. (2010). The age at onset of attention deficit hyperactivity disorder. *American Journal of Psychiatry, 167*(1), 14–16.

Kim, J., Mutyala, B., Agiovlasitis, S., & Fernhall, B. (2011). Health behaviors and obesity among US children with attention deficit hyperactivity disorder by gender and medication use. *Preventive Medicine, 52*(3–4), 218–222.

Korpi, M. (2008). *Guiding your teenager with special needs through the transition from school to adult life: Tools for parents*. London, UK: Jessica Kingsley.

Krahn, T. M., & Fenton, A. (2012). Funding priorities: Autism and the need for a more balanced research agenda in Canada. *Public Health Ethics, 5*(3), 296–310.

Lam, G. Y. H. (2016). *Self-determination during school-to-adulthood transition in young adults with autism spectrum disorder from the United States and Hong Kong*. University of South Florida Scholar Commons: Graduate Theses and Dissertations. https://scholarcommons.usf.edu/etd/6290/

Leigh, J. P., & Du, J. (2015). Brief report: Forecasting the economic burden of autism in 2015 and 2025 in the United States. *Journal of Autism and Developmental Disorders, 45*(12), 4135–4139.

Lunsky, Y., De Oliveira, C., Wilton, A., & Wodchis, W. (2019). High health care costs among adults with intellectual and developmental disabilities: A population-based study. *Journal of Intellectual Disability Research, 63*(2), 124–137.

Maenner, M. J., Shaw, K. A., Baio, J., Washington, A., Patrick, M., DiRienzo, M., & Dietz, P. M. (2020). Prevalence of autism spectrum disorder among

children aged 8 years—Autism and developmental disabilities monitoring network, 11 sites, United States, 2016. *MMWR. Surveillance Summary, 69*(4), 1–12.
Malcolm-Smith, S., Hoogenhout, M., Ing, N., Thomas, K. G. F., & de Vries, P. (2013). Autism spectrum disorders—Global challenges and local opportunities. *Journal of Child & Adolescent Mental Health, 25*(1), 1–5.
Martinez, D. C., Conroy, J. W., & Cerreto, M. C. (2012). Parent involvement in the transition process of children with intellectual disabilities: The influence of inclusion on parent desires and expectations for postsecondary education. *Journal of Policy and Practice in Intellectual Disabilities, 9*(4), 279–288. https://doi.org/10.1111/jppi.12000
Masuch, T. V., Bea, M., Alm, B., Deibler, P., & Sobanski, E. (2019). Internalized stigma, anticipated discrimination and perceived public stigma in adults with ADHD. *ADHD Attention Deficit and Hyperactivity Disorders, 11*(2), 211–220.
Matza, L. S., Paramore, C., & Prasad, M. (2005). A review of the economic burden of ADHD. *Cost Effectiveness and Resource Allocation, 3*(5). https://doi.org/10.1186/1478-7547-3-5
Maulik, P. K., Mascarenhas, M. N., Mathers, C. D., Dua, T., & Saxena, S. (2011). Prevalence of intellectual disability: A meta-analysis of population-based studies. *Research in Developmental Disabilities, 32*, 419–436.
McGarty, A. M., Downs, S. J., Melville, C. A., & Harris, L. (2018). A systematic review and meta-analysis of interventions to increase physical activity in children and adolescents with intellectual disabilities. *Journal of Intellectual Disability Research, 62*(4), 312–329.
McGarty, A. M., & Melville, C. A. (2018). Parental perceptions of facilitators and barriers to physical activity for children with intellectual disabilities: A mixed methods systematic review. *Research in Developmental Disabilities, 73*, 40–57.
McKenzie, J. A. (2013). Models of intellectual disability: Towards a perspective of (poss) ability. *Journal of Intellectual Disability Research, 57*(4), 370–379.
McKenzie, K., Milton, M., Smith, G., & Ouellette-Kuntz, H. (2016). Systematic review of the prevalence and incidence of intellectual disabilities: Current trends and issues. *Current Developmental Disorders Reports, 3*, 104–115.
Metcalfe, J., & Mischel, W. (1999). A hot/cool-system analysis of delay of gratification: Dynamics of willpower. *Psychological Review, 106*(1), 3–19.
Milton, D. E. (2012). On the ontological status of autism: The 'double empathy problem'. *Disability & Society, 27*(6), 883–887.

Muller-Heyndyk, R. (2018, February 15). Neurodiversity not a priority for nine out pf 10 businesses. *HR Magazine*. https://www.hrmagazine.co.uk/article-details/neurodiversity-not-a-priority-for-nine-out-of-10-businesses

National Institute of Mental Health. (2013). *Attention-deficit/hyperactivity disorder*. National Institutes of Health. https://www.nimh.nih.gov/health/topics/attention-deficit-hyperactivity-disorder-adhd/index.shtml

Neece, C. L., Kraemer, B. R., & Blacher, J. (2009). Transition satisfaction and family well being among parents of young adults with severe intellectual disability. *Intellectual and Developmental Disabilities, 47*(1), 31–43.

Nichols, C., Block, M. E., Bishop, J. C., & McIntire, B. (2019). Physical activity in young adults with autism spectrum disorder: Parental perceptions of barriers and facilitators. *Autism, 23*(6), 1398–1407. https://doi.org/10.1177/1362361318810221

Nicolaidis, C. (2012). What can physicians learn from the neurodiversity movement? *AMA Journal of Ethics, 14*(6), 503–510.

Nicolaidis, C., Raymaker, D., McDonald, K., Dern, S., Ashkenazy, E., Boisclair, C., … Baggs, A. (2011). Collaboration strategies in nontraditional community-based participatory research partnerships: Lessons from an academic-community partnership with autistic self-advocates. *Progress in Community Health Partnerships: Research, Education, and Action, 5*(2), 143–150.

Obrusnikova, I., & Miccinello, D. L. (2012). Parent perceptions of factors influencing after-school physical activity of children. *Adapted Physical Activity Quarterly, 29*, 6380. https://doi.org/10.1123/apaq.29.1.63

Pan, C. Y., & Frey, G. C. (2006). Physical activity patterns in youth with autism spectrum disorders. *Journal of Autism and Developmental Disorders, 36*(5), 597–606.

Pellicano, E., Dinsmore, A., & Charman, T. (2014). What should autism research focus upon? Community views and priorities from the United Kingdom. *Autism: The International Journal of Research and Practice, 18*(7), 756–770. https://doi.org/10.1177/1362361314529627

Pellicano, L., Mandy, W., Bölte, S., Stahmer, A., Lounds Taylor, J., & Mandell, D. S. (2018). A new era for autism research, and for our journal. *Autism, 22*(2), 82–83.

Phillips, A. C., & Holland, A. J. (2011). Assessment of objectively measured physical activity levels in individuals with intellectual disabilities with and without Down's syndrome. *PLoS One, 6*(12).

Rehabilitation Act of 1973, Pub. L. No. 93–112, 87 Stat. 394 (Sept. 26, 1973).

Robertson, S. M. (2009). Neurodiversity, quality of life, and autistic adults: Shifting research and professional focuses onto real-life challenges. *Disability Studies Quarterly, 30*(1). https://doi.org/10.18061/dsq.v30i1.1069
Rosen, T. E., Mazefsky, C. A., Vasa, R. A., & Lerner, M. D. (2018). Co-occurring psychiatric conditions in autism spectrum disorder. *International Review of Psychiatry, 30*(1), 40–61.
Roux, A. M., Shattuck, P. T., Rast, J. E., Rava, J. A., & Anderson, K. A. (2015). *National autism indicators report: Transition into young adulthood.* Philadelphia, PA: Life Course Outcomes Research Program, AJ Drexel Autism Institute, Drexel University.
Runswick-Cole, K. (2014). 'Us' and 'them': The limits and possibilities of a 'politics of neurodiversity' in neoliberal times. *Disability & Society, 29*(7), 1117–1129.
Sargent, K. (2019, December 6). Designing for neurodiversity and inclusion. *Work Design Magazine.* https://www.workdesign.com/2019/12/designing-for-neurodiversity-and-inclusion/
Sarrett, J. C. (2016). Biocertification and neurodiversity: The role and implications of self-diagnosis in autistic communities. *Neuroethics, 9*(1), 23–36.
Schalock, R. L., Luckasson, R. A., Shogren, K. A., Borthwick-Duffy, S., Bradley, V., Buntinx, W. H., … Yeager, M. H. (2007). The renaming of mental retardation: Understanding the change to the term intellectual disability. *Intellectual and Developmental Disabilities, 45*(2), 116–124. https://doi.org/10.1352/1934-9556(2007)45[116:TROMRU]2.0.CO;2
Segal, M., Eliasziw, M., Phillips, S., Bandini, L., Curtin, C., Kral, T. V., & Must, A. (2016). Intellectual disability is associated with increased risk for obesity in a nationally representative sample of US children. *Disability and Health Journal, 9*(3), 392–398.
Shields, N., & Synnot, A. (2016). Perceived barriers and facilitators to participation in physical activity for children with disability: A qualitative study. *BMC Pediatrics, 16*(1), 9.
Shtayermman, O. (2009). An exploratory study of the stigma associated with a diagnosis of Asperger's syndrome: The mental health impact on the adolescents and young adults diagnosed with a disability with a social nature. *Journal of Human Behavior in the Social Environment, 19*(3), 298–313.
Silberman, S. (2013, April 16). Neurodiversity rewires conventional thinking about brains. *WIRED.* https://www.wired.com/2013/04/neurodiversity/
Silberman, S. (2015, September 23). Our neurodiverse world: Autism is a valuable part of humanity's genetic legacy. *Slate Magazine.* Retrieved April 27,

2020, from https://slate.com/technology/2015/09/the-neurodiversity-movement-autism-is-a-minority-group-neurotribes-excerpt.html

Singer, J. (2017). *NeuroDiversity: The birth of an idea*. Author.

Singer, J. (2020a, March 6). That troublesome adjective "Neurodiverse". *Neurodiversity* 2.0. https://neurodiversity2.blogspot.com/2020/03/that-troublesome-adjective-neurodiverse.html

Singer, J. (2020b, March). What is neurodiversity? *Neurodiversity*. https://neurodiversity2.blogspot.com/p/what.html?m=1

Singh, J., Illes, J., Lazzeroni, L., & Hallmayer, J. (2009). Trends in US autism research funding. *Journal of Autism and Developmental Disorders, 39*(5), 788–795.

Stanish, H. I., Curtin, C., Must, A., Phillips, S., Maslin, M., & Bandini, L. G. (2017). Physical activity levels, frequency, and type among adolescents with and without autism spectrum disorder. *Journal of Autism and Developmental Disorders, 47*(3), 785–794.

Stanish, H. I., & Mozzochi, M. (2000). Participation of preschool children with developmental delay during gross motor activity sessions. *Research Quarterly for Exercise and Sport, 71*(Suppl. 1), A111–A112.

Taylor, J. L., & Mailick, M. R. (2014). A longitudinal examination of 10-year change in vocational and educational activities for adults with autism spectrum disorders. *Developmental psychology, 50*(3), 699–708.

Taylor, J. L., & Seltzer, M. M. (2011). Employment and post-secondary educational activities for young adults with autism spectrum disorders during the transition to adulthood. *Journal of Autism and Developmental Disorders, 41*(5), 566–574.

The Individuals with Disabilities Education Act. 20USC §1400 (2004). Interagency Autism Coordinating Committee. (2011). *IACC strategic plan for autism spectrum disorder research* (p. 66). Washington, DC: US Department of Health and Human Services.

The United Nations. (2006). Convention on the rights of persons with disabilities. *Treaty Series, 2515*, 3.

Thomas, R., Sanders, S., Doust, J., Beller, E., & Glasziou, P. (2015). Prevalence of attention-deficit/hyperactivity disorder: A systematic review and meta-analysis. *Pediatrics, 135*(4), e994–e1001.

Tincani, M., Travers, J., & Boutot, A. (2009). Race, culture, and autism spectrum disorder: Understanding the role of diversity in successful educational interventions. *Research and Practice for Persons with Severe Disabilities, 34*(3–4), 81–90.

Trainor, A. A., Lindstrom, L., Simon-Burroughs, M., Martin, J. E., & Sorrells, A. M. (2008). From marginalized to maximized opportunities for diverse youth with disabilities: A position paper of the division on career development and transition. *Career Development for Exceptional Individuals, 31*(1), 56–64.

Tyrer, F., Dunkley, A. J., Singh, J., Kristunas, C., Khunti, K., Bhaumik, S., & Gray, L. J. (2019). Multimorbidity and lifestyle factors among adults with intellectual disabilities: A cross-sectional analysis of a UK cohort. *Journal of Intellectual Disability Research, 63*(3), 255–265.

US Senate Committee on Health, Education, Labor, and Pensions. (2010). *Rosa's Law: Report (to accompany S. 2781).* US GPO.

Walker, N. (2014, September 27). Neurodiversity: Some basic terms & definitions. *Neurocosmopolitanism.* https://neurocosmopolitanism.com/neurodiversity-some-basic-terms-definitions/

Wang, L., Mandell, D. S., Lawer, L., Cidav, Z., & Leslie, D. L. (2013). Healthcare service use and costs for autism spectrum disorder: A comparison between Medicaid and private insurance. *Journal of Autism and Developmental Disorders, 43*(5), 1057–1064.

Warner, G., Cooper, H., & Cusack, J. (2019). *A review of the autism research funding landscape in the United Kingdom.* London, UK: Autistica.

Wehmeyer, M. L., & Abery, B. (2013). Self-determination and choice. *Intellectual and Developmental Disabilities, 51*(5), 399–411.

Whalen, C. K., Jamner, L. D., Henker, B., Delfino, R. J., & Lozano, J. M. (2002). The ADHD spectrum and everyday life: Experience sampling of adolescent moods, activities, smoking, and drinking. *Child Development, 73*(1), 209–227.

WHOQOL Group. (1995). The World Health Organization quality of life assessment (WHOQOL): position paper from the World Health Organization. *Social science & medicine, 41*(10), 1403–1409.

Wilson, S., Hicks, B. M., Foster, K. T., McGue, M., & Iacono, W. G. (2015). Age of onset and course of major depressive disorder: Associations with psychosocial functioning outcomes in adulthood. *Psychological Medicine, 45*(3), 505–514. https://doi.org/10.1017/S0033291714001640

Wing, L. (1996). Autistic spectrum disorders: No evidence for or against an increase in prevalence. *British Medical Journal, 312*, 327–328.

Wing, L., & Gould, J. (1979). Severe impairments of social interaction and associated abnormalities in children: Epidemiology and classification. *Journal of Autism and Developmental Disorders, 9*(1), 11–29.

Wolgemuth, J. R., Agosto, V., Lam, G. Y. H., Riley, M. W., Jones, R., & Hicks, T. (2016). Storying transition-to-work for/and youth on the autism spectrum in the United States: A critical construct synthesis of academic literature. *Disability & Society, 32*(6), 777–797.

World Health Organization. (2007). *Neurological disorders. Public health challenges*. World Health Organization.

World Health Organization. (2018). *ICD-11 for mortality and morbidity statistics*. Geneva, Switzerland: Author.

Wouters, M., Evenhuis, H. M., & Hilgenkamp, T. I. (2019). Physical activity levels of children and adolescents with moderate-to-severe intellectual disability. *Journal of Applied Research in Intellectual Disabilities, 32*(1), 131–142.

Yarmolovsky, J., Szwarc, T., Schwartz, M., Tirosh, E., & Geva, R. (2017). Hot executive control and response to a stimulant in a double-blind randomized trial in children with ADHD. *European Archives of Psychiatry and Clinical Neuroscience, 267*(1), 73–82.

16

Obesity and Metabolic Conditions

Kathleen Davis, Elias Mpofu, Theresa Abah, and Ami Moore

Introduction

Overweight and obesity are defined as "abnormal or excessive body fat accumulation that presents a risk to health" (World Health Organization (WHO), 2020a). People who are overweight or obese have increased adipose (fat) tissue, leading to increased risk for chronic disease (such as cardiovascular, respiratory, liver, and kidney disease; diabetes; and others); obesity in particular is also associated with increased risk for early mortality (Haidar & Cosman, 2011; Greenberg, 2013). Increased body

K. Davis (✉)
Texas Woman's University, Denton, TX, USA
e-mail: kdavis10@twu.edu

E. Mpofu
University of North Texas, Denton, TX, USA

University of Sydney, Sydney, NSW, Australia

University of Johannesburg, Johannesburg, South Africa
e-mail: Elias.Mpofu@unt.edu

E. Mpofu (ed.), *Sustainable Community Health*,
https://doi.org/10.1007/978-3-030-59687-3_16

fat is also associated with metabolic syndrome, a constellation of medical issues thought to relate to increased centrally distributed body fat and fat cell dysfunction (NHLBI, 2016). An individual must present with at least three from among these symptoms (central adiposity, glucose abnormalities, dyslipidemia, hypertriglyceridemia, and/or hypertension) to be diagnosed with metabolic syndrome (Punthakee, Goldenberg, & Katz, 2018). Overweight and obesity are critical public health issues due to their associated disease burden, cost, increased mortality, and their high and increasing rates across the globe.

Learning Objectives

After reading this chapter, the reader should be able to:

1. Define obesity and metabolic health conditions.
2. Outline the history of obesity and metabolic conditions research and practice, emphasizing current best practices.
3. Discuss current and emerging sustainable community-focused approaches for preventing and managing obesity and metabolic conditions.
4. Examine the cultural, professional, and legislative issues that influence the control and mitigation of obesity and metabolic conditions, addressing health disparities, equity and social justice concerns.
5. Summarize the role and significance of interdisciplinary approaches to managing obesity and metabolic conditions
6. Discuss the issues for research and practice in community-focused approaches to manage obesity and other metabolic conditions

The number of people dying from complications from obesity is approximately 2.8 million globally, a situation declared an epidemic by the W.H.O. (2020c). It is estimated that in the next three decades, obesity and overweight will claim about 92 million lives and reduce life expectancy by

T. Abah
University of Sacramento, Sacramento, CA, USA
e-mail: t.abah@csus.edu

A. Moore
University of North Texas, Denton, TX, USA
e-mail: Ami.Moore@unt.edu

three years by 2050 (Sassi, Devaux, Cecchini, & Rusticelli, 2009). In 2014, the McKinsey Institute estimated that 5% of all global deaths were due to obesity. One estimate of the global economic impact of obesity indicates a cost of about $2 trillion annually, second only to smoking and armed conflict (at about $2.1 trillion each) (McKinsey Global Institute, 2014). However, as indicated, most obesity cost models estimate only direct and indirect costs but do not account for the cost of prevention efforts (Sassi et al., 2009), which will be essential to mitigating the impact of this disease.

Globally, the burden of obesity and related metabolic conditions is substantial and increasing at an alarming rate around the globe (U.S. Centers for Disease Control and Prevention (U.S. C.D.C.), 2017a, 2017b; Seidell & Halberstadt, 2015) in 2016, over 1.9 billion U.S. adults were overweight, including about 650 million adults who were obese (about 34% of the global population) (Hales, Carroll, Fryar, & Ogden, 2017). Global rates of overweight among adults range from 38.5% to 39.2% among men and women, respectively, and global rates of obesity range from 11.1% to 15.1% among men and women, respectively (Development Initiatives Poverty Research, 2019). In fact, the number of people in the world who are currently overweight or obese is almost 2.5 times the number who are undernourished (McKinsey Global Institute, 2014). The global prevalence of obesity increased by about one-third between 2000 and 2014 (McKinsey Global Institute, 2014). Kelly, Yang, Chen, Reynolds, and He (2008) indicated that over one billion people (about 20%) of the world population would be obese by 2030.

An additional 340 million children and adolescents aged 5 to 19 years are also overweight or obese (Flegal, Kit, Orpana, et al., 2013; Hales et al., 2017). Among children, stunting due to long-term undernutrition coexists with overweight in the same children in many countries (Development Initiatives Poverty Research, 2019). While children have lower rates of overweight compared to adults both globally and within the US, increasingly, children may have lower quality diets, including high-fat, high-sugar, energy-dense, high-salt diets, thus increasing their risk for both overweight/obesity and metabolic diseases such as hypertension, insulin resistance, and psychological disorders (Boutayeb, 2006; WHO, 2018). Programs attempting to prevent obesity in children, even longer-term, extensive programs, have also sometimes failed to elicit any improvements (DeHenauw et al., 2015).

Developed nations tend to have higher obesity rates than those that are developing (Omran, 2005; Hales, Carroll, Fryar, & Ogden, 2020), although prevalence in developing countries is on the rise, associated with the paradoxical effect of economic development and income growth (Bhurosy & Jeewon, 2014; Zukiewicz-Sobczak et al., 2014). The paradox is that as economies begin to improve, people in the emerging economies of the developing world begin to consume more easily available, highly processed foods of minimal nutritional value, resulting in overweight and obesity (Loring & Robertson, 2014; Zukiewicz-Sobczak et al., 2014; Petersen, Pan, & Blanck, 2019; Templin, Hashiguchi, Thomson, Dieleman, & Bendavid, 2019).

Several community health-based, obesity prevention programs based on various health behavior theories have been developed and tested, which aim to prevent obesity or mitigate its effects by improving dietary quality and physical activity levels in communities. Other approaches have focused on schools, daycare and after-school programs, health care-based interventions, media education, and even church-based programs. Typical stakeholders in these efforts may include local governments and nongovernmental organizations, transportation authorities, public health and safety officials, community organizers, clergy, and urban planning officials.

Professional and/or Legal Definitions of Obesity and Metabolic Syndromes

As discussed, obesity and overweight are terms used to describe a higher than usual level of body fat, which is associated with increased health risk. Both the W.H.O. and the U.S. C.D.C. (2017a, 2017b) define overweight and obesity using **body mass index (BMI)**, which is an estimation of excess fat made by comparing weight to height in meters squared (W.H.O., 2020b). Adults with a BMI of 18.5–24.9 are considered normal weight, while those with a BMI of 25–29.9 are considered overweight. Those with a BMI of greater than 30 are considered obese. The U.S. C.D.C. has further categorized obesity into three subgroups (see Table 16.1).

Some of the professional associations dedicated to comprehensive research on evidence-based approaches and methods for treating obesity include:

Table 16.1 BMI classification

BMI range	Weight classification
<18.5	Underweight
18.5–24.9	Normal weight
25.0–29.9	Overweight
30–34.9	Class 1 obesity
35–39.9	Class 2 obesity
≥40	Class 3/morbid obesity

1. Obesity Medicine Association (OMA)—an organization of physicians, physician assistants, nurse practitioners, and other health care providers in the US, employing scientific-based, individualized, and comprehensive approaches to treat obese people (OMA, 2020).
2. The Obesity Society (TOS)—located in the US and Canada, the organization is focused on understanding the causes, prevention strategies, and treatment of obesity (www.obesity.org).
3. Obesity Action Coalition (OAC)—a not-for-profit organization dedicated to advocating for individuals affected by obesity disease to access better services and improve their health (www.obesityaction.org).

As previously noted, obesity is associated with the development of metabolic syndrome (Micciolo et al., 2010; Moller & Kaufman, 2005). Metabolic syndrome occurs when a constellation of metabolic effects occurs together, such as central obesity (obesity around the waist); diabetes, insulin resistance, or impaired fasting glucose tolerance; dyslipidemia; and/or hypertension (Huang, 2009). This cluster of metabolic conditions (metabolic syndrome) contributes to the onset of cardiovascular disease, which is one of the leading causes of death in developed nations worldwide and the leading cause of death due to obesity (Zimmet, Magliano, Matsuzawa, Alberti, & Shaw, 2005; Barkowski & Frishman, 2008). Obesity is also associated with increased school and job discrimination, higher socioeconomic burden, and earlier mortality (Puhl & Brownell, 2001; U.S. C.D.C, 2020). For these reasons, obesity poses a critical public health challenge for which community-level interventions would provide more robust health outcomes than individual-level programs.

The etiology of overweight and obesity is hotly debated, with most researchers faulting the energy imbalance that has occurred with increasing sedentary behaviors and larger portions, but others favoring social transmission theories, alterations in metabolism, or other theories (Christakis & Fowler, 2007; Hruby & Hu, 2015; Archer, Lavie, & Hill, 2018). Most obesity researchers agree that obesity is caused by energy imbalance (intake of energy from food and drink that exceeds expenditure) (Hruby & Hu, 2015). This change in average energy balance over time has been attributed to an "obesogenic" environment, one that promotes excess consumption and less physical activity. This modern shift toward an obesogenic environment is associated with social and economic changes, leading to "growing availability of abundant, inexpensive, and often nutrient-poor food, industrialization, mechanized transportation, urbanization" (Zukiewicz-Sobczak et al., 2014; Hruby & Hu, 2015). Social and environmental changes that promote greater energy consumption and less expenditure of energy interact with individual attitudes and behaviors affecting energy intake, physical activity, sedentary behaviors, and sleep, all of which are thought to influence obesity risk (Hruby & Hu, 2015). The physical/built environment, such as land use mix and walkability is also implicated in overweight and obesity, although the evidence is mixed (Mackenbach et al., 2014; see also Chap. 3, this volume).

Socioeconomic factors such as income status, education level, and belonging to a marginalized population (such as being of certain races or ethnicities) are associated with increased risk for obesity (Loring & Robertson, 2014; Petersen et al., 2019; Templin et al., 2019). In addition, obesity appears to occur in social networks, with those living together at similar risk for obesity (Christakis & Fowler, 2007). These social influence factors interact with genetic risk for obesity, which contributes from 40% to 70% of the risk for obesity (Sicat, 2018). An understanding of obesity risk requires social, economic, environmental, and genetic considerations.

History of Research and Practice in Obesity and Metabolic Syndrome

In ancient times, Hippocrates' (460 BC–370 BC) work anticipated obesity as a medical condition, resulting from an unhealthy balance between four bodily fluids: phlegm, blood, yellow bile, and black bile (Christopoulou-Aletra & Papavramidou, 2004; see Table 16.2).

Table 16.2 Historical evolution of obesity research

460 BC–370 BC	Hippocrates—defined obesity as a composition of four fluids (blood, phlegm, yellow bile, and black bile). Any extras were considered obesity (Christopoulou-Aletra & Papavramidou, 2004)
1885	Penny scale was introduced by Germany and the US as a measurement and evaluation criteria for weight (Schwartz et al. 1986–1996)
Early 1900s	Healthy weights became the criteria for eligibility for insurance policies, termed "ideal" weights. It was linked to longevity (Medico-Actuarial Mortality Investigation, 1912)
1911–1935	Dublin and Lotha developed the first tables of "ideal" weight measurement and evaluation scale computed based on gender, height, and weight (Dublin & Lotha, 1937)
1959	The Build and Blood Pressure Study conducted by 26 insurance companies in the US replaced the term "ideal" weight with desirable weight, suggesting an association between weight and morbidity, especially cardiovascular diseases (1959)
1973	Participants at an international conference updated the desired weight table to include acceptable range of weights for particular height, which was later converted to body mass index. Men had a range of 20.1–25.0 and women 18.7–23.8 (Bray, 1975)
1980	U.S. Department of Agriculture (USDA) and the Department of Health and Human Services (HHS) classified obesity and overweight as nutritional-related disorders caused by the accumulation of extra fat. The BMI table was used, with adjustments for weights and heights (1985)
1985	The USDA and HHS issued a range of weights and heights table for men and women, which was standardized to a single measurement by 1990, where healthy BMI was less than 25, unhealthy BMI was greater or equal to 25 (1985)
1990–2000s	The World Health Organization (WHO) Expert Committee on physical status modified and reclassified the body mass index (BMI) with cut off points of 25, 30, and 40 based on weight for height squared

Source: Centers for Diseases Prevention. Accessed 21 Apr 2020 at http://www.cdc.gov/growthcharts/

Hippocrates linked "surplus fluids" in the body to infertility and early mortality, including the association to different health outcomes among populations. Much later, landmark studies such as the U.S. Medico-Actuarial Mortality Investigation (Association of Life Insurance Medical Directors, 1912), the Dublin and Lotha study (1937), and the Build and

Blood Pressure Study (1959) culminated in the adoption of tables listing body weights for a given height associated with better mortality outcomes, which were to be used for insurance policies in the US (Dublin & Lotha, 1937; Schwartz, 1986). Individuals with weights between 20% and 25% above the "ideal" were considered to have unhealthy weights and ineligible to enroll in insurance policies (Metropolitan Life Insurance Company (MLIC), 1942). By the late 1970s, the weight tables were replaced with the MLIC table of weights and heights using data collated over two decades on the mortality rates of insured people from 1950 to 1972 in the Build and Blood Pressure Study of 1979.

These tables were criticized during the latter part of the twentieth century because they were too prescriptive of healthy weights for particular heights, making them unrealistic for many people (Weigley, 1984). Beginning in the early 1980s, the US Department of Agriculture (USDA) and the Department of Health and Human Services (HHS) adopted a standardized body mass index (BMI) calculation to estimate excess adiposity, which has been criticized for being inaccurate for estimating body fat, particularly in people who exercise regularly and athletes (Nutall, 2015). Other indicators of adiposity-related health risks include waist circumference, waist-to-hip ratio, and weight-to-height ratio (Maffetone & Laursen, 2020). A recent article by Park et al. proposes a new calculation to estimate cardiometabolic risk: the weight-adjusted-waist index, which these researchers believe best predicts cardiovascular disease-related mortality (2018).

Ongoing debates on overweight and obesity influences on cardiometabolic risk in global populations suggest caution in the sole reliance on BMI as an indicator (Maffetone & Laursen, 2020). Trending research seeks to develop a fail-safe index that does not underestimate risk for non-White populations such as Asians, Chinese, Africans, and Latin Americans, in which up to 40% or more of those who are normal weight according to BMI may have excess body fat (Maffetone & Laursen, 2020). Whatever the best method is for identifying obesity in individuals or populations, it is clear that the obesity problem is large and increasing.

Historically, obesity treatment programs have focused primarily on approaches that target the individual, with caloric restriction being the primary approach (Garner & Wooley, 1991; Archer et al., 2018).

However, as early as the 1970s, some rejected these approaches for their high failure rates, advocating behavior-based, non-diet approaches instead (Garner & Wooley, 1991).

As previously noted, with economic development and rising incomes, obesity has become a global health concern (Boutayeb, 2006). With this increase in global obesity, research and practice in obesity control in vulnerable populations has trended toward advocating lifestyle changes and health education awareness rather than pharmacological interventions (Roberto et al., 2015) or surgical approaches (weight loss surgery or bariatric surgery) (English & Williams, 2018), which are less available in developing country settings.

As discussed in the next section, lifestyle and education-oriented interventions involving self-monitoring of food intake, physical activity, and other behavior change hold great promise for reducing existing obesity (Heymsfield et al., 2018), when used together with prevention approaches which emphasize a healthy diet and active living (U.S. C.D.C., 2019). When implemented, these lifestyle approaches may promote weight loss of 5–10% and maintenance of that weight loss, which is closely associated with reduction in cardiometabolic risk (Heymsfield et al., 2018). In addition, lifestyle behavioral approaches for obesity control and management increasingly utilize mobile health (mHealth) technologies to provide real-time support with education, behavioral modification, interactive self-guided features, online support blogs, and other informational social interactions (Bonomi & Westerterp, 2012; see also Chap. 10, this volume). Technologies with activity monitoring capabilities support physical activity engagement, increasing the chances for sustainable obesity and metabolic conditions management.

Current research in addressing the global obesity pandemic involves newer, systems-oriented interventions, which seek to address the obesity epidemic in partnership with schools, community groups, local health authorities, and more (see also Chap. 5, this volume). A systems approach to overcoming the obesity epidemic will involve both continued research to understand the individual and societal causes of the condition as well as multidisciplinary collaboration to identify the best approaches to adopt when community-based intervention strategies are designed (The Community Guide, 2017).

Pertinent Sustainable Community Health-Oriented Approaches

Current community-level obesity and metabolic disease intervention programs include those that more traditionally seek to influence dietary intake, physical activity, or other aspects of lifestyle. In contrast to approaches focused on individuals or families, these community-based interventions may also work to change policies and practices to make adopting healthier behaviors easier. We consider current community-oriented obesity control and management practices aligned to social ecological model (SEM), policy diffusion model-based activities, and whole systems approaches (WSAs).

Social ecological models (SEMs) consider health behaviors at intrapersonal, interpersonal, organizational, community, and public policy levels that impact weight status (U.S. C.D.C., 2017a; Sallis, Owen, & Fisher, 2008). (See also Gittelsohn, Kim, He, & Pardilla, 2013; Harrison et al., 2011). These include nutrition, physical activity, and lifestyle approaches, public policy initiatives, multifactorial approaches, and creating healthier food environments within communities with the input and involvement of all parties involved; involving the media to promote healthy diets and community education on the risk factors associated with overweight; and built environment polices for obesity prevention and control.

Nutrition, Physical Activity, and Lifestyle Approaches

Both nutrition and physical activity interventions can modify risk for obesity and metabolic conditions, reducing morbidity and mortality in populations by improving dietary quality, modifying cardiometabolic risk, and promoting weight loss (Wing et al., 2011; Woodcock, Franco, Orsini, & Roberts, 2011; Schwingshackl, Missbach, Dias, König, & Hoffmann, 2014, see Chap. 5, this volume). Many such interventions aim to reduce obesity and overweight in communities through nutrition education, self-monitoring, portion control, meal planning, and social support to encourage eating healthier foods, while reducing intake of less healthy foods. The interventions focus on reducing caloric intake and sugar-sweetened beverage (SSB) intake, improving snack quality, increasing fruit and vegetable intake, and promoting moderate exercise (NHLBI, 2015). Physical activity-oriented interventions often involve walking or other forms of exercise. Walking at a moderate pace of three miles (about 6000 steps) per hour expends sufficient energy to meet the definition of moderate-intensity physical activity (Ainsworth et al., 2000). Combined nutrition and physical activity approaches may result in greater obesity and weight reduction than using either alone (Johns, Hartmann-Boyce, Jebb, & Aveyard, 2014). A weight loss of 5% or greater is associated with lower risk for cardiovascular disease (Wing et al., 2011), while moderate physical activity of 30 minutes a session, five times a week is associated with a 19% reduction in mortality (Woodcock et al., 2011). The effects are stronger and more sustained when combined with healthy diet (Schwingshackl et al., 2014).

Yet, most community-based nutrition and physical activity interventions have been implemented with small groups rather than at the community level. Some examples include ACHIEVE (Daumit et al., 2013), STRIDE (Green et al., 2015), and IN SHAPE (Bartels et al., 2015). ACHIEVE involved a randomized controlled trial of a group weight management and physical activity program in overweight and obese patients with serious mental illness in various community rehabilitation centers (Daumit et al., 2013). It was effective in helping participants achieve and maintain a healthier weight during an 18-month period (Daumit et al., 2013). STRIDE was a six-month, randomized controlled trial in a community setting of overweight and obese patients with serious mental illness and increased risk of diabetes, who were prescribed a calorie-restricted, Dietary Approaches to Stop Hypertension (DASH) diet and physical activity (Green et al., 2015). STRIDE participants were successful at achieving and maintaining a healthier weight and better glucose control. All three of these interventions were successful and took place among a specific community of adults with serious mental illness.

Community-based, lifestyle interventions involve a combination of diet, exercise, and/or behavior modification to prevent or treat obesity (Galani & Schneider, 2007; Barte et al., 2010; Olsen & Nesbitt, 2010; Bonomi & Westerterp, 2012). Those programs incorporating a dietary component have led to significant weight loss when compared with no treatment (Bonomi & Westerterp, 2012). Lifestyle interventions with a dietary component likely achieve their effects on weight loss by reducing intake of certain types of high-calorie foods, such as those high in saturated fat and/or by increasing physical activity. Moreover, such lifestyle interventions have improved plasma lipid concentrations, insulin sensitivity, and blood pressure, even in the absence of weight loss or body composition change, which may reflect the positive impact of improved dietary quality (higher in fruits, vegetables, and whole grains) and higher levels of physical activity (Bell et al., 2007).

As an example, Lifestyle Interventions for Expectant Moms (LIFE-Moms) (Lifestyle Interventions for Expectant Moms, n.d.; Peaceman et al., 2018) aimed to reduce excessive gestational weight gain among women. This intervention program targeted diet, physical activity, and behavioral and support strategies (weekly coaching phone calls and texts, individual counseling sessions, etc.). Pregnant women with BMI greater or equal to 25 took part in the program. Interventions significantly reduced the percent of women with excess gestational weight gain among this community of pregnant, adult women. (Peaceman et al., 2018).

Policy Diffusion Approaches

These seek to achieve obesity control and prevention through community access to recreational areas and infrastructure that provide an enabling environment to exercise and to stay active; information available to help individuals make healthy lifestyle choices; knowledge of food healthy supply chains and food markets; and more (see Research Box 16.1). They also include efforts to develop urban forests (Renner, 2019) and increase community access to healthy foods via incentive programs for retailers and requiring menu labeling at the point of decision in food markets such as chain restaurants (which has since been a law in the US since 2018) (US Food and Drug Administration (U.S. F.D.A.), 2020). Policy-oriented obesity control and management efforts also include increasing availability of potable water in public.

The US CDC partnership with the National Center for Chronic Disease Prevention and Health Promotion is an example of a policy

Research Box: 16.1

The Central California Regional Obesity Prevention Program: Changing Nutrition and Physical Activity Environments in California's Heartland. (Schwarte et al., 2010).

Background: In the US, California's Central Valley is one of the country's leading agricultural regions, but it also has high rates of poverty, hunger, overweight, obesity, and poor quality food environments. It is a region of food deserts and a large population of migrant workers, many of whom are undocumented.

Methods: The Central California Obesity Prevention Program (CCROPP) aimed to create a community-driven, whole systems approach to prevent obesity by changing the food and physical activity environments. It did this by addressing policy, partnering with community organizations and public health departments, and engaging community members actively in the process. It did this after an extensive review of the literature and by adopting a logic model that involved applying systems theory to social systems.

Results: The interventions within this large region led to establishment of new farmers markets, increased collaborations among schools, communities, health departments, farmers, and WIC. Self-sustaining walking groups were formed in some areas, one of which installed a new walking path at a local park. Other communities made structural improvements at parks, left school exercise areas open after school hours, passed breastfeeding policy changes, and created strategic plans for cities focused on health.

Conclusion and implications: The CCROPP showed that food and physical environment changes are possible even in areas of high poverty and few resources. However, further research is needed to evaluate the long-term impact of these initiatives and what additional measures are needed to prevent and reduce obesity within communities.

What Do You Think?

1. How do food markets influence community health?
2. How might communities differ in their obesity prevention, control and management approaches depending on community-specific characteristics?
3. How might policy initiatives address those potential cross-community differences while promoting equitable improvements in health?

initiative for obesity control and prevention (U.S. C.D.C., 2017a) aimed to "to increase the capacity of state health departments and their partners to work with and through communities to implement effective responses to obesity in populations that are facing health disparities" (U.S. C.D.C.,

2017a), such as supporting communities in plans for safe places for physical activity and nutrition health.

It is not likely that any single policy change would contribute significantly to reduced obesity, but when combined with multiple policy changes and other community-based efforts, change is possible.

Multifactorial Approaches

Multifactorial approaches to prevent obesity and metabolic conditions through health systems or structural reconfigurations, community-focused activity design, or targeting environmental and personal factors may help reduce health disparity and enable equity and social justice for people with obesity and metabolic conditions. For instance, community-level, multifactorial approaches may focus on healthier foods and beverages in schools or other communities, limiting access to less healthy food and beverage options in a variety of settings, promoting reduced screen time among individuals, and providing technology-supported coaching or counseling interventions for individuals (The Community Guide, 2017).

The Cherokee Choices program is an example of a community-based multifactorial approach intervention to improve the health of Cherokee Indians living in a rural area of North Carolina, USA (Bachar et al., 2006). This program involved three main components: elementary school mentoring (addressing inter- and intrapersonal factors and organizational factors within school settings); worksite wellness for adults (again, addressing inter, intrapersonal, and organizational factors within working adult communities); and a church-based health promotion program (addressing similar areas). The tribe also incorporated a social marketing program with TV ads and a documentary to support their efforts. Participants within the various programs met dietary and physical activity goals, reduced body fat, and were committed to the program. The program is important in that it follows established models, such as the SEM, but also involved community needs assessments that got stakeholder input into what was needed before the program was established, increasing its likelihood of success (Bachar et al., 2006).

The Healthy Eating Active Communities (HEAC: Samuels et al., 2010), the Community Health Initiative (CHI: Cheadle et al., 2010; Ross et al., 2010), and Tribal Health and Resilience in Vulnerable Environments (THRIVE: Jernigan et al., 2017, 2018; Jernigan, Salvatore, Williams, et al., 2019) are examples of other multifactorial approaches. The HEAC is a collaborative partnership located in six low-income communities in California, involving 11 school districts, a network of local organizations, and a coalition of state-level advocacy organizations, as well as local public health departments (Samuels et al., 2010). The HEAC is aimed to make changes in food and physical activity environments (such as changes in foods and beverages available in vending machines in various sectors, changes in neighborhood retail food offerings, and changes in equipment available in schools) while educating local authorities such as city councils on the importance of incorporating health considerations into development plans, and funding programs to educate local youth (Samuels et al., 2010). It resulted in many positive changes in the food and physical activity environments.

CHI is a program that originated in 2005 with Kaiser Permanente (an integrated health care delivery system) and involved just three sites in Colorado (Schwartz, Kelly, Cheadle, Pulver, & Solomon, 2018). It has since expanded to include 32 communities in Colorado, nine in northern California, nine in southern California, six in the Pacific Northwest, and two additional in each Maryland and Georgia, USA (Schwartz et al., 2018). Some of the CHI programs include local stores receiving more fresh fruits and vegetables; evidence-based prevention strategies such as promoting walking or biking to schools; and integrating health considerations into planning and development decisions (Cheadle et al., 2010). While changes attributed to this program have been modest overall, an analysis that evaluated the relationship between dose (intensity of the program) and results found more encouraging results (Schwartz et al., 2018).

THRIVE (Jernigan et al., 2017, 2019) was created to improve US Tribal food environments and control obesity. THRIVE used a participatory research process to increase healthy foods and reduce pricing of these foods in tribal stores. Formative research, used in step one, assessed tribal community food environments and health outcomes. The second phase, intervention development, examined convenience stores and created healthy retail product, pricing, promotion, and placement strategies. The last phase focused on intervention implementation and evaluation of the perception of healthier stores among both the intervention and control groups of participants (Jernigan et al., 2018). Positive outcomes included higher shopping frequency of purchasing fruits, vegetables, and other healthy items (Jernigan et al., 2018).

Whole Systems Approaches (WSAs)

WSAs are attempts to track and measure the extent to which community-based programs adhere to best practices in attempting to develop healthier practices and environments (Garside, Pearson, Hunt, Moxham, & Anderson, 2010; Bagnall et al., 2019). Ten features of a WSA program to address public health problems such as obesity include:

1. **Identifying a system:** Programs should first recognize all elements of a complex, adaptive system that affects the public health issue of concern.
2. **Capacity building:** Programs should have an explicit goal to support communities and organizations in the complex system.
3. **Creativity and innovation:** Programs should develop mechanisms to support local innovation to address the public health problem.
4. **Relationships:** Programs should use appropriate methods to develop effective relationships between participating organizations.
5. **Engagement:** Programs should use appropriate methods to help individuals, organizations, and economic sectors engage everyone involved in developing and delivering programs.
6. **Communication:** Programs should develop mechanisms to support good communication between the members of each system.
7. **Embedded action and policies:** Programs should describe in a transparent way the practices that will promote public health within organizations of the system.

8. **Robust and sustainable:** Programs should develop clear strategies to provide adequate resources for existing and new projects.
9. **Facilitative leadership:** Programs should provide strong strategic support and appropriate and adequate resources for all levels of interventions.
10. **Monitoring and evaluation:** Programs should have preplanned methods for formative and process evaluations in order to enhance the effectiveness of the interventions (Bagnall et al., 2019).

The evidence on the efficacy of WSAs is emerging (Bagnall et al., 2019). The Central California Regional Obesity Prevention Project (CCROPP: Schwarte et al., 2010) is a WSA-type program in eight agricultural, Central Valley counties to create farmers' markets in food desert communities. In addition, schoolyards became open for community members after school (Schwarte et al., 2010). A similar program is the Australian Romp and Chomp, a program that targeted all children under age 5 in the City of Geelong in Australia and the borough of Queensville in Victoria, Australia (de Silva-Sanigorski et al., 2010) promoting healthy eating and healthy play, resulting in lower prevalence of obesity in the children (de Silva-Sanigorski et al., 2010). Shape Up Somerville is another WSA-style project involving three diverse communities in Massachusetts, USA. Implementation of the program aimed to prevent and control obesity in school children showed evidence of efficacy and cost-effectiveness (Coffield et al., 2019). (Economos et al., 2013; Coffield et al., 2019). The WHO European Healthy Cities Network (de Leeuw, Tsouros, Dyakova, & Green, 2014) is yet another WSA involving 100 cities in 30 countries across Europe by addressing health inequalities that would impact obesity control and management (de Leeuw et al., 2014). Thus far the evidence for WSA approaches is mixed and outcomes may vary by context and level of adherence to WSA principles. See Discussion Box 16.1: Accessing Obesity Treatment for a discussion of some of the barriers to accessing treatment that should be addressed by using WSA principles.

Discussion Box 16.1: Accessing Obesity Treatment

Multidisciplinary approaches to managing obesity among individuals within communities are a best community health practice. However, they are very limited in their feasibility and availability, especially among members of low-income communities. While programs to improve the food and physical activity environments and to educate the public on the importance of healthy diet and exercise are very important in reducing the risk for obesity, they presume implementation of community health policies, which would not be the case across various socioeconomic groups and in different environments.

What Do You Think?

1. What types of policy changes need to occur to make obesity treatment more accessible to marginalized populations and communities?
2. What would be an appropriate universal community health policy that could reduce community health disparities in weight control?

Cultural, Legislative, and Professional Issues That Impact Community Health Approaches

Cultural, legislative, and professional issues may positively or negatively impact community health approaches to address high rates of obesity and metabolic disease. Cultural issues that influence educational attainment, SES status, gender equity, and treatment of marginalized populations can perpetuate disparities in obesity and metabolic disease. Legislative and policy issues that may positively or negatively influence efforts include ease and cost of implementation, flexibility to implement policies differently in areas that differ in various respects, and other issues. Professional issues often relate to lack of communication or understanding among professions addressing the same issues.

Cultural Influences

As described in earlier sections, there are disparities in obesity in the US and globally. In the US, these disparities are present by geographic region, SES, race/ethnicity, and rural versus urban areas (U.S. C.D.C., n.d.).

Between 2011 and 2014, an analysis of obesity rates in rural versus metropolitan areas in the US showed that obesity rates for rural residents was higher by about 15%. More than 39% of adults in nonmetropolitan areas had obesity compared to 33% of those in metropolitan countries. There are also disparities in obesity prevalence by race/ethnicity. NHANES data from 2015 to 2016 showed that Hispanics and Blacks had a higher obesity rate of 50.6% and 54.8% compared to Whites with 38% in the US (Lee, Warren, Liu, Foti, & Selvin, 2019). Even among children, obesity rates were higher among Hispanics (47%) and Blacks (46.8) compared to Whites (37.9%). Some of the factors associated with higher rates of rural obesity include institutional and systemic poverty; underinvestment in community health programs; and the consequent limited resources to implement policies that address the social determinants of health (Lundeen et al., 2018). Known contributors to these community disparities in obesity prevention and control include poor awareness about nutrition information and food deserts around residential areas (American Heart Association, 2004; Moller & Kaufman, 2005).

Among the Organization for Economic Co-operation and Development countries (OECD) (a group of 37 country economic partners), disparities in obesity rates by educational attainment is higher among women compared to men. Factors driving the higher rate of obesity among women may be attributed to low education levels and effects of socioeconomics on life course (Devaux & Sassi, 2015). These ongoing disparities, which are largely driven by poverty and lack of educational attainment, have been resistant to change. Thus, it is challenging to address obesity while equal access to educational and socioeconomic tools to battle obesity in the most at-risk populations remains limited.

Legislative Influences

Many developed countries have obesity control and prevention-related policies and programs targeted to vulnerable populations like children and older adults of low socioeconomic backgrounds. Examples in the US include the Special Supplemental Nutrition Program for Women, Infants, and Children (WIC) (which serves approximately 53% of all eligible infants) and the Child and Adult Care Food Program (CACEFP) (which serves about 3.5 million children in childcare by providing meals to one-third of all children in childcare centers). Other programs include the National School Lunch and School Breakfast programs (which provide 31 million children food twice a day, five days a week across 100,000 public and private schools) and the Supplemental Nutrition Assistance Program (SNAP) (which provides low-income individuals and families with assistance to buy healthy foods).

Policy strategies to tackle obesity in OECD countries (Finland, Belgium, and Chile) include policies that increase prices of potentially unhealthy foods, food labeling, and mandatory inclusion of physical activity in schools and primary care settings (Cecchini & Warin, 2016). In Sweden and Denmark, nutrients are listed on the front of the pack to help people make better and healthier food choices. Other interventions aimed at reducing obesity include advertisement restrictions on radio and TV, especially those enticing to children, including bans of advertisements on SSBs. Bans of ads on SSBs have also occurred in Chile, Poland, Spain, and Turkey. See Discussion Box 16.2 for a further discussion of issues related to SSBs.

Discussion Box 16.2: Sugar-Sweetened Beverage (SSB) Issues

Within developed countries, SSBs are a major contributor to excess calories. When consumed in addition to the other foods one needs, they provide excess calories without the accompanying nutrients that other foods provide and may lead to obesity. When they replace other sources of nutrition in the diet, SSB may lead to nutrient imbalances and inadequate micronutrient intake. The increasing size of SSB portions has led to front-of-package labeling or point-of-sale labeling, bans on SSB advertising, and/or SSB taxation.

What Do You Think?

1. What factors would influence a population to adopt policies to label SSBs on the front of package to mitigate risk of obesity?
2. What other food labeling would be helpful to obesity prevention and control?

However, obesity prevention and control policies are politically controversial and have not been well supported by the food industry, which has invested in its historic trade practices marketing highly processed foods, while national governments have been lukewarm regarding implementing nutrition-safe food market regulations (Lyn, Heath, & Dubhashi, 2019, see also Chap. 3, this volume). Adoption and implementation of population-wide obesity control and management policies and applying multilevel systems approaches with coordinated efforts by governments, industry, communities, and individuals would go a long way toward reducing and preventing obesity using sustainable community health approaches (Malik, Willett, & Hu, 2013).

Professional Issues

Professionals in obesity control and management (such as doctors, physician assistants, nurses, dietitians, and more) contend with fragmented health care services (Lawrence & Kisely, 2010). They also may be less well trained to treat obesity in special populations such as in those with psychiatric disabilities, who have higher rates of obesity compared to those from the general population (Jonikas et al., 2015). This lack of training in obesity management for special populations would explain why professions are less likely to advise people with developmental disabilities to adopt lifestyle changes aimed to lower their (risk for) obesity (Sciamanna, Tate, Lang, & Wing, 2000; Phelan, Nallari, Darroch, & Wing, 2009). The limitations in preparation among health care providers with people with psychiatric and developmental disabilities are suggested by the fact that intervention effects tend to be below clinically significant levels (about 5%) (Olker, Parrott, Swarbrick, & Spagnolo, 2016) with less than 40% of participants improving cardiovascular health (Bartels et al., 2013). Moreover, professionals seem less knowledgeable on sustainable resourcing of interventions with special populations (Patel, Asch, & Volpp, 2015), increasing risk for relapse to pre-intervention conditions. While professionals increasingly utilize mHealth technologies with people with or at risk for obesity (Galbraith-Emami & Lobstein, 2013), they are less familiar with community activity engagement practices for their clients to adopt and maintain healthy behavior change.

Related Disciplines Influencing Community-Oriented Health Aspects

Several disciplines work on the issue of obesity. These include, but are not limited to medicine, nutrition, kinesiology, psychology, epidemiology, biostatistics, geography, economics, political science, public health, and more. Whereas the great variety of scientists and other specialists involved means that many approaches are being discussed and tried, this variety can also lead to competitiveness and resistance to information sharing. Better coordination and understanding among disciplines will be needed if progress is to be made in reversing the obesity epidemic. Below we briefly describe the role that several disciplines have made on this important issue.

Medicine has been active in many aspects relating to obesity, including categorizing obesity as a disease (American Medical Association, 2013), thereby reducing somewhat the stigma associated with obesity and its treatment. Medical organizations are active in evaluating the evidence for clinic-based treatment, including individual and group treatment, and issuing guidelines for evidence-based treatment. For example, in 2013, the American College of Cardiology together with the American Heart Association and The Obesity Society issued guidelines for managing overweight and obesity in adults (Jensen et al., 2013). The American Association of Clinical Endocrinologists (AACE) issued its own guidelines of care in 2016 (Garvey et al., 2016). The American Academy of Pediatrics issued guidelines for pediatric weight management in 2007 (Spear et al., 2007), including guidance on when and how to incorporate the input of other health professionals such as dietitians and behavior specialists. These are just some of the guidelines that have been issued, which have included bariatric surgery guidelines and those from various international medical organizations.

The field of nutrition and dietetics has also been actively involved in obesity treatment by providing guidelines on nutrition health and safety (see Chap. 5, this volume). Professional nutrition associations such as the Academy of Nutrition and Dietetics have also issued position and practice papers to guide care (Raynor & Champagne, 2016) as well as evidence reviews for best practices in nutrition treatment for individuals

(Academy of Nutrition and Dietetics Evidence Analysis Library, 2019). They have also issued recommendations for preventing pediatric obesity particularly with emphasis on nutrition in schools and childcare settings (Hoelscher, Kirk, Ritchie, & Cunningham-Sabo, 2013).

The fields of kinesiology and exercise physiology have contributed to our understanding of how physical activity factors impact obesity risk (Donnelly et al., 2009). These professionals have also studied which forms of exercise are most helpful in combatting obesity, including recommendations that establish aerobic/cardiovascular exercise as important in preventing weight gain, promoting weight loss, and necessary to prevent regaining lost weight (Donnelly et al., 2009). Such bodies of professionals dedicated to studying physical activity as the American College of Sports Medicine (ACSM) have also published guidelines related to exercise and weight management, which aerobic exercise as the form of exercise is most likely to help those who have lost weight maintain the weight loss (Donnelly et al., 2009). This field has also been instrumental in discovering that those who have lost significant weight typically have to maintain very high levels of physical activity post weight loss to help maintain the lost weight (Donnelly et al., 2009).

Psychologists and other behavioral specialists have also been integral in obesity treatment, identifying and managing psychological comorbidities related to obesity, and raising awareness about bias related to obesity. The American Psychological Association (APA) released a Clinical Practice Guideline for the Treatment of Obesity and Overweight in Children and Adolescents in 2018, which included recommendations for family-based multicomponent interventions (consistent with recommendations of other groups) (Llabre et al., 2018). It also emphasized the need to avoid stigmatizing the condition and the people with the condition (Llabre et al., 2018).

Public health professionals have been highly active in raising the profile of obesity and disseminating information about its risks (American Public Health Association, 2007), advocating for policy changes to combat it, and developing community-based programs to address it (Blumenthal, Hendi, & Marsillo, 2002). Public health professionals often work with the most disadvantaged and marginalized populations. Therefore, they work extensively on the issue of addressing obesity disparities.

The field of epidemiology has been instrumental in helping public health professionals, physicians, policymakers, and many more understand the size and nature of the problem of obesity (Ogden, Yanovski, Carroll, & Flegal, 2007). Their work has been cited many times in this chapter (Ogden et al., 2007; Ogden, Carroll, Kit, & Flegal, 2012; Flegal et al., 2013; Hruby & Hu, 2015; Hales et al., 2017, 2020). They have studied and disseminated information regarding factors associated with increased obesity risk. They have collected and analyzed data sets within countries and globally to help track progress on addressing this epidemic. In addition, recently professionals in systems bioinformatics have begun to work together with epidemiologists and other professionals to help elucidate the underlying genetic relationships to disease, including metabolic diseases (Oulas et al., 2019).

Scientists from the social sciences have also entered the realm of research into obesity. For example, political scientists have studied how obesity has impacted policy and interest in types of policies as well as how policy has influenced health outcomes (Kersh & Monroe, 2002). They have also evaluated the unintended consequences of policies and how easily (or not) policies may be transferred from one area to another (Kersch, 2009). They have also assessed equity issues with regard to policies and their implementation (Fox & Horowitz, 2013). Even economists have been actively involved in obesity research, evaluating the cost of obesity and its comorbidities on various levels of economies (Ananthapavan, Sacks, Moodie, & Carter, 2014). They have also evaluated the cost-effectiveness of interventions that touch policy or that are implemented within communities. Geographers have also been involved as mapping of obesity trends and food deserts have become important to understand the problem (Science X: Phys.org, 2013).

Issues for Research and Other Forms of Scholarship

This discussion of obesity and metabolic disease, including its etiology, prevention, treatment (including community-based approaches), and barriers to progress, has made plain that several areas of research need to be addressed to reverse this epidemic. Among these needs include the need for better data; a systems approach to understand the data; and targeted interventions, including community-level and policy-level interventions that address the needs of marginalized populations. Moreover, there is need for a better understanding of how obesity control policy recommendations may affect individuals and communities. Another major research need is for high-quality studies of long-term interventions at the individual and community levels.

Further research is needed to unpack the social determinants of obesity in community populations with vulnerability (Roth, Foraker, Payne, & Embi, 2014). The evidence would enable the design and implementation of targeted interventions for sustainable obesity prevention and control based on the specific community health disparities. A one-size-fits-all approach to obesity prevention and control would be wasteful and unproductive.

An example of efforts to map factors that relate to both poverty and obesity for targeted interventions include the US Interactive Food Access Research Atlas (USDA Economic Research Service, 2019). This online map allows better identification at a granular level of specific areas where low income, low vehicle access, and low grocery store access exist, thereby reducing access to affordable, nutritious food. Indonesia has a similar project to map and track food insecurity (World Food Programme, 2015). However, many countries lack such good quality data and certainly do not have it mapped. Without mapping these areas in detail, it is not possible to know where the greatest resources are needed. Thus, we need more community-level epidemiological data for the design and implementation of sustainable community health addressing known and emerging risks for obesity.

Wider use of community health informatics approaches may help identify the data tools with the most yield by community characteristics (Gittner, Kilbourne, Vadapalli, Khan, & Langston, 2017; see also Chap. 10, this volume). Data types and sources are not equal in their ability to guide community-level policy interventions for obesity prevention and control. For instance, data that sample interconnected social and economic deprivation, such as lack of community infrastructure, heat stress, and higher levels of pollution and food insecurity may suggest a community setting-specific approach to obesity prevention and control (Gittner et al., 2017). As these type of data accrue globally, it would be possible to model interventions for adoption by governments, community organizations, and communities in their efforts to address the difficult, complex issues affecting obesity. Health systems-level studies are also needed to better understand the settings in which specific obesity prevention and control would be most effective. For instance, there is evidence to suggest that local programs may be most successful if they are multi-pronged and if the strategies at every level reinforce each other: identifying intervention targets or policy goals, engaging community members in identifying causes of obesity locally, as well as designing policies or interventions that appeal to groups outside of public health (Vitaliano et al., 2005; Barnhill et al., 2018). Thus, there is need to think globally and act locally in efforts to combat risks for obesity at the community level. Other, marginalized populations at higher risk of obesity and its comorbidities include people with neurodevelopmental, psychiatric, and musculoskeletal disorders (Lawrence & Kisely, 2010). Finally, most obesity interventions in the past have been short term and have targeted the individual. As discussed in the section on multifactorial interventions and WSAs, there are more community-based interventions and those that use WSAs today than ever before. However, we need more studies that adhere to the WSA features, that measure change rigorously and consistently, that evaluate cost-effectiveness and sustainability of the solutions, and that endure for longer periods of time.

Summary and Conclusion

Obesity is a critical public health issue in nearly every corner of the world. Throughout the world, even in areas of high deprivation, obesity now coexists with undernutrition. Obesity currently exceeds malnutrition (undernutrition) throughout the world. Sustainable ways to prevent and treat obesity and reduce its complications at the community level should prioritize policy approaches for creating healthy community and neighborhood environments, multilevel approaches to address obesity both at the community and individual levels, those that promote healthy food environments, health care systems, school environments, physical activity communities, and foodservice programs. Cultural, professional, and/or legislative dispensations influence the practice evidence on community-level interventions to reduce obesity and its related metabolic issues. Despite much hand-wringing, discussion, exhortation, and blaming, few efforts have been made by national governments to address overweight and obesity risk in a sustainable way. Research should focus on customizing obesity prevention and control interventions to local communities, while drawing from the available and emerging global scholarship.

Self-Check Questions

1. Define obesity and metabolic conditions and outline their prevalence in the US and globally.
2. Outline the landmark studies and practices in obesity and metabolic conditions management in the last half century, highlighting their significance.
3. What are the leading approaches to community-level interventions for managing obesity, and the strength of evidence for them?
4. How have cultural, professional, and legal issues influenced the management of obesity and metabolic conditions globally?
5. What are the roles of various disciplines in addressing obesity and metabolic conditions?
6. Identify research and practice gaps in managing obesity and metabolic conditions and suggests ways to address those gaps in knowledge.

Discussion Questions

1. Think about the issue of obesity and metabolic disease disparities in the world. Based on the content in this chapter, what types of policies would be most effective for obesity prevention and control at the community level?
2. Think about a community that has high levels of obesity and few resources. Where do you think obesity treatment would have to be offered to make it accessible? What are the barriers to access? (Time? Transportation? Cost? Values? What else?). How would you remove some of the barriers to access? Are there cultural factors that you would need to take into account?
3. To what extent are lifestyle approaches translatable across communities for addressing disparities in obesity prevention and control?
4. What are some of the ways by which professionals could be prepared to work with diverse populations for obesity prevention and control?

Field-Based Experiential Learning Activity

1. One way of better understanding how to implement obesity prevention and control at the community level is to interact with individuals in communities that have higher rates of obesity. Even when individuals are not overweight or obese, dietary quality may be poor. Consider becoming a volunteer for Cooking Matters, or a similar organization in your country. https://cookingmatters.org/volunteer-with-programming

   Consider the following based on your experience:

   1.1 How would you provide Cooking Matters at the community level? With what resources and what benefits?
   1.2 How would you win various community constituents' interest in participating in Cooking Matters?

2. Access the USDA's interactive food atlas: https://www.ers.usda.gov/data-products/food-environment-atlas/go-to-the-atlas/ Search in your area. Make note of several of the factors in your area, such as socioeconomic status, access and proximity to grocery stores, restaurant availability, and at least two others. If you do not live in the US, search for data in your country online.

   2.1 How would you rate your local food environment where you live? How easy is it to attain healthy foods because of geographic/proximity issues? How about economic resources of people in your area to buy these foods?
   2.2 What do you think needs to be improved in your area?
   2.3 Are there policies that your area could adopt to help improve this issue? Is anyone or any group currently working on this issue that you are able to find?
   2.4 If you were to design a community-level obesity intervention using the WSA described in this chapter, what would be important to do first? Which stakeholders would you include? What issues would you try to address?

Key Online Resources

Medicine Net—Metabolic syndrome quiz: Test your medical IQ. https://www.medicinenetcom/metabolic_syndrome_quiz/quiz.htm
Quizlet Inc. (2020) Metabolic Syndrome. https://quizletcom/37068287/chapter-10-metabolic-syndrome-flash-cards/
Mayo Clinic Patient Care and Health Diagnosis. https://www.mayoclinic.org/diseases-conditions/obesity/diagnosis-treatment/drc-20375749
Kids Health from Nemours—Metabolic Syndrome. https://kidshealth.org/en/parents/metabolic-syndrome.html
MERCK Manual Consumer version—Obesity. https://www.merckmanuals.com/home/disorders-of-nutrition/obesity-and-the-metabolic-syndrome/obesity
National Cancer Institute—Obesity and cancer. https://www.cancer.gov/about-cancer/causes-prevention/risk/obesity/obesity-fact-sheet
Medical News Today. What is metabolically healthy obesity? https://www.medicalnewstoday.com/articles/265405#1

References

Academy of Nutrition and Dietetics Evidence Analysis Library. (2019). *Adult weight management scoping review*. Retrieved April 30, 2020, from https://www.andeal.org/topic.cfm?menu=5276&cat=4690

Ainsworth, B. E., Haskell, W. L., Whitt, M. C., Irwin, M. L., Swartz, A. M., Strath, S. J., … Jacobs, D. R. (2000). Compendium of physical activities: An update of activity codes and MET intensities. *Medicine & Science in Sports & Exercise, 32*(9), S498–S516.

American Heart Association. (2004). *Exercise and acute cardiovascular events: placing the risks into perspective: A scientific statement from the American Heart Association Council on Nutrition, Physical*. Retrieved from https://www.ahajournals.org/doi/abs/10.1161/circulationaha.107.181485

American Medical Association (AMA). (2013). *Report of the council on science and public health: Is obesity a disease* (Resolution 115-A-12). Retrieved April 30, 2020, from https://www.ama-assn.org/sites/ama-assn.org/files/corp/media-browser/public/about-ama/councils/Council%20Reports/council-on-science-public-health/a13csaph3.pdf

American Public Health Association. (2007). *Addressing obesity and health disparities through federal nutrition and agricultural policy*. Retrieved April 30, 2020, from https://www.apha.org/policies-and-advocacy/public-health-policy-statements/policy-database/2014/07/24/15/56/addressing-obesity-and-health-disparities-through-federal-nutrition-and-agricultural-policy

Ananthapavan, J., Sacks, G., Moodie, M., & Carter, R. (2014). Economics of obesity—Learning from the past to contribute to a better future. *International Journal of Environmental Research and Public Health, 11*, 4007–4025.

Archer, E., Lavie, C. J., & Hill, J. O. (2018). The contributions of 'diet', 'genes', and physical activity to the etiology of obesity: Contrary evidence and consilience. *Progress in Cardiovascular Diseases, 61*(2), 89–102.

Association of Life Insurance Medical Directors. (1912). The Actuarial Society of America. *Medico-Actuarial Mortality Investigation*, vol. 1. New York, NY.

Bachar, J. J., Lefler, L. J., Reed, L., McCoy, T., Bailey, R., & Bell, R. (2006). Cherokee choices: A diabetes prevention program for American Indians. *Preventing Chronic Disease, 3*(3), A103.

Bagnall, A.-M., Radley, D., Jones, R., Gately, P., Nobles, J., Van Dijk, M., … Sahota, P. (2019). Whole systems approaches to obesity and other complex public health challenges: A systematic review. *BMC Public Health, 19*(8), 1–14.

Barkowski, R. S., & Frishman, W. H. (2008). HDL metabolism and CETP inhibition. *Cardiology in Review, 16*(3), 154–162.

Barnhill, A., Palmer, A., Weston, C. M., Brownell, K. D., Clancy, K., Economos, C. D., … Bennett, W. L. (2018). Grappling with complex Food systems to reduce obesity: A US public health challenge. *Public Health Reports, 133*(S1), 44S–53S.

Barte, J. C. M., Ter Bogt, N. C. W., Bogers, R. P., Teixeira, P. J., Blissmer, B., Mori, T. A., & Bemelmans, W. J. E. (2010). Maintenance of weight loss after lifestyle interventions for overweight and obesity, a systematic review. *Obesity Reviews, 11*(12), 899–906.

Bartels, S. J., Pratt, S. I., Aschbrenner, K. A., Barre, L. K., Jue, K., Wolfe, R. S., … Naslund, J. A. (2013). Clinically significant improved fitness and weight loss among overweight persons with serious mental illness. *Psychiatric Services, 64*(8), 729–736.

Bartels, S. J., Pratt, S. I., Aschbrenner, K. A., Barre, L. K., Naslund, J. A., Wolfe, R., … Feldman, J. (2015). Pragmatic replication trial of health promotion coaching for obesity in serious mental illness and maintenance of outcomes. *American Journal of Psychiatry, 172*(4), 344–352.

Bell, L. M., Watts, K., Siafarikas, A., Thompson, A., Ratnam, N., Bulsara, M., … Davis, E. A. (2007). Exercise alone reduces insulin resistance in obese children independently of changes in body composition. *The Journal of Clinical Endocrinology & Metabolism, 92*(11), 4230–4235.

Bhurosy, T., & Jeewon, R. (2014). Overweight and obesity epidemic in developing countries: A problem with diet, physical activity, or socioeconomic status? *The Scientific World Journal*. https://doi.org/10.1155/2014/964236

Blumenthal, S. J., Hendi, J. M., & Marsillo, L. (2002). A public health approach to decreasing obesity. *Journal of the American Medical Association, 288*(17), 2178.

Bonomi, A. G., & Westerterp, K. R. (2012). Advances in physical activity monitoring and lifestyle interventions in obesity: A review. *International Journal of Obesity, 36*(2), 167–177.

Boutayeb, A. (2006). The double burden of communicable and non-communicable diseases in developing countries. *Transactions of the Royal Society of Tropical Medicine and Hygiene, 100*(3), 191–199.

Bray, G. E. (1975). *Obesity in Perspective: A Conference, Part 1. John E. Fogarty International Center for Advanced Study in the Health Sciences*. U.S. Government Printing Office.

Build and Blood Pressure Study. (1959). vol. 1, Society of Actuaries, Chicago, IL.

Cecchini, M., & Warin, L. (2016). Impact of Food labelling systems on Food choices and eating behaviors: A systematic review and meta-analysis of randomized studies. *Obesity Reviews, 17*(3), 201–210.

Cheadle, A., Schwartz, P. M., Rauzon, S., Beery, W. L., Gee, S., & Solomon, L. (2010). The Kaiser Permanente community health initiative: Overview and evaluation design. *American Journal of Public Health, 100*(11), 2111–2113.

Christakis, N. A., & Fowler, J. H. (2007). The spread of obesity in a large social network over 32 years. *The New England Journal of Medicine, 357*(4), 370–379.

Christopoulou-Aletra, H., & Papavramidou, N. (2004). Methods used by the Hippocratic physicians for weight reduction. *World Journal of Surgery, 28*(5), 513–517.

Coffield, E., Nihiser, A., Carlson, S., Collins, J., Cawley, J., Lee, S., & Economos, C. (2019). Shape up Somerville's return on investment: Multi-group exposure generates net-benefits in a child obesity intervention. *Preventive Medicine Reports, 16*, 100954.

Daumit, G. L., Dickerson, F. B., Wang, N. Y., Dalcin, A., Jerome, G. J., Anderson, C. A., … Oefinger, M. (2013). A behavioral weight-loss intervention in persons with serious mental illness. *New England Journal of Medicine, 368*(17), 1594–1602.

de Leeuw, E., Tsouros, A.D., Dyakova, M., & Green, G. (2014). *Healthy cities: Promoting health and equity: Evidence for local policy and practice, summary evaluation of phase V. of the WHO European healthy cities network*. World Health Organization. Retrieved April 29, 2020, from http://www.euro.who.int/__data/assets/pdf_file/0007/262492/Healthy-Cities-promoting-health-and-equity.pdf

de Silva-Sanigorski, A. M., Bell, A. C., Kremer, P., Nichols, M., Crellin, M., Smith, M., … Swinburn, B. A. (2010). Reducing obesity in early childhood: Results from Romp & Chomp, an Australian community-wide intervention program. *The American Journal of Clinical Nutrition, 91*, 831–840.

DeHenauw, S., Huybrechts, I., De Beourdeaudhuij, I., Bammann, K., Barba, G., Marild, S., … Ahrens, W. (2015). Effects of a community-oriented obesity prevention programme on indicators of body fatness in preschool and primary school children. *Main Results from the IDEFICS Study, 16*(2), 16–29.

DeVaux, M., & Sassi, F. (2015). *The labour market impacts of obesity, smoking, alcohol use and related chronic diseases*. OECD health working papers, No. 86, OECD Publishing, Paris. https://doi.org/10.1787/5jrqcn5fpv0v-en.

Development Initiatives Poverty Research, Ltd. (2019). *Global nutrition report*. Retrieved April 27, 2020, from https://globalnutritionreport.org/media/profiles/v2.1.1/pdfs/global.pdf

Donnelly, J. E., Blair, S. N., Jakicic, J., Manore, M. M., Rankin, J. W., & Smith, B. K. (2009). Appropriate physical activity intervention strategies for weight loss and prevention of weight regain for adults. *Medicine and Science in Sports and Exercise, 41*(2), 459–471.

Dublin, L. I., & Lotha, A. J. (1937). *Twenty-five years of health Progress: A study of the mortality experience among the industrial policyholders of the Metropolitan Life Insurance Company, 1911 to 1935*. New York, NY: Metropolitan Life Insurance Co..

Economos, C. D., Hyatt, R. R., Must, A., Goldberg, J. P., Kuder, J., Naumova, E. N., … Nelson, M. E. (2013). Shape up Somerville two-year results: A community-based environmental change intervention sustains weight reduction in children. *Preventive Medicine, 57*(4), 322–327.

English, W. J., & Williams, D. B. (2018). Metabolic and bariatric surgery: An effective treatment option for obesity and cardiovascular disease. *Progress in Cardiovascular Diseases, 61*(2), 253–269.

Flegal, K. M., Kit, B. K., Orpana, H., et al. (2013). Association of all-cause mortality with overweight and obesity using standard body mass index categories: A systematic review and Meta-analysis. *JAMA, 309*(1), 71–82.

Fox, A. M., & Horowitz, C. R. (2013). Best practices in policy approaches to obesity prevention. *Journal of Health Care for the Poor and Underserved, 24*(20), 168–192.

Galani, C., & Schneider, H. (2007). Prevention and treatment of obesity with lifestyle interventions: Review and meta-analysis. *International Journal of Public Health, 52*(6), 348–359.

Galbraith-Emami, S., & Lobstein, T. (2013). The impact of initiatives to limit the advertising of Food and beverage products to children: A systematic review. *Obesity Reviews, 14*(12), 960–974.

Garner, D. M., & Wooley, S. C. (1991). Confronting the failure of behavioral and Dietary treatments for obesity. *Clinical Psychology Review, 11*, 729–780.

Garside, R., Pearson, M., Hunt, H., Moxham, T., & Anderson, R. (2010). *Identifying the key elements and interactions of a whole system approach to obesity prevention*. Exeter, UK: Peninsula Technology Assessment Group (PenTAG).

Garvey, W. T., Mechanick, J. I., Brett, E. M., Garber, A. J., Hurley, D. L., Jastreboff, A. M., … Plodkowski, R. (2016). American association of clinical endocrinologists and American college of endocrinology comprehensive clinical practice guidelines for the medical care of patients with obesity. *Endocrine Practice, 22*(S3), 1–203.

Gittelsohn, J., Kim, E. M., He, S., & Pardilla, M. (2013). A Food store–based environmental intervention is associated with reduced BMI and improved psychosocial factors and Food-related behaviors on the Navajo nation. *The Journal of Nutrition, 143*(9), 1494–1500. https://doi.org/10.3945/jn.112.165266

Gittner, L. S., Kilbourne, B. J., Vadapalli, R., Khan, H. M. K., & Langston, M. A. (2017). A multifactorial obesity model developed from nationwide health exosome data and modern computational analyses. *Obesity Research & Clinical Practice, 11*(5), 522–533.

Green, C. A., Yarborough, B. J. H., Leo, M. C., Yarborough, M. T., Stumbo, S. P., Janoff, S. L., … Stevens, V. J. (2015). The STRIDE weight loss and lifestyle intervention for individuals taking antipsychotic medications: A randomized trial. *American Journal of Psychiatry, 172*(1), 71–81.

Greenberg, J. A. (2013). Obesity and early mortality in the United States. *Obesity, 21*(2), 405–412.

Haidar, Y. M., & Cosman, B. C. (2011). Obesity epidemiology. *Clinics in Colon and Rectal Surgery, 24*(4), 205.

Hales, C. M., Carroll, M.D., Fryar C. D., & Ogden C. L. (2017). *Prevalence of obesity among adults and youth: United States, 2015–2016*. NCHS Data Brief, 288. https://www.cdc.gov/nchs/data/databriefs/db288.pdf. Accessed 23 Apr 2020.

Hales, C.M., Carroll, M.D., Fryar, C.D., Ogden, C.L. (2020). *Prevalence of obesity and severe obesity among adults: United States 2017–2018*. NCHS Data Brief 360. Retrieved April 30, 2020, from https://www.cdc.gov/nchs/products/databriefs/db360.htm

Harrison, K., Bost, K. K., McBride, B. A., Donovan, S. M., Grigsby-Toussaint, D. S., Kim, J., … Jacobsohn, G. C. (2011). Toward a developmental conceptualization of contributors to overweight and obesity in childhood: The six-Cs model. *Child Development Perspectives, 5*(1), 50–58.

Heymsfield, S., Aronne, L. J., Eneli, I., Kumar, R. B., Michalsky, M., Walker, E., Wolfe, B. M., Woolford, S. J., & Yanovski, S. (2018). *Clinical perspectives on obesity treatment: challenges, gaps, and promising opportunities*. Retrieved April 28, 2020, from https://nam.edu/clinical-perspectives-on-obesity-treatment-challenges-gaps-and-promising-opportunities/

Hoelscher, D. M., Kirk, S., Ritchie, L., & Cunningham-Sabo, L. (2013). Position of the academy of nutrition and dietetics: Interventions for the prevention of pediatric overweight and obesity. *Journal of the Academy of Nutrition and Dietetics, 113*(10), 1375–1394.

Hruby, A., & Hu, F. B. (2015). The epidemiology of obesity: A big picture. *Pharmaco Economics, 33*(7), 673–689.

Huang, P. L. (2009). A comprehensive definition for metabolic syndrome. *Disease Models & Mechanisms, 2*, 231–237.

Jensen, M. D., Ryan, D. H., Apovian, C. M., Ard, J. D., Comuzzie, A. G., Donato, K. A., … Yanovski, S. Z. (2013). 2013 AHA/ACC/TOS guideline for the management of overweight and obesity in adults. *Circulation, 129*(25 S2), S102–S138.

Jernigan, V. B., Salvatore, A. L., Wetherill, M., et al. (2018). Using community-based participatory research to develop healthy retail strategies in native American-owned convenience stores: The THRIVE study. *Preventive Medicine Reports, 11*, 148–153.

Jernigan, V. B., Salvatore, A. L., Williams, M., Wetherill, M., et al. (2019). A healthy retail intervention in native American convenience stores: The THRIVE community-based participatory Research study. *American Journal of Public Health, 109*(1), 132–139.

Jernigan, V. B. B., Wetherill, M., Hearod, J., Jacob, T., Salvatore, A. L., Cannady, T., … Wiley, A. (2017). Cardiovascular disease risk factors and health outcomes among American Indians in Oklahoma: The THRIVE study. *Journal of Racial and Ethnic Health Disparities, 4*(6), 1061–1068.

Johns, D. J., Hartmann-Boyce, J., Jebb, S. A., Aveyard, P., & Group, B. W. M. R. (2014). Diet or exercise interventions vs combined behavioral weight management programs: A systematic review and meta-analysis of direct comparisons. *Journal of the Academy of Nutrition and Dietetics, 114*(10), 1557–1568.

Jonikas, J. A., Cook, J. A., et al. (2015). Associations Between Gender and Obesity Among Adults with Mental Illnesses in a Community Health Screening Study. *Community Ment Health J., 52*(4), 406–415.

Kelly, T., Yang, W., Chen, C. S., Reynolds, K., & He, J. (2008). Global burden of obesity in 2005 and projections to 2030. *International Journal of Obesity, 32*(9), 1431–1437. https://doi.org/10.1038/ijo.2008.102

Kersch, R. (2009). The politics of obesity: A current assessment and look ahead. *The Millbank Quarterly, 87*(1), 295–316.

Kersh, R., & Monroe, J. (2002). The politics of obesity: Seven steps to government action. *Health Affairs, 21*(6). Retrieved April 30, 2020, from https://www.healthaffairs.org/doi/full/10.1377/hlthaff.21.6.142

Lawrence, D., & Kisely, S. (2010). Inequalities in healthcare provision for people with severe mental illness. *Journal of Psychopharmacology* (Oxford, England). https://doi.org/10.1177/1359786810382058

Lee, A. K., Warren, B., Liu, C., Foti, K., & Selvin, E. (2019). Number and characteristics of US adults meeting prediabetes criteria for diabetes prevention programs: NHANES 2007–2016. *Journal of General Internal Medicine, 34*(8), 1400–1402.

Lifestyle Interventions for Expectant Moms. (n.d.). Welcome Page. Retrieved Retireived October 27, 2020, from https://lifemoms.bsc.gwu.edu/

Llabre, M. M., Ard, J. D., Bennett, G., Brantley, P. J., Fiese, B., Gray, J., … Wilfley, D. (2018). *Clinical practice guideline for multicomponent behavioral treatment of obesity and overweight in children and adolescents.* American Psychological Association. Retrieved April 30, 2020, from https://www.apa.org/obesity-guideline/clinical-practice-guideline.pdf

Loring, B., & Robertson, A. (2014). *Obesity and inequities: Guidance for addressing inequities in overweight and obesity.* World Health Organization Regtional Office for Europe. Retrieved April 30, 2020, from http://www.euro.who.int/__data/assets/pdf_file/0003/247638/obesity-090514.pdf

Lundeen, E. A., Park, S., Pan, L., O'Toole, T., Matthews, K., Blanck, H. M. (2018). Obesity Prevalence Among Adults Living in Metropolitan and Nonmetropolitan Counties — United States, 2016. *MMWR Morb Mortal Wkly Rep, 67*, 653–658. https://doi.org/10.15585/mmwr.mm6723a1externalicon

Lyn, R., Heath, E., & Dubhashi, J. (2019). Global implementation of obesity prevention policies: A review of progress, politics, and the path forward. *Current Obesity Reports, 8*(4), 504–516.

Mackenbach, J. D., Rutter, H., Compernolle, S., Glonti, K., Oppert, J.-M., Charreire, H., … Lakerveld, J. (2014). Obesogenic environments: A systematic review of the association between the physical environment and adult weight status: The SPOTLIGHT project. *BMC Public Health, 14*(233).

Maffetone, P. B., & Laursen, P. B. (2020). *Revisiting the global overfat pandemic, 8*(51). Retrieved April 28, 2020 from https://www.ncbi.nlm.nih.gov/pmc/articles/PMC7052125/

Malik, V. S., Willett, W. C., & Hu, F. B. (2013). Global obesity: Trends, risk factors and policy implications. *Nature Reviews Endocrinology, 9*(1), 13–27.

McKinsey Global Institute. (2014). *Overcoming obesity: an initial economic analysis.* Retrieved April 27, 2020 from https://www.mckinsey.com/~/media/McKinsey/Business%20Functions/Economic%20Studies%20TEMP/Our%20Insights/How%20the%20world%20could%20better%20fight%20obesity/MGI_Overcoming_obesity_Full_report.ashx

Metropolitan Life Insurance Company. (1942). New weight standards for men and women. *Statistical Bulletin-Metropolitan Life Insurance Company, 23*, 6–8.

Micciolo, R., Di Francesco, V., Fantin, F., Canal, L., Harris, T. B., Bosello, O., & Zamboni, M. (2010). Prevalence of overweight and obesity in Italy (2001-2008): Is there a rising obesity epidemic? *Annals of Epidemiology, 20*(4), 258–264. https://doi.org/10.1016/j.annepidem.2010.01.006

Moller, D. E., & Kaufman, K. D. (2005). Metabolic syndrome: A clinical and molecular perspective. *Annual Review of Medicine, 56*(1), 45–62. https://doi.org/10.1146/annurev.med.56.082103.104751

National Heart, Lung, and Blood Institute (NHLBI). (2015). *In brief: Your guide to loweing your blood pressure with DASH* (NIH publication no. 06–5834). Bethesda, MD.

NHBLI. (2016). *What is metabolic syndrome?* Retrieved April 12, 2020 from https://www.nhlbi.nih.gov/health/health-topics/topiocs/ms

Nutall, F. Q. (2015). Body mass index, obesity, BMI, and health: A critical review. *50*(3):117–128. Retrieved April 28, 2020 from https://journals.lww.com/nutritiontodayonline/Fulltext/2015/05000/Body_Mass_Index__Obesity,_BMI,_and_Health_A.5.Aspx

Obesity Medicine Association. (2020). *What is Obesity Medicine?* Retireived October 27, 2020, from https://obesitymedicine.org/

Ogden, C. L., Carroll, M. D., Kit, B. K., & Flegal, K. M. (2012). Prevalence of obesity and trends in body mass index among US children and adolescents, 1999-2010. *JAMA, 307*(5), 483–490.

Ogden, C. L., Yanovski, S. Z., Carroll, M. D., & Flegal, K. M. (2007). The epidemiology of obesity. *Gastroenterology, 132*, 2087–2102.

Olker, S. J., Parrott, J. S., Swarbrick, M. A., & Spagnolo, A. B. (2016). Weight management interventions in adults with a serious mental illness: A meta-analytic review. *American Journal of Psychiatric Rehabilitation, 19*(4), 370–393.

Olsen, J. M., & Nesbitt, B. J. (2010). Health coaching to improve healthy lifestyle behaviors: An integrative review. *American Journal of Health Promotion, 25*(1), e1–e12.

Omran, A. R. (2005). The epidemiologic transition: A theory of the epidemiology of population change. *The Millbank Quarterly, 83*(4), 731–757.

Oulas, A., Minadakis, G., Zachariou, M., Sokratous, K., Bourdakou, M. M., & Spyrou, G. M. (2019). Systems bioinformatics: Increasing precision of computational diagnostics and therapeutics through network-based approaches. *Briefings in Bioinformatics, 80*(3), 806–824.

Park, Y., Kim, N. H., Kwon, T. Y., & Kim, S. G. (2018). A novel adiposity index as an integrated predictor of cardiometabolic disease morbidity and mortality. *Scientific Reports, 8*(16753), 1–8. Retrieved April 28, 2020, from https://www.nature.com/articles/s41598-018-35073-4.pdf

Patel, M. S., Asch, D. A., & Volpp, K. G. (2015). Wearable devices as facilitators, not drivers, of health behavior change. *JAMA, 313*(5), 459–460.

Peaceman, A. M., Clifton, R. G., Phelan, S., Gallagher, D., Evans, M., Redman, L. M., … Cahill, A. G. (2018). Lifestyle interventions limit gestational weight gain in women with overweight or obesity: LIFE-moms prospective Meta-analysis. *Obesity, 26*(9), 1396–1404. https://doi.org/10.1002/oby.22250

Petersen, R., Pan, L., & Blanck, H. M. (2019). Racial and ethnic disparities in adult obesity in the United States: CDC's tracking to inform local and state action. *Preventing Chronic Disease: Public Health Research, Practice, & Policy, 16*. Retrieved April 30, 2020, from https://www.cdc.gov/pcd/issues/2019/18_0579.htm

Phelan, S., Nallari, M., Darroch, F. E., & Wing, R. R. (2009). What do physicians recommend to their overweight and obese patients. *The Journal of the American Board of Family Medicine, 22*(2), 115–122.

Puhl, R., & Brownell, K. D. (2001). Bias, discrimination, and obesity. *Obesity Research, 9*(788), 805.

Punthakee, Z., Goldenberg, R., & Katz, P. (2018). Definition, classification and diagnosis of diabetes, prediabetes and metabolic syndrome. *Canadian Journal of Diabetes, 42*, S10–S15.

Raynor, H. A., & Champagne, C. M. (2016). Position of the academy of nutrition and dietetics: Interventions for the treatment of overweight and obesity in adults. *Journal of the Academy of Nutrition and Dietetics, 116*(1), 129–147.

Renner, R. (2019). Atlanta's Food Forest will provide fresh fruit, nuts, and herbs to forage. *City Lab*. Retrieved November 11, 2019 from https://www.citylab.com/environment/2019/06/urban-food-forest-local-agriculture-atlanta-fresh-produce/590869/

Roberto, C. A., Swinburn, B., Hawkes, C., Huang, T. T. K., Costa, S. A., Ashe, M., … Brownell, K. D. (2015). Patchy progress on obesity prevention: Emerging examples, entrenched barriers, and new thinking. *The Lancet*. Lancet Publishing Group. https://doi.org/10.1016/S0140-6736(14)61744-X

Rogers, J. M., Ferrari, M., Mosely, K., Lang, C. P., & Brennan, L. (2017). Mindfulness-based interventions for adults who are overweight or obese: A meta-analysis of physical and psychological health outcomes. *Obesity Reviews, 18*(1), 51–67.

Ross, R. K., Standish, M., Flores, G. R., Jhawar, M. K., Baxter, R. J., Solomon, L. S., … Schwartz, P. M. (2010, November 1). Community approaches to preventing obesity in California. *American Journal of Public Health, 100*(11), 2023–2025. https://doi.org/10.2105/AJPH.2010.198820

Roth, C., Foraker, R. E., Payne, P. R., & Embi, P. J. (2014). Community-level determinants of obesity: Harnessing the power of electronic health records for retrospective data analysis. *BMC Medical Informatics and Decision Making, 14*(36) Retrieved April 29, 2020, from https://www.ncbi.nlm.nih.gov/pmc/articles/PMC4024096/

Sallis, J. F., Owen, N., & Fisher, E. B. (2008). Ecological models of health behavior. In Glanz et al. (Eds.), *Health Behavior and Health Education*. San Francisco, CA: Wiley and Sons.

Samuels, S. E., Craypo, L., Boyle, M., Crawford, P. B., Yancey, A., & Flores, G. (2010). The California Endowment's healthy eating, active communities program: A midpoint review. *American Journal of Public Health, 100*(11), 2114–2123.

Sassi, F., Devaux, M., Cecchini, M., & Rusticelli, E. (2009). The obesity epidemic: Analysis of past and projected future trends in selected OECD countries. *OECD Health Working Papers, 45*, 1–81.

Schwarte, L., Samuels, S. E., Capitman, J., Ruwe, M., Boyle, M., & Flores, G. (2010). The Central California regional obesity prevention program: Changing nutrition and physical activity environments in California's heartland. *American Journal of Public Health, 100*(11), 2124–2128.

Schwartz, H. (1986). *Never satisfied: A cultural history of diets, fantasies, and fat.* New York, NY: The Free Press.

Schwartz, P. M., Kelly, C., Cheadle, A., Pulver, A., & Solomon, L. (2018). The Kaiser Permanente Community health initiative: A decade of implementing and evaluating community change. *American Journal of Preventive Medicine, 54*(5 S2), S105–S109.

Schwingshackl, L., & Hoffmann, G. (2015). Diet quality as assessed by the healthy eating index, the alternate healthy eating index, the dietary approaches to stop hypertension score, and health outcomes: A systematic review and meta-analysis of cohort studies. *Journal of the Academy of Nutrition and Dietetics, 115*(5), 780–800.

Schwingshackl, L., Missbach, B., Dias, S., König, J., & Hoffmann, G. (2014). Impact of different training modalities on glycaemia control and blood lipids in patients with type 2 diabetes: A systematic review and network meta-analysis. *Systems Reviews, 3*, 130.

Sciamanna, C. N., Tate, D. F., Lang, W., & Wing, R. R. (2000). Who reports receiving advice to lose weight: Results from a multistate survey. *Archives of Internal Medicine, 160*(15), 2334–2339.

Science X: Phys.org. (2013). *Mapping food deserts.* Retrieved April 30, 2020, from https://phys.org/news/2011-03-food.html

Seidell, J. C., & Halberstadt, J. (2015). The global burden of obesity and the challenges of prevention. *Annals of Nutrition and Metabolism, 66*, 7–12. https://doi.org/10.1159/000375143

Shipan, C. R., & Volden, C. (2008). The mechanisms of policy diffusion. *American Journal of Political Science, 52*(4), 840–857.

Sicat, J. (2018). Defining Obesity's interplay among environment, behavior, and genetics. *Obesity Medicine*. Retrieved April 27, 2020, from https://obesitymedicine.org/obesity-and-genetics/

Spear, B. A., Barlow, S. E., Ervin, C., Ludwig, D. S., Saelens, B. E., Schetzina, K. E., & Taveras, E. M. (2007). Recommendations for treatment of child and Adolecent overweight and obesity. *Pediatrics, 120*(S4), S254–S288.

Templin, T., Hashiguchi, T. C. O., Thomson, B., Dieleman, J., & Bendavid, E. (2019). The overweight and obesity transition from the wealthy to the poor in low- and middle-income countries: A survey of household data from 103 countries. *PLoS Medicine*. Retrieved April 30, 2020, from https://journals.plos.org/plosmedicine/article?id=10.1371/journal.pmed.1002968

The Community Guide (2017). *What works: Obesity prevention and control.* Retrieved from: https://www.thecommunityguide.org/sites/default/files/assets/What-Works-Factsheet-Obesity.pdf

U.S. C.D.C. (2015). National Center for Health Statistics. *Health, United States, 2015, Table 58.* National Health and Nutrition Examination Survey. https://www.cdc.gov/nchs/data/hus/2015/058.pdf. Accessed 22 Apr 2020.

U.S. C.D.C. (2017a). *Health equity resource toolkit for state practitioners addressing obesity disparities.* https://www.cdc.gov/nccdphp/dnpao/state-local-programs/health-equity/pdf/toolkit.pdf

U.S. C.D.C. (2017b). *CDC's childhood obesity research demonstration (CORD) project 2.0.* Centers for Disease Control and Prevention, April 1, 2017. https://www.cdc.gov/obesity/strategies/healthcare/cord2.html. Accessed 23 Apr 2020.

U.S. C.D.C. (2018). *Overweight & obesity.* CDC, April 2018. https://www.cdc.gov/obesity/

U.S. C.D.C. (2019, October 23). *Strategies to prevent and manage obesity.* Retrieved April 28, 2020 from https://www.cdc.gov/obesity/strategies/index.html

U.S. C.D.C. (2020). *Healthy weight: The health effects of overweight and obesity.* Retrieved April 27, 2020, from https://www.cdc.gov/healthyweight/effects/index.html

U.S. C.D.C. (n.d.). *Data & statistics | Overweight & obesity.* Retrieved January 15, 2020, from https://www.cdc.gov/obesity/data/index.html

U.S. Department of Agriculture and U.S. Department of Health and Human Services. (1985). *Nutrition and your health: Dietary guidelines for Americans.* Home and garden bulletin no. 232, US Government Printing Office, Washington, DC.

U.S. Food and Drug Adminstration (F.D.A). (2020). *Menu labeling requirements.* Retrieved April 30, 2020, from https://www.fda.gov/food/food-labeling-nutrition/menu-labeling-requirements

U.S.D.A. Economic Research Service. (2019). *Food environment atlas.* Retrieved April 14, 2020, from https://www.ers.usda.gov/data-products/food-environment-atlas/go-to-the-atlas/

Vitaliano, P. P., Yi, J., Phillips, P. E. M., Echeverria, D., Young, H., & Siegler, I. C. (2005). Psychophysiological mediators of caregiver stress and differential cognitive decline. *Psychology and Aging, 20*(3), 402–411. https://doi.org/10.1037/0882-7974.20.3.402

W.H.O. (2018). *Obesity and overweight: Fact sheet world health organization*, Geneva, Switzerland. Retreived April 12, 2020 from http://www.who.int/mediacentre/factsheets/fs311/cn/

W.H.O. | Controlling the global obesity epidemic. (n.d.). Retrieved January 15, 2020, from https://www.who.int/nutrition/topics/obesity/en/

W.H.O. Expert Consultation. (2004). Appropriate body-mass index for Asian populations and its implications for policy and intervention strategies. *The Lancet, 363*(9403), 157–163.

Weigley, E. S. (1984). Average? ideal? desirable? A brief of overview of height-weight tables in the United States. *Journal of the American Dietetic Association, 84*(4), 417–423.

Wing, R. R., Lang, W., Wadden, T. A., Safford, M., Knowler, W. C., Bertoni, A. G., … Look AHEAD Research Group. (2011). Benefits of modest weight loss in improving cardiovascular risk factors in overweight and obese individuals with type 2 diabetes. *Diabetes Care, 34*(7), 1481–1486.

Woodcock, J., Franco, O. H., Orsini, N., & Roberts, I. (2011). Non-vigorous physical activity and all-cause mortality: Systematic review and meta-analysis of cohort studies. *International Journal of Epidemiology, 40*(1), 121–138.

World Food Programme. (2015). *Indonesia-food security and vulnerability atlas 2015*. Retrieved April 19, 2020, from https://www.wfp.org/publications/indonesia-food-security-and-vulnerability-atlas-2015.

World Health Organization. (2020a). *Health topics: Obesity*. Retrieved April 27, 2020, from https://www.who.int/topics/obesity/en/

World Health Organization (WHO). (2020b). *Health topics fact sheets: Overweight and obesity*. Retrieved April 27, 2020, from https://www.who.int/news-room/fact-sheets/detail/obesity-and-overweight

World Health Organization. (2020c). *Facts in Pictures*. Retrieved October 27, 2020, from https://www.who.int/news-room/facts-in-pictures/detail/6-facts-on-obesity

Zimmet, P., Magliano, D., Matsuzawa, Y., Alberti, G., & Shaw, J. (2005). The metabolic syndrome: A global public health problem and a new definition. *Journal of Atherosclerosis and Thrombosis, 12*(6), 295–300.

Zukiewicz-Sobczak, W., Wróblewska, P., Zwolinski, J., Chmielewska-Badora, J., Adamczuk, P., Krasowska, E., … Silny, W. (2014). Obesity and poverty paradox in developed countries. *Annals of Agricultural and Environmental Medicine, 21*(3), 590–594.

17

Indigenous Community Health

Seth Oppong, Kendall R. Brune, and Elias Mpofu

Introduction

What defines good health and wellbeing? The processes and tools for it vary across communities based on their heritages (World Health Organization [WHO], 2010; United Nations [UN], 2016). Surprisingly, the health understandings of indigenous communities or first nations have historically been marginalized since the colonial conquest period, mostly by western nations (WHO, 2010; Williams, Potestio, & Austen-Wiebe, 2019). There are approximately 370 million designated

S. Oppong (✉)
University of Botswana, Gaborone, Botswana
e-mail: Oppongs@ub.ac.bw

K. R. Brune
Meharry Medical College, Nashville, TN, USA

E. Mpofu
University of North Texas, Denton, TX, USA

University of Sydney, Sydney, NSW, Australia

University of Johannesburg, Johannesburg, South Africa
e-mail: Elias.Mpofu@unt.edu

E. Mpofu (ed.), *Sustainable Community Health*,
https://doi.org/10.1007/978-3-030-59687-3_17

Indigenous peoples residing in about 90 countries around the world (UN, 2016). Indigenous people's habitats include Africa, Asia, Arctic Regions, Central and North America, the Pacific Region, the Russian Federation, South America, and the Caribbean.

Learning Objectives

On studying this chapter, the reader should be able to:

1. Define indigenous communities in historical and contemporary worlds for a better understanding of the variety and diversity of Indigenous peoples' experiences.
2. Trace the history of research and practice in indigenous community health, highlighting emerging contemporary themes.
3. Discuss primary themed practices in indigenous community health practices important for sustainable health in those communities.
4. Analyze the influences of culture, professional, and legal precedent on current and emerging indigenous community health practices.
5. Propose areas for research and practice to develop for sustainable health in indigenous communities.

Indigenous communities may experience health neglect by state and federal governments from modifiable social determinants of health. Within the mainstream Western healthcare worldview, social determinants of health are factors that account for the health disparities among members of a given society (NEJM Catalyst, 2017; Horrill, McMillan, Schultz, & Thompson, 2018). Key sociohistorical determinants of health among indigenous groups can be classified into distal (such as history and socioeconomic contexts), intermediate (such as community resources and capacities), and proximal determinants (such as health behaviors, the physical and micro-social environment) (Reading & Wien, 2009; Horrill et al., 2018).

While a lot has been documented about the health disparities affecting Indigenous peoples around the world (Australian Bureau of Statistics, 2012; Gruen, Weeramanthri, & Bailie, 2002; Marrone, 2007; Reading & Wien, 2009; Shah, Gunraj, & Hux, 2003; Waterworth, Pescud, Braham, Dimmock, & Rosenberg, 2015; WHO, 2007, 2010; Crombie, Irvine, Elliott, & Wallace, 2005; Brown, 2018; Murray, Kulkarni, & Ezzati, 2005; Thresia, 2018), relatively less is known about Indigenous people's health beliefs and practices that are essential for their public health support programs (Mpofu, 2006; Mpofu, Peltzer, & Bojuwoye, 2011). For instance, following a systematic literature review, Marrone (2007) documented the following healthcare disparities among indigenous populations in North America, Australia, and New Zealand:

- Limited access to ambulatory, acute, and specialized healthcare due to residing in remote rural locations
- Higher rates of chronic diseases, such as hypertension, obesity, and type 2 diabetes than the majority population
- A shorter life expectancy and higher mortality rates compared with other ethnic groups
- Higher rates of cancer, obesity, and smoking
- Higher prevalence estimates of current common mental disorders

Thus, the poorer health outcomes among indigenous populations are well documented (see also *African Commission's Working Group on Indigenous Populations/Communities,* 2007, Anderson, Robson, Connolly, et al., 2016). What has received scant attention is indigenous community health assets that have sustained indigenous communities for millennia, and some of which are belatedly recognized by modern medicine as complementary or alternative medicine (Twumasi, 1981; Phillips, Hyma, & Ramesh, 1992; Mpofu, 2006; Obomsawin, 2007).

Indigenous medicine has strong community member buy-in and at an exceedingly lower cost than with modern medicine (Twumasi, 1981; Phillips et al., 1992; White, 2015). Among its benefits is the recognition of the interdependency of the natural environment (traditional lands, territories, and natural resources) and social wellbeing mutually constitutive of physical and spiritual wellbeing (Struthers, Eschiti, & Patchell, 2004; Mpofu et al., 2011). Nonetheless, health sustenance outcomes among indigenous communities are diverse, on some indicators rather than others, and are explained by geography, degree of interaction with modern society, genetic predispositions, access to public health services, and so on (Kipuri & Sørensen, 2008; Maina, Kim, Rutherford, et al., 2014).

The UN Convention on the rights of indigenous communities (UN, 2008) recognizes the self-determination of health outcomes by indigenous communities. When indigenous communities self-determine their health outcomes, they decide, control, and pursue their own self-selected health goals and pathways (Chandler & Lalonde, 1998; Reading & Wien, 2009). They also practice cultural traditions and customs to promote health and wellness (Mpofu, 2006; see articles 24 and 25, UN, 2008).

The UN *Declaration on the Rights of Indigenous Peoples* (UN, 2008) affirmed that "indigenous peoples are equal to all other peoples, while recognizing the right of all peoples to be different, to consider themselves different, and to be respected as such" (p. 1) and reaffirmed that "indigenous peoples, in the exercise of their rights, should be free from discrimination of any kind" (p. 2). The United Nations has also expressed concern that "indigenous peoples have suffered from historic injustices as a result of, inter alia, their colonization and dispossession of their lands, territories and resources, thus preventing them from exercising, in particular, their right to development in accordance with their own needs and interests" (p. 2). Nonetheless, sociohistorical determinants of health disproportionately affect Indigenous people (Kolahdooz, Nader, Yi, & Sharma, 2015). Thus, it is important to give attention to the sociohistorical determinants of health of indigenous populations in their own right.

Professional and Legal Definitions of Health and Wellbeing in Indigenous Communities

But what do we mean exactly by Indigenous peoples and/or communities? According to the United Nations (UN, 2016, p. 4), indigenous communities to refer to:

> peoples and nations are those which, having a historical continuity with pre-invasion and pre-colonial societies that developed on their territories, consider themselves distinct from other sectors of the societies now prevailing on those territories, or parts of them. They form at present non-dominant sectors of society and are determined to preserve, develop and transmit to future generations their ancestral territories, and their ethnic identity, as the basis of their continued existence as peoples, in accordance with their own cultural patterns, social institutions and legal system.

WHO (2012) reports that, among Africans, good health is understood to be constitutive of the ability to work and move around in additon to "the emotional, psychological, economic, mental and spiritual aspects of health" (p. 12). This view is consistent with the representations of health among the Māori indigenous people of New Zealand who consider health to be defined by four interconnected dimensions of *te taha wairua* (spiritual health), *te taha hinengaro* (mental health), *te taha tinana* (physical health), and *te taha whanau* (family health) (Durie, 1985; New Zealand Ministry of Health, 2017). Like the Māori, the American Indians/Alaska Natives "hold a holistic perspective on health based on a balance of the interrelationships of body, mind, spirit, and the environment" (Horowitz, 2012, p. 25). Similarly, the Aboriginal and Torres Strait Islander people in Australia as well as Papua New Guinea regard *wellbeing to be* determined by "*lifestyle* and *relationship with others and the community*" that "*involves a balance with regard to holistic dimensions*" while illness meant "being *really sick*"; 2. And needing to "*take action …seeking treatment* and *undertaking preventative measures*"; and 3. "*an imbalance involving holistic dimensions* including physical, spiritual, social, and environmental" (Boulton-Lewis, Pillay, & Wilss, 2001, para. 19–24).

Among indigenous people, health is essentially a social process and outcome. For instance, the Aboriginal and Torres Strait Islander people in Australia associate good health with "happiness … and sustaining cultural identity, community, and family life that provides a source of strength against adversity, poverty, neglect, and other challenges of life" (Dudgeon & Walker, 2015, p. 278; see also Australian National Mental Health Commission, 2013; Boulton-Lewis et al., 2001; Dudgeon & Walker, 2015; Maher, 1999). The Canadian First Nations "embrace a holistic concept of health that reflects physical, spiritual, emotional and mental dimensions" (Reading & Wien, 2009, p. 3). First Nations Health Council (FNHC) of Canada promotes a holistic perspective to health and wellbeing "in which individual human beings own their health and wellness journeys…" and it is constituted by "four dimensions of wellness—physical, mental, emotional, and spiritual health—and acknowledges the influence of factors including values and supports, where we come from, and the social determinants of health" as stated by Gallagher (2019, p. 5).

Some of the recurring themes among different conceptions of health and wellness (or wellbeing) held by indigenous populations in different regions of the world are the emphasis on (1) the multidimensionality of health and wellness including economic dimensions that go beyond allopathic medicine, (2) a dynamic balance among the interconnected dimensions of health and wellness, (3) spirituality and connection to the universe and ancestors, (4) the view that disruptions to ways of life undermine health and wellness, and (5) the fact that treatment must recognize and respect ways of life. By contrast, the prevailing mainstream or modern medicine perspective about health is biomedical or allopathic in nature (Horrill et al., 2018; Gallagher, 2019), in which health is synonymous with the absence of disease at the individual level (Twumasi, 1975, 1981; Yuill, Crinson, & Duncan, 2011; Gallagher, 2019). Indigenous communities hold beliefs about health and wellbeing that are closer and inclusive.

History of Research and Practice in Indigenous Health

Published research and practice in indigenous health is bound up in the history of colonization (Coates, 2004; Laycock, Walker, Harrison, & Brands, 2011; Gray & Oprescu, 2016; Axelsson, Kukutai, & Kippen, 2016b). Much of it is misrepresentation of indigenous community health competencies by omission or commission.

In the colonial period, Indigenous people's health assets were marginalized through being discounted "through the lens of Western prejudice" (Gray & Oprescu, 2016, p. 461). Some research on indigenous community health has been patronizing, and carried out "without permission, consultation or involvement from Indigenous people, with the primary benefit being to the researcher." (Gray & Oprescu, 2016, p. 461; see also Australian Government Department of Families, Community Services and Indigenous Affairs, 2006; Bainbridge et al., 2015; Laycock et al., 2011). From this erosion of the indigenous knowledge base, indigenous communities have experienced avoidable intergenerational health degradations (Coates, 2004). Not surprisingly, indigenous communities harbor mistrust of contemporary health research and practice, preferring their own healthcare services (Gray & Oprescu, 2016).

As stated earlier, mainstream health services are tied to modern medicine and, often with gross disregard for the ways of life of indigenous populations (Axelsson et al., 2016b), resulting in substantial avoidable mortality in indigenous populations (Marrone, 2007; Reading & Wien, 2009; Shah et al., 2003; Waterworth et al., 2015; WHO, 2007, 2010; Crombie et al., 2005; Brown, 2018; Murray et al., 2005; Thresia, 2018). Many of the health disparities affecting indigenous communities are from the direct or indirect outcomes of health knowledge marginalization by colonization, resulting in the past and current poorer health outcomes (Horrill et al., 2018; Marrone, 2007: Reading & Wien, 2009).

In recent times, indigenous communities seek to self-determine their health outcomes as recognized by UN conventions (see Table 17.1). To foreground our discussion of current and prospective approaches to sustainable health among indigenous communities, we refer to the UN Convention on the Rights of Indigenous Communities (UNCRIC, 2008) as follows:

Article 24

1. Indigenous peoples have the right to their traditional medicines and to maintain their health practices, including the conservation of their vital medicinal plants, animals, and minerals. Indigenous individuals also have the right to access, without any discrimination, to all social and health services.
2. Indigenous individuals have an equal right to the enjoyment of the highest attainable standard of physical and mental health. States shall take the necessary steps with a view to achieving progressively the full realization of this right. (p. 9)

Article 25

Indigenous peoples have the right to maintain and strengthen their distinctive spiritual relationship with their traditionally owned or otherwise occupied and used lands, territories, waters and coastal seas and other resources and to uphold their responsibilities to future generations in this regard. (p. 10)

Table 17.1 History of advocacy for indigenous peoples at the United Nations

1923–25	First International Involvement. In 1923, Haudenosaunee Chief Deskaheh traveled to Geneva to speak to the League of Nations and was not allowed so returned home in 1925. Maori religious leader, T.W. Ratana, also traveled with his entourage to London to petition King George about the breaking of the Treaty of Waitangi (between Maori in New Zealand in 1840) but was denied access; he later traveled to Geneva to petition the League of Nations in 1925 and was again denied access.
1981	The Martínez Cobo Study provided the definition and criteria for identifying indigenous communities.
1982	Working Group on Indigenous Populations (WGIP)
1989	International Labour Organization (ILO) Convention 169 (C169), *Indigenous and Tribal Peoples Convention, 1989 (No. 169)*
1993	International Year of the World's Indigenous People
1994	International Decade of the World's Indigenous Peoples
2000	Permanent Forum on Indigenous Issues
2001	Special Rapporteur on the Rights of Indigenous Peoples
2005	Second International Decade of the World's Indigenous Peoples
2007	Expert Mechanism on the Rights of Indigenous Peoples (EMRIP)
2007	UN Declaration on the Rights of Indigenous Peoples (UNDRIP)
2014	World Conference on Indigenous Peoples (WCIP)

Source: United Nations (n.d.)

Human rights-based approaches to health sustenance go beyond simply making the required infrastructure for care available (Horrill et al., 2018) to prioritizing indigenous knowledge systems in caring for Indigenous populations (UNCRIC, 2008). They affirm the right of communities to their self-determination of wellbeing (Webb, 2012; Murphy, 2014) and applying community-driven interventions. Indigenous populations themselves have self-advocated for these rights (Wachira & Karjala, 2012; Gray & Oprescu, 2016) and in partnership with allies of indigenous communities such as Survival International, the *Gesellschaft für bedrohte Völker*, the Foundation for Aboriginal and Islander Research Action in Australia, Indigenous Peoples of Africa Coordinating Committee, the Working Group for Indigenous Minorities in Southern Africa, and the International Work Group for Indigenous Affairs (Kemner, 2014; Pelican & Maruyama, 2015; Gray & Oprescu, 2016).

For instance, the advocacy by the Foundation for Aboriginal and Islander Research Action in Australia in the 1970s led to the development of protocols for acceptable indigenous health research (Gray & Oprescu, 2016), while the Indigenous Peoples of Africa Coordinating Committee and the Working Group for Indigenous Minorities in Southern Africa have been instrumental to promoting indigenous ways to health and wellbeing in Namibia, Botswana, Angola, and South Africa (Pelican & Maruyama, 2015). Moreover, academic programs for the study of indigenous populations are on the increase around the world. Examples include programs at the University of Western Australia, Georgetown University, McMaster University, the University of Kansas, University of Alberta, University of Sydney, and University of Wollongong. These academic programs have contributed to capacity building for indigenous health research and practice as well as indigenous community development. Collectively, these developments have promoted the broadening of the allopathic model of health to include other representations of health held by Indigenous peoples for their health sustainability.

Current and Prospective Practices in Indigenous Community Health Sustenance

Indigenous knowledge systems and worldviews prioritize participatory holistic health (Laycock et al., 2011; Gray & Oprescu, 2016), with co-planning, co-creation, co-implementation, and co-evaluation of health improvement (Horowitz, 2012; Donatuto, Campbell, & Gregory, 2016; Gray & Oprescu, 2016; Fijal & Beagan, 2019; Gallagher, 2019; Williams et al., 2019). Indigenous community-oriented health practices are aimed at health programming addressing avoidable health disparities and injustices at the policy level (Donatuto et al., 2016; Horrill et al., 2018; Williams et al., 2019; Dudgeon, Bray, & Walker, 2020). We discuss several current approaches to the sustainable health and wellbeing of indigenous populations, with examples.

Community-Driven Interventions

These are strength-based approaches with an emphasis on community control (Pattoni, 2012; Murphy, 2014; Dudgeon et al., 2020). Strength-based approaches allow "meaningful self-determination ....to improved levels of indigenous physical and mental health, and, conversely... [since] control and domination by others is a contributing factor to ill-health and elevated levels of mortality in indigenous communities" (Murphy, 2014, p. 1). Strength-based, community-driven approaches allow community members to decide what is constitutive of health and wellness for them (Dudgeon et al., 2020). Strength-based approaches to indigenous community health build on their collective knowledge, capacities, skills, connections, and potentials in the community (Pattoni, 2012). They also engage families and the clan as part of healthcare for cultural security in the delivery of health services (see Case Illustration 17.1).

Interface with Modern Medicine We present on two case examples of community-driven, asset-based healthcare approaches for indigenous communities interfacing mainstream western medicine: the "*Casa de la Mujer Indígena*" (hereafter referred to as *Casas:* Pelcastre-Villafuerte et al., 2014) and the Chickasaw Nation's healthcare initiatives.

Chickasaw Nation's Health Initiatives The Chickasaw Nation's first healthcare clinic opened in 1968, in Tishomingo Oklahoma, US, as a part-time facility. It was the first Indian health facility of any kind to be located within the Chickasaw Nation's identified lands. Today, the Chickasaw Nation Medical Center (CNMC) serves 112,000 emergency patients per year, operating a 52-bed inpatient Critical Access Hospital and treating over 1.2 million outpatient visits per year (Office of the Governor of the Chickasaw Nation, 2013). The Chickasaw Nation Department of Health operates a hospital, five clinics, eight pharmacies, a diabetes care center, an emergency medical services building, four nutrition services centers, eight WIC offices, and five wellness centers throughout South Central Oklahoma. The Chickasaw Nation has an inpatient and outpatient pharmacy delivering over 1.8 million prescriptions annually.

Case Illustration 17.1: Discretionary Use of Western Medicine in Indigenous Community

Mr. Minjarra is a 23-year-old, Aboriginal Australian with intellectual disability from an indigenous community home about six hours away from the hospital. He was admitted for inpatient care following a left hip surgery. On admission to inpatient care, Mr. Minjarra refused to participate in therapy with the nurses, who believed the refusal to be due to his communication difficulties and/or lack of understanding of the treatment procedures secondary to his intellectual disability. However, his family was of a different view and said Mr. Minjarra was a very compliant individual. An Aboriginal Liaison person engaged Mr. Minjarra and his family for a way forward. It turned out on review of his pain medication that it was inadequate, and Mr. Minjarra was in pain, but not telling it to the nurses. The Aboriginal liaison officer explained the treatment procedure to Mr. Minjarra and family in their own language, which they accepted. Mr. Minjarra took the prescription pain medication, which worked. His family expressed a need for short-term accommodation in the local community for his outpatient treatment, using a walking frame. They also said he would rather stay in the hospital until he can walk without the frame, and indicated concern that if he was discharged back to the community, he would viewed as a disabled person for using the walking frame. By considering their viewpoint in the matter and allowing for further extension, Mr. Minjarra was eventually discharged without a frame.

What Do You Think?

1. Consider the cultural issues pertinent to Mr. Minjarra's healthcare important to the treatment team. Explain the issues you identify in relation to cultural safety or comfort in treatment adherence.
2. How may the cultural influences on healthcare adherence be similar or different for a person from an indigenous community you are familiar with?

The Casas *Casas* is a community-based project designed to deliver culturally appropriate health and wellness services to indigenous women in Mexico. This project was initiated by the Inter-Sectorial Programme for Indigenous Women's Healthcare under the supervision of Mexico's Federal Office of Representation for the Development of Indigenous Peoples (Pelcastre-Villafuerte et al., 2014). Within this project, the indigenous community, non-governmental organizations (NGOs), and governmental institutions work together to create a physical space to perform the following functions:

- To provide health education and basic healthcare through *promotoras*. A *promotora* is a lay Hispanic/Latino community member who has received specialized training to provide basic health education in the community, though he or she is not a professional healthcare provider (Elder, Ayala, Parra-Medina, & Talavera, 2009).
- To serve as a referral desk to mainstream health facilities with an emphasis on reproductive healthcare.
- To create physical space to enable *promotoras*, traditional birth attendants, healthcare professionals, and NGOs working on health, reproductive rights, and domestic violence in indigenous communities meet to collaborate on local actions.

A process and outcome evaluation of *Casas* suggests the program has had some success but faces some critical challenges that needed to be addressed (Pelcastre-Villafuerte et al., 2014). For instance, poor record management, lack of administrative procedures, and lack of clarity about sources of funding were some of the challenges that required immediate attention to improve the implementation of the *Casas* project.

Partnerships for Sustainable Indigenous Health

Examples of a partnership for sustainable indigenous community health include the community-government partnership in Ghana called Community-Based Health Planning and Services and the American Indian Health System Joint Venture Construction Program. We describe each of these below.

Community-Government Partnership in Ghana The Ghana's Community-Based Health Planning and Services (CHPS) Initiative began in 1999 and was relaunched in 2017 as CHPS Plus or CHPS+ (Ghana Health Service [GHS], 2002; Kweku et al., 2020; Nyonator, Awoonor-Williams, Phillips, Jones, & Miller, 2005). The purpose of the initiatives is to relocate primary healthcare away from district health facilities to the communities to ensure immediate access to basic primary healthcare. The implementation of CHPS Initiative follows the steps outlined below:

- The District Health Management Team initiates a planning process to identify the most remote and deprived communities in their district.
- A preliminary community needs analysis is conducted and a "community entry" process involving a dialogue between healthcare professionals and the community leaders is launched.
- Leadership responsibilities are clarified and the selected communities are encouraged to mobilize community resources to construct village clinics known as Community Health Compounds (CHCs).
- Upon successful completion of a CHC, a nurse (known as a Community Health Officer [CHO]) is posted to the CHC. The CHOs become community-based frontline health workers who visit households, provide community health services, and conduct CHC clinics.

The CHPS initiative is funded by the Government of Ghana providing for Community Health Training Colleges and community health workers in the CHC model.

American Indian Health System Joint Venture Construction Program According to the Indian Health Service (IHS, n.d.), the Section 818 of the Indian Healthcare Improvement Act, P.L. 94–437, empowers the IHS

> …to establish joint venture projects under which Tribes or Tribal organizations would acquire, construct, or renovate a healthcare facility and lease it to the IHS, at no cost, for a period of 20 years. Participants in this competitive program are selected from among eligible applicants who agree to provide an appropriate facility to IHS. The facility may be an inpatient or outpatient facility. The Tribe must use Tribal, private or other available (non-IHS) funds to design and construct the facility. In return the IHS will submit requests to Congress for funding for the staff, operations, and maintenance of the facility per the Joint Venture Agreement. (para. 3)

Under the Indian Health System Joint Venture Construction Program, *Indian Country Today* (2020) reports that "Since 1991, more than 25 tribes have partnered to provide more than 30 facilities, from health centers to hospitals, increasing access to quality healthcare services for their communities." (para. 3). Thus, the IHS provides additional funding for staffing and operational costs to tribes who fund construction of health facilities (Office of the Governor of the Chickasaw Nation, 2013). One outcome of the Indian Health System Joint Venture Construction Program is the Chickasaw Nation Medical Center (CNMC). Indeed, CNMC was the first hospital to be completed under the Indian Health System Joint Venture Construction Program (Office of the Governor of the Chickasaw Nation, 2013). The Chickasaw Nation also built the Chickasaw Nation Medical Center (CNMC) in 2010, CNMC is nearly triple the size of the previous Carl Albert Indian Health Facility. The Chickasaw Nation committed US$ 148 million of tribal funds to the design, construction, and furnishing of CNMC. The success of the first CNMC campus has led to similar initiatives. For instance, the Chickasaw Nation broke ground for the construction of a new health support building on July 29, 2016 for the Apila Center on CNMC campus (Office of the Governor of the Chickasaw Nation, 2016). Similarly, in May 2016, the Chickasaw Nation began the construction of the medical center for a Veterans Lodge to enable Chickasaw veterans to relax, gather, and enjoy camaraderie. Earlier, in 2013, the Chickasha House also opened on the CNMC campus to house families and caretakers of long-term patients at CNMC; Chikasha House is reported to have accommodated more than 1700 families as of July 2016 (Office of the Governor of the Chickasaw Nation, 2016).

Cultural, Professional, and Legislative Influences on the Health of Indigenous Communities

The disruptions to indigenous ways of life in the process of colonization and its precursor, racism (Axelsson, Kukutai, & Kippen, 2016a; Pihama & Lee-Morgan, 2019), have contributed to health disparities affecting indigenous communities. The intergenerational marginalization experiences of indigenous people by (1) income, (2) education, (3) employment, (4) housing, and (5) community infrastructure further negatively impact their utilization of public healthcare services (Reading & Wien, 2009; Kolahdooz et al., 2015).

Culture Influences Indigenous people around the world would limit their use of modern medicine to levels commensurate with their level of need (Garvey, 2008; Westerman, 2010), until their symptoms have become chronic and severe and engage in modern medicine for shorter time frames (Mpofu et al., 2011). This would result in health disparities from postponing and prematurely dropping out of public health services (Dell'Osso, Glick, Baldwin, & Altamura, 2013). For instance, Australian Aboriginal and Torres Strait Islander (ATSI) underutilize mental health services (Vos, Barker, Stanley, & Lopez, 2007), yet they experience psychological distress at a rate 2.7 times higher than that of other Australians (Australian Bureau of Statistics, [ABS], 2010; Australian Health Ministers' Advisory Council [AHMAC], 2015). The underutilization suggests that current mental health programs may not be appropriately designed or implemented for this population. In 2014, almost 50% of ATSI people lived in outer regional, remote, and very remote areas (AIHW, 2013). Rural and remote community indigenous communities experience healthcare access difficulties due to issues of affordability, location, travel distance, and transportation (see also Chap. 11, this volume). Rural and remote community medical facilities are seriously understaffed around the world. In the Australian case, major cities have approximately 22 full-time employed psychiatrists per 100,000 people compared to three per 100,000 in outer regional and remote areas (MHWAC, 2008). These access limitations are in addition to the fact that modern health services may lack in cultural security or safety, discouraging indigenous community members from using them (Mpofu, Chronister, Johnson, & Denham, 2012; Westerman, 2010).

Racism and colonization have both created enduring conditions for a vicious cycle of social exclusion for members of indigenous communities (Reading & Wien, 2009). This social exclusion has been the result of structural violence that perpetrates and perpetuates social injustice and inequalities. They manifest in the form of institutional discrimination and stereotyping that makes members of the dominant groups in the society insensitive to, unappreciative of, or unconcerned about the needs of the indigenous populations. This may not always be deliberate, but may simply be the product of lack of exposure or contact with indigenous populations. This may also be the product of the demonization and dehumanization of indigenous populations to which members of the dominant groups have been subliminally exposed (Oppong, 2020).

Professional Issues Healthcare providers are obliged to respect the customs and traditions of indigenous populations when providing care to them (Mpofu et al., 2011). This realization on the part of healthcare providers and health policymakers will go a long way to help them define for themselves what an appropriate and quality care is from the perspective of the indigenous populations rather than using the allopathic representations of standard of care. In other words, the standard of care at the national and facility levels has to be reformulated in accordance with indigenous representations of health and wellness. Though the First Nations of Canada and the Aboriginals of Australia have formalized their representations of health and wellness/wellbeing, barriers to quality care still exist (Marrone, 2007; Horrill et al., 2018; Wilk, Cooke, Stranges, & Maltby, 2018; Gallagher, 2019), and in part from a lack of a workforce trained in indigenous health.

Legislative Influences There is wide variability across the globe in legal protections for indigenous community rights to healthcare. Of those with legal enablers, the American Indian and Alaskan Native Tribes (AI/AN) are among the exceptions. The AI/AN has a federally funded Indian Health Service, providing federal health services to them. This relationship, established in 1787, is based on Article I, Section 8 of the Constitution of the United States, and has been given form and substance by numerous treaties, laws, Supreme Court decisions, and Executive Orders. Nonetheless, there are disparities among the AI/AN from the fact that all tribes are not created equal, as they vary in size, sophistication, and geography. Although there are more than 573 federally recognized tribes, equity in the services they receive is skewed toward the larger tribes. This would be the same for other tribal groups around the globe.

Related Disciplines on Indigenous Community Health

Disciplines germane to promoting sustainable indigenous community health include indigenous studies, anthropology, community psychology, health psychology, rural sociology, medical sociology, public health, community health nursing, and healthcare administration. **Indigenous Studies (IS)** is a field of study concerned with the past, present, and future of indigenous populations with a particular focus on one or two Indigenous peoples. The indigenous Studies discipline tends to cover the cultures, social organizations, languages, and histories of indigenous populations as well as their struggles for self-determination. Although IS does not always directly focus on indigenous health, the knowledge generated and shared through this field of study is useful for promoting sustainable indigenous community health. In the few cases where there is a focus on Indigenous Health as found at the University of Wollongong and University of Sydney, the IS professional is equipped with a better understanding of the link between identity and health. When involved in health programming, IS professionals contribute to ensuring that culturally and socially appropriate quality care is delivered through meaningful collaborations; the IS professionals make the needed cultural competence available to the multidisciplinary health planning teams.

Anthropology, particularly social and cultural branches of anthropology which focus on the pattern of human behavior and culture respectively, contributes useful knowledge for the promotion of sustainable indigenous community health. Similarly, the subfield of **medical anthropology**, with its emphasis on the interrelationships among health, illness, and culture, can make significant contributions to promoting indigenous community health. Given its emphasis on understanding people within their context and the deployment of tools for such purposes, anthropology provides an equally better understanding of the interactions between culture and behavior. Thus, indigenous community health stands to benefit from anthropology through understanding the importance of culture, appropriate community entry and engagements, and the design of programs that builds on the existing ways of life.

Community Psychology involves the study of people within their social environment and how to improve their wellbeing within their context. This implies that community psychology applies the understanding of the interactions between behavior and culture directly to the improvement of health and wellbeing at the community level. The implication is that indigenous community health benefits from the participatory models developed in community psychology for the initiating, planning, designing, implementing, and evaluating community interventions aimed at improving wellness.

Health Psychology is also a related discipline that can contribute to sustainable indigenous community health. Health psychologists apply psychological knowledge and methods to the prevention and management of physical diseases and disorders as well as promote general wellbeing. Thus, health psychologists bring to the table expertise about the psychological and behavioral processes in health and wellness that can help address modifiable risk factors of ill-health at the individual level. This will ensure that there are long-term desirable changes in behavior that sustain good health.

Rural Sociology and **Medical Sociology** are two subfields of sociology that contribute to promoting sustainable indigenous community health. **Rural sociology** is concerned with rural people and places with varying emphasis on food production, environment, natural resources, social disruptions, culture, and rural health. Thus, rural sociology can contribute to sustainable indigenous community health through the knowledge of how rural communities are organized as well as rural health. **Medical sociology**, on the other hand, concentrates on health, organization of health services, and healthcare utilization issues including health disparities. This implies that medical sociologists can help elucidate the organization of health services within rural communities, and this knowledge is valuable to organizing for partnerships for sustainable indigenous community health.

Public Health protects and promotes the health of people or populations and their communities. This discipline contributes competencies in good sanitation practices, personal hygiene, control of both communicable and noncommunicable diseases, and effective organization of health services. Indigenous community health equally benefits from knowledge about how to promote health and wellbeing at the community level rather than at the individual level.

Closely related to public health is **community or public health nursing.** Community health nurses (CHNs) conduct primary healthcare and nursing practice in a community setting by way of providing curative care, preventive care, interventions, and health education. CHNs are crucial to the promotion of sustainable indigenous community health and are often the ones directly involved in working with the indigenous communities to deliver care. Thus, CHNs constitute an essential human resource for sustainable indigenous community health.

Healthcare Administration (also known as health administration, hospital management, or healthcare management) is concerned with planning, organizing, coordinating, monitoring and leading public health systems, healthcare systems, hospitals, and hospital networks at all levels of care, primary, secondary, and tertiary levels of care. Healthcare administrators contribute the needed skills in operating a single or a network of healthcare facilities across a vast geographical area to ensure and assure access to the right care required. Healthcare administrators are key to promoting sustainable indigenous community health.

Issues for Research and Other Forms of Scholarship on Indigenous Health

It is important to understand historical harm to the health of indigenous communities from the colonization experiences to be able to discover and nurture their cultural assets for health and wellbeing. Colonization has been a progressive, invasive process on all aspects of indigenous community life, giving rise to contested health realities and memories of themselves (Bulhan, 2015; Oppong, 2019). Research to advance the health and wellness of indigenous communities invariably involves examining the restorative process of mending the "fractured" self through re-embedding the "dislocated" self onto its authentic cultural base to ensure harmony of the self with the universe (Oppong, 2014; Horrill et al., 2018), without denying the apparent benefits of modern medicine (Mpofu et al., 2011).

Specifically, there is need for research on strategies for enhancing health and wellbeing in indigenous communities through (1) increased varieties of services provided, (2) increased number of healthcare providers, (3) increased awareness of available healthcare services, and (4) increased funding for medical transportation are potential facilitators to healthcare (Marrone, 2007; Horrill et al., 2018). Though some studies are being conducted to identify such strategies (see Research Box 17.1), more research is needed to scale up evidence-informed practices.

Practice-led evidence with indigenous communities is needed to highlight the efficacy of the following: utilization of mobile clinics that visit inaccessible communities periodically and construction of community-based clinics to provide basic healthcare (Kweku et al., 2020). Moreover, evidence is needed on health support programs in indigenous communities for (1) creation of "socially accepting, safe, and inviting spaces to reduce social distance," (2) decolonization of health education, research, and practices, and

Research Box 17.1: Improving Handwashing with Soap in Remote Aboriginal Communities. (McDonald, Cunningham, & Slavin, 2015)

Background

In 2007, Northern Territory (NT) Government Environmental Health Officers (EHOs) developed a No Germs on Me (NGoM) Social Marketing Campaign to promote handwashing with soap to reduce the incidence of diarrheal, respiratory (lung and ear), and skin infections among children living in remote NT Aboriginal communities. During the first phase of the NGoM Program, eight TV commercials were produced and televised. The EHOs completed an evaluation of the NGoM Program in 2008–2009, reporting limited success. In 2013, when additional funding became available to expand the NGoM program, three new television commercials were developed to target adult viewers. These TV commercials were produced with inputs from the Aboriginal people living in remote communities and were also filmed in regional and remote locations featuring Aboriginal people from those areas. From May 11, 2014, to June 8, 2014, five TV channels (four "free-to-air" and one satellite) intensively televised the three commercials. The broadcast covered a vast geographical area including remote and rural communities across the NT and Western Australia, northern South Australia, and central and far west Queensland and New South Wales.

(*continued*)

(continued)

Method

An evaluation of the new TV commercials added to NGoM was conducted using an evaluation design informed by the Theory of Planned Behavior (TPB). A pre- and post-intervention study design was applied and took place in six remote Aboriginal communities representing three different geographical regions. A mixed methods approach was used in which interviews were conducted at the research sites while a questionnaire based on the principal components of the TPB was also developed for quantitative data collection. Data were completed in the six Aboriginal communities immediately before and on completion of four weeks (from May 11, 2014 to June 8, 2014) intensive televising of the three new commercials.

Results

It was found that access in homes to TV varied across the six communities. All but one community reported less than a 100% access to TV in their homes. The access for the five communities ranged between 49% and 83%. Most of participants said they saw one or more of the commercials being evaluated. Most of the targeted audience also found the content of the commercials acceptable and comprehensible. In terms of intentions, a majority reported they would buy more soap, toilet paper, and facial tissues, if these were not so expensive in their communities.

Conclusions and Implications

Sociohistorical determinants of health make it challenging to promote handwashing with soap among adults and children as well as maintaining clean faces such that these behaviors become habits in indigenous communities. This study also highlighted the fact that social marketing programs such as NGoM can accentuate health inequalities if the media to access health information are expensive. For instance, in the NGoM program, the TV sets required to gain access to the health information were expensive. The findings can be leveraged for an evidence-based approach to planning NGoM or similar programs in the future.

What Do You Think?

1. How did the involvement of the Aboriginal communities in the production of the new TV commercials enhance the chances of success of NGoM Program?
2. What were the goals of the commercials and how successful were they in meeting those goals?
3. How does theory-informed evaluation practice improve health evaluation research with indigenous communities?

(3) coordinated policy implementation between different levels of government to eliminate jurisdictional ambiguity (Horrill et al., 2018, p. 9).

Public health professionals working with indigenous communities lack in cultural competence and education (Mpofu et al., 2012). Indigenous epistemologies of health must permeate all practices with them. In this regard, providers of health services must not only be understood as physical wellbeing, but also as constitutive of social, economic, mental, and spiritual wellbeing with a greater emphasis on their interconnectedness. Allopathic practice alone will almost always never be sufficient for care in indigenous populations as indigenous communities would seek spiritual healing before, during, and/or after being attended to by an allopathic medical practitioner on any health issue (Mensah, 2005; Yawar, 2001, Twumasi, 1975, 1981). This is because the indigenous populations perceive the care received from allopathic practitioners as only part of the care they are in search of and, therefore, will source from appropriate indigenous healers to complement the allopathic care with whatever is deemed missing (see Discussion Box 17.1).

Based on the ongoing discussion about indigenous health and wellness disparities, we offer the following suggestions for engaging in strengths-focused and sustainable health programming with Indigenous peoples:

Discussion Box 17.1: Coping with Workplace Accident Using Spirituality

In the wake of a fire outbreak in Tema (Ghana) in 2009 during which four lives were lost and several properties destroyed, the Managing Director (MD) of the affected company hired psychologists to deliver psychosocial solutions (trauma management and safety training). In addition, the MD engaged a traditional priest to make a libation to appease the gods on whose land the company was located as well as a pastor to pray for deliverance (Oppong, 2011). Thus, the conceptions of health and wellness in occupational health are, in many ways, similar to the general conceptions of health at the societal level.

What Do You Think?

1. What are the implications of indigenous representations of health for the management of occupational health and safety?
2. What is the significance of spirituality to occupational health practices with indigenous communities?

- Planning for sustainable health programs should begin with an understanding of indigenous conceptions of health.
- Collaboration between allopathic health practitioners and indigenous communities should characterize any health program. Health programs should be co-designed, co-implemented, and co-evaluated with indigenous populations.
- Health programs should be community-based and where possible, community-initiated programs should receive priority in support and implementation. This is to say that governments should not reinvent the wheel and must work with and through existing indigenous structures.
- Regardless of the orientation of the allopathic medical professional, an appreciation of and for the spiritual nature of the indigenous community is a sine qua non for successful practice in indigenous communities.
- Health programs should be designed to draw on the individual and collective resources inherent in the indigenous communities.
- Health programs should lead to the creation of "socially accepting, safe, and inviting spaces to reduce social distance" (Horrill et al., 2018, p. 9).
- Healthcare providers who work with indigenous communities should receive appropriate training in culturally competent caregiving. This can be achieved through decolonizing health education, research, and practices.
- Health programs should be developed with an understanding that indigenous populations have the right to self-determination and the right to ask for and receive the type of care they prefer which closely aligns with their conceptions of health and wellbeing.
- Health programs developers should innovate the delivery of healthcare to indigenous populations. This may include the use of mobile clinics to deliver care in the remotest parts of the country or constructing shelters to accommodate persons coming from hard-to-reach areas while they are receiving care (see Case Illustration 17.2). However, health programs developers are called upon to think of new ways to use what works already in the indigenous communities to deliver care without causing disruptions to their ways of life.

Case Illustration 17.2: Using Maternal Shelters to Improve Maternal and Neonatal Care

Garissa is located in the north-eastern part of Kenya, an area characterized by higher-than-average neonatal mortality in 2003 (UN, 2016). About half of all deaths in developing countries are avoidable if there is access to emergency care, often caused by situations such as obstetric complications. To improve maternal and neonatal health in the area, a maternal shelter was constructed on the premises of the provincial hospital in Garissa. Maternal shelters are waiting homes for pregnant women who live far away from the hospital and those with high-risk pregnancies; the pregnant women stay at the maternal shelters to enable healthcare workers to monitor them to increase speed and access to life-saving emergency care. The purpose of maternal shelters is, therefore, to save the lives of both mothers and newborns in hard-to-reach areas. The maternal shelter located at Garissa was constructed with support from UNICEF. Mr. Mohamed, a husband to one of the beneficiaries of the service had this to say: "Had my wife not come to the shelter to deliver at this hospital, she could have died. The two previous deliveries were difficult and she almost died from excessive bleeding." His wife had a Caesarean section and it required a life-saving blood transfusion procedure. Mr. Mohamed said he was more than happy to travel an 800 km journey to visit his wife at the shelter (UN, 2016, p. 026).

What Do You Think?

1. 1. How is public infrastructure a major influence to the health of indigenous communities? 2. Propose some solutions to assisting indigenous community maternal and child care needs minimizing avoidable risks from limited medical services in their localities?

Summary and Conclusion

Due to the differences in worldviews, indigenous populations tend to have representations of health grounded in collective wellness rather than disease causation and treatment. Moreover, indigenous communities endorse (1) the multidimensionality of health and wellness, including an economic dimension that goes beyond allopathic medicine and health and wellness as a way of life. Health services for indigenous communities should be consonance with the indigenous health conceptions toward restorative processes of health and wellbeing premised on respectful relationships with nature and the spirit world. To optimize the health of indigenous communities, there is a need to strengthen their health self-determination through participation in the design, implementation, and evaluation of sustainable community health programs for their benefit.

References

African Commission on Human and Peoples' Rights (ACHPR) and International Work Group for Indigenous Affairs (IWGIA). (2007, September). *Report of the African Commission's working group on indigenous populations/communities: Research and information visit to the Republic of Congo.* https://www.achpr.org/public/Document/file/English/achpr38_misrep_specmec_indpop_congo_2005_eng.pdf

Anderson, I., Robson, B., Connolly, M., et al. (2016). Indigenous and tribal peoples' health (The Lancet–Lowitja Institute Global Collaboration): A population study. *The Lancet, 388*(10040), 131–157. https://doi.org/10.1016/S0140-6736(16)00345-7

Australian Bureau of Statistics. (2010). *Measures of Australia's progress.* ABS cat. No. 1370.0. Australian Government.

Australian Bureau of Statistics. (2012, October). *The health and welfare of Australia's Aboriginal and Torres Strait Islander peoples.* ABS cat. No. 4704.0. Australian Government.

Australian Department of Health and Ageing. (2013). *National Aboriginal and Torres Strait Islander suicide prevention strategy.* https://www1.health.gov.au/internet/main/publishing.nsf/Content/1CE7187EC4965005CA25802800127B49/$File/Indigenous%20Strategy.pdf

Australian Government Department of Families, Community Services and Indigenous Affairs. (2006). *Aboriginal and Torres Strait Islander views on research in their communities.* Occasional Paper No. 16. https://www.dss.gov.au/sites/default/files/documents/05_2012/op16.pdf

Australian Health Ministers' Advisory Council. (2015). *Aboriginal and Torres Strait Islander health performance framework 2014 report.* Australian Health Ministers' Advisory Council.

Australian National Mental Health Commission. (2013). *The mental health and social and emotional wellbeing of Aboriginal and Torres Strait Islander peoples, families and communities* [Supplementary paper]. https://www.mentalhealthcommission.gov.au/getmedia/f014d128-ab8a-4a40-b3d2-9a4668a8fd93/Mental-Health-Report-Card-on-Aboriginal-and-Torres-Strait-Islander

Axelsson, P., Kukutai, T., & Kippen, R. (2016a). Indigenous wellbeing and colonisation: Editorial. *Journal of Northern Studies, 10*(2), 7–18.

Axelsson, P., Kukutai, T., & Kippen, R. (2016b). The field of Indigenous health and the role of colonisation and history. *Journal of Population Research, 33*, 1–7. https://doi.org/10.1007/s12546-016-9163-2

Bainbridge, R., Tsey, K., McCalman, J., Kinchin, I., Saunders, V., Lui, F. W., … Lawson, K. (2015). No one's discussing the elephant in the room: Contemplating questions of research impact and benefit in Aboriginal and Torres Strait Islander Australian health research. *BMC Public Health, 15*(696). https://doi.org/10.1186/s12889-015-2052-3

Boulton-Lewis, G., Pillay, H., & Wilss, L. (2001, July 16–18). Conceptions of health held by Aboriginal, Torres Strait Islander, and Papua New Guinea health science students. A paper presented at the *Higher Education Close Up Conference 2*, Lancaster University, UK. http://www.leeds.ac.uk/educol/documents/00001739.htm

Brown, M. O. (2018). *The health disparities in Africa: Impact of colonial institutions on Indigenous mortality from the 20th century* (Unpublished Master's thesis). School of Economics and Management, Lund University, Sweden.

Bulhan, H. A. (2015). Stages of colonialism in Africa: From occupation of land to occupation of being. *Journal of Social and Political Psychology, 3*(1), 239–256. https://doi.org/10.5964/jspp.v3i1.143

Chandler, M. J., & Lalonde, C. (1998). Cultural continuity as a hedge against suicide in Canada's first nations. *Transcultural Psychiatry, 35*, 191–219. https://doi.org/10.1177/136346159803500202

Coates, K. S. (2004). *A global history of indigenous peoples: Struggle and survival.* Houndmills/Basingstoke/Hampshire, UK/New York, NY: Palgrave Macmillan.

Crombie, I. K., Irvine, L., Elliott, L., & Wallace, H. (2005). *Closing the health inequalities gap: An international perspective.* WHO Regional Office for Europe. http://www.euro.who.int/__data/assets/pdf_file/0005/124529/E87934.pdf

Dell'Osso, B., Glick, I., Baldwin, D., & Altamura, A. (2013). Can long-term outcomes be improved by shortening the duration of untreated illness in psychiatric disorders? A conceptual framework. *Psychpathology, 46*, 14–21.

Donatuto, J., Campbell, L., & Gregory, R. (2016). Developing responsive indicators of indigenous community health. *International Journal of Environmental Research and Public Health, 13*(9), 899. https://doi.org/10.3390/ijerph13090899

Dudgeon, P., Bray, A., & Walker, R. (2020). Self-determination and strengths-based Aboriginal and Torres Strait Islander suicide prevention: An emerging evidence-based approach. In A. Page & W. Stritzke (Eds.), *Alternatives to suicide: Beyond risk and toward a life worth living* (pp. 237–256). Academic Press. https://doi.org/10.1016/B978-0-12-814297-4.00012-1

Dudgeon, P., & Walker, R. (2015). Decolonising Australian psychology: Discourses, strategies, and practice. *Journal of Social and Political Psychology, 3*(1), 276–297. https://doi.org/10.5964/jspp.v3i1.126

Durie, M. H. (1985). A Maori perspective of health. *Social Science & Medicine, 20*(5), 483–486. https://doi.org/10.1016/0277-9536(85)90363-6

Elder, J. P., Ayala, G. X., Parra-Medina, D., & Talavera, G. A. (2009). Health communication in the Latino community: Issues and approaches. *Annual Review of Public Health, 30*(1), 227–251. https://doi.org/10.1146/annurev.publhealth.031308.100300

Fijal, D., & Beagan, B. L. (2019). Indigenous perspectives on health: Integration with a Canadian model of practice. *Canadian Journal of Occupational Therapy, 86*(3), 220–231. https://doi.org/10.1177/0008417419832284

Gallagher, J. (2019). Indigenous approaches to health and wellness leadership: A BC first nations perspective. *Healthcare Management Forum, 32*(1), 5–10. https://doi.org/10.1177/0840470418788090

Garvey, D. (2008). *Review of the social and emotional wellbeing of Indigenous Australian peoples – Considerations, challenges and opportunities.* http://www.healthinfonet.ecu.edu.au/sewb_review

Ghana Health Service. (2002). *Community-based health planning and services (CHPS) initiative.* https://www.ghanahealthservice.org/downloads/ServicestoClients.pdf

Gray, M. A., & Oprescu, F. I. (2016). Role of non-Indigenous researchers in Indigenous health research in Australia: A review of the literature. *Australian Health Review, 40*, 459–465. https://doi.org/10.1071/AH15103

Gruen, R. L., Weeramanthri, T. S., & Bailie, R. S. (2002). Outreach and improved access to specialist services for indigenous people in remote Australia: The requirements for sustainability. *Journal of Epidemiology and Community Health, 56*, 517–521.

Horowitz, S. (2012). American Indian health: Integrating traditional native healing practices and western medicine. *Alternative and Complementary Therapies, 18*(1), 24–30. https://doi.org/10.1089/act.2012.18103

Horrill, T., McMillan, D. E., Schultz, A. S. H., & Thompson, G. (2018). Understanding access to healthcare among Indigenous peoples: A comparative analysis of biomedical and postcolonial perspectives. *Nursing Inquiry, 25*(3), e12237. https://doi.org/10.1111/nin.12237

Indian Country Today. (2020, May). *Indian health service selects five projects for joint venture health care facilities.* https://indiancountrytoday.com/the-press-pool/indian-health-service-selects-five-projects-for-joint-venture-health-care-facilities-xgDiDaYq30Ks1PiF38Ihew

Indian Health Service. (n.d.). *Programs.* https://www.ihs.gov/dfpc/programs/#jvcp

Kemner, J. (2014). Fourth world activism in the first world: The rise and consolidation of European solidarity with Indigenous peoples. *Journal of Modern European History, 12*(2), 262–279. https://doi.org/10.17104/1611-8944_2014_2_262

Kipuri, N., & Sørensen, C. (2008). *Poverty, pastoralism and policy in Ngorongoro: Lessons learned from the Ereto I Ngorongoro Pastoralist Project with implications for pastoral development and the policy debate.* https://pubs.iied.org/pdfs/12548IIED.pdf

Kolahdooz, F., Nader, F., Yi, K. J., & Sharma, S. (2015). Understanding the social determinants of health among Indigenous Canadians: Priorities for health promotion policies and actions. *Global Health Action, 8.* https://doi.org/10.3402/gha.v8.27968

Kweku, M., Amu, H., Awolu, A., Adjuik, M., Ayanore, M. A., Manu, E., et al. (2020). Community-based health planning and services plus programme in Ghana: A qualitative study with stakeholders in two systems learning districts on improving the implementation of primary health care. *PLoS One, 15*(1), e0226808. https://doi.org/10.1371/journal.pone.0226808

Laycock, A., Walker, D., Harrison, N., & Brands, J. (2011). *Researching Indigenous health: A practical guide for researchers.* The Lowitja Institute. Carlton South, Australia.

Maher, P. (1999). A review of 'traditional' Aboriginal health beliefs. *Australian Journal of Rural Health, 7,* 229–236. https://doi.org/10.1046/j.1440-1584.1999.00264.x

Maina, W. K., Kim, A. A., Rutherford, G. W., et al. (2014). Kenya AIDS indicator surveys 2007 and 2012: Implications for public health policies for HIV prevention and treatment. *Journal of Acquired Immune Deficiency Syndromes, 66*(Suppl 1), S130–S137. https://doi.org/10.1097/QAI.0000000000000123

Marrone, S. (2007). Understanding barriers to health care: A review of disparities in health care services among indigenous populations. *International*

Journal of Circumpolar Health, 66(3), 188–198. https://doi.org/10.3402/ijch.v66i3.18254
McDonald, E., Cunningham, T., & Slavin, N. (2015). Evaluating a handwashing with soap program in Australian remote Aboriginal communities: A pre and post intervention study design. *BMC Public Health, 15*(1188). https://doi.org/10.1186/s12889-015-2503-x
Mensah, F. A. (2005, March 25–27). *Beyond medicine: Spiritual basis of illnesses and healing in Africa.* A paper presented at the African conference on the theme "African Health and Illness", The University of Texas at Austin, USA.
Mpofu, E. (2006). Majority world health care traditions intersect indigenous and complementary and alternative medicine. *International Journal of Disability, Development and Education, 53*(4), 375–379.
Mpofu, E., Chronister, J., Johnson, E., & Denham, G. (2012). Aspects of culture influencing rehabilitation and persons with disabilities. In P. Kennedy (Ed.), *The Oxford Handbook of rehabilitation psychology* (pp. 543–553). Oxford, UK: Oxford University Press.
Mpofu, E., Peltzer, K., & Bojuwoye, O. (2011). Indigenous healing practices in sub-Saharan Africa. In E. Mpofu (Ed.), *Counseling people of African ancestry* (pp. 3–21). Cambridge, UK/New York, NY: Cambridge University Press.
Murphy, M. (2014). *Self-determination and indigenous health: Is there a connection?* https://www.e-ir.info/2014/05/26/self-determination-and-indigenous-health-is-there-a-connection/
Murray, C. J., Kulkarni, S., & Ezzati, M. (2005). Eight Americas: New perspectives on U.S. health disparities. *American Journal of Preventive Medicine, 29*(5 Suppl 1), 4–10. https://doi.org/10.1016/j.amepre.2005.07.031
NEJM Catalyst. (2017). *Social determinants of health (SDOH).* https://catalyst.nejm.org/doi/full/10.1056/CAT.17.0312
New Zealand Ministry of Health. (2017). *Māori health models – Te Whare Tapa Whā.* https://www.health.govt.nz/our-work/populations/maori-health/maori-health-models/maori-health-models-te-whare-tapa-wha
Nyonator, F. K., Awoonor-Williams, J. K., Phillips, J. F., Jones, T. C., & Miller, R. A. (2005). The Ghana community-based health planning and services initiative for scaling up service delivery innovation. *Health Policy and Planning, 20*(1), 25–34. https://doi.org/10.1093/heapol/czi003
Obomsawin, R. (2007). Traditional medicine for Canada's first peoples. *Indigenous Sites.* https://lfs-indigenous.sites.olt.ubc.ca/files/2014/07/RayObomsawin.traditional.medicine-1.pdf

Office of the Governor of the Chickasaw Nation. (2013, October). *State of Chickasaw nation great and getting better.* Chickasaw.net. https://chickasaw.net/News/Press-Releases/Release/State-of-Chickasaw-Nation-Great-and-Getting-Better-1463.aspx

Office of the Governor of the Chickasaw Nation. (2016, July). *Chickasaw nation breaks ground on Apila Center.* Chickasaw.net. https://www.chickasaw.net/News/Press-Releases/Release/Chickasaw-Nation-breaks-ground-on-Apila-Center-2065.aspx

Oppong, S. (2011). *Health & safety: Theory and practice in the oil and gas sector.* Saarbrücken, Germany: VDM Publishing House Ltd..

Oppong, S. (2014). Psychology, economic policy design, and implementation: Contributing to the understanding of economic policy failures in Africa. *Journal of Social and Political Psychology, 2*(1), 183–196. https://doi.org/10.5964/jspp.v2i1.306

Oppong, S. (2019). Overcoming obstacles to a truly global psychological theory, research and praxis in Africa. *Journal of Psychology in Africa, 29*(4), 292–300. https://doi.org/10.1080/14330237.2019.1647497

Oppong, S. (2020). When something dehumanizes, it is violent but when it elevates, it is not violent. *Theory & Psychology, 30*(3), 468–472. https://doi.org/10.1177/0959354320920942

Pattoni, L. (2012). Strengths-based approaches for working with individuals. *Insight* 16. https://www.iriss.org.uk/resources/insights/strengths-based-approaches-working-individuals

Pelcastre-Villafuerte, B., Ruiz, M., Meneses, S., Amaya, C., Márquez, M., Taboada, A., & Careaga, K. (2014). Community-based health care for indigenous women in Mexico: A qualitative evaluation. *International Journal for Equity in Health, 13*(2). https://doi.org/10.1186/1475-9276-13-2

Pelican, M., & Maruyama, J. (2015). The indigenous rights movement in Africa: Perspectives from Botswana and Cameroon. *African Study Monographs, 36*(1), 49–74.

Phillips, D. R., Hyma, B., & Ramesh, A. (1992). A comparison of the use of traditional and modern medicine in primary health centres in Tamil Nadu. *GeoJournal, 26*, 21–30. https://doi.org/10.1007/BF00159434

Pihama, L., & Lee-Morgan, J. (2019). Colonization, education, and Indigenous peoples. In E. McKinley & L. Smith (Eds.), *Handbook of Indigenous Education.* Springer. https://doi.org/10.1007/978-981-10-3899-0_67

Reading, C. L., & Wien, F. (2009). *Health inequalities and social determinants of Aboriginal Peoples' health*. https://www.nccah-ccnsa.ca/docs/social%20determinates/NCCAH-Loppie-Wien_Report.pdf

Shah, B., Gunraj, N., & Hux, J. E. (2003). Markers of access to and quality of primary care for aboriginal people in Ontario, Canada. *American Journal of Public Health, 93*, 798–802.

Struthers, R., Eschiti, V. S., & Patchell, B. (2004). Traditional indigenous healing: Part I. *Complementary Therapies in Nursing & Midwifery, 10*(3), 141–149.

Thresia, C. U. (2018). Health inequalities in South Asia at the launch of sustainable development goals: Exclusions in health in Kerala, India need political interventions. *International Journal of Health Services, 48*(1), 57–80. https://doi.org/10.1177/0020731417738222

Twumasi, P. A. (1975). *Medical Systems in Ghana: A study in medical sociology*. Accra, Ghana: Ghana Publishing Corporation.

Twumasi, P. A. (1981). Colonialism and international health: A study in social change in Ghana. *Social Science & Medicine. Part B: Medical Anthropology, 15*(2), 147–151. https://doi.org/10.1016/0160-7987(81)90037-5

United Nations. (2008). *United nations declaration on the rights of indigenous peoples*. https://www.un.org/esa/socdev/unpfii/documents/DRIPS_en.pdf

United Nations. (2016). *State of the world's indigenous peoples: Indigenous peoples' access to health services*. https://www.un.org/esa/socdev/unpfii/documents/2016/Docs-updates/SOWIP_Health.pdf

United Nations. (n.d). *Indigenous peoples at the United Nations*. https://www.un.org/development/desa/indigenouspeoples/about-us.html

Vos, T., Barker, B., Stanley, L., & Lopez, A. (2007). *The burden of disease and injury in Aboriginal and Torres Strait Islander peoples 2003*. University of Queensland.

Wachira, G. M., & Karjala, T. (2012). Advocacy for indigenous peoples' rights in Africa: Dynamics, methods and mechanisms. In R. Laher & K. Sing'Oei (Eds.), *Indigenous People in Africa: Contestations, empowerment and group rights* (pp. 104–123). Pretoria, South Africa: Africa Institute of South Africa.

Waterworth, P., Pescud, M., Braham, R., Dimmock, J., & Rosenberg, M. (2015). Factors influencing the health behaviour of Indigenous Australians: Perspectives from support people. *PLoS One, 10*(11), e0142323. https://doi.org/10.1371/journal.pone.0142323

Webb, J. (2012). Indigenous people and the right to self-determination. *Journal of Indigenous Policy, 13*, 75–102.

Westerman, T. G. (2010). Engaging Australian Aboriginal youth in mental health services. *Australian Psychologist, 45*, 212–222.

White, P. (2015). The concept of diseases and health care in African traditional religion in Ghana. *HTS Teologiese Studies/Theological Studies, 71*(3), art. No. 2762. https://doi.org/10.4102/hts.v71i3.2762

Wilk, P., Cooke, M., Stranges, S., & Maltby, A. (2018). Reducing health disparities among indigenous populations: The role of collaborative approaches to improve public health systems. *International Journal of Public Health, 63*, 1–2. https://doi.org/10.1007/s00038-017-1028-8

Williams, K., Potestio, M. L., & Austen-Wiebe, V. (2019). Indigenous health: Applying truth and reconciliation in Alberta health services. *Canadian Medical Association Journal, 191*(Suppl 1), S44–S46. https://doi.org/10.1503/cmaj.190585

World Health Organization. (2007). *Health inequities in the South-East Asia region.* https://www.who.int/social_determinants/media/health_inequities_searo_122007.pdf

World Health Organization. (2010). *Indigenous health – Australia, Canada, Aotearoa New Zealand and the United States – Laying claim to a future that embraces health for us all.* World Health Report – Background Paper, No. 33. https://www.who.int/healthsystems/topics/financing/healthreport/IHNo33.pdf

World Health Organization. (2012). *Health systems in Africa: Community perceptions and perspectives.* https://www.afro.who.int/sites/default/files/2017-06/english%2D%2D-health_systems_in_africa%2D%2D-2012.pdf

Yawar, A. (2001). Spirituality in medicine: What is to be done? *Journal of the Royal Society of Medicine, 94*(10), 529–533. https://doi.org/10.1177/014107680109401013

Yuill, C., Crinson, I., & Duncan, E. (2011). *Key concepts in health studies.* London, UK: Sage.

Part V

Epilogue

18

The Futures of Sustainable Community Health

Stanley Ingman and Elias Mpofu

Introduction

As the world approaches the mid-twenty-first century, it lives to realize the extent to which aspirational global Sustainable Development Goals (SDGs) and national and regional health systems provide to expectation. For the most part, SDGs have been aspiration and forward-looking aimed to "the needs of the present without compromising the ability of future generations to meet their own needs" (World Commission on Environment and Development, 2014, p. 4). There is increasing appreciation of the interconnectedness of health and development well into

S. Ingman (✉)
University of North Texas, Denton, TX, USA
e-mail: Stan.Ingman@unt.edu

E. Mpofu
University of North Texas, Denton, TX, USA

University of Sydney, Sydney, NSW, Australia

University of Johannesburg, Johannesburg, South Africa
e-mail: Elias.Mpofu@unt.edu

E. Mpofu (ed.), *Sustainable Community Health*,
https://doi.org/10.1007/978-3-030-59687-3_18

the future in that there can be no development without population health while population health is a reliable indicator of future development (Hertzman, 1999; Mills, 2014). With the broad endorsement of the interdependency of the three pillars of social, economic, and environment sustainability to community health, future debates will consider how implementation of these pillars in specific health systems translate into population health outcomes.

Learning Objectives

By the end of the chapter, the reader should be able to:

1. Identify and justify priorities for futuristic, sustainable community health.
2. Explain how the social and economic sustainability of community health are grounded in environmental sustainability.
3. Distinguish between wellness and health-focused approaches to sustainable community health.
4. Evaluate the place of health in all as a sustainable community health policy and practice.
5. Propose indicators and outcomes for tracking the performance of sustainable community health systems.
6. Estimate the relative value of mathematical modeling in the planning and design of sustainable community health systems.

Conceivably, policies and practices for sustainable community health will align with population health needs for a particular context, optimizing access, relevance, and equity. Regardless of context, sustainable community health systems of the future will prioritize natural environment safety, disease prevention and health promotion, efficient health system management, affordable financial resources and investments in innovative funding models, and health technologies, public housing and transportation, cross-sector workforce development for wellness, and inclusive wellbeing policies (Fineberg, 2012; Morrison, Petticrew, & Thomson, 2003; Thomson, Morrison, & Petticrew, 2007). However, while sustainable health has wide endorsement as an ideology, questions remain regarding the practical commitment to related policies and practices. Practically, national and state policies around the globe have not supported sustainability or community public health in a serious manner

(Frenk & Moon, 2013; Mpofu, 2015). In fact, for some 30 or more years, sustainability has been a voice in the wilderness and confined to some cities, some nonprofits, and some academics. Moreover, there is a lack of consensus on essential indicators and outcomes of sustainable health across settings and populations (Hunter & Fineberg, 2015).

The Future of the Foundations of Sustainable Community Health

By the end of the century, the world as we know it may be unrecognizable from changes to the lived natural environment secondary to climate changes, population demographic shifts, capital flows and investments, as well as technology innovation altering the entire way in which humans relate to each other and the environment (McMichael, 2013). This would bring to the forefront aspects of the social, economic, and environmental sustainabilities for health, to which local, national, and regional governments must be responsive for population health gains.

Environmental Suitability Futures

Present trends suggest an acceleration of global warming due to the absence or delay in adopting alternative energy resources (Rahmstorf, Foster, & Cahill, 2017). First, heat waves associated with global warming compromise the livability of human habits, increasing the demand for hospital care services for vulnerable populations like children, older adults, people with significant disability, and remote and rural community populations, placing strain on health care budgets. Rising sea levels from global warning will reduce habitable land, flooding some cities and causing increased risk for water system contamination from industrial waste. Global warming will also escalate energy demands from use of industrial and residential local climate control systems, escalating depletion of the natural resources required for the energy production. Pollution of water and air resources increases the risk of contamination of the human food systems, which would lead to costly, avoidable infections (He et al., 2015). Natural environment degradation contributes to an estimated 16% of the global burden of disease associated with chronic obstructive pulmonary diseases, cardiovascular diseases, or asthma (WHO, 2013).

Environmental decline leads to escalation of health conditions requiring medical treatments, which typically come with medical waste such as disposable items (biohazardous agents, such as heavy metals and radioactive isotopes), contributing to environmental degradation (Carnero, 2015; Conrardy, Hillanbrand, Myers, & Nussbaum, 2010). Reuse of medical devices is a sustainable health practice, were it not for the fact that "medical devices are designed for single use and should not be used twice or more times" (Thomke, Bigler, & Lehmann, 2013. p. 1). Use of pharmaceuticals continues to increase and is likely to persist in the near future; their disposal contributes significantly to heavy pollution of the environment (Kümmerer, 2010). Development and increasing use of non-pharmaceutical therapies would reduce environmental contamination and related health risks.

Moreover, the continued reliance on fossil fuels beyond the mid-twenty-first century harms human habitats from air pollution (Kampa & Castanas, 2008; WHO, 2012), which is associated with lung cancer and other cardiopulmonary mortality (Brauer et al., 2016). Air pollution-related diseases represent 7.6% of total global deaths (approximately 4.2 million deaths), of which 59% of these deaths occur in east and south Asia (Cohen et al., 2017). Global warming also comes with a rapid loss in biodiversity, which is essential for a balanced human and natural environment system (Liddicoat, Waycott, & Weinstein, 2016; Oliver et al., 2015). Healthy biodiversity provides for good air quality and freshwater, reducing the risks of infectious disease spread while promoting physical and mental health (Keune, Martens, Kretsch, & Prieur-Richard, 2013).

While the Paris climate accord held much promise, there has been low commitment to the terms of accord by the world's leading producers of carbon emissions, shortchanging its likely benefits to sustainable health by the mid-twenty-first century (Mahapatra & Ratha, 2017). Evidently, the environmental pillar of sustainable health continues to be under threat for the near future, despite being the backbone of the economic and social sustainability pillars. A one-health approach that integrates health systems "across the tree of life, including, but not limited to, wildlife, livestock, crops, and humans" might be a serviceable model for sustainable community health systems of the future (Wallace et al., 2015, p. 68).

There is growing interest by industry, governance, and academia regarding investing in new technologies for environmental health sustainability (Zapico, Brandt, & Turpeinen, 2010). Example initiatives include use of environmental health scans for information to reduce health inequalities and disparities by providing accurate measures for required improvements needed in a community (Graham, Evitts, & Thomas-MacLean, 2008; Martenies, Milando, Williams, & Batterman, 2017). For instance, increased wind and solar farms would decrease the dependence on fossil fuels, which in turn would reduce the greenhouse gas effect from the burning of carbon-based products. These initiatives are critical to sustainable community health in the twenty-first century.

Economic Sustainability Futures

The financial pillar of health systems is often touted as the backbone of sustainability (Karanikolos et al., 2013; WHO, 2007). While health services financing is clearly important to any health system, its singular significance is questionable, given the evidence that countries with the largest health care budgets like the US have marginal population health outcomes compared to others, such as Japan, which spend considerably less on health services (Kontis et al., 2017). At the same time, countries with poorly funded universal health care services also have poor population health outcomes (Mills, 2014). Nations such as Denmark, Germany, and Sweden are examples that it is possible to have a strong economy with a strong welfare state and a strong ecological preservation program (Mackenbach, Karanikolos, & McKee, 2013). Even a relatively poor country like Costa Rica is an example that poorer nations can create a better balance between economic distribution and ecological preservation resulting in sustainable health and social care systems for the citizens (Gindling & Trejos, 2005).

Nonetheless, public budgets in poorer countries fail to provide even the barest of health care systems, and particularly to the rural poor and those working in the informal sector, which is more than 80% of the population in developing countries (Mills, 2014; Rahman & Smith, 2000). The poor in developing countries often have to make hard choices between spending their limited resources on food and basic shelter over much needed medical care. Innovative communal health insurance, perhaps modeled around cooperative health insurance membership with minimum monthly payment and with reinsurance by a local author like the city council, might make health care affordable for the rural and informal sector people in developing countries (Preker, Langenbrunner, & Jakab, 2002). Local, state, and national governments could commit to providing targeted subsidies to co-pay for the premiums of low-income community health cooperatives, safeguarding their financial solvency and technical management support capacity of the local schemes (Dror & Preker, 2002; Preker et al., 2002). As a part of universal health coverage, local, state, and federal governments and other stakeholders should work to ensure access to basic health care for all, prioritizing vulnerable populations.

Social Sustainability Futures

Evidence shows that a whole community health system, attending to health-related social needs can improve health outcomes and reduce costs (Morrison et al., 2003; Mpofu, 2015). The relevance of the social contract to community wellbeing has long historical roots. As an example, in the fourteenth-century Europe under feudalism, farmers ruled by lords accepted the social contract, or what John Rawls (2009) would call the "distributive justice" model, where a landlord with his army of knights would protect farmers or peasants (Baker, 2012). In return, peasants were bound to surrender one-half of production (e.g., crops and animals raised) to the manor or ruler and his vassals in the feudal "mini" state (Baker, 2012; Voice, 2011). This represents an early "sustainable community" with a heavy focus on security. Most peasants likely perceived the manor to provide for their wellbeing compared to living in the forest with little or no protection. Without detailing the complex history of the historical transformation of feudalism, it is fair to assert that feudalism was rejected over time, and capitalism emerged as the dominant socioeconomic model, thus a new social contract emerged. Nonetheless, it is essential to link citizens with information regarding their community in order to address distributive justice and to instigate and enforce sustainable environmental health practices.

In the last 100 or more years, nations have been modifying capitalism to make it a more humane (as in social welfare states) and revised social contract between ruler and those ruled, oriented toward investment in population health (Abrahamson, 2010). Some might call the new models the welfare state or social capitalism. In the US, we have adopted some part of the European welfare state ideas (e.g., social security, Medicare, Medicaid, EPA, HUD, FDA, NIH), and many more national governmental interventions to help capitalism function better for more people. However, there is always the serious question of: who profits from these "reforms"—the general population or the elites or upper classes?

Three issues have challenged the notion that all is well with our current blend of capitalism and socialist programs: climate crisis, inequalities, and the COVID-19 pandemic of 2020. In this context, how should population health sustainability be redefined? Applied social scientists and activists are seeking to encourage or stimulate debates around such questions. This volume is one attempt to answer and chart a path forward to community health sustainability.

In 1975, Stanley Ingman (lead author, this chapter) and Tony Thomas attempted to stimulate the debate around community or "sustainable" health under the banner of "Topias" and "Utopias" in health to characterize current realities of how communities are organized (Ingman & Thomas, 2011; see also Follesdal & Pogge, 2006; Rawls, 2009). Following the work of Karl Mannheim from the 1920s, they defined "utopias" as any viable plan to replace "topias," or the dominant existing social order (see also Rawls, 2009), thereby conceiving utopias of community health to be within the reach of future generations. Subsequently, a conference hosted in 1995 on debates around topias and utopias as social innovations for the health and wellbeing of the aging population (Follett, 2012) and supported by the *Sustainable Communities Review* journal (http://scrjournal.org/) aimed to publish research on building a more sustainable future around the globe. Both efforts were in the spirit of challenging the existing order or "topias" in community health and to provide alternatives to the existing order, or "utopias" in community health.

Policies and Practices Health policy has three levels: international, national, and local level (Weiss, Isaac, Parkar, Chowdhury, & Raguram, 2001; WHO, 2008). International-level health policy creates regulations that guide communities in preventing and responding to acute public health risks that have the potential to cross borders and threaten people worldwide. In the context of community health, health policy not only focuses on medical care but also includes any action that affects health such as water sanitation, smoking ban, pharmaceutical patent laws, air pollution restrictions, and so on. National or state health policies provide a framework to regions, cities, and municipalities around which they should formulate and implement health policies in their local domains. They typically are public health-oriented and determine the health initiatives that the national government would provide funding and other support for. Local health polices, while framed on the national health policies, seek to address the specific health needs of the local health communities in their diversity (e.g., by city, municipality, neighborhood, population segments).

As previously noted, while sustainability is a widely endorsed policy (WHO, 2008), surprisingly, sustainability is rarely a design feature of health systems design around the world, and tends to bend to the winds of short-term political expediencies at the cost of population health (Fischer, 2015; Frenk & Moon, 2013). Yet, "establishing a sustainable system requires a governance structure that can guide and oversee development and assign responsibility for making progress. This can help define the key services and roles, their expected benefits and who should deliver them, as well as the barriers that may exist to their being fulfilled" (WHO, 2017, p. 4). If existing governance systems become more unjust, they may be less accountable for population health from a lack of commitment to health social justice. Some argue that the poor global response to the COVID-19 crisis by the US was from a weak social justice and public health system.

The lack of practical commitment to sustainability goals in community health systems design is a major loss to long-term population health, and is certainly a betrayal to the aspirational goal to "meet the[health] needs of the present without compromising the ability to meet future needs" (Roberts, & World Health Organization, 1998, p. 5). Practical institutionalization of sustainability across all community health systems is a futuristic, aspirational goal. Barriers to sustainability include the inertia from present community health systems acting in silos and often in competition with each other for the same resources, and in some cases, the politicization of population health with election cycles determining the life cycle of initiatives.

The success of sustainable community health practices in the twenty-first century is dependent on the adoption and implementation of inclusive population health approaches that are adaptive to sociodemographic changes, such as the aging society, urbanization, and neurodiversity, as well as opportunistic pandemics. Presently, approaches to addressing the health and wellness of various population groups tends to be piecemeal and reactive, often from political pressure by advocacy groups rather than grand futuristic policy design (Frenk & Moon, 2013). Moreover, sustainable health policies are those that focus on wellness principally, understanding that well individuals and communities make for healthy communities. To achieve community wellness requires cross-sector collaborations of human service agencies, including nonmedical agencies.

A wellness approach prioritizes prevention over treatment with the long-term benefit of a healthy population contributing to the economic and social health of the community. Wellness is about both the health promotion and protections for improved general health (not only healthcare) conditions. Achieving population wellness requires political commitment to health as a human right, citizenry education, commitment to self-managing of own health (Beirão, Patrício, & Fisk, 2017; Russo, Moretta Tartaglione, & Cavacece, 2019), and a healthcare workforce educated about environmentally protection and safety practices (Crisp & Chen, 2014), in addition to community-oriented cross-sector collaborations for population wellbeing (Glasgow, Goldstein, Ockene, & Pronk, 2004; Guglielmin, Muntaner, O'Campo, & Shankardass, 2018; Mpofu, 2015). This requires health and wellbeing value co-creation with the stakeholders from across the community, engaging them in actual decision-making and ongoing forums for permanent partnerships for future development.

Policies for global sustainable development are part of the solution given the interconnectedness of today's world, which is trending toward stronger networking across all human service sectors in the years to come, including environmental management. Conceivably, national and regional health agencies will increasingly engage in cross learning and co-learning (as evidenced by the cross-sector collaborations demands to mitigate the COVID-19). Thus, a one-size-fits-all community health systems would not be a goal and the fact that demonstratively "radically different health systems are imaginable" (World Economic Forum, 2013, p. 3), the richness in diversity of sustainable community health system qualities around the globe will be a resource for mutual development. *The Economist* (2020) published an editorial on May 16, 2020 titled "Goodbye Globalisation: The dangerous lure of self-sufficiency." However, their editorial projects a lurch toward more nationalism and less globalization so that "poorer countries will find it harder to catch up and, in the rich world; life will be more expensive and less free.... Moreover, a fractured world will make solving global problems harder, including finding a vaccine and securing an economic recovery... this logic is no longer fashionable." It is essential to think global while acting local in implementing sustainable health services.

As best practices, we suggest the implementation of health-in-all policies and practices for sustainable community health through benchmarking both environmental wellbeing (climate crisis related) (e.g., housing and business energy efficiency, and transportation), and social wellbeing (distributive justice or inequalities issues). Access to resources such as food security and adequate public housing may decrease the population risks of developing chronic conditions, increasing the ability of communities manage health conditions, which in turn would reduce avoidable health care utilization health care costs.

Health-in-all policies would build agriculture, education, the environment, fiscal policies, housing, and transport systems for health (Muntaner et al., 2011; Ståhl, Wismar, Ollila, Lahtinen, & Leppo, 2006; Young & Lambie, 2007). Health-in-all policies are a means to develop policies across sectors with the explicit goal of improving health for all (Guglielmin et al., 2018).

For over 100 years, social reformers have advised against a primarily acute-care-oriented approach to developing health systems as compared to a wellness approach (Eddy, Bibeau, Glover, Hunt, & Westerfield,

1989; Mirza, Mirza, Chung, & Sundaram, 2016). The trending of health-in-all approaches suggests a great future for inclusive health and participatory health systems for wellness rather than disease management (Baum & Sanders, 1995; Gostin, 2012; Guglielmin et al., 2018). An example of participatory community health is engaging trained community health workers or health advocates in dispensing simple medicines, providing antenatal care, and conducting blood tests under supervision. In fact, initiatives aimed to enhance community health by integrating community resources to clinical care, while addressing nonmedical factors related to health, is an old idea from the early 1900s when peer health aides offered activities and social benefits to local residents alongside clinical care services (Kark & Kark, 1983). The relative advantage of contemporary and future health systems is the phenomenal abundance of the tools for indicators and outcomes of health systems.

Indicators and Outcomes Indicators and outcomes are critical to telling how successful we have been on the long journey toward sustainable communities and healthier living. One of the barriers to population health management is the availability of information on social determinants to identify obstacles impeding efforts to improving community health outcomes (Barten, Mitlin, Mulholland, Hardoy, & Stern, 2007; Sheiham, 2000). Moreover, building health systems for wellness across the life course "will require …the collection, recording and linkage of health and administrative information, which is currently often condition- or intervention-based" rather than wellness oriented (WHO, 2017, p. 6). Information on trajectories of community wellness in the form of community health metrics should be a part of any sustainable health initiative. For instance, community environmental health metrics provide a comprehensive insight into natural environment factors affecting the health of a community. These metrics monitor the health of environmental media, contaminants present in individuals (biomonitoring), and the health effects caused by structural layouts of communities (Jakubowski & Frumkin, 2010; Lobdell, Murphy, & Calderon, 2007). By their nature, community environmental metrics profiles will be unique to each community, and the best tools and indicators will vary depending on the needs and characteristics of that community.

The fact that sustainability of community health systems may be incidental to community health, rather than being a long-term in-built strategic advantage is a lost opportunity to federal, state, and local governments to develop sustainability indicators for evidence-based implementation. The adage of what gets measured applies here. As we trend toward the middle of the twenty-first century, the future of sustainability of community health systems will require use of metrics, not just for diseases and hospital systems performances, for wellness (Barten et al., 2007; WHO, 2013). The development of sustainability health metrics cannot be left to chance or convenience. It will be critically important to the world's health systems what metrics of sustainability are used for each of the social, economic, and environment pillars across settings.

Indicator and outcomes data would be helpful for modeling of long-term sustainability of community health systems. Community Operational Research is a proven community health modeling approach in both the developed and developing country settings (Jackson, 2004; Midgley, Johnson, & Chichirau, 2018). Indicator and outcomes data are useful for health systems development planning to enhance efficiency and cost-effectiveness, when communicating with funding bodies, and explaining decision-making at community level (Rahman & Smith, 2000). Use of indicator and outcomes data that include environmental metrics and epidemiological data from across multiple domains (e.g., architecture, planning, parks and recreation, etc.) and applying interdisciplinary approaches to health system design would make for sustainability.

Environmental justice health practices strive to identify environmental factors that disproportionately affect disadvantaged populations (Clark, Millet, & Marshall, 2014; Kelly-Reif & Wing, 2016). This differs from simple monitoring of environmental metrics, such as air and water pollution, by focusing on the differences amongst different populations. A common metric for measuring environmental justice is the Environmental Justice Screening Method EJSM (Sadd, Pastor, Morello-Frosch, Scoggins, & Jesdale, 2011). This metric combines 23 indicators for mapping the cumulative effects of environmental stressors into an impact score to identify neighborhoods experiencing environmental injustice for remediation.

Summary and Conclusion

Two pillars of sustainable community health are in crisis: the environmental pillar from the climate change crisis and the economic pillar from the economic inequality crisis. Without long-term solutions to these two crises, sustainable community health seems a far cry. With climate change, the planet is warming up more. "Heat island" episodes will arise in major cities in the US and around the world. Rising seas will mean coastal cities will experience flooding. The decreasing biodiversity due to natural environment degradation will result in health catastrophes, in addition to pollutants to air, soil, and water harming the human food chain and causing avoidable infections. More diseases, common closer to the equator, are moving north and south as the planet warms. The widening of the economic divides between and within nations, accompanied with escalating health costs and under-investment in health-in-all and health-for-all distracts from the population health in the long term. Progress toward sustainable health will need to address these emerging challenges. Comprises to the environmental and economic sustainability of community health systems will damage their social sustainability through disempowering community members to participate in promoting and safeguarding their own health. Yet, "one-size-fits-all" approaches do not necessarily affect all populations equally and, in some cases, can widen existing disparities in community health systems.

For sustainability of community health systems, environmental, economic, and social resources and supports must underpin health policies and practices. The requisite health-in-all and for all sustainability enablers will need to be identified and developed. Moreover, it is critical to identify within local, state, national, and international health governance the responsibility for system development and to enact inclusive planning, defining the roles of community and state/federal and other stakeholders for identifying the approaches that will work optimally in the local health systems based on assets mapping for implementing health-in-all and health-for-all policies.

In seeking to develop sustainable health systems, we need to ask if a policy or practice would add to environmental, economic, and social sustainability for all rather than only for people with the necessary resources. At the same time, we need to create and implement health systems that are prevention and wellness-oriented, efficient, cost-effective, and easy to access and utilize by community members to minimize exacerbating health disparities. These would be health systems that are life situations focused to promote wellness rather than with a disease focus, relatively affordable and easy to disseminate and use.

References

Abrahamson, P. (2010). European welfare states beyond neoliberalism: Toward the social investment state. *Development and Society, 39*(1), 61–95.

Baker, E. (2012). *Social contract, essays by Locke, Hume and Rousseau*. Redditch, UK: Read Books Ltd.

Barten, F., Mitlin, D., Mulholland, C., Hardoy, A., & Stern, R. (2007). Integrated approaches to address the social determinants of health for reducing health inequity. *Journal of Urban Health, 84*(1), 164–173.

Baum, F., & Sanders, D. (1995). Can health promotion and primary health care achieve Health for All without a return to their more radical agenda? *Health Promotion International, 10*(2), 149–160.

Beirão, G., Patrício, L., & Fisk, R. P. (2017). Value co-creation in service ecosystems. *Journal of Service Management, 28*(2), 227–249.

Brauer, M., Freedman, G., Frostad, J., Van Donkelaar, A., Martin, R. V., Dentener, F., … Balakrishnan, K. (2016). Ambient air pollution exposure estimation for the global burden of disease 2013. *Environmental Science & Technology, 50*(1), 79–88.

Carnero, M. C. (2015). Assessment of environmental sustainability in health care organizations. *Sustainability, 7*(7), 8270–8291.

Clark, L. P., Millet, D. B., & Marshall, J. D. (2014). National patterns in environmental injustice and inequality: Outdoor NO_2 air pollution in the United States. *PLoS One, 9*(4), e94431.

Cohen, A. J., Brauer, M., Burnett, R., Anderson, H. R., Frostad, J., Estep, K., … Feigin, V. (2017). Estimates and 25-year trends of the global burden of disease attributable to ambient air pollution: An analysis of data from the Global Burden of Diseases Study 2015. *The Lancet, 389*(10082), 1907–1918.

Conrardy, J., Hillanbrand, M., Myers, S., & Nussbaum, G. F. (2010). Reducing medical waste. *AORN Journal, 91*(6), 711–721.

Crisp, N., & Chen, L. (2014). Global supply of health professionals. *New England Journal of Medicine, 370*(10), 950–957.

Dror, D. M., & Preker, A. S. (Eds.). (2002). *Social reinsurance: A new approach to sustainable community health financing*. Washington, DC: The World Bank.

Eddy, J. M., Bibeau, D. L., Glover, E. D., Hunt, B. P., & Westerfield, R. C. (1989). Wellness perspectives part 1: History, philosophy and emerging trends. *Wellness Perspectives, Research, Theory and Practices, 6*, 3–19.

Fineberg, H. V. (2012). A successful and sustainable health system—How to get there from here. *New England Journal of Medicine, 366*(11), 1020–1027.

Fischer, M. (2015). Fit for the future? A new approach in the debate about what makes healthcare systems really sustainable. *Sustainability, 7*(1), 294–312.

Follesdal, A., & Pogge, T. (Eds.). (2006). *Real world justice: Grounds, principles, human rights, and social institutions* (Vol. 1). Dordrecht, Netherlands: Springer Science & Business Media.

Follett, K. (2012). *World without end*. New York, NY: Penguin.

Frenk, J., & Moon, S. (2013). Governance challenges in global health. *New England Journal of Medicine, 368*(10), 936–942.

Gindling, T. H., & Trejos, J. D. (2005). Accounting for changing earnings inequality in Costa Rica, 1980–99. *Journal of Development Studies, 41*(5), 898–926.

Glasgow, R. E., Goldstein, M. G., Ockene, J. K., & Pronk, N. P. (2004). Translating what we have learned into practice: Principles and hypotheses for interventions addressing multiple behaviors in primary care. *American Journal of Preventive Medicine, 27*(2), 88–101.

Gostin, L. O. (2012). A framework convention on global health: Health for all, justice for all. *JAMA, 307*(19), 2087–2092.

Graham, P., Evitts, T., & Thomas-MacLean, R. (2008). Environmental scans: How useful are they for primary care research? *Canadian Family Physician, 54*(7), 1022–1023.

Guglielmin, M., Muntaner, C., O'Campo, P., & Shankardass, K. (2018). A scoping review of the implementation of health in all policies at the local level. *Health Policy, 122*(3), 284–292.

He, Z., Shentu, J., Yang, X., Baligar, V. C., Zhang, T., & Stoffella, P. J. (2015). Heavy metal contamination of soils: Sources, indicators and assessment. *Journal of Environmental Science, 9*, 17–18.

Hertzman, C. (1999). Population health and human development. In D. P. Keating & C. Hertzman (Eds.), *Developmental health and the wealth of nations: Social, biological, and educational dynamics* (pp. 21–40). New York, NY: The Guilford Press.

Hunter, D. J., & Fineberg, H. V. (2015). Convergence to common purpose in global health. In *Readings in Global Health: Essential Reviews from the New England Journal of Medicine* (pp. 289–293). New York, NY: Oxford University Press.

Ingman, S. R., & Thomas, A. E. (Eds.). (2011). *Topias and Utopias in health: Policy studies*. Berlin, Germany: Walter de Gruyter.

Jackson, M. C. (2004). Community operational research: Purposes, theory and practice. In *Community operational research* (pp. 57–74). Boston, MA: Springer.

Jakubowski, B., & Frumkin, H. (2010). Environmental metrics for community health improvement. *Preventing Chronic Disease, 7*(4), A76. Retrieved from https://www.ncbi.nlm.nih.gov/pubmed/20550834

Kampa, M., & Castanas, E. (2008). Human health effects of air pollution. *Environmental Pollution, 151*(2), 362–367.

Karanikolos, M., Mladovsky, P., Cylus, J., Thomson, S., Basu, S., Stuckler, D., … McKee, M. (2013). Financial crisis, austerity, and health in Europe. *The Lancet, 381*(9874), 1323–1331.

Kark, S. L., & Kark, E. (1983). An alternative strategy in community health care: Community-oriented primary health care. *Israel Journal of Medical Sciences, 19*(8), 707–713.

Kelly-Reif, K., & Wing, S. (2016). Urban-rural exploitation: An underappreciated dimension of environmental injustice. *Journal of Rural Studies, 47*, 350–358.

Keune, H., Martens, P., Kretsch, C., & Prieur-Richard, A. H. (2013). The natural relation between biodiversity and public health: An ecosystem services perspective. In *Ecosystem services* (pp. 181–189). San Diego, CA: Elsevier.

Kontis, V., Bennett, J. E., Mathers, C. D., Li, G., Foreman, K., & Ezzati, M. (2017). Future life expectancy in 35 industrialised countries: Projections with a Bayesian model ensemble. *The Lancet, 389*(10076), 1323–1335.

Kümmerer, K. (2010). Pharmaceuticals in the environment. *Annual Review of Environment and Resources, 35*, 57–75.

Liddicoat, C., Waycott, M., & Weinstein, P. (2016). Environmental change and human health: Can environmental proxies inform the biodiversity hypothesis for protective microbial–human contact? *Bioscience, 66*(12), 1023–1034.

Lobdell, D., Murphy, P., & Calderon, R. (2007). Environmental public health indicators. *Epidemiology, 18*(Suppl), S155. https://doi.org/10.1097/01.ede.0000276805.08216.cf

Mackenbach, J. P., Karanikolos, M., & McKee, M. (2013). The unequal health of Europeans: Successes and failures of policies. *The Lancet, 381*(9872), 1125–1134.

Mahapatra, S. K., & Ratha, K. C. (2017). Paris climate accord: Miles to go. *Journal of International Development, 29*(1), 147–154.

Martenies, S. E., Milando, C. W., Williams, G. O., & Batterman, S. A. (2017). Disease and health inequalities attributable to air pollutant exposure in Detroit, Michigan. *International journal of environmental research and public health, 14*(10), 1243.

McMichael, A. J. (2013). Globalization, climate change, and human health. *New England Journal of Medicine, 368*(14), 1335–1343.

Midgley, G., Johnson, M. P., & Chichirau, G. (2018). What is community operational research? *European Journal of Operational Research, 268*(3), 771–783.

Mills, A. (2014). Health care systems in low-and middle-income countries. *New England Journal of Medicine, 370*(6), 552–557.

Mirza, F., Mirza, A., Chung, C. Y. S., & Sundaram, D. (2016, November). Sustainable, holistic, adaptable, real-time, and precise (SHARP) approach towards developing health and wellness systems. In *International conference on future network systems and security* (pp. 157–171). Cham, Switzerland: Springer.

Morrison, D. S., Petticrew, M., & Thomson, H. (2003). What are the most effective ways of improving population health through transport interventions? Evidence from systematic reviews. *Journal of Epidemiology & Community Health, 57*(5), 327–333.

Mpofu, E. (2015). *Community-oriented health services: Practices across disciplines.* New York; NY: Springer.

Muntaner, C., Borrell, C., Ng, E., Chung, H., Espelt, A., Rodriguez-Sanz, M., … O'Campo, P. (2011). Politics, welfare regimes, and population health: Controversies and evidence. *Sociology of Health & Illness, 33*(6), 946–964.

Oliver, T. H., Isaac, N. J., August, T. A., Woodcock, B. A., Roy, D. B., & Bullock, J. M. (2015). Declining resilience of ecosystem functions under biodiversity loss. *Nature Communications, 6*, 10122.

Preker, A. S., Langenbrunner, J., & Jakab, M. (2002). Rich-poor differences in health care financing. *Social Reinsurance, 21*, 3–51.

Rahman, S. U., & Smith, D. K. (2000). Use of location-allocation models in health service development planning in developing nations. *European Journal of Operational Research, 123*(3), 437–452.

Rahmstorf, S., Foster, G., & Cahill, N. (2017). Global temperature evolution: Recent trends and some pitfalls. *Environmental Research Letters, 12*(5), 054001.

Rawls, J. (2009). *A theory of justice*. Cambridge, MA: Harvard University Press.

Roberts, J. L., & World Health Organization. (1998). *Terminology: A glossary of technical terms on the economics and finance of health services* (No. EUR/ICP/CARE 94 01/CN01). Copenhagen, Denmark: WHO Regional Office for Europe.

Russo, G., Moretta Tartaglione, A., & Cavacece, Y. (2019). Empowering patients to co-create a sustainable healthcare value. *Sustainability, 11*(5), 1315.

Sadd, J. L., Pastor, M., Morello-Frosch, R., Scoggins, J., & Jesdale, B. (2011). Playing it safe: Assessing cumulative impact and social vulnerability through an environmental justice screening method in the South Coast Air Basin, California. *International Journal of Environmental Research and Public Health, 8*(5), 1441–1459.

Sheiham, A. (2000). Improving oral health for all: Focusing on determinants and conditions. *Health Education Journal, 59*(4), 351–363.

Ståhl, T., Wismar, M., Ollila, E., Lahtinen, E., & Leppo, K. (2006). Health in all policies. In *Prospects and potentials*. Helsinki, Finland: Finnish Ministry of Social Affairs and Health.

The Economist. (2020, May 16). Editorial "Goodbye Globalization: A More Nationalistic and Self-Sufficient Era Beckons. It Won't be wicker – or safer" p. 7.

Thomke, R., Bigler, H. P., & Lehmann, P. (2013). *U.S. Patent No. 8,382,804*. Washington, DC: U.S. Patent and Trademark Office.

Thomson, H., Morrison, D., & Petticrew, M. (2007). The health impacts of housing-led regeneration: A prospective controlled study. *Journal of Epidemiology & Community Health, 61*(3), 211–214.

Voice, P. (2011). *Rawls explained: From fairness to Utopia* (pp. 41–48). Chicago, IL: Open Court.

Wallace, R. G., Bergmann, L., Kock, R., Gilbert, M., Hogerwerf, L., Wallace, R., & Holmberg, M. (2015). The dawn of structural one health: A new science tracking disease emergence along circuits of capital. *Social Science & Medicine, 129*, 68–77.

Weiss, M. G., Isaac, M., Parkar, S. R., Chowdhury, A. N., & Raguram, R. (2001). Global, national, and local approaches to mental health: Examples from India. *Tropical Medicine & International Health, 6*(1), 4–23.

World Commission on Environment and Development. (2014). *Report on our common future*. Retrieved on May 17, 2020 from online: http://www.un-documents.net/our-common-future.pdf

World Economic Forum. (2013). *Human Capital Report 2013*. Geneva, Switzerland: World Economic Forum.

World Health Organization. (2007). *Everybody's business—Strengthening health systems to improve health outcomes: WHO's framework for action*. Geneva, Switzerland: Author.

World Health Organization. (2008). *International health regulations (2005)*. Geneva, Switzerland: Author.

World Health Organization (WHO). (2012). *Burden of disease from household air pollution for 2012*. Summary of Results. Retrieved on May 17, 2020 from http://www.who.int/phe/health_topics/outdoorair/databases/FINAL_HAP_AAP_BoD_24March2014.pdf

World Health Organization. (2013). *Air Quality and Health, Fact Sheet N°313* [Online]. Retrieved on July 10, 2020 from http://www.who.int/mediacentre/factsheets/fs313/en/

World Health Organization (WHO). (2017). *Global strategy on aging and health*. Geneva, Switzerland: Author.

Young, M. E., & Lambie, G. W. (2007). Wellness in school and mental health systems: Organizational influences. *The Journal of Humanistic Counseling, Education and Development, 46*(1), 98–113.

Zapico, J. L., Brandt, N., & Turpeinen, M. (2010). Environmental metrics. *Journal of Industrial Ecology, 14*(5), 703.

Index[1]

[1] Note: Page numbers followed by 'n' refer to notes.

E. Mpofu (ed.), *Sustainable Community Health*,
https://doi.org/10.1007/978-3-030-59687-3

9783030596866